Numerical Methods for Scientific and Engineering Computation

M. K. JAIN
S. R. K. IYENGAR
R. K. JAIN

Department of Mathematics
Indian Institute of Technology Delhi
India

A HALSTED PRESS BOOK

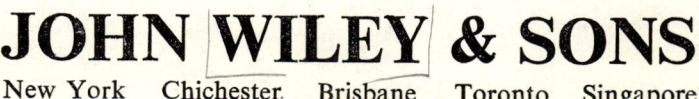

New York   Chichester   Brisbane   Toronto   Singapore

Copyright © 1985 WILEY EASTERN LIMITED
NEW DELHI

Published in the Western Hemisphere
by Halsted Press, A Division of
John Wiley & Sons, Inc., New York

Library of Congress Cataloging in Publication Data

Jain, M. K. (Mahinder Kumar), 1932–
  Numerical methods for scientific and engineering computation.

  "A Halsted Press book."
  Bibliography: P.
  Includes index.
  1. Numerical analysis. I. Iyenger, S. R. K.
II. Jain, Rajendra K., 1951–   III. Title.
QA297. J28 1984   519.4   84–22024

ISBN 0-470-20143-6

Printed in India at Urvashi Press, Meerut.

Preface

This book has grown from the lectures which we and our colleagues in the Department of Mathematics have delivered at the Indian Institute of Technology Delhi. Both undergraduate and postgraduate students of various engineering disciplines and M.Sc. mathematics (1st and 2nd year) students have attended these lectures. The material has also been covered in various summer schools organised by the Department of Mathematics at this Institute. Further, the book is based on the curriculum recommended by the workshop of the Engineering College teachers held at IIT Delhi, sponsored by the Ministry of Education, Government of India.

Also, the book has a number of BIT problems which have appeared in the examinations of Course I and partly Course II of Numerical Analysis at the Universities and Institutes of Technology in the Scandinavian countries for the years 1964–1983. Evidently, the book will meet the requirements of the students taking a first course in Numerical Analysis at most of the international universities and Institutes of Technology.

The book has seven chapters. Chapter 1 provides an introduction to computer arithmetic, errors and machine computation. In Chapter 2, the direct and iterative methods for finding the roots of transcendental and polynomial equations are given. A brief section on the choice of an iterative method is discussed. Chapter 3 contains the direct and iterative methods for the solution of a system of linear algebraic equations. The error analysis and convergence of iterative methods are also given. Various methods for finding eigenvalues and eigenvectors are included. Chapter 4 gives the derivation of interpolating polynomials and approximating functions. At the end, a section on the choice of a method is given. Chapter 5 provides methods for numerical differentiation and integration. Extrapolation procedures are discussed in detail. Chapters 6 and 7 have been adapted from the text entitled *Numerical Solution of Differential Equations*, 2nd edition, written by one of the present authors, published by Wiley Eastern Ltd., 1983. Chapter 6 includes a detailed treatment of single step and multistep methods for solving first order initial value problems. The difference and shooting methods for solving two point second order boundary value problems are discussed. Chapter 7 contains numerical methods for the solution of parabolic, hyperbolic and elliptic partial differential equations.

Chapters 2–7 are followed by problems, including BIT problems. A number of examples have been solved in each chapter to enable the student to understand the concepts described in the text. Flowcharts of the frequently used numerical methods for computer program are given in Appendix 1. Answers to the problems are also listed at the end of the book.

We wish to express our sincere thanks to Prof. C. E. Froberg for allowing the use of BIT problems. Thanks are due to Mr. D. R. Joshi and Miss Neelam Dhody for typing the manuscript and to Mr. Ranjit Kumar for his assistance. Finally, we are thankful to the authorities of the Indian Institute of Technology, Delhi for providing necessary facilities and encouragement.

New Delhi M. K. JAIN
October, 1984 S. R. K. IYENGAR
R. K. JAIN

Contents

Preface .... *iii*

CHAPTER 1. High Speed Computation **1**

    1.1 Introduction   1
    1.2 Computer Arithmetic   1
        Binary number system
        Octal and hexadecimal system
        Floating-point arithmetic
    1.3 Errors   7
        Significant digits and numerical instability
    1.4 Machine Computation   12
    1.5 Computer Software   16

CHAPTER 2. Transcendental and Polynomial Equations **18**

    2.1 Introduction   18
        Initial approximations
    2.2 The Bisection Method   22
    2.3 Iteration Methods Based on First Degree Equation   23
        Secant method
        The Newton-Raphson method
    2.4 Iteration Methods Based on Second Degree Equation   27
        The Muller method
        The Chebyshev method
        Multipoint iteration method
    2.5 Rate of Convergence   32
        Secant method
        The Newton-Raphson method
    2.6 Iteration Methods   34
        First order method
        Second order method
        High order methods
        Acceleration of convergence
        Efficiency of a method
        Methods for multiple roots
        Methods for complex roots
    2.7 Polynomial Equations   43

The Birge-Vieta method
The Bairstow method
Graeffe's root squaring method

    2.8 Choice of an Iterative Method and Implementation 56
          Flowchart for a zero finder
          Problems

CHAPTER 3. System of Linear Algebraic Equations and Eigenvalue Problems   69

    3.1 Introduction 69

    3.2 Direct Methods 72
          Cramer rule
          Gauss elimination method
          Gauss-Jordan elimination method
          Triangularization method
          Cholesky method
          Partition method

    3.3 Error Analysis 88
          Iterative improvement of the solution

    3.4 Iteration Methods 92
          Jacobi iteration method
          Gauss-Seidel iteration method
          Successive over-relaxation method
          Convergence analysis
          Iterative method for A^{-1}

    3.5 Eigenvalues and Eigenvectors 102
          Jacobi method for symmetric matrices
          Given's method for symmetric matrices
          Householder's method for symmetric matrices
          Rutishauser methodf or arbitrary matrices
          Power method

    3.6 Choice of a Method 121
          Problems

CHAPTER 4. Interpolation and Approximation   135

    4.1 Introduction 135

    4.2 Lagrange and Newton Interpolations 137
          Linear interpolation
          Higher order interpolation

    4.3 Finite Difference Operators 145

    4.4 Interpolating Polynomials Using Finite Differences 148

    4.5 Hermite Interpolation 153

    4.6 Piecewise and Spline Interpolation 155

    4.7 Bivariate Interpolation 162
          Lagrange bivariate interpolation
          Newton's bivariate interpolation for equispaced points

- 4.8 Approximation 164
- 4.9 Least Squares Approximation 166
- 4.10 Uniform Approximation 174
  Uniform (minimax) polynomial approximation
  Chebyshev polynomial approximation and Lanczos economization
- 4.11 Rational Approximation 181
- 4.12 Choice of a Method 183
  Problems

CHAPTER 5. Differentiation and Integration 196

- 5.1 Introduction 196
- 5.2 Numerical Differentiation 196
  Methods based on interpolation
  Methods based on finite differences
  Methods based on undetermined coefficients
- 5.3 Optimum Choice of Step-length 205
- 5.4 Extrapolation Methods 208
- 5.5 Partial Differentiation 211
- 5.6 Numerical Integration 213
- 5.7 Methods Based on Interpolation 214
- 5.8 Methods Based on Undetermined Coefficients 219
  Gauss-Legendre integration method
  Lobatto integration method
  Radau integration method
  Gauss-Chebyshev integration methods
  Gauss-Laguerre integration methods
  Gauss-Hermite integration methods
- 5.9 Composite Integration Methods 227
  Trapezoidal rule
  Simpson's rule
- 5.10 Romberg Integration 231
- 5.11 Double Integration 234
  Trapezoidal method
  Simpson's method
  Problems

CHAPTER 6. Ordinary Differential Equations 243

- 6.1 Introduction 243
- 6.2 Numerical Methods 247
  Euler method
  Backward Euler method
  Mid-point method
- 6.3 Single Step Methods 258
  Taylor series method

Runge-Kutta methods
Implicit Runge-Kutta methods

6.4 **Multistep Methods**   270
Determination of a_i and b_i
Convergence of multistep methods
Predictor-corrector methods

6.5 **Stability Analysis**   280
Singlestep methods
Multistep methods

6.6 **Boundary Value Problems**   289
Difference methods
Boundary value problem $u'' = f(x, u)$
Convergence of difference schemes
Shooting method
Problems

CHAPTER 7. Partial Differential Equations   314

7.1 Introduction   314

7.2 Difference Methods   318

7.3 Parabolic Equations   319
One space dimension
Convergence and stability analysis
Two space dimensions

7.4 Hyperbolic Equations   341
One space dimension
Two space dimensions
First order equation
System of equations

7.5 Elliptic Equations   359
Dirichlet problem
Neumann problem
Mixed problem
Problems

Answers and Hints to the Problems   385

Index   403

CHAPTER 1

High Speed Computation

1.1 INTRODUCTION

With the advent of the modern high speed electronic digital computers, the numerical methods have been successfully applied to study problems in mathematics, engineering, computer science and physical sciences such as biophysics, physics, atmospheric sciences and geosciences. The art and science of preparing and solving scientific and engineering problems have undergone considerable changes. This is due to the following two reasons:
   (i) The mathematical problem is to be reduced to a form amenable to machine solution.
   (ii) Several million operations are performed per minute on a high speed computer. This makes difficult to check the intermediate results for possible build up of large errors during calculations.
In what follows, we examine these two aspects in detail.

1.2 COMPUTER ARITHMETIC

The basic arithmetic operations performed by the computer are addition, subtraction, multiplication and division. The decimal numbers are first converted to the machine numbers consisting of the only digits 0 and 1 with a **base** or **radix** depending on the computer. If the base is two, eight or sixteen, the number system is called **binary, octal** or **hexadecimal** respectively. The decimal number system has the base 10. The decimal integer number 4987 actually means

$$(4987)_{10} = 4 \times 10^3 + 9 \times 10^2 + 8 \times 10^1 + 7 \times 10^0 \tag{1.1}$$

which represents a polynomial in the base 10. Similarly a fractional decimal number 0.6251 means

$$(0.6251)_{10} = 6 \times 10^{-1} + 2 \times 10^{-2} + 5 \times 10^{-3} + 1 \times 10^{-4} \tag{1.2}$$

which is a polynomial in 10^{-1}.

Combining (1.1) and (1.2), we may write the number 4987.6251 in decimal system as

$$(4987.6251)_{10} = 4 \times 10^3 + 9 \times 10^2 + 8 \times 10^1 + 7 \times 10^0$$
$$+ 6 \times 10^{-1} + 2 \times 10^{-2} + 5 \times 10^{-3} + 1 \times 10^{-4} \tag{1.3}$$

where the subscripts denote the base of the number system.

Thus a number $N = (d_{n-1}d_{n-2}\ldots d_0 \cdot d_{-1}d_{-2}\ldots d_{-m})$ in decimal system can always be expressed in the form

$$(N)_{10} = d_{n-1}10^{n-1} + d_{n-2}10^{n-2} + \ldots + d_1 10^1 + d_0 10^0 \\ + d_{-1}10^{-1} + d_{-2}10^{-2} + \ldots + d_{-m}10^{-m} \qquad (1.4)$$

where $d_{n-1}, d_{n-2}, \ldots, d_{-m}$ are any digits between 0 and 9.

1.2.1 Binary Number System

Binary number system has a base 2 with digits 0 and 1 called **bits** and any number N can be written as

$$(N)_2 = b_{n-1}b_{n-2}\ldots b_1 b_0 \cdot b_{-1}b_{-2}\ldots b_{-m} \qquad (1.5)$$

where $b_{n-1}, b_{n-2}, \ldots, b_{-m}$ are binary bits 0 or 1 and the point is called the binary point. The corresponding decimal number is easily calculated by using the formula

$$(N)_{10} = b_{n-1}2^{n-1} + b_{n-2}2^{n-2} + \ldots + b_1 2^1 + b_0 2^0 \\ + b_{-1}2^{-1} + b_{-2}2^{-2} + \ldots + b_{-m}2^{-m} \qquad (1.6)$$

Example 1.1 Find the decimal number corresponding to the binary number $(111.011)_2$.

We have

$$(111.011)_2 = 1\times 2^2 + 1\times 2^1 + 1\times 2^0 + 0\times 2^{-1} + 1\times 2^{-2} + 1\times 2^{-3}$$
$$= (7.375)_{10}$$

We now consider the conversion of the integer N in decimal system to the binary number $b_{n-1}b_{n-2}\ldots b_1 b_0$. We write

$$b_{n-1}2^{n-1} + b_{n-2}2^{n-2} + \ldots + b_1 2^1 + b_0 = N \qquad (1.7)$$

The last binary digit b_0 in (1.7) is zero if and only if N is even. The second digit b_1 is zero if and only if $(N - b_0)/2$ is even, and so on. Thus, we have

$$N_0 = N$$

$$N_{k+1} = \frac{N_k - b_k}{2}, \quad k = 0, 1, 2, \ldots \qquad (1.8)$$

until $N_k = 0$, where

$$b_k = \begin{cases} 1, & \text{if } N_k \text{ is odd} \\ 0, & \text{if } N_k \text{ is even} \end{cases}$$

Example 1.2 Convert $(58)_{10}$ to the corresponding binary number.

We have

$$N_0 = 58, \qquad\qquad b_0 = 0$$
$$N_1 = (58 - 0)/2 = 29, \qquad b_1 = 1$$
$$N_2 = (29 - 1)/2 = 14, \qquad b_2 = 0$$

$$N_3 = (14 - 0)/2 = 7, \quad b_3 = 1$$
$$N_4 = (7 - 1)/2 = 3, \quad b_4 = 1$$
$$N_5 = (3 - 1)/2 = 1, \quad b_5 = 1$$
$$N_6 = (1 - 1)/2 = 0$$

Thus the binary number is 111010.

Next, we convert the fraction N in decimal system to binary fraction $.b_{-1}b_{-2}\ldots b_{-m}$.

We write

$$b_{-1}2^{-1} + b_{-2}2^{-2} + \ldots + b_{-m}2^{-m} = N, \quad 0 < N < 1 \tag{1.9}$$

where $b_{-1}, b_{-2}, \ldots, b_{-m}$ are 0 or 1.

It is easily seen that b_{-1} is 1 if and only if $2N > 1$ and 0 if and only if $2N \leqslant 1$. Similarly, using (1.9), we may determine b_{-k} recursively as follows:

$$N_1 = N$$

$$b_{-k} = \begin{cases} 1, & \text{if } 2N_k > 1 \\ 0, & \text{if } 2N_k \leqslant 1 \end{cases}$$

$$N_{k+1} = 2N_k - b_{-k}, \quad k = 1, 2, \ldots \tag{1.10}$$

Example 1.3 Convert $(.859375)_{10}$ to the corresponding binary fraction.

We have

k	b_{-k}	N_{k+1}
0		.859375
		$\times 2$
1	1	.718750
		$\times 2$
2	1	.437500
		$\times 2$
3	0	.875000
		$\times 2$
4	1	.750000
		$\times 2$
5	1	.500000
		$\times 2$
6	1	.000000

The required binary fraction becomes

$$(.859375)_{10} = (.110111)_2$$

4 Numerical Methods

Example 1.4 Convert $(0.7)_{10}$ to the corresponding binary fraction. We have

k	b_{-k}	N_{k+1}
0		.7
		$\times 2$
1	1	.4
		$\times 2$
2	0	.8
		$\times 2$
3	1	.6
		$\times 2$
4	1	.2
		$\times 2$
5	0	.4
		$\times 2$
6	0	.8
		$\times 2$
7	1	.6
		$\times 2$
8	1	.2
		$\times 2$
9	0	.4

Thus we obtain

$$(.7)_{10} = (.101100110\ldots)_2$$

which is a never ending sequence. If only 7 bits are retained in the binary fraction then the corresponding decimal number becomes

$$(.1011001)_2 = 1 \times 2^{-1} + 0 \times 2^{-2} + 1 \times 2^{-3} + 1 \times 2^{-4} + 0 \times 2^{-5}$$
$$+ 0 \times 2^{-6} + 1 \times 2^{-7}$$
$$= .6953125$$

which is not exactly the same as the given number. The difference

$$.7 - .6953125 = .0046875$$

is the round-off error.

1.2.2 Octal and Hexadecimal System

The octal system has base 8 and uses the digits 0, 1, 2, 3, 4, 5, 6, 7. Similarly, the hexadecimal system has base 16 and uses the digits 0 to 9 and A, B, C, D, E, F to represent 10, 11, 12, 13, 14, 15 respectively.

Decimal numbers can be converted to octal or hexadecimal number in a similar manner. The conversion between binary and octal or between binary and hexadecimal is simple due to the relationship between the bases, $8 = 2^3$ for octal and $16 = 2^4$ for hexadecimal. We convert a binary number to an octal number by grouping the binary bits in groups of three to the right and left of the binary point by adding sufficient zeros to complete the groups and replacing each group of three bits by its octal equivalent. Similarly to convert a binary number to a hexadecimal number, we form groups of four of the binary bits and replace it by the corresponding digit in the hexadecimal system. The hexadecimal system is sometimes referred to as hex.

Example 1.5 Convert the binary number
$$1101001.1110011$$
to the octal and the hexadecimal systems.
We have
$$(1101001.1110011)_2 = (001101001.111001100)_2$$
$$= (151.714)_8$$
and
$$(1101001.1110011)_2 = (01101001.11100110)_2$$
$$= (69.E6)_{16}$$

1.2.3 Floating Point Arithmetic

The first step in the computation with digital computers is to convert the decimal numbers to another number system with the base (say) β understandable to that particular computer and then to store these converted numbers in computer memory. The memory of the digital computer is divided into separate cells called **words**. Each word can hold the same number of digits called bits, with respect to its base plus a sign. Negative numbers are stored as absolute values plus a sign or in complement form. The number of digits which can be stored in a computer word is called its word length. The word length varies from one computer to another. The numbers in the computer word can be stored in two forms:

(i) Fixed-point form.
(ii) Floating-point form.

In **fixed-point** form a t digit number is assumed to have its decimal point at the left-hand end of the word. This implies that all numbers are assumed to be less than 1 in magnitude. The fixed-point number with base β and t digits word length may be written as

$$\pm \sum_{k=1}^{p} a_k \beta^{-k}$$

where $0 \leqslant a_k < \beta$.

If y is any real number and y^* is its machine representation then the error in y^* is at most $a_{p+1}\beta^{-(p+1)}$ or

$$|y - y^*| \leqslant \beta^{-(p+1)} \tag{1.11}$$

To avoid the difficulty of keeping every number less than 1 in magnitude during computation, most computers use **floating-point** representation for a real number. A floating-point number is characterized by four parameters, the base β, the number of digits t, and the exponent range (m, M).

Definition 1.1 A floating-point number is a number represented in the form

$$.d_1 d_2 \ldots d_t \times \beta^e \qquad (1.12)$$

where $d_1, d_2, \ldots, d_t$ are integers and satisfy $0 \leqslant d_i < \beta$ and the exponent e is such that $m \leqslant e \leqslant M$.

The fractional part $.d_1 d_2 \ldots d_t$ is called the **mantissa** and it lies between $+1$ and -1. The number 0 is written as

$$+.00 \ldots 0 \times \beta^e \qquad (1.13)$$

Definition 1.2 A non-zero floating-point number (1.12) is in **normal form** if the value of the mantissa lies in the interval $(-1, -.1/\beta]$ or in the interval $[.1/\beta, 1)$.

Example 1.6 Subtract the following two floating-point numbers 0.36143447×10^7 and 0.36132346×10^7.

We have

$$\begin{array}{r} 0.36143447 \times 10^7 \\ -0.36132346 \times 10^7 \\ \hline 0.00011101 \times 10^7 \end{array}$$

The result is a floating-point number, but not a normalized floating-point number due to the presence of the three leading zeros. Shifting the fractional part three places to the left, we get the result 0.11101×10^4 which is a normalized floating-point number.

Definition 1.3 A non-zero floating-point number (1.12) is in **t-digit-mantissa standard form** if it is normalized and its mantissa consists of exactly t digits.

If a number x has the representation in the form

$$x = .d_1 d_2 \ldots d_t d_{t+1} \ldots \times \beta^e \qquad (1.14)$$

then the floating-point number $\mathrm{fl}(x)$ in t-digit-mantissa standard form can be obtained in the following two ways:

(i) *Chopping.* Here we neglect $d_{t+1}, d_{t+2}, \ldots$ in (1.14) and obtain

$$\mathrm{fl}(x) = .d_1 d_2 \ldots d_t \times \beta^e \qquad (1.15)$$

(ii) *Rounding.* Here the fractional part in (1.14) is written as

$$.d_1 d_2 \ldots d_t d_{t+1} + \tfrac{1}{2}\beta \qquad (1.16)$$

and the first t digits are taken to write the floating-point number.

Example 1.7 Find the sum of $.123 \times 10^3$ and $.456 \times 10^2$ and write the result in three-digit mantissa.

The number of the smaller magnitude is adjusted so that its exponent is same as that of the number of larger magnitude. We have

$$.123 \times 10^3$$
$$.0456 \times 10^3$$
$$\overline{.1686 \times 10^3} = \begin{cases} .168 \times 10^3, & \text{for chopping} \\ .169 \times 10^3, & \text{for rounding} \end{cases}$$

1.3 ERRORS

A computer has a finite word length and so only a fixed number of digits are stored and used during computation. This would mean that even in storing an exact decimal number in its converted form in the computer memory, an error is introduced. This error is machine dependent and is called **machine epsilon**. After the computation is over, the result in the machine form (with base β) is again converted to decimal form understandable to the users and some more error may be introduced at this stage. We now discuss the effect of the errors on the results.

The quantity,

True value − Approximate value

is called the **error**. In order to determine the accuracy in an approximate solution to a problem, either we find the bound of the

$$\textbf{Relative error} = \frac{|\text{Error}|}{|\text{True value}|} \tag{1.17}$$

or of the

$$\textbf{Absolute error} = |\text{Error}| \tag{1.18}$$

Neglecting a blunder or mistake, the errors may be classified into the following types:

Definition 1.4 The **inherent error** is that quantity which is already present in the statement of the problem before its solution.

The inherent error arises either due to the simplified assumptions in the mathematical formulation of the problem or due to the physical measurements of the parameters of the problem.

Definition 1.5 The **round-off error** is the quantity R which must be added to the finite representation of a computed number in order to make it the true representation of that number.

Thus, if x is the computed number given by (1.14)

$$x = .d_1 d_2 \ldots d_t d_{t+1} \ldots \times \beta^e$$

then the relative error (1.17) for t-digit mantissa standard form representation of x becomes

$$\frac{|x - \text{fl}(x)|}{|x|} \leqslant \begin{cases} \beta^{1-t} & \text{for chopping} \\ \tfrac{1}{2}\beta^{1-t} & \text{for rounding} \end{cases} \tag{1.19}$$

Thus the bound on the relative error of a floating-point number is reduced by half when rounding is used than chopping. It is for this reason that on most computers rounding is used. We write

$$\text{fl}(x) = x(1 + \delta) \tag{1.20}$$

where $\delta = \delta(x)$, some number depending on x, is called the relative round-off error in fl(x). The number δ is called the **machine epsilon** and is denoted by *EPS*. From (1.19) we have

$$|\delta(x)| = EPS = \begin{cases} \beta^{1-t} & \text{for chopping} \\ \tfrac{1}{2}\beta^{1-t} & \text{for rounding} \end{cases} \tag{1.21}$$

Definition 1.6 The **truncation error** is the quantity T which must be added to the true representation of the quantity in order that the result be exactly equal to the quantity we are seeking to generate.

This error is a result of the approximate formulas used which are generally based on truncated series. The Taylor series with a remainder is an invaluable tool in the study of the truncation error.

Example 1.8 Obtain a second degree polynomial approximation to

$$f(x) = (1 + x)^{1/2}, \quad x \in [0, 0.1]$$

using the Taylor series expansion about $x = 0$. Use the expansion to approximate $f(.05)$ and bound the truncation error.

We have

$$f(x) = (1 + x)^{1/2}, \qquad f(0) = 1$$
$$f'(x) = \tfrac{1}{2}(1 + x)^{-1/2}, \qquad f'(0) = \tfrac{1}{2}$$
$$f''(x) = -\tfrac{1}{4}(1 + x)^{-3/2}, \qquad f''(0) = -\tfrac{1}{4}$$
$$f'''(x) = \tfrac{3}{8}(1 + x)^{-5/2}$$

Thus the Taylor series expansion with remainder term may be written as

$$(1 + x)^{1/2} = 1 + \frac{x}{2} - \frac{x^2}{8} + \frac{1}{16}\frac{x^3}{[(1 + \xi)^{1/2}]^5}, \quad 0 < \xi < 0.1$$

The truncation term is given by

$$T = (1 + x)^{1/2} - \left(1 + \frac{x}{2} - \frac{x^2}{8}\right)$$

$$= \frac{1}{16}\frac{x^3}{[(1 + \xi)^{1/2}]^5}$$

We have

$$f(.05) = 1 + \frac{.05}{2} - \frac{(.05)^2}{8} = .10246875 \times 10^1$$

The bound of the truncation error, for $x \in [0, .1]$ is

$$|T| \leq \frac{(.1)^3}{16[(1 + \xi)^{1/2}]^5} \leq \frac{(.1)^3}{16} = .625 \times 10^{-4}$$

1.3.1 Significant Digits and Numerical Instability

When a number x is written in normalized floating point form with t digits in base β as given in (1.15), we say that the number has t **significant digits**. The leading digit d_1 is called the most significant digit. In other words a number x^* is an approximation to x to t significant digits if

$$\frac{|x - x^*|}{|x|} \leq \tfrac{1}{2}\beta^{1-t} \tag{1.22}$$

in the base β. As an example, $x^* = .3$ agrees with $1/3$ to one significant digit and $x^* = .3333$ agrees with $1/3$ to four significant digits in the base $\beta = 10$. Suppose now, that x^* and y^* are approximations to x and y to t significant digits and we wish to calculate the number $z = x - y$. Then $z^* = x^* - y^*$ is an approximation to z which is also correct to t significant digits unless x^* and y^* have one or more leading digits same. In this case, there will be cancellation of digits during subtraction and z^* will not be accurate to t significant digits. For example, if we have

$x^* = .178693 \times 10^1$

$y^* = .178439 \times 10^1$

each correct to six digits in decimal system, then

$z^* = .000254 \times 10^1$

is correct to only three significant digits.

It may be noted that the relative error in x is $\tfrac{1}{2} \times 10^{-5}/.178693 \times 10^1 \approx 2.8 \times 10^{-6}$, while that in $x - y$ is $(\tfrac{1}{2} \times 10^{-5} + \tfrac{1}{2} \times 10^{-5})/.000254 \times 10^1 \approx 3.9 \times 10^{-3}$. When this number z^* is used in further arithmetic calculations, error may considerably increase. Thus we find that the subtraction of two nearly equal numbers causes a considerable loss of significant digits and may greatly magnify the error in later calculations.

A similar loss in significant digits occurs when a number is divided by a small divisor. For example consider

$$f(x) = \frac{1}{1 - x^2}$$

and we want to calculate $f(x)$ for $x = .9$. The exact value of $f(x)$ for $x = 0.9$ correct to six digits is

$f(x) = .527053 \times 10^1$

If x is approximated to $x^* = .900005$ that is an error is introduced in the sixth decimal place, we obtain the value

$f(x^*) = .526454 \times 10^1$

Therefore an error in the sixth place in x has caused an error in the third place in $f(x)$. Thus, when a number is divided by a very small number (or multiplied by a very large number) there is loss of significant digits and magnification of errors in the result.

It is noticed, therefore, that every arithmetic operation performed during computation, gives rise to some error, which once generated may decay or grow in subsequent calculations. In some cases, errors may grow so large as to make the computed result totally redundant and we call such a procedure **numerically unstable**. In some cases it can be avoided by changing the calculation procedure, which avoids subtractions of nearly equal numbers or division by a small number or by retaining more digits in the mantissa.

While finding numerical solution of problems of scientific nature, we often encounter problems with inherent instability or induced instability. Inherent instability may arise due to the ill-conditionedness of the problem. We cannot avoid inherent instability by changing the method of solution. It is the property of the problem itself. Sometimes, we can avoid this instability by a suitable reformulation of the problem. For example, consider the **Wilkinson's** example of finding the zeros of a polynomial. The polynomial

$$p_{20}(x) = (x-1)(x-2)\ldots(x-20)$$
$$= x^{20} - 210x^{19} + \ldots + 20! \qquad (1.23)$$

has the zeros $1, 2, \ldots, 20$. Let the coefficient of x^{19} be changed from -210 to $-(210 + 2^{-23})$. This is a very small absolute change, of magnitude 10^{-7} approximately. Most computers neglect this small change which occurs after 23 binary bits. If the solution of the new equation is now obtained, we find that the smaller roots are obtained with good accuracy, while the roots of larger magnitude are changed by a large amount. The largest change occurs in the roots 16 and 17. They are now obtained as the complex pair $16.73\ldots \pm i2.81\ldots$ whose magnitude is 17 approximately. The change is very substantial and it is due to the inherent instability or ill-conditionedness of the polynomial. If the problem cannot be reformulated, then the method that we are using must at least provide some information about the degree of ill-conditioning.

The second type of instability that we encounter is the induced instability which arises usually because of the wrong choice of the method of solution. The problem is often well conditioned in this case. Induced instability can be avoided by a suitable modification or change of the method of solution. For example, to evaluate the integral

$$I_n = \int_0^1 \frac{x^n}{x+6} \, dx, \quad n = 1, 2, \ldots, 10$$

we may use the recurrence relation

$$I_n = \frac{1}{n} - 6I_{n-1}, \quad n = 1, 2, \ldots, 10 \qquad (1.24)$$

where

$$I_0 = \int_0^1 \frac{dx}{x+6} = \ln(7/6) = 0.15415$$

Using the recurrence relation (1.24), we obtain

$I_1 \approx 0.07510,\quad I_2 \approx 0.04940,\quad I_3 \approx 0.03693$
$I_4 \approx 0.02842,\quad I_5 \approx 0.02948,\quad I_6 \approx -0.01021$
$I_7 \approx 0.20412,\quad I_8 \approx -1.09972,\; I_9 \approx 6.70943$
$I_{10} \approx -40.15658$

The exact value is $I_{10} = 0.01449$. This explosion has occurred because of the induced instability. This problem is well conditioned and accurate solutions can be obtained by choosing a suitable method.

We may write the recurrence relation (1.24) as

$$I_{n-1} = \frac{1}{6}\left(\frac{1}{n} - I_n\right), \quad n = 10, 9, \ldots, 1 \tag{1.25}$$

Since I_n decreases as n increases, choose $I_{10} = 0$. We obtain

$I_9 \approx 0.01666,\quad I_8 \approx 0.01574,\quad I_7 \approx 0.01821$
$I_6 \approx 0.02077,\quad I_5 \approx 0.02432,\quad I_4 \approx 0.02928$
$I_3 \approx 0.03679,\quad I_2 \approx 0.04942,\quad I_1 \approx 0.07510$
$I_0 \approx 0.15415$

The exact value is $I_0 = 0.15415$.

The criteria when to stop an infinite sequence of computation should be carefully given to the computer. To illustrate this point, consider finding the sum of the series

$$1 - \tfrac{1}{2} + \tfrac{1}{3} - \tfrac{1}{4} + \cdots$$

If the computer is asked to stop when the absolute value of the next term is less than the tolerable error, then nothing would happen and we get accurate solution. However, if the same criterion is applied to the series

$$1 + \tfrac{1}{2} + \tfrac{1}{3} + \tfrac{1}{4} + \cdots$$

the computer would give finite solution, while the sum is infinite.

Example 1.9 Find the smaller root of the equation
$$x^2 - 400x + 1 = 0$$
using four digit arithmetic.

For the equation $ax^2 - bx + c = 0$, $b > 0$, the smaller root is given by

$$x = \frac{b - \sqrt{b^2 - 4ac}}{2a}$$

Here
$a = 1\ \ = .1000 \times 10^1$
$b = 400 = .4000 \times 10^3$
$c = 1\ \ = .1000 \times 10^1$
$b^2 - 4ac = .1600 \times 10^6 - .4000 \times 10^1$
$\qquad\quad\ = .1600 \times 10^6$ (to four digit accuracy).
$\sqrt{b^2 - 4ac} = .4000 \times 10^3$

Substituting in the above formula we get $x = .0000$.

However, if we rewrite the above formula in the form
$$x = \frac{2c}{b + \sqrt{b^2 - 4ac}}$$
we get
$$x = \frac{.2000 \times 10^1}{.4000 \times 10^3 + .4000 \times 10^3}$$
$$= \frac{.2000 \times 10^1}{.8000 \times 10^3}$$
$$= .0025$$
which is the exact root.

Example 1.10 Compute the midpoint of the numbers
$$a = 4.568 \text{ and } b = 6.762$$
using the four digit arithmetic.
If we take the midpoint as the mean of the numbers, we have
$$c = \frac{a+b}{2} = \frac{.4568 \times 10^1 + .6762 \times 10^1}{2}$$
$$= \frac{.1133 \times 10^2}{2} = .5660 \times 10^1$$
However, if we use the formula
$$c = a + \frac{b-a}{2}$$
we get
$$c = .4568 \times 10^1 + \frac{.6762 \times 10^1 - .4568 \times 10^1}{2}$$
$$= .4568 \times 10^1 + .1097 \times 10^1$$
$$= .4665 \times 10^1$$
which is the correct result.

1.4 MACHINE COMPUTATION

To obtain meaningful results for a given problem using computers, there are five distinct phases:
1. Choice of a method
2. designing the algorithm
3. flow charting
4. programming
5. computer execution

A method is defined to be a mathematical formula for finding the solution of a given problem. There may be more than one method available to solve the same problem. We should choose the method which suits the given problem best. The inherent assumptions and limitations of the method must be studied carefully.

Once the method has been decided, we must describe a complete and unambiguous set of computational steps to be followed in a particular sequence to obtain the solution. This description is called an **algorithm**. It may be emphasized that the computer is concerned with the algorithm and not with the method. The algorithm tells to the computer where to start, what information to use, what operations to be carried out and in which order, what information to be printed and when to stop.

Example 1.11 Design an algorithm to find the real roots of the equation

$$ax^2 + bx + c = 0, \, a, b, c \text{ real}$$

for 10 sets of values of a, b, c, using the method

$$x_1 = \frac{-b+e}{2a}, \, x_2 = \frac{-b-e}{2a} \tag{1.26}$$

where $e = \sqrt{(b^2 - 4ac)}$.

The following computational steps are involved:

Step 1: Set $I = 1$.
Step 2: Read a, b, c.
Step 3: Check: is $a = 0$? if yes, print wrong data and go to step 9, otherwise go to next step.
Step 4: Calculate $d = b^2 - 4ac$.
Step 5: Check: is $d < 0$? if yes, print complex roots and go to step 9, otherwise go to next step.
Step 6: Calculate $e = \sqrt{(b^2 - 4ac)}$
Step 7: Calculate x_1 and x_2 using method (1.26).
Step 8: Print x_1 and x_2.
Step 9: Increase I by 1.
Step 10: Check: Is $I \leqslant 10$? if yes, go to step 2, otherwise go to next step.
Step 11: Stop.

On execution of the above eleven steps or **instructions** in the same order, the problem is completely solved. These eleven steps constitute the algorithm of the method (1.26).

An algorithm has five important features:
1. finiteness : an algorithm must terminate after a finite number of steps.
2. definiteness: each step of an algorithm must be clearly defined or the action to be taken must be unambiguously specified.
3. inputs: an algorithm must specify the quantities which must be read before the algorithm can begin. In the algorithm of example 1.11 the three input quantities are a, b, c.
4. outputs: an algorithm must specify the quantities which are to be outputted and their proper place. In the algorithm of example 1.11 the two output quantities are x_1, x_2.

5. **effectiveness**: an algorithm must be effective which means that all operations are executable. For example, in the algorithm of example 1.11, we must avoid the case $a = 0$, as division by zero is not possible. Similarly, if $b^2 - 4ac < 0$, some alternate path must be defined to avoid finding the square root of a negative number.

A **flow chart** is a graphical representation of a specific sequence of steps (algorithm) to be followed by the computer to produce the solution of a given problem. It makes use of the flow chart symbols to represent the basic operations to be carried out. The various symbols are connected by arrows to indicate the flow of information and processing. While drawing a flow chart any logical error in the formulation of the problem or applying the algorithm can be easily seen and corrected. Some of the symbols used in drawing flow charts are given in Table 1.1.

Table 1.1 Flow chart Symbols

Flow Chart Symbols	Meaning
$C = A + B$	A processing symbol such as addition or subtraction of two numbers and movement of data in computer memory.
is $D < 0$?	A decision taking symbol. Depending on the answer yes or no, a particular path is chosen.
Read A, B	An input symbol, specifying quantities which are to be read before processing can proceed.
Print A, B	An output symbol, specifying quantities which are to be outputted.
Start or End	A terminating symbol, indicating start or end of the flow chart. This symbol is also used as a connector.

Example 1.12 Draw a flow chart to find real roots of the equation $ax^2 + bx + c = 0$, a, b, c real, using the method

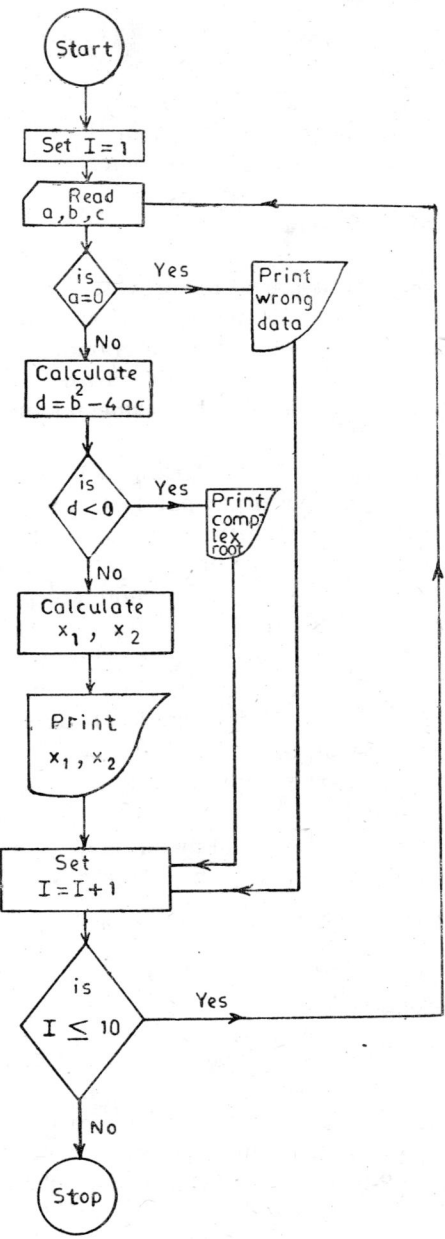

Fig. 1.1 Flow chart for finding real roots of the quadratic equation

$$x_1 = \frac{-b + \sqrt{b^2 - 4ac}}{2a}$$

$$x_2 = \frac{-b - \sqrt{b^2 - 4ac}}{2a}$$

for ten sets of values of a, b, c.

The flow chart is given in Fig. 1.1. The flow chart can be easily translated into any high level language, for example FORTRAN, ALGOL, BASIC etc. and can be executed on the computer.

1.5 COMPUTER SOFTWARE

The purpose of computer software is to provide a useful computational tool for users. The writing of computer software requires a good understanding of numerical analysis and art of programming. A good computer software must satisfy certain criteria of **self-starting, accuracy** and **reliability, minimum number of levels, good documentation, ease of use** and **portability**.

A computer software should be self-starting as far as possible. A numerical method very often involves some parameters whose values are determined by the properties of the problem to be solved. For example in finding the roots of an equation one or more initial approximations to the root have to be given. The program will be more acceptable, if it can be made automatic in the sense that the program will select the initial approximations itself rather than requiring the user to specify them.

Accuracy and reliability are measures of the performance of an algorithm on all similar problems. Once an error criterion is fixed, it should produce solutions of all similar problems to that accuracy. The program should be able to prevent and handle most of the exceptional conditions like division by zero, infinite loops etc.

The structure of the program should avoid many levels. For example, many programs, used to find roots of an equation have three levels:

    Program calls zero-finder (parameters, function)

    zero-finder calls function

    function subprogram.

More the number of levels in the program, more time is wasted in interlinking and transfer of parameters.

Good documentation and easy to use are two very important criteria. The program must have some comment lines or comment paragraphs at various places giving explanation and clarification of the method used and steps involved. A good documentation should clarify, what kind of problems can be solved using this software, what parameters are to be supplied, what accuracy can be achieved, which method has been used and other relevant details.

The criterion of portability means that the software should be made independent of the computer being used as far as possible. Since most machines have different hardware configuration, complete independency of the machine may not be possible. However, the aim at the time of writing a computer software should be that the same program should be able to run on any machine with minimal modifications. Machine dependent constants, for example machine error EPS, must be avoided or automatically generated. A standard dialect of the programming language should be used rather than using a local dialect.

CHAPTER 2

Transcendental and Polynomial Equations

2.1 INTRODUCTION

A problem of great importance in applied mathematics and engineering is that of determining the roots of an equation of the form

$$f(x) = 0 \tag{2.1}$$

The function $f(x)$ may be given explicitly, for example

$$f(x) = P_n(x) = x^n + a_1 x^{n-1} + \ldots + a_{n-1} x + a_n$$

a polynomial of degree n in x, or $f(x)$ may be known only implicitly as a transcendental function.

Definition 2.1 A number ξ is a solution of $f(x) = 0$ if $f(\xi) \equiv 0$. Such a solution ξ is called a **root** or a **zero** of $f(x) = 0$.

Geometrically, a root of the equation (2.1) is the value of x at which the graph of $y = f(x)$ intersects the x-axis.

Definition 2.2 If we can write (2.1) as

$$f(x) = (x - \xi)^m g(x) = 0$$

where $g(x)$ is bounded and $g(\xi) \neq 0$, then ξ is called a **multiple root** of multiplicity m.

For $m = 1$, the number ξ is said to be a **simple root**.

There are generally two types of methods used to find the roots of the equation (2.1).

(i) *Direct methods.* These methods give the exact value of the roots in a finite number of steps. Further, the methods give all the roots at the same time. For example, a direct method gives the root of a linear or first degree equation

$$a_0 x + a_1 = 0, \quad a_0 \neq 0 \tag{2.2}$$

as

$$x = -\frac{a_1}{a_0}$$

Similarly, the roots of the quadratic equation

$$a_0 x^2 + a_1 x + a_2 = 0, \quad a_0 \neq 0 \tag{2.3}$$

are given by
$$x = \frac{-a_1 \pm \sqrt{(a_1^2 - 4a_0 a_2)}}{2a_0}$$

(ii) *Iterative methods.* These methods are based on the idea of successive approximation, i.e. starting with one or more initial approximations to the root, we obtain a sequence of approximants or iterates $\{x_k\}$, which in the limit converges to the root. The methods may give only one root at a time. For example, to solve the quadratic equation (2.3) we may choose any one of the following iteration methods:

(a) $x_{k+1} = -\dfrac{a_2 + a_0 x_k^2}{a_1}, \quad k = 0, 1, 2, \ldots$

(b) $x_{k+1} = -\dfrac{a_2}{a_0 x_k + a_1}, \quad k = 0, 1, 2, \ldots$

(c) $x_{k+1} = -\dfrac{a_2 + a_1 x_k}{a_0 x_k}, \quad k = 0, 1, 2, \ldots$ \hfill (2.4)

The convergence of the sequence $\{x_k\}$ to a number ξ, the root of the equation (2.3) depends on the rearrangement (2.4) and the choice of starting approximation x_0.

Definition 2.3 A sequence of iterates $\{x_k\}$ is said to converge to the root ξ, if
$$\lim_{k \to \infty} |x_k - \xi| = 0$$

If $x_k, x_{k-1}, \ldots, x_{k-m+1}$ are m approximations to the root, then a multipoint iteration method is defined as
$$x_{k+1} = \phi(x_k, x_{k-1}, \ldots, x_{k-m+1}) \tag{2.5}$$

The function ϕ is called the multipoint iteration function.

For $m = 1$, we get the one point iteration method
$$x_{k+1} = \phi(x_k) \tag{2.6}$$

Thus, given one or more initial approximations to the root, we require a suitable iteration function ϕ for a given function $f(x)$, such that the sequence of iterates obtained from (2.5) or (2.6) converges to the root ξ. In practice, except in rare cases, it is not possible to find ξ which satisfies the given equation exactly. We, therefore, attempt to find an approximate root ξ^* such that either
$$|f(\xi^*)| < \epsilon$$
or
$$|x_{k+1} - x_k| < \epsilon \tag{2.7}$$
where x_k and x_{k+1} are two consecutive iterates and ϵ is the prescribed **error tolerance**.

2.1.1 Initial Approximations

Initial approximations to the root are often known from the physical considerations of the problem. Otherwise graphical methods are generally used to obtain initial approximations to the root. Since the value of x, at which the graph of the equation $y = f(x)$ intersects the x-axis, gives the root of $f(x) = 0$, any value in the neighbourhood of this point may be taken as an initial approximation to the root (see Fig. 2.1a, b). If the equation

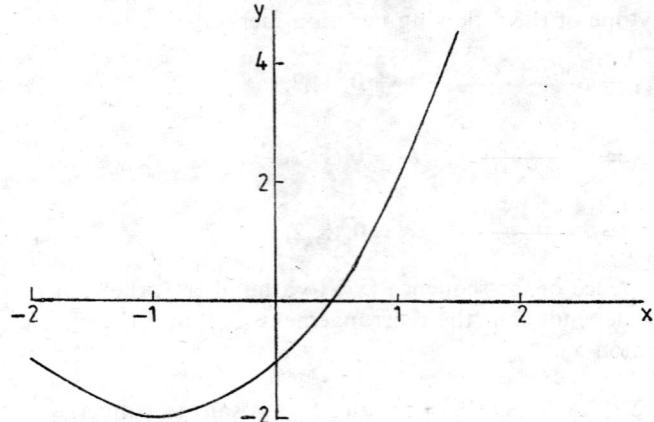

Fig. 2.1(a) Graph of $y = x^2 + 2x - 1$

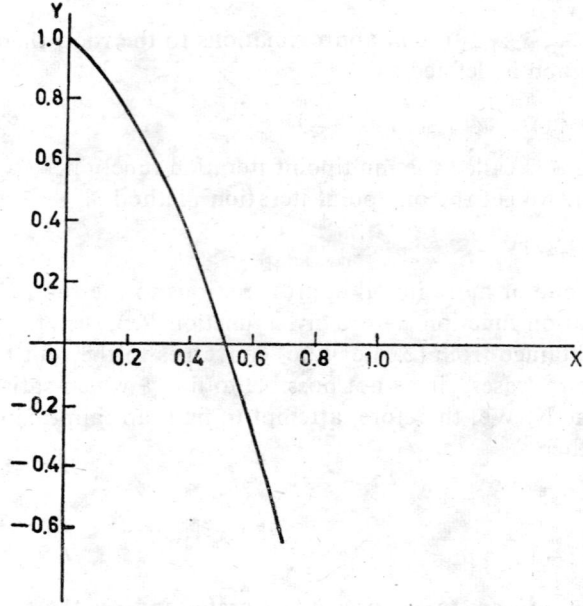

Fig. 2.1(b) Graph of $y = \cos x - xe^x$

$f(x) = 0$ can be conveniently written in the form $f_1(x) = f_2(x)$, then the point of intersection of the graphs of the equations $y = f_1(x)$ and $y = f_2(x)$ gives the root of $f(x) = 0$ and therefore any value in the neighbourhood of this point can be taken as an initial approximation to the root (see Fig. 2.1c).

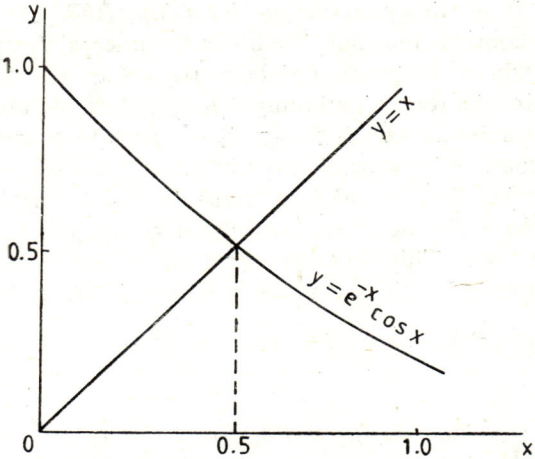

Fig. 2.1(c) Graph of $y = x$ and $y = e^{-x} \cos x$

Another commonly used method to obtain the initial approximation to the root is based upon the **Intermediate Value Theorem**, which states:

THEOREM 2.1 *If $f(x)$ is a continuous function on some interval $[a, b]$ and $f(a) f(b) < 0$, then the equation $f(x) = 0$ has at least one real root or an odd number of roots in the interval (a, b).*

We can set up a table of values of $f(x)$ for various values of x and obtain a suitable initial approximation to the root.

Example 2.1 Obtain an initial approximation to a root of the equation

$$f(x) = \cos x - xe^x = 0$$

We prepare a table of the values of the function $f(x)$ for various values of x. We get

Table 2.1 Values of $f(x)$

x	0	0.5	1	1.5	2
$f(x)$	1	0.0532	-2.1780	-6.6518	-15.1942

From the table we find that the equation $f(x) = 0$ has at least one root in the interval (0.5, 1). The exact root correct to ten decimal places is 0.5177573637.

2.2 THE BISECTION METHOD

This method is based on the repeated application of the intermediate value theorem. If we know that a root of $f(x) = 0$ lies in the interval $I_0 = (a_0, b_0)$, we bisect I_0 at the point $m_1 = \frac{1}{2}(a_0 + b_0)$. Denote by I_1 the interval (a_0, m_1) if $f(a_0)f(m_1) < 0$ or the interval (m_1, b_0) if $f(m_1)f(b_0) < 0$. Therefore, the interval I_1 also contains the root. We bisect the interval I_1 and get a sub-interval I_2 at whose end points $f(x)$ takes the values of opposite signs and therefore contains the root. Continuing this procedure, we obtain a sequence of nested sets of sub-intervals $I_0 \supset I_1 \supset I_2 \ldots$ such that each sub-interval contains the root. After repeating the bisection process q times, we either find the root or find the interval I_q of length $(b_0 - a_0)/2^q$ which contains the root. We take the midpoint of the last sub-interval as the desired approximation to the root. This root has error not greater than one half of the length of the interval of which it is the midpoint. Thus we have

$$m_{k+1} = a_k + \tfrac{1}{2}(b_k - a_k), \quad k = 0, 1, 2, \ldots$$

where

$$(a_{k+1}, b_{k+1}) = \begin{cases} (a_k, m_{k+1}), & \text{if } f(a_k)f(m_{k+1}) < 0 \\ (m_{k+1}, b_k), & \text{if } f(m_{k+1})f(b_k) < 0 \end{cases}$$

We notice that this method uses only the end points of the interval $[a_k, b_k]$ for which $f(a_k)f(b_k) < 0$ and not the values of $f(x)$ at these end points, to obtain the next approximation to the root. The method is simple to use and the sequence of approximations always converges to the root for any $f(x)$ which is continuous in the interval that contains the root. If the permissible error is ϵ then the approximate number of iterations required may be determined from the relation

$$\frac{b_0 - a_0}{2^n} \leqslant \epsilon \quad \text{or} \quad n \geqslant \frac{\log(b_0 - a_0) - \log \epsilon}{\log 2}$$

The minimum number of iterations required for converging to a root in the interval $(0, 1)$ for a given ϵ are listed in Table 2.2.

Table 2.2 Number of Iterations

ϵ	10^{-2}	10^{-3}	10^{-4}	10^{-5}	10^{-6}	10^{-7}
n	7	10	14	17	20	24

Thus, the bisection method requires a large number of iterations to achieve a reasonable degree of accuracy for the root. It requires one function evaluation for each iteration.

Example 2.2 Perform five iterations of the bisection method to obtain the smallest positive root of the equation

$$f(x) = x^3 - 5x + 1 = 0$$

Since $f(0) > 0$ and $f(1) < 0$, the smallest positive root lies in the interval $(0, 1)$. Taking $a_0 = 0$, $b_0 = 1$, we get

$$m_1 = \tfrac{1}{2}(a_0 + b_0) = \tfrac{1}{2}(0 + 1) = 0.5$$
$$f(m_1) = -1.375 \quad \text{and} \quad f(a_0)f(m_1) < 0$$

Thus the root lies in the interval $(0, 0.5)$. The sequence of intervals are given in Table 2.3.

Table 2.3 Sequence of Intervals for the Bisection Method

k	a_{k-1}	b_{k-1}	m_k	$f(m_k)f(a_{k-1})$
1	0	1	0.5	< 0
2	0	0.5	0.25	< 0
3	0	0.25	0.125	> 0
4	0.125	0.25	0.1875	> 0
5	0.1875	0.25	0.21875	< 0

Hence, the root lies in $(0.1875, 0.21875)$. The approximate root is taken as 0.203125.

2.3 ITERATION METHODS BASED ON FIRST DEGREE EQUATION

We have already seen that if $f(x) = 0$ is a first degree equation in x then it can be readily solved. We now study the iteration methods which will produce exact results whenever $f(x) = 0$ is a first degree equation. Thus, if we approximate $f(x)$ by a first degree equation then we may write

$$f(x) = a_0 x + a_1 = 0 \tag{2.8}$$

The solution of (2.8) is given by

$$x = -a_1/a_0 \tag{2.9}$$

where $a_0 \neq 0$ and a_1 are arbitrary parameters to be determined by prescribing two appropriate conditions on $f(x)$ and/or its derivatives.

2.3.1 Secant Method

If x_{k-1} and x_k are two approximations to the root, then we determine a_0 and a_1 in (2.8) using the conditions

$$f_{k-1} = a_0 x_{k-1} + a_1$$
$$f_k = a_0 x_k + a_1$$

where $f_{k-1} = f(x_{k-1})$ and $f_k = f(x_k)$.

On solving, we obtain

$$a_0 = (f_k - f_{k-1})/(x_k - x_{k-1})$$
$$a_1 = (x_k f_{k-1} - x_{k-1} f_k)/(x_k - x_{k-1}) \qquad (2.10)$$

From the equations (2.9) and (2.10), the next approximation x_{k+1} to the root is given by

$$x_{k+1} = \frac{x_{k-1} f_k - x_k f_{k-1}}{f_k - f_{k-1}} \qquad (2.11)$$

which may also be written as

$$x_{k+1} = x_k - \frac{x_k - x_{k-1}}{f_k - f_{k-1}} f_k \qquad (2.12)$$

This is called the **secant** or the **chord method**.

Geometrically, in this method we replace the function $f(x)$ by a straight line or a chord passing through the points (x_k, f_k) and (x_{k-1}, f_{k-1}) and take the point of intersection of the straight line with the x-axis as the next approximation to the root. If the approximations are such that $f_k f_{k-1} < 0$, then the method (2.11) or (2.12) is known as the **Regula-Falsi** method. The method is shown graphically in Figure 2.2. Since (x_{k-1}, f_{k-1}), (x_k, f_k) are known before the start of the iteration, the secant method requires one function evaluation per step.

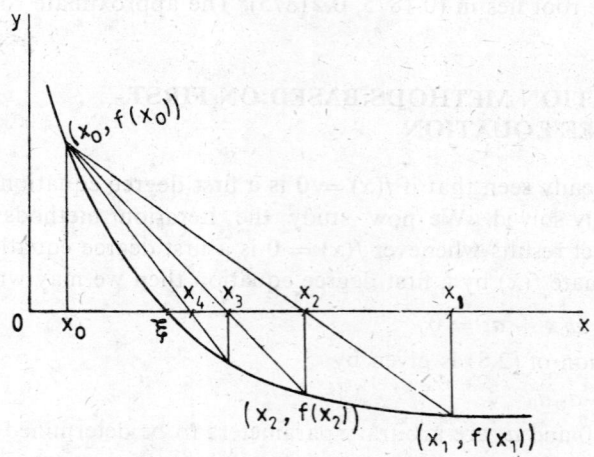

Fig. 2.2 The Regula-Falsi method

Example 2.3 Use the secant and regula-falsi methods to determine the root of the equation

$$\cos x - x e^x = 0$$

Taking the initial approximations as $x_0 = 0$, $x_1 = 1$, we have for the secant method

$$f(0) = 1, f(1) = \cos 1 - e = -2 \cdot 177979523$$

$$x_2 = x_1 - \frac{x_1 - x_0}{f_1 - f_0} f_1 = 0.3146653378$$

$$f(.3146653378) = 0.519871175$$

$$x_3 = x_2 - \frac{x_2 - x_1}{f_2 - f_1} f_2 = 0.4467281466$$

The computed results are tabulated in Table 2.4.

Table 2.4 Approximations to the Root by the Secant and the Regula-Falsi Methods

k	Secant Method		Regula-Falsi Method	
	x_{k+1}	$f(x_{k+1})$	x_{k+1}	$f(x_{k+1})$
1	0.3146653378	0.519871	0.3146653378	0.519871
2	0.4467281466	0.203545	0.4467281446	0.203545
3	0.5317058606	−0.429311(−01)	0.4940153366	0.708023(−01)
4	0.5169044676	0.259276(−02)	0.5099461404	0.236077(−01)
5	0.5177474653	0.301119(−04)	0.5152010099	0.776011(−02)
6	0.5177573708	−0.215132(−07)	0.5169222100	0.253886(−02)
7	0.5177573637	0.178663(−12)	0.5174846768	0.829358(−03)
8	0.5177573637	0.222045(−15)	0.5176683450	0.270786(−03)
10	—	—	0.5177478783	0.288554(−04)
20	—	—	0.5177573636	0.396288(−09)

The numbers within the parenthesis denote exponentiation.

2.3.2 The Newton-Raphson Method

We determine a_0 and a_1 in (2.8) using the conditions

$$f_k = a_0 x_k + a_1$$

$$f'_k = a_0 \qquad (2.13)$$

where a prime denotes differentiation with respect to x.

On substituting a_0 and a_1 from (2.13) in (2.9) and representing the approximate value of x by x_{k+1}, we obtain

$$x_{k+1} = x_k - \frac{f_k}{f'_k} \qquad (2.14)$$

This method is called the **Newton-Raphson** method. The method (2.14) may also be obtained directly from (2.12) by taking the limit $x_{k-1} \to x_k$. In the limit when $x_{k-1} \to x_k$, the chord passing through the points (x_k, f_k) and (x_{k-1}, f_{k-1}) becomes the tangent at the point (x_k, f_k). Thus, in this case the problem of finding the root of the equation (2.1) is equivalent to finding the point of intersection of the tangent to the curve, $y = f(x)$ at the point (x_k, f_k)

with the x-axis. The method is shown graphically in Figure 2.3. The Newton-Raphson method requires two evaluations f_k, f'_k for each iteration.

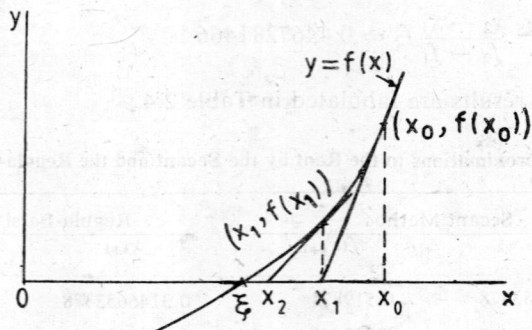

Fig. 2.3 The Newton-Raphson method

Example 2.4 Apply Newton-Raphson's method to determine a root of the equation

$$f(x) = \cos x - xe^x = 0$$

such that $|f(x^*)| < 10^{-8}$, where x^* is the approximation to the root. We write (2.14) in the form

$$x_{k+1} = x_k - \Delta x_k, \quad k = 0, 1, 2, \ldots$$

where $\Delta x_k = \dfrac{(\cos x_k - x_k e^{x_k})}{(-\sin x_k - x_k e^{x_k} - e^{x_k})}$

Starting with $x_0 = 1$, the results obtained are given in Table 2.5.

Table 2.5 Approximations to the Root by the Newton-Raphson Method

k	x_k	Δx_k	x_{k+1}	$f(x_{k+1})$
0	1.0	0.3469	0.65307940	-0.4606
1	0.65307940	0.1217	0.53134337	$-0.4180(-1)$
2	0.53134337	0.1343(-1)	0.51790991	$-0.4641(-3)$
3	0.51790991	0.1525(-3)	0.51775738	$-0.5926(-7)$
4	0 51775738	0.1948(-7)	0.51775736	$-0.2910(-10)$

Example 2.5 Show that the initial approximation x_0 for finding $1/N$, where N is a positive integer, by the Newton-Raphson method must satisfy $0 < x_0 < 2/N$, for convergence.

We write $f(x) = \dfrac{1}{x} - N = 0$.

The Newton-Raphson method becomes

$$x_{n+1} = 2x_n - Nx_n^2$$

Let us now draw the graphs of $y = x$ and $y = 2x - Nx^2$. The second curve is the parabola

$$\left(x - \frac{1}{N}\right)^2 = -\frac{1}{N}\left(y - \frac{1}{N}\right)$$

The graphs are given in Figure 2.4. The point of intersection of these two curves is the required value $1/N$. From Fig. 2.4, we find that any initial approximation outside the range $0 < x_0 < 2/N$ diverges. If $x_0 = 0$, the iteration does not converge to $1/N$ but remains zero always. This shows the importance of choosing a suitable initial approximation.

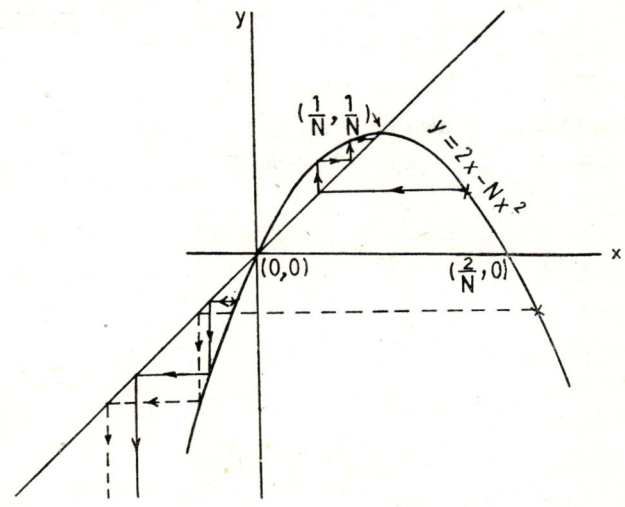

Fig. 2.4. Choice of initial approximation

2.4 ITERATION METHODS BASED ON SECOND DEGREE EQUATION

We assume for the function $f(x)$ a polynomial of degree two and write as

$$f(x) = a_0 x^2 + a_1 x + a_2 = 0, \quad a_0 \neq 0 \qquad (2.15)$$

where a_0, a_1 and a_2 are arbitrary parameters to be determined by prescribing three appropriate conditions on $f(x)$ and its derivatives.

2.4.1 The Muller Method

If x_{k-2}, x_{k-1} and x_k are three approximations to the root ξ of $f(x) = 0$, then we may determine a_0, a_1 and a_2 in (2.15) by using the conditions

$$f_{k-2} = a_0 x_{k-2}^2 + a_1 x_{k-2} + a_2$$
$$f_{k-1} = a_0 x_{k-1}^2 + a_1 x_{k-1} + a_2$$
$$f_k = a_0 x_k^2 + a_1 x_k + a_2 \qquad (2.16)$$

Eliminating a_0, a_1 and a_2 from (2.15) and (2.16), we get

$$\begin{vmatrix} f(x) & x^2 & x & 1 \\ f_{k-2} & x_{k-2}^2 & x_{k-2} & 1 \\ f_{k-1} & x_{k-1}^2 & x_{k-1} & 1 \\ f_k & x_k^2 & x_k & 1 \end{vmatrix} = 0 \qquad (2.17)$$

which we may simplify and obtain

$$f(x) = \frac{(x-x_{k-1})(x-x_k)}{(x_{k-2}-x_{k-1})(x_{k-2}-x_k)} f_{k-2} + \frac{(x-x_{k-2})(x-x_k)}{(x_{k-1}-x_{k-2})(x_{k-1}-x_k)} f_{k-1}$$
$$+ \frac{(x-x_{k-2})(x-x_{k-1})}{(x_k-x_{k-2})(x_k-x_{k-1})} f_k = 0 \qquad (2.18)$$

The equation (2.18) may also be written as

$$\frac{h(h+h_k)}{h_{k-1}(h_{k-1}+h_k)} f_{k-2} - \frac{h(h+h_k+h_{k-1})}{h_k h_{k-1}} f_{k-1}$$
$$+ \frac{(h+h_k)(h+h_k+h_{k-1})}{h_k(h_k+h_{k-1})} f_k = 0 \qquad (2.19)$$

where
$$h = x - x_k$$
$$h_k = x_k - x_{k-1}$$
$$h_{k-1} = x_{k-1} - x_{k-2}$$

We further define

$$\lambda = h/h_k, \quad \lambda_k = h_k/h_{k-1} \quad \text{and} \quad \delta_k = 1 + \lambda_k$$

and express (2.19) in the form

$$\lambda^2 \lambda_k [\lambda_k f_{k-2} - \delta_k f_{k-1} + f_k] \delta_k^{-1} + \lambda g_k \delta_k^{-1} + f_k = 0 \qquad (2.20)$$

where
$$g_k = \lambda_k^2 f_{k-2} - \delta_k^2 f_{k-1} + (\lambda_k + \delta_k) f_k$$

Dividing (2.20) throughout by $f_k \lambda^2$ and solving for $1/\lambda$ we obtain

$$\lambda \approx \lambda_{k+1} = \frac{-2 \delta_k f_k}{g_k \pm \sqrt{g_k^2 - 4 \delta_k f_k c_k}} \qquad (2.21)$$

where $c_k = \lambda_k (\lambda_k f_{k-2} - \delta_k f_{k-1} + f_k)$.

The sign in the denominator in (2.21) is chosen such that λ_{k+1} has the smallest absolute value. Thus we have

$$\lambda_{k+1} = \frac{x - x_k}{x_k - x_{k-1}}$$

or
$$x = x_k + (x_k - x_{k-1}) \lambda_{k+1} \qquad (2.22)$$

Replacing x on the left hand side of (2.22) by x_{k+1}, we obtain the method

$$x_{k+1} = x_k + (x_k - x_{k-1})\lambda_{k+1} \tag{2.23}$$

which is called the **Muller** method. The graphical representation of this method is shown in Fig. 2.5. Generally, this method converges for all initial approximations. If no better approximations are known then we choose $x_0 = -1$, $x_1 = 0$ and $x_2 = 1$. This method is an extension of the secant method. Here, the next approximation x_{k+1} is found as the zero of the second degree curve passing through the points (x_{k-2}, f_{k-2}), (x_{k-1}, f_{k-1}) and (x_k, f_k) where x_{k-2}, x_{k-1} and x_k are the initial approximations to the root.

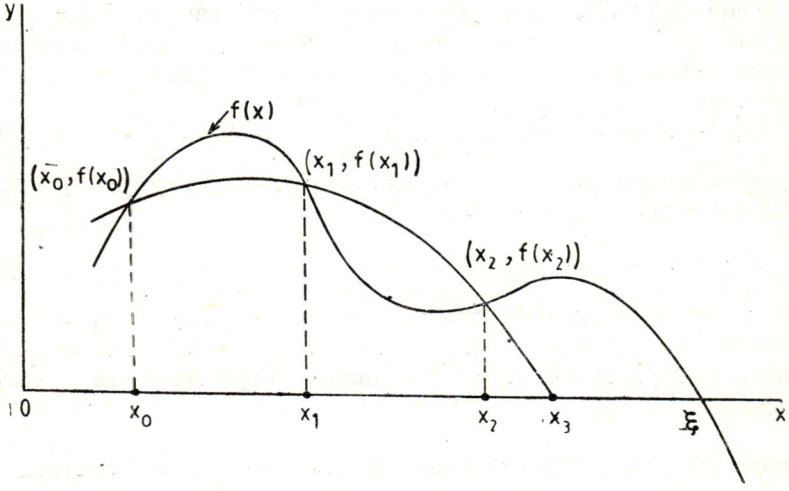

Fig. 2.5 The Muller method

Example 2.6 Perform five iterations of the Muller method to find the root of the equation

$$f(x) = \cos x - xe^x = 0$$

Computations are performed using the initial approximations $x_0 = -1.0$, $x_1 = 0.0$ and $x_2 = 1.0$. The results are given in Table 2.6.

Table 2.6 Approximations to the root by the Muller's Method

k	x_{k-2}	x_{k-1}	x_k	x_{k+1}	λ_{k+1}	$f(x_{k+1})$
2	−1.0	0.0	1.0	0.44151732	−0.5585	0.2175
3	0.0	1.0	0.44151732	0.51254636	−0.1272	0.1578(−1)
4	1.0	0.44151732	0.51254636	0.51769325	0.7246(−1)	0.1950(−3)
5	0.44151732	0.51254636	0.51769325	0.51775737	0.1246(−1)	−0.2240(−7)
6	0.51254636	0.51769325	0.51775737	0.51775736	−0.1148(−3)	−0.1455(−10)

2.4.2 The Chebyshev Method

We determine a_0, a_1 and a_2 in (2.15) using the conditions
$$f_k = a_0 x_k^2 + a_1 x_k + a_2$$
$$f'_k = 2a_0 x_k + a_1$$
$$f''_k = 2a_0 \tag{2.24}$$

On eliminating a's from (2.15) and (2.24) we obtain
$$f_k + (x - x_k)f'_k + \tfrac{1}{2}(x - x_k)^2 f''_k = 0 \tag{2.25}$$

which is the Taylor series expansion of $f(x)$ about $x = x_k$ such that the terms of order $(x - x_k)^3$ and higher powers are neglected.

The equation (2.25) is a quadratic equation and can be solved easily. Only one of the two roots converges to the correct root. In order to get the next approximation to the correct root we write (2.25) as

$$x_{k+1} - x_k = -\frac{f_k}{f'_k} - \frac{1}{2}(x_{k+1} - x_k)^2 \frac{f''_k}{f'_k} \tag{2.26}$$

We substitute for $(x_{k+1} - x_k)$ from (2.14) by $(-f_k/f'_k)$ on the right side of (2.26) and obtain

$$x_{k+1} = x_k - \frac{f_k}{f'_k} - \frac{1}{2}\frac{f_k^2}{f'^3_k} f''_k \tag{2.27}$$

which is called the **Chebyshev** method. This method requires three evaluations for each iteration. If $(x_{k+1} - x_k)$ in the right hand side of (2.26) is replaced by the Secant or Regula-falsi method, the order of the method is reduced.

Example 2.7 Using Chebyshev method, find the root of the equation
$$f(x) = \cos x - xe^x = 0$$
correct to six decimal places.

We write (2.27) in the form
$$x_{k+1} = x_k - \Delta x_k - \Delta^* x_k$$

where $\Delta x_k = \dfrac{f_k}{f'_k}$

$$\Delta^* x_k = \frac{1}{2}\frac{f_k^2 f''_k}{f'^3_k}$$

Choosing the initial approximation $x_0 = 1.0$ we obtain the successive iterates as given in Table 2.7.

Table 2.7 Successive Iterates for the Root by the Chebyshev Method

k	x_k	Δx_k	$\Delta^* x_k$	x_{k+1}	$f(x_{k+1})$
0	1.0	0.3469	0.8335(-1)	0.56973366	-0.1651
1	0.56973366	0.4982(-1)	0.2016(-2)	0.51789543	$-0.4201(-3)$
2	0.51789543	0.1381(-3)	0.1596(-7)	0.51775738	$-0.1455(-10)$

2.4.3 Multipoint Iteration Methods

We write the equation (2.25) as

$$x_{k+1} - x_k = -\frac{f_k}{f'_k + \tfrac{1}{2}(x_{k+1} - x_k)f''_k}$$

$$\simeq -\frac{f_k}{f'(x_k + \tfrac{1}{2}(x_{k+1} - x_k))} \qquad (2.28)$$

and again replace $(x_{k+1} - x_k)$ by $(-f_k/f'_k)$ on the right side of (2.28) to get the next approximation to the root ξ. We have

$$x_{k+1} = x_k - \frac{f_k}{f'(x_k - \tfrac{1}{2} f_k/f'_k)}, \quad k = 0, 1, 2, \ldots \qquad (2.29)$$

which may be called a **multipoint iteration** method.

For computation purposes we may write (2.29) as the two-stage method

$$x^*_{k+1} = x_k - \frac{1}{2}\frac{f_k}{f'_k}$$

$$x_{k+1} = x_k - \frac{f_k}{f'^*_{k+1}}, \quad k = 0, 1, 2, \ldots$$

where $f'^*_{k+1} = f'(x^*_{k+1})$.

We may also write (2.25) in the form

$$x_{k+1} - x_k = -\frac{1}{f'_k}\left(f(x_k + (x_{k+1} - x_k)) - (x_{k+1} - x_k)f'_k\right) \qquad (2.30)$$

and replace $(x_{k+1} - x_k)$ by $(-f_k/f'_k)$ on the right side of (2.30), to get the multipoint iteration method

$$x_{k+1} = x_k - \frac{f_k}{f'_k} - \frac{f(x_k - f_k/f'_k)}{f'_k} \qquad (2.31)$$

This method may also be expressed as the two-stage method

$$x^*_{k+1} = x_k - \frac{f_k}{f'_k}$$

$$x_{k+1} = x^*_{k+1} - \frac{f^*_{k+1}}{f'_k}$$

where $f^*_{k+1} = f(x^*_{k+1})$.

These methods require three evaluations for each iteration as in Chebyshev method, but we avoid evaluating the second derivative which may be advisable in many cases.

Example 2.8 Perform three iterations of the multipoint iteration method (2.31), to find the root of the equation

$$f(x) = \cos x - xe^x = 0$$

Starting with the initial approximation $x_0 = 1$ we get the following results as given in Table 2.8.

Table 2.8 Approximations to the Root by the Method (2.31)

k	x_k	x_{k+1}^*	x_{k+1}	$f(x_{k+1})$
0	1.0	0.6530794035	0.5797057848	-0.1984
1	0.5797057848	0.5207904077	0.5180441619	$-0.8726(-03)$
2	0.5180441619	0.5177574325	0.5177573637	$-0.1006(-09)$
3	0.5177573637	0.5177573637	0.5177573637	$0.1388(-16)$

2.5 RATE OF CONVERGENCE

We now study the rate at which the iteration method converges if the initial approximation to the root is sufficiently close to the desired root.

Definition 2.4 An iterative method is said to be of **order** p or has the rate of **convergence** p, if p is the largest positive real number for which there exists a finite constant $C \neq 0$ such that

$$|\epsilon_{k+1}| \leq C|\epsilon_k|^p \qquad (2.32)$$

where $\epsilon_k = x_k - \xi$ is the error in the kth iterate.

The constant C is called the **asymptotic error** constant and usually depends on derivatives of $f(x)$ at $x = \xi$.

2.5.1 Secant Method

We assume that ξ is a simple root of $f(x) = 0$ and substitute $x_k = \xi + \epsilon_k$ in (2.12) and obtain

$$\epsilon_{k+1} = \epsilon_k - \frac{(\epsilon_k - \epsilon_{k-1}) f(\xi + \epsilon_k)}{f(\xi + \epsilon_k) - f(\xi + \epsilon_{k-1})} \qquad (2.33)$$

Expanding $f(\xi + \epsilon_k)$ and $f(\xi + \epsilon_{k-1})$ in Taylor's series about the point ξ and noting that $f(\xi) = 0$, we get

$$\epsilon_{k+1} = \epsilon_k - \left[\epsilon_k + \tfrac{1}{2} \epsilon_k^2 \frac{f''(\xi)}{f'(\xi)} + \ldots \right] \left[1 + \tfrac{1}{2} (\epsilon_{k-1} + \epsilon_k) \frac{f''(\xi)}{f'(\xi)} + \ldots \right]^{-1}$$

or

$$\epsilon_{k+1} = \frac{1}{2} \frac{f''(\xi)}{f'(\xi)} \epsilon_k \epsilon_{k-1} + O(\epsilon_k^2)$$

or

$$\epsilon_{k+1} = C \epsilon_k \epsilon_{k-1} \qquad (2.34)$$

where $C = \dfrac{1}{2} \dfrac{f''(\xi)}{f'(\xi)}$ and higher powers of ϵ_k are neglected.

The relation of the form (2.34) is called the **error equation**. Keeping in view the definition of the rate of convergence, we seek a relation of the form

$$\epsilon_{k+1} = A\epsilon_k^p \qquad (2.35)$$

where A and p are to be determined.

From (2.35) we have

$$\epsilon_k = A\epsilon_{k-1}^p \quad \text{or} \quad \epsilon_{k-1} = A^{-1/p}\epsilon_k^{1/p}$$

Substituting the values of ϵ_{k+1} and ϵ_{k-1} in (2.34) we obtain

$$\epsilon_k^p = CA^{-(1+1/p)}\epsilon_k^{1+1/p}$$

Comparing the powers of ϵ_k on both sides, we get

$$p = 1 + \frac{1}{p}.$$

which gives

$$p = \tfrac{1}{2}(1 \pm \sqrt{5}) \qquad (2.36)$$

Neglecting the minus sign, we find that the rate of convergence for the Secant method (2.12) is $p = 1.618$.

2.5.2 The Newton-Raphson Method

On substituting $x_k = \xi + \epsilon_k$ in (2.14), expanding $f(\xi + \epsilon_k), f'(\xi + \epsilon_k)$ in Taylor's series about the point ξ, we obtain after simplification the error equation

$$\epsilon_{k+1} = \frac{1}{2}\frac{f''(\xi)}{f'(\xi)}\epsilon_k^2 + 0(\epsilon_k^3)$$

On neglecting ϵ_k^3 and higher powers of ϵ_k, we get

$$\epsilon_{k+1} = C\epsilon_k^2 \qquad (2.37)$$

where

$$C = \frac{1}{2}\frac{f''(\xi)}{f'(\xi)}$$

Thus the Newton-Raphson method has second order convergence.

It can be verified that the rate of convergence of the Muller method is 1.84. The Chebyshev and multipoint methods have third order convergence.

Example 2.9 Determine α_1 and α_2 so that the order of the iterative method

$$x_{k+1} = x_k - \alpha_1 W_1(x_k) - \alpha_2 W_2(x_k)$$

where $W_1(x_k) = f(x_k)/f'(x_k)$ and $W_2(x_k) = f(x_k)/f'(x_k + \beta W_1(x_k))$, $\beta \neq 0$, for finding a root of the equation $f(x) = 0$ becomes as high as possible.

34 Numerical Methods

We substitute $x_k = \xi + \epsilon_k$ and $f(\xi) = 0$ in $W_1(x_k)$ and $W_2(x_k)$ to get

$$W_1(\xi + \epsilon_k) = \frac{f(\xi + \epsilon_k)}{f'(\xi + \epsilon_k)} = \epsilon_k - \frac{1}{2}\frac{f''(\xi)}{f'(\xi)}\epsilon_k^2 + 0(\epsilon_k^3)$$

$$W_2(\xi + \epsilon_k) = \frac{f(\xi + \epsilon_k)}{f'(\xi + \epsilon_k + \beta W_1(\xi + \epsilon_k))}$$

$$= \epsilon_k + \frac{1}{2}\frac{f''(\xi)}{f'(\xi)}(1 - 2(1 + \beta))\epsilon_k^2 + 0(\epsilon_k^3)$$

Using these values in the iteration formula and equating the coefficients of ϵ_k and ϵ_k^2 to zero, we obtain

$$\alpha_1 + \alpha_2 = 1$$
$$\alpha_1 + \alpha_2(1 + 2\beta) = 0$$

If $\beta \neq 0$, then we have

$$\alpha_1 = \frac{1 + 2\beta}{2\beta}, \quad \alpha_2 = -\frac{1}{2\beta}$$

Thus, the order of the iteration method is three for arbitrary $\beta \neq 0$.

2.6 ITERATION METHODS

Some of the iteration methods considered in the previous sections are special cases of the following

$$x_{k+1} = \phi(x_k), \quad k = 0, 1, 2, \ldots \tag{2.38}$$

The function $\phi(x)$ is known as an **iteration function**. The convergence of an iteration method depends on the suitable choice of the function $\phi(x)$, and x_0, a suitable initial approximation to the root.

Suppose x_k converges to the value ξ so that we have

$$\xi = \phi(\xi) \tag{2.39}$$

Subtracting (2.39) from (2.38), we obtain

$$x_{k+1} - \xi = \phi(x_k) - \phi(\xi) \tag{2.40}$$

which may also be written as

$$\epsilon_{k+1} = a_1 \epsilon_k + a_2 \epsilon_k^2 + 0(\epsilon_k^3), \quad k = 0, 1, 2, \ldots \tag{2.41}$$

where

$$\epsilon_k = x_k - \xi, \quad a_1 = \phi'(\xi), \quad a_2 = \tfrac{1}{2}\phi''(\xi)$$

We now consider the following cases:

2.6.1 First Order Method

Here $a_1 \neq 0$ and the equation (2.41) becomes

$$\epsilon_{k+1} = a_1 \epsilon_k, \quad k = 0, 1, 2, \ldots \tag{2.42}$$

From (2.42), we have

$$\epsilon_k = a_1^k \epsilon_0$$

If $|a_1| < 1$ and ϵ_0 is not very large then the iteration method (2.38) converges. The method has the first order or linear convergence. Thus each iteration adds approximately the same number of significant figures to the accuracy. Further, if x_0 is in the neighbourhood of the root ξ, and $a_1 > 0$, then a convergent iteration method provides a **staircase** solution as shown in Fig. 2.6, and it provides a spiral solution as shown in Fig. 2.7, if $a_1 < 0$.

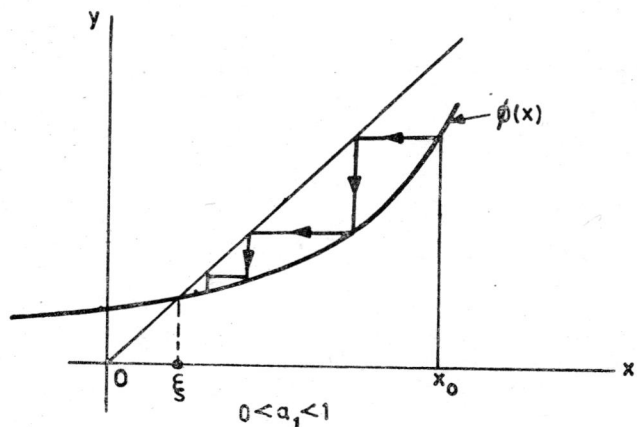

Fig. 2.6 Stair case solution

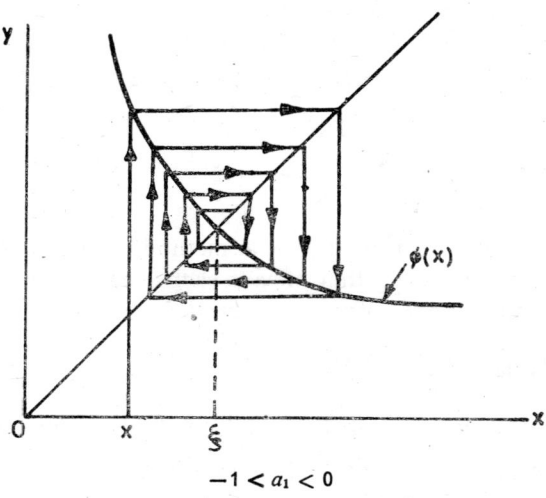

Fig. 2.7 Spiral case solution

A sufficient condition on $\phi(x)$ for the convergence is given in the following theorem:

36 Numerical Methods

THEOREM 2.2 *If $\phi(x)$ is a continuous function in the interval $[a, b]$ that contains the root and $|\phi'(x)| \leq c < 1$ in this interval, then for any choice of $x_0 \in [a, b]$, the sequence $\{x_k\}$ determined from*

$$x_{k+1} = \phi(x_k), \quad k = 0, 1, 2, \ldots$$

converges to the root ξ of $x = \phi(x)$.

Proof. The exact solution ξ satisfies the equation

$$\xi = \phi(\xi)$$

We have

$$\xi - x_{k+1} = \phi(\xi) - \phi(x_k), \quad k = 0, 1, 2, \ldots$$

Using the Mean Value Theorem, we get

$$\xi - x_{k+1} = (\xi - x_k)\phi'(\bar{\xi}_k), \quad x_k < \bar{\xi}_k < \xi$$

or

$$\epsilon_{k+1} = \epsilon_k \phi'(\bar{\xi}_k)$$
$$= \epsilon_{k-1} \phi'(\bar{\xi}_k) \phi'(\bar{\xi}_{k-1})$$
$$= \epsilon_0 \phi'(\bar{\xi}_k) \phi'(\bar{\xi}_{k-1}) \ldots \phi'(\bar{\xi}_0)$$

where $x_0 < \bar{\xi}_0 < \xi$, $x_1 < \bar{\xi}_1 < \xi, \ldots, x_{k-1} < \bar{\xi}_{k-1} < \xi$.

If $|\phi'(\bar{\xi}_r)| \leq c$, $r = 0, 1, 2, \ldots, k$, then

$$|\epsilon_{k+1}| \leq |\epsilon_0| c^{k+1}$$

If $c < 1$, the right hand side goes to zero as k becomes large. Thus, the iteration method converges if $|\phi'(x)| \leq c < 1$. This condition is same as the Lipschitz condition and c is the Lipschitz constant.

2.6.2 Second Order Method

Here $a_1 = 0$, $a_2 \neq 0$ and the equation (2.41) becomes

$$\epsilon_{k+1} = a_2 \epsilon_k^2, \quad k = 0, 1, 2, \ldots \tag{2.43}$$

which can also be written as

$$\epsilon_k = a_2^{2^k - 1} \epsilon_0^{2^k}$$

If $|a_2| \simeq 1$, ϵ_0 is small, then the iteration method (2.38) converges and has second order convergence, that means each successive iteration approximately doubles the number of significant figures of accuracy.

2.6.3 High Order Methods

Definition 2.5 The iteration method (2.38) is said to be of the p-th order if

$$\phi'(\xi) = \phi''(\xi) = \ldots = \phi^{(p-1)}(\xi) = 0, \; \phi^{(p)}(\xi) \neq 0$$

where ξ is the solution of $x = \phi(x)$.

The equation (2.41) for a p-th order method becomes

$$\epsilon_{k+1} = \frac{1}{p!} \phi^{(p)}(\xi) \epsilon_k^p + O(\epsilon_k^{p+1}) \tag{2.44}$$

Transcendental and Polynomial Equations 37

Thus, the number of significant figures of accuracy at each step is approximately p-times the number of significant figures of accuracy of the previous step.

2.6.4 Acceleration of the Convergence

The linear convergence of the iterative method (2.38) can be improved with the help of the **Aitken** Δ^2 method. The error in two successive approximations, using (2.42), may be written as

$$\epsilon_{k+1} = a_1 \epsilon_k$$

$$\epsilon_{k+2} = a_1 \epsilon_{k+1} \tag{2.45}$$

Eliminating a_1 in (2.45) and solving for ξ, we get

$$\xi = \frac{x_k x_{k+2} - x_{k+1}^2}{x_{k+2} - 2x_{k+1} + x_k} = x_k - \frac{(x_{k+1} - x_k)^2}{x_{k+2} - 2x_{k+1} + x_k} \tag{2.46}$$

The number ξ gives an improved value of the approximation x_{k+2}. The computational procedure can be written as follows. Choose an initial approximation x_0 and calculate $x_1 = \phi(x_0)$, $x_2 = \phi(x_1)$, and determine the sequence $\{x_k\}$ from the following:

$$x_{3k+2}^* = x_{3k} - \frac{(\phi(x_{3k}) - x_{3k})^2}{\phi(\phi(x_{3k})) - 2\phi(x_{3k}) + x_{3k}}, \quad k = 0, 1, 2, \ldots$$

Example 2.10 Perform two iterations of the linear iteration method followed by one iteration of the Aitken Δ^2 method to find the root of the equation

$$f(x) = \cos x - xe^x = 0$$

Take the initial approximation $x_0 = 0.0$. Repeat the process three times.

We write the given equation in the form

Table 2.9 Approximations to the Root by the Aitken Δ^2 Method

k	x_k	x_{k+1}	x_k^*
0	0.0	0.50000000	
1	0.50000000	0.52661096	
2	0.52661096		0.52810686
3	0.52810686	0.51222771	
4	0.51222771	0.52059979	
5	0.52059979		0.51770956
6	0.51770956	0.51778227	
7	0.51778227	0.51774438	
8	0.51774438		0.51775736

38 *Numerical Methods*

$$x_{k+1} = \phi(x_k), \quad k = 0, 1, 2, \ldots$$

where

$$\phi(x) = x + \tfrac{1}{2}(\cos x - xe^x)$$

The results obtained with the initial approximation $x_0 = 0.0$ are given in Table 2.9.

2.6.5 Efficiency of a Method

Definition 2.6 The efficiency index of an iterative method in the sense of **Traub** is defined by the equation

$$E^* = p^{1/n}$$

where p is the order of the method and n is the total number of function and derivative evaluations at each step of the iteration. Obviously, if the value of the index is larger, then the method is more efficient. Table 2.10 gives the efficiency index of the methods discussed in the previous sections.

Table 2.10 Comparison of Efficiency Index

Method	n	p	E^*
Secant	1	1.62	1.62
Newton-Raphson	2	2	1.41
Chebyshev	3	3	1.44
Muller	1	1.84	1.84
Multipoint	3	3	1.44

Alternatively, if the cost of evaluating $f(x)$ and $f'(x)$ are taken as 1 and w respectively, then for the Newton-Raphson method we have

$$E = \log E^* = \frac{\log 2}{1 + w}$$

where $n = 1 + w$ is the total cost of evaluation at each iteration. For a multipoint method with one function and two derivative evaluations, we have

$$E = \frac{\log 3}{1 + 2w}$$

If the multipoint method is to be superior to the Newton-Raphson method, then we require

$$\frac{\log 3}{1 + 2w} > \frac{\log 2}{1 + w}, \quad \text{i.e.,} \quad w < 1.4$$

Similarly, if the Newton-Raphson method is to be superior to the Secant method, we require

$$\frac{\log 2}{1+w} > \log 1.618, \text{ i.e., } w < 0.44$$

Thus, the cost of evaluating the derivatives may also decide the method that may be used.

2.6.6 Methods for Multiple Roots

If ξ is a multiple root of multiplicity m of the equation (2.1) then we have from definition 2.2,

$$f(\xi) = f'(\xi) = \ldots = f^{(m-1)}(\xi) = 0 \text{ and } f^{(m)}(\xi) \neq 0 \qquad (2.47)$$

It can be easily verified that all the iteration methods discussed in previous sections have only linear rate of convergence when $m > 1$. For example, the error equation in the Newton-Raphson method (2.14) becomes

$$\epsilon_{k+1} = \left(1 - \frac{1}{m}\right)\epsilon_k + \frac{1}{m^2(m+1)} \epsilon_k^2 \frac{f^{(m+1)}(\xi)}{f^{(m)}(\xi)} + O(\epsilon_k^3)$$

If $m \neq 1$, then we obtain

$$\epsilon_{k+1} = \left(1 - \frac{1}{m}\right)\epsilon_k + O(\epsilon_k^2)$$

which shows that the method has only linear rate of convergence. However, when the multiplicity of the root is known in advance we can modify the methods by introducing parameters dependent on the multiplicity of the root to increase their order of convergence. For example, consider the Newton-Raphson method in the form

$$x_{k+1} = x_k - \alpha \frac{f_k}{f'_k} \qquad (2.48)$$

where α is an arbitrary parameter to be determined. If ξ is a multiple root of multiplicity m, we obtain from (2.48), the error equation

$$\epsilon_{k+1} = \left(1 - \frac{\alpha}{m}\right)\epsilon_k + \frac{\alpha}{m^2(m+1)} \epsilon_k^2 \frac{f^{(m+1)}(\xi)}{f^{(m)}(\xi)} + O(\epsilon_k^3)$$

If the method (2.48) is to have quadratic rate of convergence, then the coefficient of ϵ_k must vanish, which gives

$$1 - \frac{\alpha}{m} = 0 \quad \text{or} \quad \alpha = m$$

Thus the method

$$x_{k+1} = x_k - m \frac{f_k}{f'_k} \qquad (2.49)$$

has quadratic rate of convergence for determining a multiple root of multiplicity m.

If the multiplicity of the root is not known in advance, then we use the following procedure. It is known that if $f(x) = 0$ has a root ξ of multiplicity m, then $f'(x) = 0$ has the same root ξ and is of multiplicity $m - 1$. Hence

$g(x) = f(x)/f'(x)$ has a simple root ξ. We can now use the Newton-Raphson method

$$x_{k+1} = x_k - \frac{g(x_k)}{g'(x_k)}$$

to find the approximate value of the multiple root ξ. Simplifying, we have

$$x_{k+1} = x_k - \frac{f_k f'_k}{f'^2_k - f_k f''_k} \tag{2.50}$$

It can easily be verified that this method has second order convergence.

The Secant method can also be generalized to find a multiple root. We can apply the Secant method on $g = f/f'$, which has a simple root. In this case the Secant method becomes

$$x_{k+1} = \frac{x_{k-1} g_k - x_k g_{k-1}}{g_k - g_{k-1}} = \frac{x_{k-1} f_k f'_{k-1} - x_k f_{k-1} f'_k}{f_k f'_{k-1} - f_{k-1} f'_k} \tag{2.51}$$

However, this method (2.51) can be made derivative free, if we use the approximation

$$g(x) \simeq G(x) = -\frac{f^2(x)}{f(x - f(x)) - f(x)}$$

When $G(x)$ is substituted in the Secant method, we have

$$x_2 = x_1 - (x_0 - x_1) \frac{G_1}{G_0 - G_1} \tag{2.52}$$

where $G_1 = G(x_1)$ and $G_0 = G(x_0)$. The order of convergence is again 1.618. Since two function evaluations are required, the efficiency index is 1.272. While finding a multiple root, roundoff errors are avoided by using multi-precision arithmetic. The above method also gives an estimate of the order of multiplicity. The multiplicity m is obtained by computing

$$m \simeq \frac{x_2 - x_1}{G_2 - G_1}$$

at each step of iteration.

Example 2.11 Find the multiple root of the equation

$$f(x) = 27x^5 + 27x^4 + 36x^3 + 28x^2 + 9x + 1 = 0$$

using the methods (2.14), (2.50), (2.51) and (2.52).

The root lies in $(-1, 0)$. For the Newton-Raphson methods (2.14) and (2.50) the initial approximation is taken as -1.0. For the Secant method (2.51), choose $x_0 = 0$ and $x_1 = -1$. The modified Secant method (2.52) diverged when $x_0 = 0$ and $x_1 = -1$. Another set of initial values $x_0 = 0$, $x_1 = -0.5$ are chosen for the method (2.52). The computational results are given in Table 2.11. The exact root is $-1/3$.

Table 2.11 Approximations to the Multiple Roots Using Different Methods

k	Method (2.14)		Method (2.50)	
	x_{k+1}	f_{k+1}	x_{k+1}	f_{k+1}
1	−0.81818181	−0.514(1)	−0.18518518	0.908(−1)
2	−0.67866701	−0.162(1)	−0.33266153	0.909(−8)
3	−0.57467968	−0.505(0)	−0.33333324	0.139(−15)
4	−0.49945937	−0.155(0)	−0.33333320	−0.208(−15)
5	−0.44643164	−0.468(−1)	−0.33333313	−0.153(−15)
6	−0.40976142	−0.141(−1)	−0.33333303	−0.180(−15)
20	−0.33360180	−0.581(−9)	—	—

k	Method (2.51)		Method (2.52)	
	x_{k+1}	f_{k+1}	x_{k+1}	f_{k+1}
1	−0.37931034	−0.300(−2)	−0.13673044	0.209(0)
2	−0.32280294	0.348(−4)	−0.34455964	−0.427(−4)
3	−0.33343798	−0.344(−10)	−0.34089774	−0.130(−4)
4	−0.33333355	0.416(−16)	−0.33336414	−0.877(−12)
5	−0.33335693	−0.394(−12)	−0.33336408	−0.872(−12)
6	−0.33334668	−0.712(−13)	−0.33336410	−0.874(−12)

2.6.7 Methods for Complex Roots

The equation (2.1) in which the independent variable is a complex variable $z = x + iy$ becomes

$$f(z) = 0 \tag{2.53}$$

which can also be written as

$$f(z) = u(x, y) + iv(x, y) = 0 \tag{2.54}$$

where $u(x, y)$ and $v(x, y)$ are the real and imaginary parts of $f(z)$ respectively.

Thus, the problem of finding the roots of the single equation (2.53) is equivalent to determining the roots of two simultaneous equations

$$u(x, y) = 0$$
$$v(x, y) = 0 \tag{2.55}$$

We assume that (x_k, y_k) is an initial approximation to a solution of (2.55) and the exact solution is $(x_k + \Delta x, y_k + \Delta y)$.

Then we have
$$u(x_k + \Delta x, y_k + \Delta y) = 0$$
$$v(x_k + \Delta x, y_k + \Delta y) = 0 \qquad (2.56)$$

The Taylor series expansions about (x_k, y_k) are
$$u(x_k + \Delta x, y_k + \Delta y) = u(x_k, y_k) + [\Delta x\, u_x(x_k, y_k)$$
$$+ \Delta y\, u_y(x_k, y_k)] + \ldots = 0$$
$$v(x_k + \Delta x, y_k + \Delta y) = v(x_k, y_k) + [\Delta x\, v_x(x_k, y_k)$$
$$+ \Delta y\, v_y(x_k, y_k)] + \ldots = 0 \qquad (2.57)$$

where suffixes with respect to x and y represent partial differentiation.

If all terms of second and higher degrees in Δx and Δy are neglected then we obtain

$$u(x_k, y_k) + \Delta x\, u_x(x_k, y_k) + \Delta y\, u_y(x_k, y_k) = 0$$
$$v(x_k, y_k) + \Delta x\, v_x(x_k, y_k) + \Delta y\, v_y(x_k, y_k) = 0$$

Solving for Δx and Δy, we get

$$\Delta x = -\frac{u(x_k, y_k)v_y(x_k, y_k) - v(x_k, y_k)u_y(x_k, y_k)}{J}$$

$$\Delta y = -\frac{u_x(x_k, y_k)v(x_k, y_k) - u(x_k, y_k)v_x(x_k, y_k)}{J}$$

where

$$J = \begin{vmatrix} u_x & u_y \\ v_x & v_y \end{vmatrix}_k$$
$$= u_x(x_k, y_k)v_y(x_k, y_k) - u_y(x_k, y_k)v_x(x_k, y_k)$$

is the **Jacobian** of the functions u and v evaluated at (x_k, y_k). Thus an improved approximation (x_{k+1}, y_{k+1}) to the exact solution is given by

$$x_{k+1} = x_k - J^{-1}(u(x_k, y_k)v_y(x_k, y_k) - v(x_k, y_k)u_y(x_k, y_k))$$
$$y_{k+1} = y_k - J^{-1}(u_x(x_k, y_k)v(x_k, y_k) - u(x_k, y_k)v_x(x_k, y_k)),$$
$$k = 0, 1, 2, \ldots \qquad (2.58)$$

This is the Newton-Raphson method for two variables and has **quadratic** convergence. Let us assume now that $f(z)$ is an analytic function of z. Then the functions $u(x, y)$ and $v(x, y)$ satisfy the **Cauchy-Riemann** equations

$$u_x = v_y$$
$$v_x = -u_y \qquad (2.59)$$

and the equations (2.58) become

$$x_{k+1} = x_k - \frac{u(x_k, y_k)u_x(x_k, y_k) + v(x_k, y_k)v_x(x_k, y_k)}{u_x^2(x_k, y_k) + v_x^2(x_k, y_k)}$$

$$y_{k+1} = y_k - \frac{v(x_k, y_k)u_x(x_k, y_k) - u(x_k, y_k)v_x(x_k, y_k)}{u_x^2(x_k, y_k) + v_x^2(x_k, y_k)}$$

$$k = 0, 1, 2, \ldots \quad (2.60)$$

Alternatively, the Newton-Raphson method (2.14) for the equation (2.53) may be defined by

$$z_{k+1} = z_k - \frac{f(z_k)}{f'(z_k)}, \quad k = 0, 1, 2, \ldots \quad (2.61)$$

It is easy to verify that (2.61) is equivalent to (2.60) when $f(z)$ is an analytic function of z. The initial approximation z_0 in (2.61) must be a complex number and complex arithmetic is used throughout.

Example 2.12 Obtain the complex roots of the equation

$$f(z) = z^3 + 1 = 0$$

correct to eight decimal places. Compare with the exact values of the roots $(1 \pm i\sqrt{3})/2$.

If we use the method (2.61), then with the initial approximation $z_0 = (0.25, 0.25)$ and using complex arithmetic we obtain the successive iterates as given in Table 2.12.

Table 2.12 Approximations to the Complex Root by the Newton-Raphson Method

k	z_k	z_{k+1}	$f(z_{k+1})$
0	(0.25, 0.25)	(0.16666667, 2.83333333)	(0.9687, 0.3125(−1))
1	(0.16666667, 2.83333333)	(0.15220505, 1.89374026)	(−0.3009(1), −0.2251(2))
2	(0.15220505, 1.89374026)	(0.19263553, 1.27724322)	(−0.6340, −0.6660(1))
3	(0.19263553, 0.27724322)	(0.31932197, 0.91041889)	(0.6438(−1), −0.1941(1))
4	(0.31932197, 0.91041889)	(0.49252896, 0.83063199)	(0.2385, −0.4761)
5	(0.49252896, 0.83063199)	(0.49983161, 0.86738607)	(0.1000, 0.3140(−1))
6	(0.49983161, 0.86738607)	(0.49999870, 0.86602675)	(−0.3284(−2), −0.2484(−2))
7	(0.49999870, 0.86602675)	(0.50000000, 0.86602540)	(−0.1548(−5), −0.5414(−5))

Obviously, the approximation to the second root is $(0.5, -0.8660254)$.

2.7 POLYNOMIAL EQUATIONS

The methods discussed in previous sections can be directly applied to obtain the roots of a polynomial equation of degree n

$$P_n(x) = x^n + a_1 x^{n-1} + \ldots + a_{n-1} x + a_n = 0 \quad (2.62)$$

where $a_1, \ldots, a_n$ are real numbers. However, if we are interested in determining all the roots (real and complex) of a polynomial equation then it is advantageous from computational view point to use the direct methods of solution.

It is useful to know the following information about the roots:
  (i) The exact number of real and complex roots along with their multiplicity.
  (ii) The interval in which each root lies.

Definition 2.7 When in a polynomial (terms being written in order) a $+$ sign follows a $+$ sign or a $-$ sign follows a $-$ sign, a continuation or a permanence of signs is said to occur. But if a $+$ sign follows a $-$ sign or a $-$ sign follows a $+$ sign then a change of sign is said to occur. For example, the polynomial

$$P_5(x) = 8x^5 + 12x^4 - 10x^3 + 17x^2 - 18x + 5 = 0 \tag{2.63}$$

has four changes and one continuation of signs.

Descartes' rule of signs

The number of positive real roots of $P_n(x) = 0$ cannot exceed the number of sign changes in $P_n(x)$ and the number of negative real roots of $P_n(x) = 0$ cannot exceed the number of sign changes in $P_n(-x)$.

Thus the polynomial (2.63) has maximum of four positive roots and one negative root. This rule does not give the exact number of real roots but gives only their upper bounds. The exact number of real roots of a polynomial can be found by **Sturm's** theorem.

Let $f(x)$ be a given polynomial of degree n and let $f_1(x)$ represent its first order derivative. Denote by $f_2(x)$ the remainder of $f(x)$ divided by $f_1(x)$ taken with the reverse sign and $f_3(x)$ for the remainder of $f_1(x)$ divided by $f_2(x)$ with the reverse sign and so on, until a constant is arrived at. We thus obtain a sequence of functions

$$f(x), f_1(x), f_2(x), \ldots, f_n(x)$$

which are called the **Sturm functions** or the **Sturm sequence**.

THEOREM (STURM) 2.3 *The number of real roots of the equations $f(x) = 0$ on $[a, b]$ equals the difference between the number of changes of sign in the Sturm sequence at $x = a$ and $x = b$, provided that $f(a) \neq 0, f(b) \neq 0$.*

If $f(x) = 0$ has a multiple root, we can as before obtain the Sturm sequence $f(x), f_1(x), \ldots, f_l(x)$, where $f_l(x)$ will not be a constant in this case. Since $f_l(x)$ gives the greatest common divisor of $f(x)$ and $f'(x)$, the multiplicity of the root of $f(x) = 0$ is one more than that of $f_l(x)$. We obtain a new Sturm sequence by dividing all the functions $f(x), f_1(x), \ldots, f_l(x)$ by $f_l(x)$. Using this sequence we can find the number of real roots of the equation $f(x) = 0$ on $[a, b]$ in the same way, without taking their multiplicity into account. While obtaining the sequence, any constant common factor in any of Sturm functions $f_i(x)$, $i = 0, 1, 2, \ldots, l$ can be neglected.

Since a polynomial of degree n has exactly n roots, the number of complex roots equals ($n -$ number of real roots), where a real root of multiplicity r is to be counted r times.

If $\xi_1, \xi_2, \ldots, \xi_n$ are real and distinct roots of (2.62) then we may write

$$P_n(x) = (x - \xi_1)(x - \xi_2) \ldots (x - \xi_n) = 0 \tag{2.64}$$

However, if we assume that $\xi_1, \xi_2, \ldots, \xi_s$ are real and distinct roots with multiplicity $\nu_1, \nu_2, \ldots, \nu_s$ respectively then the equation (2.64) takes the form

$$P_n(x) = (x - \xi_1)^{\nu_1}(x - \xi_2)^{\nu_2} \ldots (x - \xi_s)^{\nu_s} = 0 \tag{2.65}$$

where $\nu_1 + \nu_2 + \ldots + \nu_s = n$.

It is easily verified by substitution that if either ξ_1, ξ_2 are a complex pair or ξ_1 and ξ_2 are real and multiplied together and all other roots ξ_j, $j = 3, 4, \ldots, n$ are real and distinct then (2.64) becomes

$$P_n(x) = (x^2 + px + q)(x - \xi_3) \ldots (x - \xi_n) \tag{2.66}$$

Thus it is obvious that the methods of finding roots of a polynomial equation will determine either linear factors $(x - p)$ or quadratic factors $x^2 + px + q$.

2.7.1 The Birge-Vieta Method

In this method we seek to determine a real number p such that $(x - p)$ is a factor of the polynomial equation (2.62). If we divide $P_n(x)$ by the factor $(x - p)$ then we get a quotient $Q_{n-1}(x)$ of degree $(n - 1)$

$$Q_{n-1}(x) = x^{n-1} + b_1 x^{n-2} + \ldots + b_{n-2} x + b_{n-1} \tag{2.67}$$

and a remainder R. Thus we have

$$P_n(x) = (x - p)Q_{n-1}(x) + R \tag{2.68}$$

The value of R depends on p. Starting with an initial approximation p_0 to p, we use some iterative method to improve the value of p such that

$$R(p) = 0 \tag{2.69}$$

This is a single equation in one unknown and the Newton-Raphson method or any other iterative method can be applied to improve the assumed value p_0. The Newton-Raphson method for (2.69) becomes

$$p_{k+1} = p_k - \frac{P_n(p_k)}{P'_n(p_k)}, \quad k = 0, 1, 2, \ldots \tag{2.70}$$

For polynomial equations, the computation of $P_n(p_0)$ and $P'_n(p_0)$ can be systematized with the help of synthetic division. On comparing the coefficients of like powers of x on both sides of (2.68), we get

$$a_1 = b_1 - p \qquad b_1 = a_1 + p$$
$$a_2 = b_2 - pb_1 \qquad b_2 = a_2 + pb_1$$
$$\vdots \qquad\qquad \vdots$$
$$a_k = b_k - pb_{k-1} \qquad b_k = a_k + pb_{k-1}$$
$$\vdots \qquad\qquad \vdots$$
$$a_n = R - pb_{n-1} \qquad R = a_n + pb_{n-1} \qquad (2.71)$$

Let us introduce a quantity b_n and define the following recurrence relations

$$b_k = a_k + pb_{k-1}, \quad k = 1, 2, \ldots, n \qquad (2.72)$$

From the equation (2.68) we have

$$P_n(p) = R = b_n \qquad (2.73)$$

To determine $P_n'(p)$, we differentiate (2.72) with respect to p and obtain

$$\frac{db_k}{dp} = b_{k-1} + p\frac{db_{k-1}}{dp} \qquad (2.74)$$

If we put

$$\frac{db_k}{dp} = c_{k-1} \qquad (2.75)$$

then the equation (2.74) becomes

$$c_{k-1} = b_{k-1} + pc_{k-2}$$

which can also be written as

$$c_k = b_k + pc_{k-1}, \quad k = 1, 2, \ldots, n-1 \qquad (2.76)$$

Differentiating (2.73) and using (2.75), we get

$$P_n'(p) = \frac{dR}{dp} = \frac{db_n}{dp} = c_{n-1} \qquad (2.77)$$

The Newton-Raphson method in the above notations becomes

$$p_{k+1} = p_k - \frac{b_n}{c_{n-1}}, \quad k = 0, 1, 2, \ldots \qquad (2.78)$$

This method is often called the **Birge-Vieta** method.

The coefficients c_k are determined from b_k in a similar way as b_k are obtained from a_k. The calculations of the coefficients b_k and c_k can be carried out as given below:

	1	a_1	a_2	$\ldots$	a_{n-2}	a_{n-1}	a_n
p		p	pb_1	$\ldots$	pb_{n-3}	pb_{n-2}	pb_{n-1}
	1	b_1	b_2	$\ldots$	b_{n-2}	b_{n-1}	$b_n = R$
p		p	pc_1	$\ldots$	pc_{n-3}	pc_{n-2}	
	1	c_1	c_2	$\ldots$	c_{n-2}	$c_{n-1} = \dfrac{dR}{dp}$	

Transcendental and Polynomial Equations 47

Example 2.13 Use synthetic division and perform two iterations by Birge-Vieta method to find the smallest positive root of the equation

$$x^4 - 3x^3 + 3x^2 - 3x + 2 = 0$$

We take the initial approximation $p_0 = 0.5$ and obtain

	1	−3	3	−3	2
0.5		0.5	−1.25	0.875	−1.0625
	1	−2.5	1.75	−2.125	0.9375 = b_4
		0.5	−1.00	0.375	
	1	−2.0	0.75	−1.750 = c_3	

$$p_1 = p_0 - \frac{b_4}{c_3} = 0.5 + \frac{0.9375}{1.750} = 1.0356$$

	1	−3	3	−3	2
1.0356		1.0356	−2.0343	1.0001	−2.0711
	1	−1.9644	0.9657	−1.9999	−0.0711 = b_4
		1.0356	−0.9619	0.0039	
	1	−0.9288	0.0038	−1.9960 = c_3	

$$p_2 = p_1 - \frac{b_4}{c_3} = .999979$$

The exact root is 1.0.

2.7.2 The Bairstow Method

The Bairstow method extracts a quadratic factor of the form $x^2 + px + q$ from the polynomial (2.62), which may give a pair of complex roots or a pair of real roots. If we divide the polynomial (2.62) by the quadratic factor $x^2 + px + q$, then we obtain a quotient polynomial $Q_{n-2}(x)$ of degree $n - 2$ and a remainder term which is a polynomial of degree one, i.e., $Rx + S$.
Thus

$$P_n(x) = (x^2 + px + q)Q_{n-2}(x) + Rx + S \qquad (2.79)$$

where

$$Q_{n-2}(x) = x^{n-2} + b_1 x^{n-3} + \ldots + b_{n-3}x + b_{n-2}$$

The problem is then to find p and q, such that

$$R(p, q) = 0, \quad S(p, q) = 0 \qquad (2.80)$$

48 Numerical Methods

The above equations are two simultaneous equations in two unknowns p and q. Suppose that (p_0, q_0) is an initial approximation and that $(p_0 + \Delta p, q_0 + \Delta q)$ is the true solution. Following the Newton-Raphson method, derived in section 2.8, we obtain

$$\Delta p = -\frac{RS_q - SR_q}{R_p S_q - R_q S_p}$$
$$\Delta q = -\frac{R_p S - RS_p}{R_p S_q - R_q S_p} \qquad (2.81)$$

where R_p, R_q, S_p, S_q are the partial derivatives of R and S with respect to p and q evaluated at p_0, q_0.

The coefficients b_i, R and S can be determined by comparing the like powers of x in (2.79). We obtain

$$\begin{aligned}
a_1 &= b_1 + p & b_1 &= a_1 - p \\
a_2 &= b_2 + pb_1 + q & b_2 &= a_2 - pb_1 - q \\
&\vdots & &\vdots \\
a_k &= b_k + pb_{k-1} + qb_{k-2} & b_k &= a_k - pb_{k-1} - qb_{k-2} \\
&\vdots & &\vdots \\
a_{n-1} &= R + pb_{n-2} + qb_{n-3} & R &= a_{n-1} - pb_{n-2} - qb_{n-3} \\
a_n &= S + qb_{n-2} & S &= a_n - qb_{n-2}
\end{aligned} \qquad (2.82)$$

We now introduce the recursion formula

$$b_k = a_k - pb_{k-1} - qb_{k-2}, \quad k = 1, 2, \ldots, n \qquad (2.83)$$

where $b_0 = 1, b_{-1} = 0$.

Comparing the last two equations with those of (2.82), we get

$$\begin{aligned} R &= b_{n-1} \\ S &= b_n + pb_{n-1} \end{aligned} \qquad (2.84)$$

The partial derivatives R_p, R_q, S_p and S_q can be determined by differentiating (2.83) with respect to p and q.

We have

$$-\frac{\partial b_k}{\partial p} = b_{k-1} + p\frac{\partial b_{k-1}}{\partial p} + q\frac{\partial b_{k-2}}{\partial p}; \quad \frac{\partial b_0}{\partial p} = \frac{\partial b_{-1}}{\partial p} = 0$$
$$-\frac{\partial b_k}{\partial q} = b_{k-2} + p\frac{\partial b_{k-1}}{\partial q} + q\frac{\partial b_{k-2}}{\partial q}; \quad \frac{\partial b_0}{\partial q} = \frac{\partial b_{-1}}{\partial q} = 0 \qquad (2.85)$$

Putting

$$\frac{\partial b_k}{\partial p} = -c_{k-1}, \quad k = 1, 2, \ldots, n$$

in the first equation of (2.85), we find

$$c_{k-1} = b_{k-1} - pc_{k-2} - qc_{k-3} \qquad (2.86)$$

Furthermore, if we write
$$c_{k-2} = -\frac{\partial b_k}{\partial q}$$
then the second equation of (2.85) gives
$$c_{k-2} = b_{k-2} - pc_{k-3} - qc_{k-4}$$
Thus, we get a recurrence relation for the determination of c_k from b_k,
$$c_k = b_k - pc_{k-1} - qc_{k-2}, \quad k = 1, 2, \ldots, n-1$$
where $c_0 = 1$, $c_{-1} = 0$.

We obtain
$$R_p = -c_{n-2}, \quad S_p = b_{n-1} - c_{n-1} - pc_{n-2}$$
$$R_q = -c_{n-3}, \quad S_q = -(c_{n-2} + pc_{n-3})$$

Substituting the above values in (2.81) and simplifying, we get
$$\Delta p = -\frac{b_n c_{n-3} - b_{n-1} c_{n-2}}{c_{n-2}^2 - c_{n-3}(c_{n-1} - b_{n-1})}$$

$$\Delta q = -\frac{b_{n-1}(c_{n-1} - b_{n-1}) - b_n c_{n-2}}{c_{n-2}^2 - c_{n-3}(c_{n-1} - b_{n-1})}$$

The improved values of p_0 and q_0 are
$$p_1 = p_0 + \Delta p$$
$$q_1 = q_0 + \Delta q$$

For computing b_k's and c_k's we use the following scheme:

	1	a_1	a_2	$\ldots$	a_{n-2}	a_{n-1}	a_n
$-p$		$-p$	$-pb_1$	$\ldots$	$-pb_{n-3}$	$-pb_{n-2}$	$-pb_{n-1}$
$-q$			$-q$	$\ldots$	$-qb_{n-4}$	$-qb_{n-3}$	$-qb_{n-2}$
	1	b_1	b_2	$\ldots$	b_{n-2}	b_{n-1}	b_n
$-p$		$-p$	$-pc_1$	$\ldots$	$-pc_{n-3}$	$-pc_{n-2}$	
$-q$			$-q$	$\ldots$	$-qc_{n-4}$	$-qc_{n-3}$	
	1	c_1	c_2	$\ldots$	c_{n-2}	c_{n-1}	

When p and q have been obtained to the desired accuracy, the polynomial $Q_{n-2}(x) = P_n(x)/(x^2 + px + q) = x^{n-2} + b_1 x^{n-4} + \ldots + b_{n-2}$ is called the deflated polynomial. The coefficients b_i, $i = 1, 2, \ldots, n-2$ are known from the synthetic division procedure. The next quadratic factor is obtained using this deflated polynomial.

Example 2.14 Perform one iteration of the Bairstow method to extract a quadratic factor $x^2 + px + q$ from the polynomial
$$x^4 + x^3 + 2x^2 + x + 1 = 0$$

Use the initial approximations $p_0 = 0.5$, $q_0 = 0.5$.
Starting with $p_0 = 0.5$, $q_0 = 0.5$ we obtain

1	1	2	1	1
-0.5	-0.5	-0.25	-0.625	-0.0625
-0.5		-0.5	-0.25	-0.625
1	0.5	1.25	$0.125 = b_3$	$0.3125 = b_4$
	-0.5	0.0	-0.375	
		-0.5	0.0	
1	$0.0 = c_1$	$0.75 = c_2$	$-0.25 = c_3$	

$$\Delta p = -\frac{b_4 c_1 - b_3 c_2}{c_2^2 - c_1(c_3 - b_3)} = 0.1667$$

$$\Delta q = -\frac{b_3(c_3 - b_3) - b_4 c_2}{c_2^2 - c_1(c_3 - b_3)} = 0.5$$

Therefore,

$$p_1 = p_0 + \Delta p = 0.6667$$
$$q_1 = q_0 + \Delta q = 1.0$$

The exact values of p and q are 1.0.

2.7.3 Graeffe's Root Squaring Method

This is a direct method and it is used to find the roots of a polynomial with real coefficients. The roots may be real and distinct, real and equal or complex roots.

We separate the roots of the equation (2.62), by forming another equation with the help of root squaring process, whose roots are very high powers of the roots of the equation (2.62). Separating the even and odd powers of x in (2.62) and squaring we get

$$(x^n + a_2 x^{n-2} + a_4 x^{n-4} + \ldots)^2$$
$$= (a_1 x^{n-1} + a_3 x^{n-3} + \ldots)^2$$

Simplifying we obtain

$$x^{2n} - (a_1^2 - 2a_2)x^{2n-2} + (a_2^2 - 2a_1 a_3 + 2a_4)x^{2n-4} - \ldots$$
$$+ (-1)^n a_n^2 = 0$$

Substituting z for $-x^2$ we have

$$z^n + b_1 z^{n-1} + \ldots + b_{n-1} z + b_n = 0 \tag{2.87}$$

where
$$b_1 = a_1^2 - 2a_2$$
$$b_2 = a_2^2 - 2a_1a_3 + 2a_4$$
$$\vdots$$
$$b_n = a_n^2 \qquad (2.88)$$

Thus all the b_k's are known in terms of a_k's. The roots of the equation (2.87) are $-\xi_1^2, -\xi_2^2, \ldots, -\xi_n^2$ where $\xi_1, \xi_2, \ldots, \xi_n$ are roots of (2.62). The coefficients b_k's can be obtained as follows:

1	a_1	a_2	a_3	$\ldots$	a_n
1	a_1^2	a_2^2	a_3^2	$\ldots$	a_n^2
	$-2a_2$	$-2a_1a_3$	$-2a_2a_4$	$\ldots$	
		$+2a_4$	$+2a_1a_5$		
			$\vdots$	$\vdots$	
1	b_1	b_2	b_3	$\ldots$	b_n

The $(k+1)$th column in the above table can be obtained as follows:

The terms alternate in sign starting with a positive sign. The first term is the square of the $(k+1)$th coefficient a_k. The second term is twice the product of the nearest neighbouring coefficients a_{k-1} and a_{k+1}. The third is twice the product of the next neighbouring coefficients a_{k-2} and a_{k+2}. This procedure is continued until there are no available coefficients to form the cross products.

This procedure can be repeated m times and we obtain the equation
$$x^n + B_1 x^{n-1} + B_2 x^{n-2} + \ldots + B_{n-1} x + B_n = 0 \qquad (2.89)$$
whose roots
$$R_1, R_2, R_3, \ldots, R_n$$
are the 2^mth power of the roots of the equation (2.62) with opposite signs,
$$R_i = -\xi_i^{2^m}, \quad i = 1, 2, \ldots, n$$
If we assume that
$$|\xi_1| > |\xi_2| \ldots > |\xi_n|$$
then
$$|R_1| \gg |R_2| \ldots \gg |R_n|$$

Thus, if the roots of (2.62) differ in magnitude, then the 2^mth power of the roots are widely separated for large m. We have
$$-B_1 = \Sigma R_i \simeq R_1$$
$$B_2 = \Sigma R_i R_j \simeq R_1 R_2$$

$$-B_3 = \Sigma R_i R_j R_k \simeq R_1 R_2 R_3$$
$$\vdots$$
$$(-1)^n B_n = R_1 R_2 \ldots R_n$$

which give

$$R_i = -\frac{B_i}{B_{i-1}}, \quad i = 1, 2, \ldots, n$$

where $B_0 = 1$.

We have

$$|R_i| = \frac{|B_i|}{|B_{i-1}|} = |\xi_i|^{2^m}$$

or

$$\log |\xi_i| = 2^{-m}(\log |B_i| - \log |B_{i-1}|), \quad i = 1, 2, \ldots, m \tag{2.90}$$

This determines the absolute values of the roots and substitution in the original equation (2.62) will give the sign of the roots. The squaring process is stopped when another squaring process produces new coefficients that are almost the squares of the corresponding coefficients B_k's i.e., when the cross product terms become negligible in comparison to square terms. Thus the vanishing of all the cross product terms in the squaring process can be used as an indication that the roots have been widely separated.

After few squarings, if the magnitude of the coefficient B_k is half the square of the magnitude of the corresponding coefficient in the previous equation, then it indicates that ξ_k is a double root. We can find this double root by using the following procedure. We have

$$R_k \simeq -\frac{B_k}{B_{k-1}} \quad \text{and} \quad R_{k+1} \simeq -\frac{B_{k+1}}{B_k}$$

$$R_k R_{k+1} \simeq R_k^2 \simeq \left|\frac{B_{k+1}}{B_{k-1}}\right|$$

$$\therefore \quad |R_k^2| = |\xi_k|^{2(2^m)} = \left|\frac{B_{k+1}}{B_{k-1}}\right|$$

This gives the magnitude of the double root. Substituting in the given equation, we can find its sign. This double root can also be found directly since R_k and R_{k+1} converge to the same root after sufficient squarings. Usually, this convergence to the double root is slow. By making use of the above observation, we can save a number of squarings.

If ξ_k and ξ_{k+1} form a complex pair, then this would cause the coefficients of x^{n-k} in the successive squarings to fluctuate both in magnitude and sign. If $\xi_k, \xi_{k+1} = \beta_k \exp(\pm i\phi_k)$ is the complex pair, then the coefficients would fluctuate in magnitude and sign by an amount $2\beta_k^m \cos m\phi_k$. A complex pair can be spotted by such an oscillation. For m sufficiently large, β_k can be determined from the relation

$$\beta_k^{2(2^m)} \simeq \left|\frac{B_{k+1}}{B_{k-1}}\right|$$

and ϕ is suitably determined from the relation

$$2\beta_k^m \cos m\phi_k \simeq \frac{B_{k+1}}{B_{k-1}}$$

If the equation has only one complex pair, then we can first determine all the real roots. The complex pair can be written as $\xi_{k,\,k+1} = p \pm iq$. The sum of the roots then gives

$$\xi_1 + \xi_2 + \ldots + \xi_{k-1} + 2p + \xi_{k+2} + \ldots + \xi_n = -a_1$$

This determines p. We also have $|\beta_k|^2 = p^2 + q^2$. Since $|\beta_k|$ is already determined, this equation gives q.

Example 2.15 Find all the roots of the polynomial

$$x^3 - 6x^2 + 11x - 6 = 0$$

using the Graeffe's root squaring method.

The coefficients of the successive root squarings are given in Table 2.13.

Table 2.13 Coefficients in the Root Squarings by Graeffe's Method

m	2^m				
0	1	1	-6	11	-6
		1	36	121	36
			-22	-72	
1	2	1	14	49	36
		1	196	2401	1296
			-98	-1008	
2	4	1	98	1393	1296
		1	9604	1940449	1679616
			-2786	-254016	
3	8	1	6818	1686433	1679616
		1	46485124	2.8440562(12)	2.8211099(12)
			-3372866	$-2.2903243(10)$	
4	16	1	43112258	2.8211530(12)	2.8211099(12)

Successive approximations to the roots are given in the Table 2.14. The exact roots of the equation are 3, 2, 1.

Table 2.14 Approximations to the Roots

m	α_1	α_2	α_3
1	3.7417	1.8708	0.8571
2	3.1463	1.9417	0.9821
3	3.0144	1.9914	0.9995
4	3.0003	1.9998	1.0000

Example 2.16 Find all the roots of the polynomial
$$x^3 - 4x^2 + 5x - 2 = 0$$
using the Graeffe's root squaring method.

The coefficients in the successive root squarings are tabulated in the Table 2.15.

Table 2.15 Coefficients in the Root Squarings by Graeffe's Method

m	2^m				
0	1	1	−4	5	−2
		1	0.1600(2)	0.2500(2)	0.4000(1)
			−0.1000(2)	−0.1600(2)	
1	2	1	0.6000(1)	0.9000(1)	0.4000(1)
		1	0.3600(2)	0.8100(2)	0.1600(2)
			−0.1800(2)	−0.4800(2)	
2	4	1	0.1800(2)	0.3300(2)	0.1600(2)
		1	0.3240(3)	0.1089(4)	0.2560(3)
			−0.0660(3)	−0.0576(4)	
3	8	1	0.2580(3)	0.5130(3)	0.2560(3)
		1	0.6656(5)	0.2632(6)	0.6554(5)
			−0.0103(5)	−0.1321(6)	
4	16	1	0.6653(5)	0.1311(6)	0.6554(5)
		1	0.4294(10)	0.1718(11)	0.4295(10)
			−0.2622(6)	−0.0859(11)	
5	32	1	0.4294(10)	0.0859(11)	0.4295(10)

We notice from above that the magnitude of the coefficient B_1 has become constant (upto four decimal places), whereas the magnitude of the coefficient B_2 is half the square of the magnitude of the corresponding coefficient in the previous equation. This phenomenon indicates that ξ_2 is a double root. We obtain the magnitude of the roots as

$$|\xi_1|^{32} = \left|\frac{B_1}{B_0}\right| = (0.4294)10^{10} \text{ or } |\xi_1| = 1.99999$$

$$|\xi_2|^{64} = |\xi_3|^{64} = \left|\frac{B_3}{B_1}\right| = 1.0002 \text{ or } |\xi_2| = |\xi_3| = 1.000003$$

All the roots are positive. The exact roots are 2, 1, 1.

However, if we want to get the double root by the direct procedure, it would take a large number of squarings. In the present example, we find directly $|\xi_2| = 1.0219$ and $|\xi_3| = 0.9786$. One more squaring produces $|\xi_2| = 1.0109$, $|\xi_3| = .9892$, so that the convergence is slow. It would require few more squarings to stabilize the roots.

Example 2.17 Find all the roots of the polynomial

$$x^3 - x^2 - x - 2 = 0$$

using the Graeffe's root squaring method.

The coefficients in the successive root squarings are given in Table 2.16.

Table 2.16 Coefficients in the Root Squarings by Graeffe's Method

m	2^m				
0	1	1	-1	-1	-2
		1	0.1000(1) 0.2000(1)	0.1000(1) $-0.4000(1)$	0.4000(1)
1	2	1	0.3000(1)	$-0.3000(1)$	0.4000(1)
		1	0.9000(1) 0.6000(1)	0.9000(1) $-2.4000(1)$	0.1600(2)
2	4	1	0.1500(2)	$-0.1500(2)$	0.1600(2)
		1	0.2250(3) 0.0300(3)	0.2250(3) $-0.0048(3)$	0.2560(3)
3	8	1	0.2550(3)	0.2202(3)	0.2560(3)
		1	0.6503(5) $-0.0044(5)$	0.4849(5) $-1.3056(5)$	0.6554(5)
4	16	1	0.6459(5)	$-0.8207(5)$	0.6554(5)
		1	0.4172(10) 0.1641(6)	0.6735(10) $-0.8466(10)$	0.4295(10)
5	32	1	0.4172(10)	$-0.1731(10)$	0.4295(10)

We observe from above that the magnitude of the coefficient B_1 has become constant (upto four decimal places), whereas the magnitude of the coefficient B_2 oscillates. This indicates that ξ_1 is the real root and ξ_2 and ξ_3 are a pair of complex roots. The real root is given by

$$|\alpha_1|^{32} = (0.4172)10^{10}, \quad |\alpha_1| = 1.9982.$$

It can be verified that this root is positive. The magnitude of the complex root is obtained as

$$\beta_2^{64} \simeq \left|\frac{B_3}{B_1}\right| = 1.0295$$

which gives $\beta_2 = 1.00045$.

If we write

$$\xi_2 = p + iq \text{ and } \xi_3 = p - iq$$

then using the given equation, we get

$$\xi_1 + \xi_2 + \xi_3 = -a_1$$
$$\xi_1 + 2p = 1 \text{ or } p = -0.4991$$

We find

$$q = \sqrt{(\beta^2 - p^2)} = 0.8671.$$

Thus, the required roots of the given equation are

$$\xi_1 = 1.9982$$
$$\xi_2 = -0.4991 + 0.8671i$$
$$\xi_3 = -0.4991 - 0.8671i$$

The exact roots of the equation are $2.0, (-1.0 \pm i\sqrt{3})/2$.

2.8 CHOICE OF AN ITERATIVE METHOD AND IMPLEMENTATION

The bisection method is slow but it never fails. If the evaluation of $f(x)$ is rapid, then the use of the bisection method is strongly advised. The precision of one binary bit is gained in each iteration of the bisection method. It is better to find the bisection point using the formula $x_{k+1} = x_{k-1} + \frac{1}{2}(x_k - x_{k-1})$ rather than using the conventional formula $x_{k+1} = \frac{1}{2}(x_k + x_{k-1})$. The criterion $f(x_{k-1})f(x_k) < 0$, to ensure that $f(x_{k-1})$ and $f(x_k)$ are of opposite signs, should be carefully used. For, if $f(x_{k-1}) = -10^{-40}$ and $f(x_k) = 10^{-40}$, then their product may cause an underflow and this value is set as zero by the computer and the test fails. If the function $f(x)$ is smooth enough to have continuous derivatives of first and second order which can be evaluated easily, then the Newton-Raphson, multipoint and Chebyshev methods considerably reduce the number of iterations required for attaining the prescribed accuracy. The Secant method also requires lesser iterations than the bisection method. In the Secant method we always use x_i and x_{i-1} to generate x_{i+1}. Often, Secant method gives better results than the Regula-Falsi method (see example 2.3). The only disadvantage of the Secant method is that it may not converge sometimes. If the evaluation of $f'(x)$ is difficult, then the Secant method is a better choice than the Newton's method. The difficulty with the Newton-Raphson's method is in finding a suitable choice of the initial approximation x_0. If this choice is poor, it is possible that the root may never be obtained. Hence, frequently, some convergent method like bisection is used for the first few iterations before using the rapidly convergent Newton-Raphson method. If the evaluation of $f'(x)$ is difficult, then the use of the Secant method, preceded by a few iterations by the bisection method is often recommended.

If all the roots of a polynomial are required, then Bairstow's method is recommended. After a quadratic factor has been found, then the Bairstow's method must be applied on the deflated polynomial $P_n(x)/(x^2 + px + q)$. If the polynomial is of odd degree, we are finally left with a linear factor. If the location of some roots are known, we first find these roots to a given accuracy, then apply the Bairstow method on the deflated polynomial. It is known that if Bairstow method is applied to a polynomial with exactly one real root p and the initial trial factor vanishes at p, then all the succeeding iterates will vanish at p and hence the method cannot converge. A remedy for this problem is to apply the Bairstow method only to polynomials of even degree, by considering $xP(x)$ or $P(x)/x$ if necessary. If the roots of $P_n(x)$ are real and distinct then Graeffe's root squaring method works very well.

The bisection method cannot be applied for locating the complex zeros of analytic functions. The Secant and Newton-Raphson methods can be used in the complex plane without any change. A good initial approximation to the complex root is necessary for application of these methods. A real initial approximation cannot produce a complex root. For certain functions, it is possible that the sequence of approximations tend to a periodic state, in which the approximations cycle with a period P. Muller's method is efficient in finding complex zeros of the analytic functions. This method allows real initial approximations and proceeds to complex iterates. Once a complex zero z_1 is found, it is necessary to prevent a new iteration for finding another root, from returning to the root z_1. The new iteration should be performed on the new function (deflated polynomial)

$$f^*(z) = \frac{f(z)}{z - z_1} \qquad (2.91)$$

where z_1 is the complex zero already obtained. This procedure can be repeated after finding every root. If k roots are already determined then the new iteration can be applied on the function

$$f^*(z) = \frac{f(z)}{\prod_{i=1}^{k}(z - z_i)} \qquad (2.92)$$

58   Numerical Methods

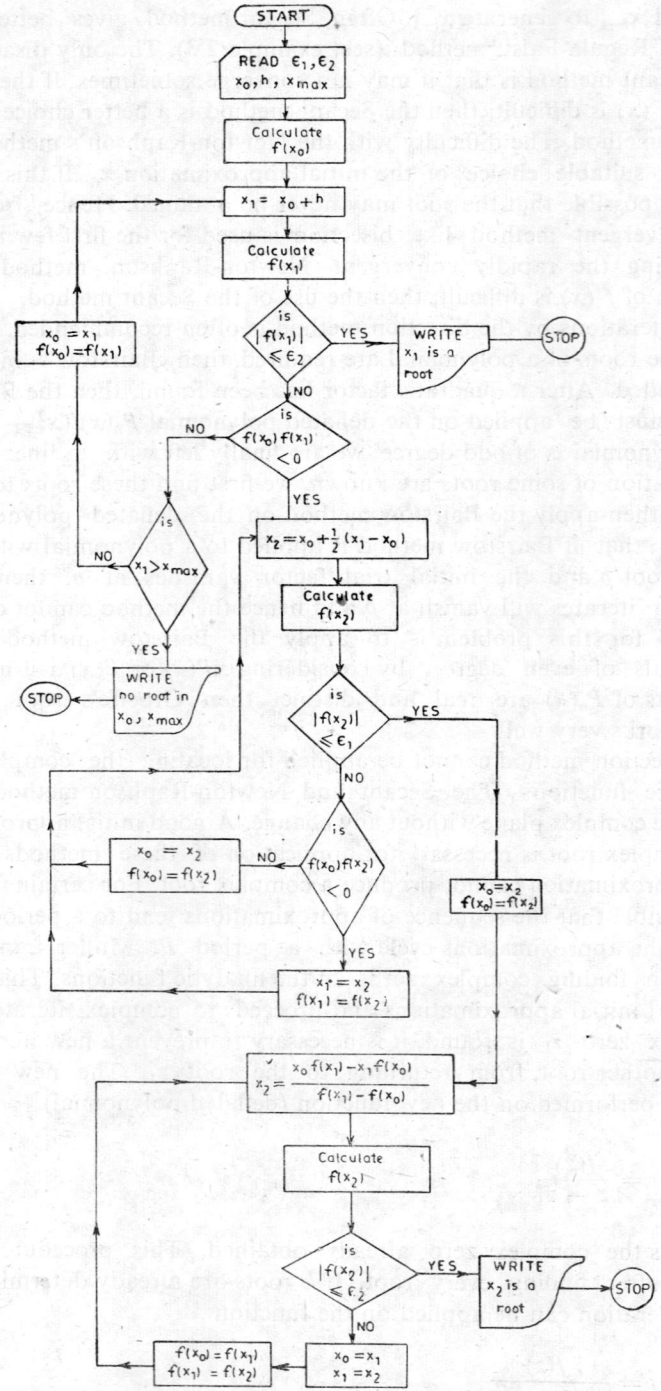

Fig. 2.8   Flow chart for the method using Incremental Search method,
Bisection method followed by the Secant method.

The Newton-Raphson method for finding the complex zero of (2.91) becomes

$$z_{n+1} = z_n - \frac{f^*(z_n)}{f^{*\prime}(z_n)} \tag{2.93}$$

The computation of $f^*(z_n)/f^{*\prime}(z_n)$ can easily be performed as follows:

$$\frac{f^{*\prime}}{f^*} = \frac{d}{dz}[\log f^*] = \frac{d}{dz}[\log f(z) - \log(z - z_1)[= \frac{f'}{f} - \frac{1}{z - z_1}$$

Hence, the computations are carried on with (2.93) where

$$\frac{f^{*\prime}(z_n)}{f^*(z_n)} = \frac{f'(z_n)}{f(z_n)} - \frac{1}{z_n - z_1}$$

2.8.1 Flow Chart for a Zero Finder

The flow chart for the method using incremental search method, Bisection method followed by the Secant method is given in Figure 2.8.

The incremental search method searches for an interval $(x_k, x_k + h)$ containing the root in $(x_0, x_{\max})$ starting with x_0 and using the increment h (which may not be small) everytime. ϵ_1 and ϵ_2 are tolerances for the Bisection and Secant methods. ϵ_1 is much larger than ϵ_2.

PROBLEMS

1. Find the interval in which the smallest positive root of the following equations lies:
   (a) $\tan x + \tanh x = 0$
   (b) $x^3 - x - 4 = 0$
   Determine the roots correct to two decimals using the bisection method.

2. Determine the order of convergence of the iterative method
   $$x_{k+1} = (x_0 f(x_k) - x_k f(x_0))/(f(x_k) - f(x_0)).$$
   for finding a root of the equation $f(x) = 0$.

3. (i) Develop a flow diagram to combine the bisection method and the Newton-Raphson method to find the roots of the equation $f(x) = 0$. Assume that subroutines are available to evaluate $f(x)$ and $f'(x)$ for any x.

   (ii) Find the iterative methods based on the Newton-Raphson method for finding $\sqrt{N}$, $1/N$, $N^{1/3}$, where N is a real number. Apply the methods to $N = 18$ to obtain the results correct to two decimals.

4. Given the following equations:
   (i) $x^4 - x - 10 = 0$
   (ii) $x - e^{-x} = 0$
   Determine the initial approximations. Use these to find the roots correct to three decimal places with the following methods:

(a) the Regula-Falsi method
(b) the Secant method
(c) the Newton-Raphson method.

5. Show that the following two sequences have convergence of the second order with the same limit $\sqrt{a}$.

$$x_{n+1} = \tfrac{1}{2}x_n(1 + a/x_n^2)$$

$$x_{n+1} = \tfrac{1}{2}x_n(3 - x_n^2/a)$$

If x_n is a suitably close approximation to $\sqrt{a}$, show that the error in the first formula for x_{n+1} is about one-third of that in the second formula, and deduce that the formula

$$x_{n+1} = \tfrac{1}{8}x_n(6 + 3a/x_n^2 - x_n^2/a)$$

gives a sequence with third-order convergence.

6. Let $x = a$ be a simple root of the equation $f(x) = 0$. We try to find the root by means of the iteration formula

$$x_{i+1} = x_i - (f(x_i))^2/(f(x_i) - f(x_i - f(x_i)))$$

Find the order of convergence and compare the convergence properties with those of Newton-Raphson's method.

(Bergen Univ., Sweden, BIT 20(1980), 262)

7. A root of the equation $f(x) = 0$ can be obtained by combining the Newton-Raphson method and the Regula-Falsi method. We start from $x_0 = \xi + \epsilon$ where ξ is the true solution of $f(x) = 0$. Further $y_0 = f(x_0)$, $x_1 = x_0 - \dfrac{f_0}{f'_0}$ and $y_1 = f_1$ are computed. Lastly, a straight line is drawn through the points (x_1, y_1) and $\left(\dfrac{x_0 + x_1}{2}, \dfrac{y_0}{2}\right)$. If ϵ is sufficiently small, the intersection of the line and the x-axis gives a good approximation to ξ. To what power of ϵ is the error term proportional? Use this method to compute the root of the equation $x^4 - x - 10 = 0$, correct to three decimal places.

8. Determine p, q and r so that the order of the iterative method

$$x_{n+1} = px_n + qa/x_n^2 + ra^2/x_n^5$$

for $a^{1/3}$ becomes as high as possible. For this choice of p, q and r, indicate how the error in x_{n+1} depends on the error in x_n.

(Lund Univ., Sweden, BIT 8(1968), 138)

9. Use the Chebyshev third order method with $f(x) = x^2 - a$ and with $f(x) = 1 - a/x^2$ to obtain the iteration methods converging to $a^{1/2}$, in the form

$$x_{k+1} = \tfrac{1}{2}\left(x_k + \dfrac{a}{x_k}\right) - \dfrac{1}{8x_k}\left(x_k - \dfrac{a}{x_k}\right)^2$$

and
$$x_{k+1} = \tfrac{1}{2}x_k\left(3 - \frac{x_k^2}{a}\right) + \tfrac{3}{8}x_k\left(1 - \frac{x_k^2}{a}\right)^2$$

Perform two iterations with these methods to find the value of $\sqrt{6}$.

10. Let the function $f(x)$ be four times continuously differentiable and have a simple zero ξ. Successive approximations x_n, $n = 1, 2, \ldots$ to ξ are computed from
$$x_{n+1} = \tfrac{1}{2}(x'_{n+1} + x''_{n+1})$$
where $x'_{n+1} = x_n - f(x_n)/f'(x_n)$
$$x''_{n+1} = x_n - g(x_n)/g'(x_n), \quad g(x) = f(x)/f'(x)$$

Prove that if the sequence $\{x_n\}$ converges to ξ, then the convergence is cubic.

(Lund Univ., Sweden, BIT 8(1968), 59)

11. Given the equation $f(x) = 0$.

   (i) Obtain an iteration method using the rational approximation
   $$f(x) = \frac{x - a_0}{b_0 + b_1 x}$$
   where the coefficients a_0, b_0 and b_1 are determined by evaluating $f(x)$ at x_k, x_{k-1} and x_{k-2}.

   (ii) Find the order of convergence for this method.

   (iii) Carry out two iterations using this method for the equation $f(x) = 2x^3 - 3x^2 + 2x - 3 = 0$ with $x_0 = 0$, $x_1 = 1$, $x_2 = 2$.

12. Determine an iteration method using the rational approximation
   $$f(x) = \frac{x - a_0}{b_0 + b_1 x + b_2 x^2}$$
   where the parameters a_0 and b's are obtained by evaluating both f and f' at x_k and x_{k-1}. Find the rate of convergence of the iteration method. Also carry out two iterations using this method for the equation
   $$x^4 - x - 10 = 0 \text{ with } x_0 = 1.8, \ x_1 = 2.0$$

13. Find the order of convergence of the **Steffensen** method
   $$x_{k+1} = x_k - \frac{f_k}{g_k}, \quad k = 0, 1, 2, \ldots$$
   $$g_k = \frac{f(x_k + f_k) - f_k}{f_k}$$
   where $f_k = f(x_k)$. Use this method to determine the nonzero root of the equation
   $$f(x) = x - 1 + e^{-2x} = 0, \text{ with } x_0 = .7$$
   correct to three decimals.

14. (i) Draw a flow diagram for the Muller method.

(ii) Perform 2 iterations with the Muller method for the following equations:

(a) $x^3 - \frac{1}{2} = 0$, $x_0 = 0$, $x_1 = 1$, $x_2 = \frac{1}{2}$

(b) $\log x - x + 3 = 0$, $x_0 = \frac{1}{4}$, $x_1 = \frac{1}{2}$, $x_2 = 1$

15. The equation $x^2 + ax + b = 0$ has two real roots α and β. Show that the iteration method

$$x_{k+1} = -(ax_k + b)/x_k$$

is convergent near $x = \alpha$ if $|\alpha| > |\beta|$ and that

$$x_{k+1} = -b/(x_k + a)$$

is convergent near $x = \alpha$ if $|\alpha| < |\beta|$.
Show also that the iteration method

$$x_{k+1} = -(x_k^2 + b)/a$$

is convergent near $x = \alpha$ if

$$2|\alpha| < |\alpha + \beta|$$

16. The equation $x^4 + x = \epsilon$, where ϵ is a small number, has a root ξ which is close to ϵ. Computation of this root is done by the expression

$$\xi = \epsilon - \epsilon^4 + 4\epsilon^7$$

(i) Find an iteration formula $x_{n+1} = F(x_n)$, $x_0 = 0$ for the computation. Show that we get the expression above after three iterations when neglecting terms of higher order.

(ii) Give a good estimate (of the form $N\epsilon^k$, where N and k are integers) of the maximal error when the root is estimated by the expression above.

(Inst. Tech. Stockholm, Sweden, BIT 9(1969), 87)

17. The equation $x = f(x)$ is solved by the iteration method $x_{k+1} = f(x_k)$, and a solution is wanted with a maximum error not greater than 0.5×10^{-4}. The first and second iterates were computed; $x_1 = 0.50000$, $x_2 = 0.52661$. How many iterations must be performed further, if it is known that $|f'(x)| \leq 0.53$ for all values of x.

18. A root of the equation $f(x) = x - F(x) = 0$ can often be determined by combining the iteration method with Regula-Falsi:
(i) With a given approximate value x_0 we compute

$$x_1 = F(x_0), \quad x_2 = F(x_1)$$

(ii) Observing that $f(x_0) = x_0 - x_1$ and $f(x_1) = x_1 - x_2$, we find a better approximation x' using Regula-Falsi on the points $(x_0, x_0 - x_1)$ and $(x_1, x_1 - x_2)$.

(iii) This last x' is taken as a new x_0 and we start from (i) all over again.

Compute the smallest root of the equation $x - 5 \log_e x = 0$ with an error less than 0.5×10^{-4} starting with $x_0 = 1.3$.

(Inst. Tech. Stockholm, Sweden, BIT 6(1966), 176)

19. The root of the equation $x = \frac{1}{2} + \sin x$ by using the iteration method

$$x_{k+1} = \tfrac{1}{2} + \sin x_k, \quad x_0 = 1$$

correct to six decimals is $x = 1.497300$. Determine the number of iteration steps required to reach the root by the linear iteration. If the Aitken Δ^2 process is used after three approximations are available, how many iterations are required.

20. Consider the iteration method

$$x_{k+1} = \phi(x_k), \quad k = 0, 1, 2, \ldots$$

for solving the equation $f(x) = 0$.
We choose the iteration function in the form

$$\phi(x) = x - \gamma_1(x)f(x) - \gamma_2(x)f^2(x) - \gamma_3(x)f^3(x)$$

where $\gamma_1(x)$, $\gamma_2(x)$ and $\gamma_3(x)$ are arbitrary functions to be determined. Find the γ's such that the iteration method has the following orders: (i) third, (ii) fourth.

Apply these methods to determine a root of the equation

$x = \tfrac{1}{5} e^x$ with $x_0 = 0.4$

correct to three decimal places.

21. The equation

$x^3 - 5x^2 + 4x - 3 = 0$

has one root near $x = 4$, which is to be computed by the iteration

$x_0 = 4$

$$x_{n+1} = \frac{3 + (k-4)x_n + 5x_n^2 - x_n^3}{k}, \quad k \text{ integer}$$

(a) Determine which value of k will give the fastest convergence.
(b) Using this value of k, iterate three times and estimate the error in x_3.

(Royal Inst. Stockholm, Sweden, BIT 11(1971), 125)

22. How should the constant α be chosen to ensure the fastest possible convergence with the iteration formula

$$x_{n+1} = \frac{\alpha x_n + x_n^{-2} + 1}{\alpha + 1}$$

(Uppsala Univ., Sweden, BIT 11(1971), 225)

23. (a) Show that the equation $\log_e x = x^2 - 1$ has exactly two real roots, $\alpha_1 = 0.45$ and $\alpha_2 = 1$.

(b) Determine for which initial approximation x_0, the iteration
$$x_{n+1} = \sqrt{1 + \log_e x_n}$$
converges to α_1 or α_2.

(c) Determine for which initial approximation x_0, the iteration
$$x_{n+1} = x_n - \frac{x_n - \exp(x_n^2 - 1)}{1 - 2x_n \exp(x_n^2 - 1)}$$
converges to a root of the equation in (a) and in case of convergence also which root.

(Uppsala Univ., Sweden, BIT 10(1970), 115)

24. A sequence $\{x_n\}_1^\infty$ is defined by
$$x_0 = 5$$
$$x_{n+1} = \frac{1}{16} x_n^4 - \frac{1}{2} x_n + 8x_n - 12$$
Show that it gives cubic convergence to $\xi = 4$.
Calculate the smallest integer n for which the inequality
$$|x_n - \xi| < 10^{-6}$$
is valid.

(Uppsala Univ., Sweden, BIT 13(1973), 493)

25. If an attempt is made to solve the equation $x = 1.4 \cos x$ by using the iteration formula
$$x_{n+1} = 1.4 \cos x_n$$
it is found that for large n, x_n alternates between the two values A and B.

(i) Calculate A and B correct to 3 decimal places.

(ii) Calculate the correct solution of the equation to 4 decimal places.

(Lund Univ., Sweden, BIT 17(1977), 115)

26. We wish to compute the root of the equation
$$e^{-x} = 3 \log_e x$$
using the formula
$$x_{n+1} = x_n + \frac{3 \log_e x_n - \exp(-x_n)}{p}$$
Show that $p = 3$ gives rapid convergence.

(Stockholm Univ., Sweden, BIT 14(1974), 254)

27. Apply the Newton-Raphson method, with $x_0 = 0.8$, the Secant method with $x_0 = 0.8$, $x_1 = 1.2$, and the Muller method with $x_0 = 0.6$, $x_1 = 0.8$, $x_2 = 1.2$ to the equation
$$f(x) = x^3 - x^2 - x + 1 = 0$$
and verify that the convergence is only of first order in each case.

Then apply the Newton-Raphson method (2.49), with $m = 2$ and verify that the convergence is of second order.

28. We consider the multipoint iteration method

$$x_{k+1} = x_k - \alpha \frac{f(x_k)}{f'\left(x_k - \beta \frac{f(x_k)}{f'(x_k)}\right)}$$

where α and β are arbitrary parameters, for solving the equation $f(x) = 0$. Determine α and β such that the multipoint method is of order as high as possible for finding ξ, a simple root of $f(x) = 0$.

29. The multiple root ξ of multiplicity two of the equation $f(x) = 0$ is to be determined. We consider the multipoint method

$$x_{k+1} = x_k - \tfrac{1}{2} \frac{f(x_k + 2f(x_k)/f'(x_k))}{f'(x_k)}$$

Show that the iteration method has third order rate of convergence. Hence solve the equation

$$9x^4 + 30x^3 + 34x^2 + 30x + 25 = 0, \quad \text{with } x_0 = -1.4$$

correct to three decimals.

30. Perform three iterations of the Newton-Raphson method (2.60) for solving the following equations
   (i) $1 + z^2 = 0$, $z_0 = \tfrac{1}{2}(1 + i)$
   (ii) $z^3 - 4iz^2 - 3e^z = 0$, $z_0 = -0.53 - 0.36i$

   Perform the same number of iterations using the method (2.61).

31. Consider the equation

   $$10z^{10} + z^5 + z - 1 = 0$$

   (i) Determine the number of roots satisfying $|z| \leq 1.1$
   (ii) Determine the number of real roots
   (iii) Compute the largest positive real root with an error less than 0.02

   (Uppsala Univ., Sweden, BIT 14(1974), 254)

32. The system of equations

   $y \cos(xy) + 1 = 0$
   $\sin(xy) + x - y = 0$

   has one solution close to $x = 1$, $y = 2$. Calculate this solution correct to 2 decimal places.

   (Umea Univ., Sweden, BIT 19(1979), 552)

33. The system of equations

   $\log_e(x^2 + y) - 1 + y = 0$
   $\sqrt{x} + xy = 0$

has one approximate solution $(x_0, y_0) = (2.4, -0.6)$.
Improve this solution and estimate the accuracy of the result.

(Lund Univ., Sweden, BIT 18(1978), 366)

34. Calculate all solutions of the system

$$x^2 + y^2 = 1.12$$

$$xy = 0.23$$

correct to three decimal places.

(Lund Univ., Sweden, BIT 20(1980), 389)

35. It is required to solve the two simultaneous equations

$$x = f(x, y), \quad y = g(x, y)$$

by means of an iterative sequence. Show that the sequence

$$x_{n+1} = f(x_n, y_n), \quad y_{n+1} = g(x_n, y_n)$$

will converge to a solution if the roots of the quadratic

$$\lambda^2 - \left(\frac{\partial f}{\partial x} + \frac{\partial g}{\partial y}\right)\lambda + \left(\frac{\partial f}{\partial x}\frac{\partial g}{\partial y} - \frac{\partial f}{\partial y}\frac{\partial g}{\partial x}\right) = 0$$

are less than unity in modulus, the derivatives being evaluated at the solution.

Obtain the condition that the iterative sequence

$$x_{n+1} = f(x_n, y_n), \quad y_{n+1} = g(x_{n+1}, y_n)$$

will converge. Show further that if $\dfrac{\partial f}{\partial x} = \dfrac{\partial g}{\partial y} = 0$ and both sequences converge, then the second sequence converges more rapidly than the first.

36. Describe how, in general, suitable values of a, b, c and d may be estimated so that the sequence of values of x and y determined from the recurrence relations

$$x_{n+1} = x_n + af(x_n, y_n) + bg(x_n, y_n)$$

$$y_{n+1} = y_n + cf(x_n, y_n) + dg(x_n, y_n)$$

will converge to a solution of

$$f(x, y) = 0$$

$$g(x, y) = 0$$

Illustrate the method by finding a suitable initial point and a recurrence relation to find the real solution of

$$y = \sin(x + y)$$

$$x = \cos(y - x)$$

37. (a) Take one step from a suitable starting point with Newton-Raphson's method applied to the system
    $$10x + \sin(x+y) = 1$$
    $$8y - \cos^2(z-y) = 1$$
    $$12z + \sin z = 1$$
    (b) Suggest some explicit method of the form $\mathbf{x}^{(k+1)} = \mathbf{F}(\mathbf{x}^{(k)})$ where no inversion is needed for $\mathbf{F}$, and estimate how many iterations are required to obtain a solution correct to two decimal places from the starting point in (a).
    (Uppsala Univ., Sweden, BIT 19(1979), 139)

38. Calculate the solution of the system of equations
    $$x^3 + y^3 = 53$$
    $$2y^3 + z^4 = 69$$
    $$3x^5 + 10z^2 = 770$$
    which is close to $x = 3$, $y = 3$, $z = 2$,
    (Stockholm Univ., Sweden, BIT 19(1979), 285)

39. Obtain the number of real roots between 0 and 3 of the equation
    $$P(x) = x^4 - 4x^3 + 3x^2 + 4x - 4 = 0,$$
    using Sturm sequences.

40. Determine the multiplicity of the root $\xi = 1$, of the polynomial equation
    $$P(x) = x^5 - 2x^4 + 4x^3 - x^2 - 7x + 5 = 0,$$
    using synthetic division. Find also $P'(2)$ and $P''(2)$.

41. Use the Birge-Vieta method to find a real root correct to three decimals of the following equations:
    (i) $x^3 - 11x^2 + 32x - 22 = 0$, $p = 1.0$
    (ii) $x^5 - x + 1 = 0$, $p = -1.5$
    (iii) $x^6 - x^4 - x^3 - 1 = 0$, $p = 1.5$

42. Carry out two iterations of the Chebyshev method, the multipoint methods (2.29), and (2.31) for finding the root of the polynomial equation $x^3 - 2 = 0$ with $x_0 = 1$, using synthetic division.

43. Given the two polynomials
    $$P(x) = x^6 - 4.8x^4 + 3.3x^2 - 0.05$$
    and
    $$Q(x, h) = x^6 - (4.8 - h)x^4 + (3.3 + h)x^2 - (0.05 - h)$$
    (a) Calculate all the roots of P.
    (b) When $h \ll 1$ the roots of Q are close to those of P. Estimate the difference between the smallest positive root of P and the corresponding root of Q.
    (Danmarks Tekniske Hojskole, Denmark, BIT 19(1979), 139)

44. Using Bairstow's method obtain the quadratic factor of the following equations (perform two iterations):
    (i) $x^4 - 3x^3 + 20x^2 + 44x + 54 = 0$, with $(p, q) = (2, 2)$
    (ii) $x^4 - x^3 + 6x^2 + 5x + 10 = 0$, with $(p, q) = (1.14, 1.42)$
    (iii) $x^3 - 3.7x^2 + 6.25x - 4.069 = 0$, with $(p, q) = (-2.5, 3)$.

45. Develop a flow diagram for Graeffe's root squaring method. Apply the method to find the roots of the following equations correct to two decimals:
    (i) $x^3 - 2x + 2 = 0$
    (ii) $x^3 + 3x^2 - 4 = 0$
    (iii) $x^4 - x^3 + 3x^2 + x - 4 = 0$.

46. Find, to two decimal places, the real and complex roots of the equation
    $$x^5 = 3x - 1.$$

47. The equation
    $$2e^{-x} = \frac{1}{x+2} + \frac{1}{x+1}$$
    has two roots greater than -1.
    Calculate these roots correct to five decimal places.
    (Inst. Tech., Lund, Sweden, BIT 21(1981), 136)

48. Find the positive root of the equation
    $$e^x = 1 + x + \frac{x^2}{2} + \frac{x^3}{6} e^{0.3x}$$
    correct to two decimal places.
    (Royal Inst. Tech., Stockholm, Sweden, BIT 21(1981), 242)

49. Assuming that Δx, in the Taylor expansion of $f(x_0 + \Delta x)$, can be approximated by $\Delta x = a_1 f(x_0) + a_2 f^2(x_0) + a_3 f^3(x_0) + \ldots$, where $a_0, a_1, a_2, \ldots$ are arbitrary parameters to be determined. Derive the Chebyshev methods of third and fourth orders for finding a root of $f(x) = 0$.

50. Consider the system of equations $f(x, y) = 0$, $g(x, y) = 0$. Let $x = x_0 + \Delta x$, $y = y_0 + \Delta y$, where (x_0, y_0) is an initial approximation to the solution. Assume
    $$\Delta x = A_1(x_0, y_0) + A_2(x_0, y_0) + A_3(x_0, y_0) + \ldots,$$
    $$\Delta y = B_1(x_0, y_0) + B_2(x_0, y_0) + B_3(x_0, y_0) + \ldots,$$
    where $A_1(x_0, y_0)$, $B_1(x_0, y_0)$ are linear in f_0, g_0, $A_2(x_0, y_0)$, $B_2(x_0, y_0)$ are quadratic in f_0, g_0 and so on. Use Taylor series method to derive iterative methods of second and third orders.

CHAPTER 3

System of Linear Algebraic Equations and Eigenvalue Problems

3.1 INTRODUCTION

Consider a system of n linear algebraic equations in n unknowns

$$a_{11}x_1 + a_{12}x_2 + \ldots + a_{1n}x_n = b_1$$
$$a_{21}x_1 + a_{22}x_2 + \ldots + a_{2n}x_n = b_2$$
$$\vdots \qquad\qquad\qquad\qquad\qquad\qquad (3.1)$$
$$a_{n1}x_1 + a_{n2}x_2 + \ldots + a_{nn}x_n = b_n$$

where a_{ij}, $(i, j = 1(1)n)$ are the known coefficients, b_i, $(i = 1(1)n)$ are the known values and x_i, $(i = 1(1)n)$ are the unknowns to be determined. We introduce the following notation and definitions.

Notation

A	square matrix of order n
a_{ij}	element in the ith row and jth column of the matrix A
A^{-1}	inverse of A
A^T	transpose of A
$\|A\|$	determinant of A
M_{ij}	minor of a_{ij} in A
A_{ij}	cofactor of a_{ij} in A
0	null matrix
I	unit matrix of order n
D	diagonal matrix of order n
L	Lower triangular matrix of order n
U	upper triangular matrix of order n
P	permutation matrix

x column vector with elements x_i, $i = 1(1)n$

$\mathbf{x}^T$ row vector with elements x_i, $i = 1(1)n$

$\rho(\mathbf{A})$ spectral radius of $\mathbf{A}$

$\|\mathbf{A}\|$ norm of $\mathbf{A}$

$\|\mathbf{x}\|$ norm of $\mathbf{x}$

Definitions The matrix $\mathbf{A}$ is said to be

nonsingular if $|\mathbf{A}| \neq 0$

symmetric if $\mathbf{A} = \mathbf{A}^T$

skew symmetric if $\mathbf{A} = -\mathbf{A}^T$

orthogonal if $\mathbf{A}^{-1} = \mathbf{A}^T$

null if $a_{ij} = 0$, $i, j = 1(1)n$

diagonal if $a_{ij} = 0$, $i \neq j$

unit matrix if $a_{ij} = 0$, $i \neq j$, $a_{ii} = 1$, $i = 1(1)n$

lower triangular if $a_{ij} = 0$, $j > i$

upper triangular if $a_{ij} = 0$, $i > j$

band matrix if $a_{ij} = 0$, for $j > i + p$ and $i > j + q$ with band width $p + q + 1$

tridiagonal if $a_{ij} = 0$, for $|i - j| > 1$

diagonally dominant if $|a_{ij}| \geq \sum_{\substack{j=1 \\ i \neq j}}^{n} |a_{ij}|$, $i = 1(1)n$

permutation matrix if it has exactly one 1 in each row and column and all other entries are 0

reducible if there exists a permutation matrix $\mathbf{P}$ such that

$$\mathbf{PAP}^T = \begin{bmatrix} \mathbf{A}_{11} & \mathbf{A}_{12} \\ \mathbf{0} & \mathbf{A}_{22} \end{bmatrix}$$

where $\mathbf{A}_{11}$ and $\mathbf{A}_{22}$ are square submatrices.

Hermitian, denoted by $\mathbf{A}^*$ or $\mathbf{A}^H$, if $\mathbf{A} = (\bar{\mathbf{A}})^T$ where $\bar{\mathbf{A}}$ is the complex conjugate of $\mathbf{A}$

Unitary if $\mathbf{A}^{-1} = (\bar{\mathbf{A}})^T$

normal if $\mathbf{AA}^* = \mathbf{A}^*\mathbf{A}$

A matrix $\mathbf{A}$ is said to be a positive definite matrix, if $\mathbf{x}^*\mathbf{Ax} > 0$ for any vector $\mathbf{x} \neq \mathbf{0}$ and $\mathbf{x}^* = (\bar{\mathbf{x}})^T$. Further, $\mathbf{x}^*\mathbf{Ax} = 0$ if $\mathbf{x} = \mathbf{0}$. If $\mathbf{A}$ is a Hermitian, strictly diagonal dominant matrix with positive real diagonal entries, then $\mathbf{A}$ is positive definite. Positive definite matrices have the following important properties.

(i) If **A** is nonsingular and positive definite, then $\mathbf{B} = \mathbf{A}^*\mathbf{A}$ is Hermitian and positive definite.

(ii) The eigenvalues of a positive definite matrix are all positive.

(iii) All the leading minors of **A** are positive.

A real matrix **B** is said to have 'property A' iff there exists a permutation matrix **P** such that

$$\mathbf{PBP}^T = \begin{bmatrix} \mathbf{A}_{11} & \mathbf{A}_{12} \\ \mathbf{A}_{21} & \mathbf{A}_{22} \end{bmatrix}$$

where $\mathbf{A}_{11}$ and $\mathbf{A}_{22}$ are diagonal matrices.

The minor M_{ij} of a_{ij} is the determinant of the $n-1 \times n-1$ submatrix obtained by deleting the ith row and jth column in **A**.

The cofactor A_{ij} is defined by

$$A_{ij} = (-1)^{i+j} M_{ij}$$

The transpose of the matrix of the cofactors of elements of **A** is called the **adjoint** matrix and is denoted by Adj (**A**)

The inverse of a matrix **A** is defined by

$$\mathbf{A}^{-1} = \frac{1}{|\mathbf{A}|} \text{Adj}(\mathbf{A})$$

The matrix [**A**|**b**] is called the **augmented** matrix. It is formed by appending the column **b** to the $n \times n$ matrix **A**.

In the matrix notation, the system (3.1) can be written as

$$\mathbf{Ax} = \mathbf{b} \tag{3.2}$$

If all b_i are zero then the system of equations (3.1) is said to be homogeneous, and if at least one of b_i is not zero then it is said to be inhomogeneous. The inhomogeneous system (3.1) has a unique solution if and only if the determinant of **A** is non zero, i.e.

$$\det(\mathbf{A}) = \begin{vmatrix} a_{11} & a_{12} & \cdots & a_{1n} \\ a_{21} & a_{22} & \cdots & a_{2n} \\ \vdots & & & \\ a_{n1} & a_{n2} & \cdots & a_{nn} \end{vmatrix} \neq 0$$

The solution of (3.1) may be written as

$$\mathbf{x} = \mathbf{A}^{-1}\mathbf{b} \tag{3.3}$$

The homogeneous system ($b_i = 0$, $i = 1(1)n$) possesses only a trivial solution

$$x_1 = x_2 = \ldots = x_n = 0$$

if $\det(\mathbf{A}) \neq 0$. We, therefore, consider the system in which a parameter λ occurs, and we determine values of λ, called **eigenvalues**, for which the system has a non trivial solution. Such a solution is called an **eigenfunction** and the entire system is called an **eigenvalue problem** or the **characteristic** value problem. The system (3.2), ($b = 0$) may then be written as

$$Ax = \lambda x$$

or

$$(A - \lambda I)x = 0 \tag{3.4}$$

In order that the equations (3.4) have a nontrivial solution $x \neq 0$ the determinant of the matrix $(A - \lambda I)$ must be zero

$$\det(A - \lambda I) = 0 \tag{3.5}$$

The equation (3.5) is called the **characteristic equation**. The n roots $\lambda_1, \lambda_2, \ldots, \lambda_n$ are called the eigenvalues of A and may be distinct or repeated, real or complex. The modulus of the largest eigenvalue is called the **spectral radius** of A. Corresponding to each eigenvalue λ_i, there exists an eigenvector x_i which is a nontrivial solution of

$$(A - \lambda_i I)x_i = 0 \tag{3.6}$$

If the n eigenvalues λ_i, $i = 1(1)n$ of A are different and x_i are the corresponding eigenvectors then any vector y, in the eigenspace can be expressed as a linear combination of the x_i

$$y = c_1 x_1 + c_2 x_2 + \ldots + c_n x_n \tag{3.7}$$

The methods of solution of the linear algebraic equations (3.2) and the methods to determine the eigenvalues and eigenvectors of the system (3.4) may broadly be classified into two types.

(i) *Direct methods*: These methods produce the exact solution after a finite number of steps (disregarding the round-off errors).

(ii) *Iterative methods*: These methods give a sequence of approximate solutions, which converge when the number of steps tend to infinity.

We now give a few direct and iterative computational methods to solve (3.2) and (3.4).

3.2 DIRECT METHODS

The system of equations (3.2)

$$Ax = b$$

can be directly solved in the following cases.

(i) $A = D$

The equations (3.1) become

$$a_{11}x_1 \qquad\qquad = b_1$$
$$\qquad a_{22}x_2 \qquad = b_2$$
$$\qquad\qquad \vdots$$
$$\qquad\qquad a_{nn}x_n = b_n$$

The solution is given by
$$x_i = \frac{b_i}{a_{ii}}, \quad i = 1(1)n \tag{3.8}$$
where $a_{ii} \neq 0$, $i = 1(1)n$.

(ii) $\mathbf{A} = \mathbf{L}$

The equations (3.1) may be written as
$$\begin{aligned} a_{11}x_1 &= b_1 \\ a_{21}x_1 + a_{22}x_2 &= b_2 \\ a_{31}x_1 + a_{32}x_2 + a_{33}x_3 &= b_3 \\ &\vdots \\ a_{n1}x_1 + a_{n2}x_2 + \ldots + a_{nn}x_n &= b_n \end{aligned} \tag{3.9}$$

Solving the first equation and then successively solving second, third and so on, we obtain
$$\begin{aligned} x_1 &= b_1/a_{11} \\ x_2 &= (b_2 - a_{21}x_1)/a_{22} \\ x_3 &= (b_3 - a_{31}x_1 - a_{32}x_2)/a_{33} \\ &\vdots \\ x_n &= (b_n - \sum_{j=1}^{n-1} a_{nj}x_j)/a_{nn} \end{aligned} \tag{3.10}$$

where $a_{ii} \neq 0$, $i = 1(1)n$.

Since the unknowns are solved by forward substitutions, this method is called the **forward substitution** method.

(iii) $\mathbf{A} = \mathbf{U}$

The system of equations (3.1) becomes
$$\begin{aligned} a_{11}x_1 + a_{12}x_2 + \ldots + a_{1n}x_n &= b_1 \\ a_{22}x_2 + \ldots + a_{2n}x_n &= b_2 \\ &\vdots \\ a_{n-1,\,n-1}x_{n-1} + a_{n-1,\,n}x_n &= b_{n-1} \\ a_{nn}x_n &= b_n \end{aligned} \tag{3.11}$$

Solving for the unknowns in the order $x_n, x_{n-1}, \ldots, x_1$, we get
$$\begin{aligned} x_n &= b_n/a_{nn} \\ x_{n-1} &= (b_{n-1} - a_{n-1,\,n}x_n)/a_{n-1,\,n-1} \\ &\vdots \\ x_1 &= (b_1 - \sum_{j=2}^{n} a_{1j}x_j)/a_{11} \end{aligned}$$

74 Numerical Methods

The unknowns are solved by the back substitution and this method is called the **back substitution** method.

Thus, the equations (3.2) are exactly solvable, if the matrix **A** in (3.2) can be transformed into any one of the forms **D**, **L** or **U**.

3.2.1 Cramer Rule

In order to determine the unknown x_1 in the equations (3.1) we multiply the equations by the cofactors A_{i1}, $i(1)n$ and add these together to get

$$(\sum_{i=1}^{n} a_{i1}A_{i1})x_1 + \sum_{j=2}^{n}(\sum_{i=1}^{n} a_{ij}A_{i1})x_j = \sum_{i=1}^{n} b_i A_{i1}$$

which gives

$$|A|x_1 = b_1 A_{11} + b_2 A_{21} + \ldots + b_n A_{n1} \tag{3.12}$$

Similarly, we may obtain

$$|A|x_2 = b_1 A_{12} + b_2 A_{22} + \ldots + b_n A_{n2}$$
$$|A|x_3 = b_1 A_{13} + b_2 A_{23} + \ldots + b_n A_{n3} \tag{3.13}$$
$$\vdots$$
$$|A|x_n = b_1 A_{1n} + b_2 A_{2n} + \ldots + b_n A_{nn}$$

Thus we have

$$x_i = \frac{|\mathbf{B}_i|}{|\mathbf{A}|}, \quad i = 1(1)n \tag{3.14}$$

where $|\mathbf{B}_i|$ is the determinant of the matrix obtained by replacing the ith column of **A** by the right hand vector **b**. This process is known as the **Cramer rule**.

The equations (3.12) and (3.13) may be written as

$$\mathbf{x} = \frac{1}{|\mathbf{A}|} \begin{bmatrix} A_{11} & A_{21} & \ldots & A_{n1} \\ A_{12} & A_{22} & \ldots & A_{n2} \\ \vdots & & & \\ A_{1n} & A_{2n} & \ldots & A_{nn} \end{bmatrix} \begin{bmatrix} b_1 \\ b_2 \\ \vdots \\ b_n \end{bmatrix} = \frac{1}{|\mathbf{A}|} \text{adj}(\mathbf{A})\mathbf{b} = \mathbf{A}^{-1}\mathbf{b}$$

which is same as (3.3). The Cramer rule provides a method of obtaining the solution immediately, but since these solutions involve $n+1$ determinants of order n they are not in a very convenient form.

Example 3.1 Solve the equations

$$x_1 + 2x_2 - x_3 = 2$$
$$3x_1 + 6x_2 + x_3 = 1$$
$$3x_1 + 3x_2 + 2x_3 = 3$$

(i) by using Cramer's rule
(ii) by determining the inverse of the coefficient matrix.

Using Cramer's rule the solution of the equations may immediately be written as

$$x_i = |B_i|/|A|, \quad i = 1(1)3$$

where

$$|B_1| = \begin{vmatrix} 2 & 2 & -1 \\ 1 & 6 & 1 \\ 3 & 3 & 2 \end{vmatrix}, \quad |B_2| = \begin{vmatrix} 1 & 2 & -1 \\ 3 & 1 & 1 \\ 3 & 3 & 2 \end{vmatrix}$$

$$|B_3| = \begin{vmatrix} 1 & 2 & 2 \\ 3 & 6 & 1 \\ 3 & 3 & 3 \end{vmatrix}, \quad |A| = \begin{vmatrix} 1 & 2 & -1 \\ 3 & 6 & 1 \\ 3 & 3 & 2 \end{vmatrix}$$

The cofactors of the elements a_{ij}, $i, j = 1(1)3$ are given by

$A_{11} = 9, \quad A_{12} = -3, \quad A_{13} = -9$
$A_{21} = -7, \quad A_{22} = 5, \quad A_{23} = 3$
$A_{31} = 8, \quad A_{32} = -4, \quad A_{33} = 0$

Thus, we have

$|A| = 9 - 6 + 9 = 12$
$|B_1| = 18 - 7 + 24 = 35$
$|B_2| = -6 + 5 - 12 = -13$
$|B_3| = -18 + 3 + 0 = -15$

$$x_1 = \frac{35}{12}, \quad x_2 = -\frac{13}{12}, \quad x_3 = -\frac{15}{12}$$

$$A^{-1} = \frac{1}{12} \begin{bmatrix} 9 & -7 & 8 \\ -3 & 5 & -4 \\ -9 & 3 & 0 \end{bmatrix}$$

$$x = \frac{1}{12} \begin{bmatrix} 9 & -7 & 8 \\ -3 & 5 & -4 \\ -9 & 3 & 0 \end{bmatrix} \begin{bmatrix} 2 \\ 1 \\ 3 \end{bmatrix} = \frac{1}{12} \begin{bmatrix} 35 \\ -13 \\ -15 \end{bmatrix}$$

3.2.2 Gauss Elimination Method

Here the unknowns are eliminated by combining equations such that the n equations in n unknowns are reduced to an equivalent upper triangular system which is then solved by back substitution. Consider the 3×3 system

$$a_{11}x_1 + a_{12}x_2 + a_{13}x_3 = b_1$$
$$a_{21}x_1 + a_{22}x_2 + a_{23}x_3 = b_2 \qquad (3.15)$$
$$a_{31}x_1 + a_{32}x_2 + a_{33}x_3 = b_3$$

In the first stage of elimination, multiply the first row in (3.15) by a_{21}/a_{11} and a_{31}/a_{11} respectively and subtract from the second and third rows. We get

$$a_{22}^{(2)}x_2 + a_{23}^{(2)}x_3 = b_2^{(2)}$$
$$a_{32}^{(2)}x_2 + a_{33}^{(2)}x_3 = b_3^{(2)} \qquad (3.16)$$

where $a_{22}^{(2)} = a_{22} - \dfrac{a_{21}}{a_{11}} a_{12}$

$$a_{23}^{(2)} = a_{23} - \dfrac{a_{21}}{a_{11}} a_{13}$$

$$a_{32}^{(2)} = a_{32} - \dfrac{a_{31}}{a_{11}} a_{12}$$

$$a_{33}^{(2)} = a_{33} - \dfrac{a_{31}}{a_{11}} a_{13}$$

$$b_2^{(2)} = b_2 - \dfrac{a_{21}}{a_{11}} b_1$$

$$b_2^{(2)} = b_3 - \dfrac{a_{31}}{a_{11}} b_1$$

In the second stage of elimination, multiply the first row in (3.16) by $(a_{32}^{(2)}/a_{22}^{(2)})$ and subtract from the second row in (3.16). We get

$$a_{33}^{(3)}x_3 = b_3^{(3)} \qquad (3.17)$$

where $a_{33}^{(3)} = a_{33}^{(2)} - \dfrac{a_{32}^{(2)}}{a_{22}^{(2)}} a_{23}^{(2)}$

$$b_3^{(3)} = b_3^{(2)} - \dfrac{a_{32}^{(2)}}{a_{22}^{(2)}} b_2^{(2)}$$

Collecting the first equation from each stage, i.e. from (3.15), (3.16) and (3.17) we obtain the system

$$a_{11}^{(1)}x_1 + a_{12}^{(1)}x_2 + a_{13}^{(1)}x_3 = b_1^{(1)}$$
$$a_{22}^{(2)}x_2 + a_{23}^{(2)}x_3 = b_2^{(2)} \qquad (3.18)$$
$$a_{33}^{(3)}x_3 = b_3^{(3)}$$

where $a_{ij}^{(1)} = a_{ij}$, $b_i^{(1)} = b_i$, $i, j = 1(1)3$.

The system (3.18) is an upper triangular system and can be solved using the back substitution method. Therefore, the Gauss elimination method gives

$$[A|b] \xrightarrow[\text{Elimination}]{\text{Gauss}} [U|C] \tag{3.19}$$

where $[A|b]$ is the augmented matrix. The elements $a_{11}^{(1)}$, $a_{22}^{(2)}$ and $a_{33}^{(3)}$ which have been assumed to be non-zero are called **pivot elements**. The elimination procedure described above to determine the unknowns is called the **Gauss elimination** method.

We now solve the system (3.1) in n unknowns by performing the Gauss elimination on the augmented matrix $[A|b]$. We put

$$b_i^{(k)} = a_{i,n+1}^{(k)}, \quad i, k = 1(1)n \tag{3.20}$$

The elements $a_{ij}^{(k)}$ with $i, j > k$ are given by

$$a_{ij}^{(k+1)} = a_{ij}^{(k)} - \frac{a_{ik}^{(k)}}{a_{kk}^{(k)}} a_{kj}^{(k)} \tag{3.21}$$

$$i = k+1, k+2, \ldots, n; \quad j = k+1, \ldots, n, n+1$$

$$a_{ij}^{(1)} = a_{ij}$$

The elimination is performed in $(n-1)$ steps, $k = 1, 2, \ldots, n-1$. In the elimination process, if any one of the pivot elements $a_{11}^{(1)}, a_{22}^{(2)}, \ldots, a_{nn}^{(n)}$ vanishes or becomes very small compared to other elements in that row then we attempt to rearrange the remaining rows so as to obtain a non-vanishing pivot or to avoid the multiplication by a large number. This strategy is called **pivoting**. The pivoting is of the following two types.

Partial Pivoting

In the first stage of elimination, the first column is searched for the largest element in magnitude and brought as the first pivot by interchanging the first equation with the equation having the largest element in magnitude. In the second elimination stage, the second column is searched for the largest element in magnitude among the $n-1$ elements leaving the first element, and this element is brought as the second pivot by an interchange of the second equation with the equation having the largest element in magnitude. This procedure is continued until we arrive at the equation (3.19). We are thus led to the following algorithm to find the pivot.

Choose j, the smallest integer for which

$$|a_{jk}^{(k)}| = \max |a_{ik}^{(k)}|, \ k \leqslant i \leqslant n$$

and interchange rows k and j.

Complete Pivoting

We search the matrix **A** for the largest element in magnitude and bring it as the first pivot. This requires not only an interchange of equations but also an interchange of the position of the variables. This leads us to the following algorithm to find the pivot.

Choose l and m as the smallest integers for which
$$|a_{lm}^{(k)}| = \max |a_{ij}^{(k)}|, \quad k \leqslant i, j \leqslant n$$
and interchange rows k and l and columns k and m..

If the matrix A is diagonally dominant or real, symmetric and positive definite then no pivoting is necessary.

Example 3.2 Solve the equations
$$10x_1 - x_2 + 2x_3 = 4$$
$$x_1 + 10x_2 - x_3 = 3$$
$$2x_1 + 3x_2 + 20x_3 = 7$$
using the Gauss elimination method.

The system is diagonally dominant and no pivoting is necessary. We have, after the first elimination stage
$$10x_1 - x_2 + 2x_3 = 4$$
$$\frac{101}{10}x_2 - \frac{12}{10}x_3 = \frac{26}{10}$$
$$\frac{32}{10}x_2 + \frac{196}{10}x_3 = \frac{62}{10}$$
second elimination stage
$$10x_1 - x_2 + 2x_3 = 4$$
$$\frac{101}{10}x_2 - \frac{12}{10}x_3 = \frac{26}{10}$$
$$\frac{20180}{1010}x_3 = \frac{5430}{1010}$$
Using back substitution, we get the solution
$$x_3 = 0.269, \; x_2 = 0.289 \text{ and } x_1 = 0.375$$

Example 3.3 Solve the equations
$$x_1 + x_2 + x_3 = 6$$
$$3x_1 + 3x_2 + 4x_3 = 20$$
$$2x_1 + x_2 + 3x_3 = 13$$
using the Gauss elimination method.

In the first step we eliminate x_1 from the last two equations and obtain
$$x_1 + x_2 + x_3 = 6$$
$$x_3 = 2$$
$$-x_2 + x_3 = 1$$

Here the pivot in the second equation is zero and so we cannot proceed as usual. We interchange the equations 2 and 3 before the second step. We obtain the upper triangular system

$$x_1 + x_2 + x_3 = 6$$
$$-x_2 + x_3 = 1$$
$$x_3 = 2$$

which has the solution

$$x_1 = 3, \quad x_2 = 1 \text{ and } x_3 = 2$$

Example 3.4 Solve the equations

$$x_1 + x_2 + x_3 = 6$$
$$3x_1 + (3+\epsilon)x_2 + 4x_3 = 20$$
$$2x_1 + x_2 + 3x_3 = 13$$

using the Gauss elimination method, where ϵ is small such that $1 \pm \epsilon^2 \approx 1$. Eliminating x_1 from the last two equations, we get

$$x_1 + x_2 + x_3 = 6$$
$$\epsilon x_2 + x_3 = 2$$
$$-x_2 + x_3 = 1$$

Here, the pivot in the second equation is ϵ which is a very small number. If we do not use pivoting, then we get

$$x_1 + x_2 + x_3 = 6$$
$$\epsilon x_2 + x_3 = 2$$
$$\left(1 + \frac{1}{\epsilon}\right) x_3 = 1 + \frac{2}{\epsilon}$$

The solution is

$$x_3 = \frac{1 + \frac{2}{\epsilon}}{1 + \frac{1}{\epsilon}}, \quad x_2 = \frac{1}{\epsilon}\left[2 - \frac{1 + \frac{2}{\epsilon}}{1 + \frac{1}{\epsilon}}\right] \quad \text{and} \quad x_1 = 6 - x_2 - x_3$$

However, this solution may be very inaccurate if ϵ is of the order of the roundoff error. This situation can be avoided if pivoting is done at the second step. In this case we have

$$x_1 + x_2 + x_3 = 6$$
$$-x_2 + x_3 = 1$$
$$(1 + \epsilon)x_3 = 2 + \epsilon$$

The solution is

$$x_3 = \frac{2 + \epsilon}{1 + \epsilon}, \quad x_2 = 1 + \frac{2 + \epsilon}{1 + \epsilon} \quad \text{and} \quad x_1 = 6 - x_2 - x_3$$

3.2.3 Gauss-Jordan Elimination Method

Here the coefficient matrix is reduced to a diagonal matrix rather than a triangular matrix. At all steps of the Gauss elimination method, the elimination is done not only in the equations below but also in the equations above, producing the solution without using the back substitution method. On the completion of the Gauss-Jordan method the equations (3.1) become

$$\begin{bmatrix} 1 & 0 & 0 & \ldots & 0 \\ 0 & 1 & 0 & \ldots & 0 \\ & & \cdot & & \vdots \\ & & \cdot & & \vdots \\ 0 & 0 & 0 & & 1 \end{bmatrix} \begin{bmatrix} x_1 \\ x_2 \\ \vdots \\ \vdots \\ x_n \end{bmatrix} = \begin{bmatrix} d_1 \\ d_2 \\ \vdots \\ \vdots \\ d_n \end{bmatrix} \quad (3.22)$$

The solution is given by

$x_i = d_i, \quad i = 1(1)n$

Hence, the Gauss-Jordan method gives

$$[A|b] \xrightarrow[\text{Jordan}]{\text{Gauss}} [I|d] \quad (3.23)$$

Generally, this method is not used for the solution of a system of equations as it is more expensive from the computation view-point than the Gauss elimination method. However, it gives a simple method for finding the inverse of a given matrix **A**. We start with the augmented matrix of **A** and the identity matrix **I** of the same order. When the Gauss-Jordan procedure is completed we obtain

$$[A|I] \xrightarrow[\text{Jordan}]{\text{Gauss}} [I|A^{-1}] \quad (3.24)$$

Example 3.5 Find the inverse of the coefficient matrix of the system

$$\begin{bmatrix} 1 & 1 & 1 \\ 4 & 3 & -1 \\ 3 & 5 & 3 \end{bmatrix} \begin{bmatrix} x_1 \\ x_2 \\ x_3 \end{bmatrix} = \begin{bmatrix} 1 \\ 6 \\ 4 \end{bmatrix}$$

by the Gauss-Jordan method with partial pivoting and hence solve the system.

Using the augmented matrix $[A|I]$ we obtain

$$\begin{bmatrix} 1 & 1 & 1 & | & 1 & 0 & 0 \\ 4 & 3 & -1 & | & 0 & 1 & 0 \\ 3 & 5 & 3 & | & 0 & 0 & 1 \end{bmatrix} \sim \begin{bmatrix} 1 & \tfrac{3}{4} & -\tfrac{1}{4} & | & 0 & \tfrac{1}{4} & 0 \\ 1 & 1 & 1 & | & 1 & 0 & 0 \\ 3 & 5 & 3 & | & 0 & 0 & 1 \end{bmatrix}$$

$$\sim \begin{bmatrix} 1 & \tfrac{3}{4} & -\tfrac{1}{4} & | & 0 & \tfrac{1}{4} & 0 \\ 0 & \tfrac{1}{4} & \tfrac{5}{4} & | & 1 & -\tfrac{1}{4} & 0 \\ 0 & \tfrac{11}{4} & \tfrac{15}{4} & | & 0 & -\tfrac{3}{4} & 1 \end{bmatrix} \sim \begin{bmatrix} 1 & \tfrac{3}{4} & -\tfrac{1}{4} & | & 0 & \tfrac{1}{4} & 0 \\ 0 & 1 & \tfrac{15}{11} & | & 0 & -\tfrac{3}{11} & \tfrac{4}{11} \\ 0 & \tfrac{1}{4} & \tfrac{5}{4} & | & 1 & -\tfrac{1}{4} & 0 \end{bmatrix}$$

$$\sim \begin{bmatrix} 1 & 0 & -\frac{14}{11} & 0 & \frac{5}{11} & -\frac{3}{11} \\ 0 & 1 & \frac{15}{11} & 0 & -\frac{3}{11} & \frac{4}{11} \\ 0 & 0 & \frac{10}{11} & 1 & -\frac{2}{11} & -\frac{1}{11} \end{bmatrix} \sim \begin{bmatrix} 1 & 0 & -\frac{14}{11} & 0 & \frac{5}{11} & -\frac{3}{11} \\ 0 & 1 & \frac{15}{11} & 0 & -\frac{3}{11} & \frac{4}{11} \\ 0 & 0 & 1 & \frac{11}{10} & -\frac{1}{5} & -\frac{1}{10} \end{bmatrix}$$

$$\sim \begin{bmatrix} 1 & 0 & 0 & \frac{7}{5} & \frac{1}{5} & -\frac{2}{5} \\ 0 & 1 & 0 & -\frac{3}{2} & 0 & \frac{1}{2} \\ 0 & 0 & 1 & \frac{11}{10} & -\frac{1}{5} & -\frac{1}{10} \end{bmatrix}$$

Therefore, the solution of the system is

$$\begin{bmatrix} x_1 \\ x_2 \\ x_3 \end{bmatrix} = \begin{bmatrix} \frac{7}{5} & \frac{1}{5} & -\frac{2}{5} \\ -\frac{3}{2} & 0 & \frac{1}{2} \\ \frac{11}{10} & -\frac{1}{5} & -\frac{1}{10} \end{bmatrix} \begin{bmatrix} 1 \\ 6 \\ 4 \end{bmatrix} = \begin{bmatrix} 1 \\ \frac{1}{2} \\ -\frac{1}{2} \end{bmatrix}$$

3.2.4 Triangularization Method

This method is also known as **decomposition** method. In this method, the coefficient matrix **A** of the system of equations (3.2) is decomposed or factorized into the product of a lower triangular matrix **L** and an upper triangular matrix **U**. We write as

$$\mathbf{A} = \mathbf{LU} \tag{3.25}$$

where
$$\mathbf{L} = \begin{bmatrix} l_{11} & 0 & 0 & \cdots & 0 \\ l_{21} & l_{22} & 0 & \cdots & 0 \\ l_{31} & l_{32} & l_{33} & \cdots & 0 \\ \vdots & & & & \vdots \\ l_{n1} & l_{n2} & l_{n3} & \cdots & l_{nn} \end{bmatrix}$$

$$\mathbf{U} = \begin{bmatrix} u_{11} & u_{12} & u_{13} & \cdots & u_{1n} \\ 0 & u_{22} & u_{23} & \cdots & u_{2n} \\ 0 & 0 & u_{33} & \cdots & u_{3n} \\ \vdots & & & & \\ 0 & 0 & 0 & \cdots & u_{nn} \end{bmatrix}$$

Using the matrix multiplication rule to multiply the matrices **L** and **U** and comparing the elements of the resulting matrix with those of **A** we obtain

$$l_{i1}u_{1j} + l_{i2}u_{2j} + \ldots + l_{in}u_{nj} = a_{ij}, \quad i, j = 1(1)n \tag{3.26}$$

where $l_{ij} = 0, j > i$ and $u_{ij} = 0, i > j$.

The system of n^2 equations involve $n^2 + n$ unknowns. Thus, there are n parameter family of solutions. To produce a unique solution it is convenient to choose either $u_{ii} = 1$ or $l_{ii} = 1, i = 1(1)n$. We may choose $u_{ii} = 1$, $i = 1(1)n$. The solution of the equations (3.26) may be written as

$$l_{ij} = a_{ij} - \sum_{k=1}^{j-1} l_{ik}u_{kj}, \ i \geqslant j$$

$$u_{ij} = (a_{ij} - \sum_{k=1}^{i-1} l_{ik}u_{kj})/l_{ii}, \ i < j \qquad (3.27)$$

$$u_{ii} = 1$$

We note that the first column of the matrix **L** is identical with the first column of the matrix **A**. That is

$$l_{i1} = a_{i1}, \ i = 1(1)n \qquad (3.28)$$

We also note that

$$u_{1j} = a_{1j}/l_{11}, \ j = 2(1)n \qquad (3.29)$$

The first column of **L** and the first row of **U** have been determined. We can now proceed to determine the second column of **L** and the second row of **U**

$$l_{i2} = a_{i2} - l_{i1}u_{12}, \ i = 2(1)n$$

$$u_{2j} = (a_{2j} - l_{21}u_{1j})/l_{22}, \ j = 3(1)n \qquad (3.30)$$

Next we find the third column of **L** followed by the third row of **U**. Thus, for the relevant indices i and j, the elements are computed in the order

$$l_{i1}, u_{1j}; l_{i2}, u_{2j}; l_{i3}, u_{3j}; \ldots, l_{i,\,n-1}, u_{n-1,\,j}; l_{nn}.$$

Having determined the matrices **L** and **U**, the system of equations (3.2) becomes

$$\mathbf{LUx} = \mathbf{b} \qquad (3.31)$$

We write (3.31) as the following two systems of equations

$$\mathbf{Ux} = \mathbf{z} \qquad (3.32)$$

$$\mathbf{Lz} = \mathbf{b} \qquad (3.33)$$

The unknowns $z_1, z_2, \ldots, z_n$ in (3.33) are determined by forward substitution and the unknowns $x_1, x_2, \ldots, x_n$ in (3.32) are obtained by back substitution.

Alternatively, we can find $\mathbf{L}^{-1}$ and $\mathbf{U}^{-1}$ to get

$$\mathbf{z} = \mathbf{L}^{-1}\mathbf{b}$$

and

$$\mathbf{x} = \mathbf{U}^{-1}\mathbf{z} \qquad (3.34)$$

The inverse of **A** can also be determined from

$$\mathbf{A}^{-1} = \mathbf{U}^{-1}\mathbf{L}^{-1} \qquad (3.35)$$

Example 3.6 Consider the equations

$$x_1 + x_2 + x_3 = 1$$
$$4x_1 + 3x_2 - x_3 = 6$$
$$3x_1 + 5x_2 + 3x_3 = 4$$

Use the decomposition method to solve the system.
We write

$$\begin{bmatrix} 1 & 1 & 1 \\ 4 & 3 & -1 \\ 3 & 5 & 3 \end{bmatrix} = \begin{bmatrix} l_{11} & 0 & 0 \\ l_{21} & l_{22} & 0 \\ l_{31} & l_{32} & l_{33} \end{bmatrix} \begin{bmatrix} 1 & u_{12} & u_{13} \\ 0 & 1 & u_{23} \\ 0 & 0 & 1 \end{bmatrix}$$

and obtain

$l_{11} = 1, l_{21} = 4, l_{31} = 3$

$u_{12} = 1, u_{13} = 1$

$l_{22} = -1, l_{32} = 2$

$u_{23} = 5$

$l_{33} = -10$

Thus we have

$$L = \begin{bmatrix} 1 & 0 & 0 \\ 4 & -1 & 0 \\ 3 & 2 & -10 \end{bmatrix}, U = \begin{bmatrix} 1 & 1 & 1 \\ 0 & 1 & 5 \\ 0 & 0 & 1 \end{bmatrix}$$

From (3.33), using forward substitution, we have

$\mathbf{z} = [1 \quad -2 \quad -\tfrac{1}{2}]^T$

From (3.32), using back substitution, we get

$\mathbf{x} = [1 \quad \tfrac{1}{2} \quad -\tfrac{1}{2}]^T$

3.2.5 Cholesky Method

This method is also known as the **square-root** method. If the coefficient matrix **A** is symmetric and positive definite then the matrix **A** can be decomposed as

$$\mathbf{A} = \mathbf{L}\mathbf{L}^T \tag{3.36}$$

where $\mathbf{L} = l_{ij}$, $i, j = 1(1)n$, $l_{ij} = 0$, $i < j$. Alternatively, **A** may be decomposed as

$$\mathbf{A} = \mathbf{U}\mathbf{U}^T \tag{3.37}$$

For (3.36), equation (3.2) becomes

$$\mathbf{L}\mathbf{L}^T\mathbf{x} = \mathbf{b} \tag{3.38}$$

which may be written as

$$\mathbf{L}^T\mathbf{x} = \mathbf{z} \tag{3.39}$$

$$\mathbf{L}\mathbf{z} = \mathbf{b} \tag{3.40}$$

The intermediate solution values z_i, $i = 1(1)n$ are obtained from (3.40) by forward substitution and the solution values x_i, $i = 1(1)n$ are determined from (3.39) by back substitution. Alternatively, we can find only one inverse L^{-1} and obtain

$$z = L^{-1}b \tag{3.41}$$

and

$$x = (L^T)^{-1}z = (L^{-1})^T z$$

This results in substantial saving of computational work. The inverse of A may be obtained from

$$A = (L^{-1})^T L^{-1} \tag{3.42}$$

The elements of the lower triangular matrix L may be obtained as

$$l_{ii} = (a_{ii} - \sum_{j=1}^{i-1} l_{ij}^2)^{1/2}, \quad i = 1(1)n$$

$$l_{ij} = \frac{1}{l_{jj}}(a_{ij} - \sum_{k=1}^{j-1} l_{jk} l_{ik}),$$

$$i = j+1, j+2, \ldots, n; \quad j = 1(1)n$$

$$l_{ij} = 0, \quad i < j \tag{3.43}$$

where the summation is taken as zero if the upper limit is less than the lower limit.

For (3.37) the elements u_{ij} are given by

$$u_{nn} = (a_{nn})^{1/2}$$

$$u_{in} = a_{in}/u_{nn}, \quad i = 1(1)n - 1$$

$$u_{ij} = (a_{ij} - \sum_{k=j+1}^{n} u_{ik} u_{jk})/u_{jj}$$

$$i = n-2, n-3, \ldots, 1; \quad j = i+1, i+2, \ldots, n-1$$

$$u_{ii} = (a_{ii} - \sum_{k=i+1}^{n} u_{ik}^2)^{1/2}, \quad i = n-1, n-2, \ldots, 1$$

$$u_{ij} = 0, \quad i > j \tag{3.44}$$

Example 3.7 Solve the equations

$$\begin{bmatrix} 4 & -1 & 0 & 0 \\ -1 & 4 & -1 & 0 \\ 0 & -1 & 4 & -1 \\ 0 & 0 & -1 & 4 \end{bmatrix} \begin{bmatrix} x_1 \\ x_2 \\ x_3 \\ x_4 \end{bmatrix} = \begin{bmatrix} 1 \\ 0 \\ 0 \\ 0 \end{bmatrix}$$

using the Cholesky method. Also determine A^{-1}.

Let

$$L = \begin{bmatrix} l_{11} & & & \\ l_{21} & l_{22} & & \\ l_{31} & l_{32} & l_{33} & \\ l_{41} & l_{42} & l_{43} & l_{44} \end{bmatrix}$$

We get from (3.43)

$l_{11} = 2$

$l_{21} = -\tfrac{1}{2}, \quad l_{22} = \sqrt{\tfrac{15}{4}}$

$l_{31} = 0, \quad l_{32} = -\sqrt{\tfrac{4}{15}}, \quad l_{33} = \sqrt{\tfrac{56}{15}}$

$l_{41} = 0, \quad l_{42} = 0, \quad l_{43} = -\sqrt{\tfrac{15}{56}}, \quad l_{44} = \sqrt{\tfrac{209}{56}}$

The solution of the system

$$\begin{bmatrix} 2 & & & O \\ -\tfrac{1}{2} & \sqrt{\tfrac{15}{4}} & & \\ 0 & -\sqrt{\tfrac{4}{15}} & \sqrt{\tfrac{56}{15}} & \\ 0 & 0 & -\sqrt{\tfrac{15}{56}} & \sqrt{\tfrac{209}{56}} \end{bmatrix} \begin{bmatrix} z_1 \\ z_2 \\ z_3 \\ z_4 \end{bmatrix} = \begin{bmatrix} 1 \\ 0 \\ 0 \\ 0 \end{bmatrix}$$

is

$$z_1 = \tfrac{1}{2}, \quad z_2 = \frac{1}{\sqrt{60}}, \quad z_3 = \frac{1}{\sqrt{840}}, \quad z_4 = \frac{1}{\sqrt{11704}}$$

The solution vector is now obtained from

$$\begin{bmatrix} 2 & -\tfrac{1}{2} & 0 & 0 \\ & \sqrt{\tfrac{15}{4}} & -\sqrt{\tfrac{4}{15}} & 0 \\ & & \sqrt{\tfrac{56}{15}} & -\sqrt{\tfrac{15}{56}} \\ O & & & \sqrt{\tfrac{209}{56}} \end{bmatrix} \begin{bmatrix} x_1 \\ x_2 \\ x_3 \\ x_4 \end{bmatrix} = \begin{bmatrix} \tfrac{1}{2} \\ \frac{1}{\sqrt{60}} \\ \frac{1}{\sqrt{840}} \\ \frac{1}{\sqrt{11704}} \end{bmatrix}$$

The back substitution gives the solution

$$x_4 = \frac{1}{209}, \quad x_3 = \frac{4}{209}, \quad x_2 = \frac{15}{209} \text{ and } x_1 = \frac{56}{209}$$

We find

$$\mathbf{L}^{-1} = \begin{bmatrix} \tfrac{1}{2} & & & \\ \frac{1}{\sqrt{60}} & \sqrt{\tfrac{4}{15}} & & O \\ \frac{1}{\sqrt{840}} & \sqrt{\tfrac{2}{105}} & \sqrt{\tfrac{15}{56}} & \\ \frac{1}{\sqrt{11704}} & \sqrt{\tfrac{2}{1463}} & \frac{15}{\sqrt{11704}} & \sqrt{\tfrac{56}{209}} \end{bmatrix}$$

$$= \begin{bmatrix} 0.5 & & & \\ 0.1291 & 0.5164 & & \\ 0.0345 & 0.1380 & 0.5176 & \\ 0.0092 & 0.0370 & 0.1387 & 0.5176 \end{bmatrix}$$

Hence, we obtain

$$\mathbf{A}^{-1} = (\mathbf{L}^{-1})^T \mathbf{L}^{-1} = \begin{bmatrix} 0.2679 & 0.0718 & 0.0191 & 0.0048 \\ 0.0718 & 0.2871 & 0.0766 & 0.0192 \\ 0.0191 & 0.0766 & 0.2871 & 0.0718 \\ 0.0048 & 0.0192 & 0.0718 & 0.2679 \end{bmatrix}$$

3.2.6 Partition Method

Partition method is used to find the inverse of a matrix. Let the coefficient matrix **A** be partitioned as

$$\mathbf{A} = \left[\begin{array}{c|c} \mathbf{B} & \mathbf{C} \\ \hline \mathbf{E} & \mathbf{D} \end{array} \right] \qquad (3.45)$$

where **B** is an $r \times r$ matrix, **C** is an $r \times s$ matrix, **E** is an $s \times r$ and **D** is an $s \times s$ matrix. Here r and s are positive integers with $r + s = n$. Further, we also assume that $\mathbf{A}^{-1}$ is partitioned similarly as

$$\mathbf{A}^{-1} = \left[\begin{array}{c|c} \mathbf{X} & \mathbf{Y} \\ \hline \mathbf{Z} & \mathbf{V} \end{array} \right] \qquad (3.46)$$

where **X, Y, Z** and **V** are matrices of the same orders as **B, C, E, D** respectively. Then we have

$$\mathbf{A}\mathbf{A}^{-1} = \left[\begin{array}{c|c} \mathbf{B} & \mathbf{C} \\ \hline \mathbf{E} & \mathbf{D} \end{array} \right] \left[\begin{array}{c|c} \mathbf{X} & \mathbf{Y} \\ \hline \mathbf{Z} & \mathbf{V} \end{array} \right] = \left[\begin{array}{c|c} \mathbf{I}_1 & \mathbf{0} \\ \hline \mathbf{0} & \mathbf{I}_2 \end{array} \right] \qquad (3.47)$$

where $\mathbf{I}_1$ and $\mathbf{I}_2$ are identity matrices of orders r and s respectively.
From (3.47) we have

$$\mathbf{BX} + \mathbf{CZ} = \mathbf{I}_1$$

$$\mathbf{BY} + \mathbf{CV} = \mathbf{0}$$

$$\mathbf{EX} + \mathbf{DZ} = \mathbf{0}$$

$$\mathbf{EY} + \mathbf{DV} = \mathbf{I}_2$$

From these equations we get
$$V = (D - EB^{-1}C)^{-1}$$
$$Y = -B^{-1}CV$$
$$Z = -(D - EB^{-1}C)^{-1}EB^{-1}$$
$$= -VEB^{-1}$$
$$X = B^{-1}(I_1 - CZ)$$
$$= B^{-1} - B^{-1}CZ \tag{3.48}$$

It is obvious that we need to find two inverse matrices B^{-1} and $(D - EB^{-1}C)^{-1}$ of orders r and s respectively.

This method is useful when a few more unknowns and consequently a few more equations are added to the original system and the solution of the new system is required. The inverse of the coefficient matrix of the new system can easily be obtained, as the inverse of the coefficient matrix B^{-1} of the original system is available.

Example 3.8 Determine the inverse of the matrix
$$\begin{bmatrix} 1 & 1 & 1 \\ 4 & 3 & -1 \\ 3 & 5 & 3 \end{bmatrix}$$
using the partition method. Hence, find the solution of the system of equations
$$x_1 + x_2 + x_3 = 1$$
$$4x_1 + 3x_2 - x_3 = 6$$
$$3x_1 + 5x_2 + 3x_3 = 4$$

Let the matrix A be partitioned as
$$A = \begin{bmatrix} 1 & 1 & | & 1 \\ 4 & 3 & | & -1 \\ -- & -- & + & -- \\ 3 & 5 & | & 3 \end{bmatrix} \text{ and } A^{-1} = \begin{bmatrix} X & | & Y \\ -- & + & -- \\ Z & | & V \end{bmatrix}$$

Now,
$$B^{-1} = \begin{bmatrix} 1 & 1 \\ 4 & 3 \end{bmatrix}^{-1} = -\begin{bmatrix} 3 & -1 \\ -4 & 1 \end{bmatrix}$$

$$D - EB^{-1}C = 3 + [3 \quad 5]\begin{bmatrix} 3 & -1 \\ -4 & 1 \end{bmatrix}\begin{bmatrix} 1 \\ -1 \end{bmatrix} = -10$$

$$V = -\frac{1}{10}$$

$$Y = \begin{bmatrix} 3 & -1 \\ -4 & 1 \end{bmatrix}\begin{bmatrix} 1 \\ -1 \end{bmatrix}\left(-\frac{1}{10}\right) = -\frac{1}{10}\begin{bmatrix} 4 \\ -5 \end{bmatrix}$$

$$Z = -VEB^{-1} = -\tfrac{1}{10} [3 \ \ 5] \begin{bmatrix} 3 & -1 \\ -4 & 1 \end{bmatrix} = -\tfrac{1}{10} [-11 \ \ 2]$$

$$X = B^{-1} - B^{-1}CZ = \begin{bmatrix} -3 & 1 \\ 4 & -1 \end{bmatrix} - \tfrac{1}{10} \begin{bmatrix} 3 & -1 \\ -4 & 1 \end{bmatrix} \begin{bmatrix} 1 \\ -1 \end{bmatrix} [-11 \ \ 2]$$

$$= \begin{bmatrix} -3 & 1 \\ 4 & -1 \end{bmatrix} - \tfrac{1}{10} \begin{bmatrix} -44 & 8 \\ 55 & -10 \end{bmatrix} = \begin{bmatrix} 1.4 & 0.2 \\ -1.5 & 0 \end{bmatrix}$$

$$A^{-1} = \left[\begin{array}{cc|c} 1.4 & 0.2 & -0.4 \\ -1.5 & 0 & 0.5 \\ \hline 1.1 & -0.2 & -0.1 \end{array} \right]$$

The solution of the given system of equations is

$$x = \begin{bmatrix} 1.4 & 0.2 & -0.4 \\ -1.5 & 0 & 0.5 \\ 1.1 & -0.2 & -0.1 \end{bmatrix} \begin{bmatrix} 1 \\ 6 \\ 4 \end{bmatrix} = \begin{bmatrix} 1 \\ 0.5 \\ -0.5 \end{bmatrix}$$

3.3 ERROR ANALYSIS

The methods discussed in the previous section involve only a finite number of arithmetic operations. For example, the number of divisions and multiplications involved in solving the system of n equations are:

Gauss elimination method $\tfrac{1}{3}(n^3 + 3n^2 - n)$

Cholesky method $\quad \tfrac{1}{6}(n^3 + 9n^2 + 2n)$

If all calculations are performed exactly, then only we can hope to find the exact solution to the system (3.2). Usually, during computation, it will be necessary to round or chop the numbers. This will introduce roundoff errors in the computation. Because of this, the methods used will produce results which will differ considerably from the exact solution. The exact solution x and the corresponding approximate solution $\hat{x}$ will satisfy respectively the equations

$$Ax = b$$
$$(A + \delta A)\hat{x} = b + \delta b \qquad (3.49)$$

where δA and δb are the changes in A and b respectively, due to roundoff error. From (3.49) we obtain

$$\hat{x} - x = (A + \delta A)^{-1}(b + \delta b) - A^{-1}b$$
$$= \{(A + \delta A)^{-1} - A^{-1}\}b + (A + \delta A)^{-1} \delta b \qquad (3.50)$$

which may be called the error equation. In order to estimate the error vector $\varepsilon = \hat{x} - x$, we recall the concept of a norm of a vector x and a matrix A.

Vector norm

The non-negative quantity $\|\mathbf{x}\|$ is a measure of the size or length of a vector satisfying:

(i) $\|\mathbf{x}\| > 0$, for $\mathbf{x} \neq \mathbf{0}$ and $\|\mathbf{0}\| = 0$
(ii) $\|c\mathbf{x}\| = |c|\,\|\mathbf{x}\|$, for an arbitrary complex number c
(iii) $\|\mathbf{x} + \mathbf{y}\| \leqslant \|\mathbf{x}\| + \|\mathbf{y}\|$

The most commonly used norms are

(i) *Absolute norm* (l_1 norm)

$$\|\mathbf{x}\|_1 = \sum_{i=1}^{n} |x_i| \tag{3.51}$$

(ii) *Euclidean norm*

$$\|\mathbf{x}\|_2 = (\mathbf{x}^*\mathbf{x})^{1/2} = \left(\sum_{i=1}^{n} |x_i|^2 \right)^{1/2} \tag{3.52}$$

(iii) *Maximum norm* (l_∞ norm)

$$\|\mathbf{x}\|_\infty = \max_{1 \leqslant i \leqslant n} |x_i| \tag{3.53}$$

Matrix norm

The matrix norm, $\|\mathbf{A}\|$, is a non-negative number which satisfies the properties:

(i) $\|\mathbf{A}\| > 0$ if $\mathbf{A} \neq \mathbf{0}$ and $\|\mathbf{0}\| = 0$
(ii) $\|c\mathbf{A}\| = |c|\,\|\mathbf{A}\|$ for an arbitrary complex number c
(iii) $\|\mathbf{A} + \mathbf{B}\| \leqslant \|\mathbf{A}\| + \|\mathbf{B}\|$
(iv) $\|\mathbf{AB}\| \leqslant \|\mathbf{A}\|\,\|\mathbf{B}\|$

The most commonly used norms are:

(i) *Frobenius or Euclidean norm*

$$F(\mathbf{A}) = \left(\sum_{i,j=1}^{n} |a_{ij}|^2 \right)^{1/2}$$

(ii) *Maximum norm*

$$\|\mathbf{A}\| = \|\mathbf{A}\|_\infty = \max_i \sum_k |a_{ik}| \quad \text{(maximum absolute row sum)}$$

$$\|\mathbf{A}\| = \|\mathbf{A}\|_1 = \max_k \sum_i |a_{ik}| \quad \text{(maximum absolute column sum)}$$

(iii) *Hilbert norm or Spectral norm*

$$\|\mathbf{A}\|_2 = \sqrt{\lambda}$$

where $\lambda = \rho(\mathbf{A}^*\mathbf{A})$. If $\mathbf{A}$ is Hermitian or real and symmetric, then

$$\lambda = \rho(\mathbf{A}^2) = \rho^2(\mathbf{A})$$

so that

$$\|\mathbf{A}\|_2 = \rho(\mathbf{A})$$

The matrix norm must be consistent with the vector norm that we are using for any vector **x** and matrix **A**, i.e.,

$$\|A x\| \leqslant \|A\| \, \|x\| \tag{3.54}$$

It may be verified that the norm

$$\|A\| = \max_{i} \sum_{j=1}^{n} |a_{ij}| \tag{3.55}$$

is consistent with maximum norm $\|x\|$.

From (3.50) we obtain

$$\|\hat{x} - x\| \leqslant \|(A + \delta A)^{-1} - A^{-1}\| \, \|b\| + \|(A + \delta A)^{-1}\| \, \|\delta b\| \tag{3.56}$$

But

$$\|(A + \delta A)^{-1}\| = \|(A + \delta A)^{-1} - A^{-1} + A^{-1}\|$$

$$\leqslant \|(A + \delta A)^{-1} - A^{-1}\| + \|A^{-1}\| \tag{3.57}$$

and

$$\|(A + \delta A)^{-1} - A^{-1}\| = \|A^{-1} - (A + \delta A)^{-1}\|$$

$$\leqslant \|A^{-1}\| \, \|I - (I + A^{-1} \delta A)^{-1}\|$$

$$= \|A^{-1}\| \, \|(I + A^{-1} \delta A)^{-1} (I + A^{-1} \delta A - I)\|$$

$$= \|A^{-1}\| \, \|(I + A^{-1} \delta A)^{-1} A^{-1} \delta A\|$$

$$\leqslant \|A^{-1}\| \, \|A^{-1} \delta A\| / (1 - \|A^{-1} \delta A\|)$$

where $\|A^{-1} \delta A\|$ is assumed to be less than 1. Hence, we obtain

$$\|\hat{x} - x\| \leqslant \frac{\|A^{-1}\|}{1 - \|A^{-1} \delta A\|} (\|A^{-1} \delta A\| \, \|b\| + \|\delta b\|) \tag{3.58}$$

Since $\|A^{-1} \delta A\| \leqslant \|A^{-1}\| \, \|\delta A\|$, the equation (3.58) may be written as

$$\|\hat{x} - x\| \leqslant \frac{\|A^{-1}\|}{(1 - \|A^{-1} \delta A\|)} (\|x\| \, \|\delta A\| + \|\delta b\|)$$

or

$$\frac{\|\hat{x} - x\|}{\|x\|} \leqslant \frac{\|A^{-1}\|}{(1 - \|A^{-1} \delta A\|)} \|A\| \left[\frac{\|x\| \, \|\delta A\|}{\|A\| \, \|x\|} + \frac{\|\delta b\|}{\|A\| \, \|x\|} \right]$$

or

$$\frac{\|\hat{x} - x\|}{\|x\|} \leqslant \frac{K(A)}{(1 - \|A^{-1} \delta A\|)} \left[\frac{\|\delta b\|}{\|b\|} + \frac{\|\delta A\|}{\|A\|} \right] \tag{3.59}$$

where the quantity $K(A) = \|A^{-1}\| \, \|A\|$ is called the **condition number** of the matrix A and is denoted by cond (A). The left-hand side in (3.59) gives the overall relative error in **x**. The first term inside the brackets on the right hand is the overall relative error in **b** and the second term is the relative error in A. If there is no error in **b** then $\|\delta b\| = 0$ and (3.59) becomes

$$\frac{\|\hat{x} - x\|}{\|x\|} \leqslant \frac{K(A)}{(1 - \|A^{-1} \delta A\|)} \frac{\|\delta A\|}{\|A\|} \tag{3.60}$$

If there is no error in A then $\|\delta A\| = 0$ and (3.59) gives

$$\frac{\|\hat{x} - x\|}{\|x\|} \leqslant K(A) \frac{\|\delta b\|}{\|b\|} \tag{3.61}$$

If $K(A)$ is small then the small changes in A or b produce only small changes in x. If $K(A)$ is large then small relative changes in A and b will produce large relative changes in x, and the system of equations $Ax = b$ is called **illconditioned**. If $K(A)$ is near unity, the system (3.2) is well conditioned.

If $\|\cdot\|$ is the spectral norm, then

$$K(A) = \|A\|_2 \|A^{-1}\|_2 = \sqrt{\frac{\lambda}{\mu}}$$

where λ and μ are the largest and smallest eigenvalues in modulus of A^*A. If A is Hermitian or real and symmetric, we have

$$K(A) = \frac{\lambda^*}{\mu^*}$$

where λ^* and μ^* are the largest and smallest eigenvalues in modulus of **A**.

Example 3.9 Find the condition number of the system

$$\begin{bmatrix} 2.1 & 1.8 \\ 6.2 & 5.3 \end{bmatrix} \begin{bmatrix} x_1 \\ x_2 \end{bmatrix} = \begin{bmatrix} 2.1 \\ 6.2 \end{bmatrix}$$

We have

$$A^*A = A^TA = \begin{bmatrix} 42.85 & 36.64 \\ 36.64 & 31.33 \end{bmatrix}$$

The eigenvalues of A^*A are the solutions of

$$\lambda^2 - 74.18\lambda + 0.0009 = 0$$

We find

$$\lambda_1 = 74.17998787 \text{ and } \lambda_2 = 0.000012132$$

The condition number is

$$K(A) = \sqrt{\frac{\lambda_1}{\lambda_2}} = 2472.73$$

Hence, this system of equations is highly illconditioned and is very sensitive to roundoff errors.

3.3.1 Iterative Improvement of the Solution

We assume that the system (3.2) is well conditioned. The error equation (3.50) for $\delta A = 0$ becomes

$$A \, \delta x = \delta b \tag{3.62}$$

where $\delta b = b - A\hat{x}$ is called the **residual** vector. The approximate solution

$\hat{\mathbf{x}}$ may be improved as follows:
$$\delta\mathbf{x} = \mathbf{A}^{-1}\,\delta\mathbf{b}$$
$$\mathbf{x} = \hat{\mathbf{x}} + \delta\mathbf{x} \tag{3.63}$$

This process of using error equation (3.63) may be carried out as far as necessary to obtain solution values to the desired accuracy.

3.4 ITERATION METHODS

A general linear iterative method for the solution of the system of equations (3.2) may be defined in the form
$$\mathbf{x}^{(k+1)} = \mathbf{H}\mathbf{x}^{(k)} + \mathbf{c}, \quad k = 0, 1, 2, \ldots \tag{3.64}$$
where $\mathbf{x}^{(k+1)}$ and $\mathbf{x}^{(k)}$ are the approximations for $\mathbf{x}$ at the $(k+1)$th and kth iterations, respectively. $\mathbf{H}$ is called the iteration matrix depending on $\mathbf{A}$ and $\mathbf{c}$ is a column vector. In the limiting case when $k \to \infty$, $\mathbf{x}^{(k)}$ converges to the exact solution
$$\mathbf{x} = \mathbf{A}^{-1}\mathbf{b} \tag{3.65}$$
and the iteration equation (3.64) becomes by substitution from (3.65)
$$\mathbf{A}^{-1}\mathbf{b} = \mathbf{H}\mathbf{A}^{-1}\mathbf{b} + \mathbf{c} \tag{3.66}$$

From (3.66), the column vector $\mathbf{c}$ is given by
$$\mathbf{c} = (\mathbf{I} - \mathbf{H})\mathbf{A}^{-1}\mathbf{b} \tag{3.67}$$

We now determine the iteration matrix $\mathbf{H}$ and the column vector $\mathbf{c}$ for a few well known iteration methods.

3.4.1 Jacobi Iteration Method

We assume that the quantities a_{ii} in (3.2) are pivot elements. The equations (3.2) may be written as
$$\begin{aligned} a_{11}x_1 &= -(a_{12}x_2 + a_{13}x_3 + \ldots + a_{1n}x_n) + b_1 \\ a_{22}x_2 &= -(a_{21}x_1 + a_{23}x_3 + \ldots + a_{2n}x_n) + b_2 \\ &\vdots \\ a_{nn}x_n &= -(a_{n1}x_1 + a_{n2}x_2 + \ldots + a_{n,n-1}x_{n-1}) + b_n \end{aligned} \tag{3.68}$$

The **Jacobi** iteration method may now be defined as
$$\begin{aligned} x_1^{(k+1)} &= -\frac{1}{a_{11}}(a_{12}x_2^{(k)} + a_{13}x_3^{(k)} + \ldots + a_{1n}x_n^{(k)} - b_1) \\ x_2^{(k+1)} &= -\frac{1}{a_{22}}(a_{21}x_1^{(k)} + a_{23}x_3^{(k)} + \ldots + a_{2n}x_n^{(k)} - b_2) \\ &\vdots \\ x_n^{(k+1)} &= -\frac{1}{a_{nn}}(a_{n1}x_1^{(k)} + a_{n2}x_2^{(k)} + \ldots + a_{n,n-1}x_{n-1}^{(k)} - b_n) \\ & \qquad k = 0, 1, 2, \ldots \end{aligned} \tag{3.69}$$

which in matrix notation becomes
$$\mathbf{x}^{(k+1)} = -\mathbf{D}^{-1}(\mathbf{L} + \mathbf{U})\mathbf{x}^{(k)} + \mathbf{D}^{-1}\mathbf{b}$$
$$= \mathbf{H}\mathbf{x}^{(k)} + \mathbf{c}, \quad k = 0, 1, 2, \ldots \tag{3.70}$$
where
$$\mathbf{H} = -\mathbf{D}^{-1}(\mathbf{L} + \mathbf{U}), \mathbf{c} = \mathbf{D}^{-1}\mathbf{b}$$
and the matrices $\mathbf{L}$ and $\mathbf{U}$ are respectively lower and upper triangular matrices with zero diagonal entries, the matrix $\mathbf{D}$ is the diagonal matrix, such that $\mathbf{A} = \mathbf{L} + \mathbf{D} + \mathbf{U}$.

3.4.2 Gauss-Seidel Iteration Method

We now use on the right hand side of (3.69), all the available values from the present iteration. We write the **Gauss-Seidel** method as

$$x_1^{(k+1)} = -\frac{1}{a_{11}}(a_{12}x_2^{(k)} + a_{13}x_3^{(k)} + \ldots + a_{1n}x_n^{(k)}) + \frac{b_1}{a_{11}}$$

$$x_2^{(k+1)} = -\frac{1}{a_{22}}(a_{21}x_1^{(k+1)} + a_{23}x_3^{(k)} + \ldots + a_{2n}x_n^{(k)}) + \frac{b_2}{a_{22}}$$

$$\vdots$$

$$x_n^{(k+1)} = -\frac{1}{a_{nn}}(a_{n1}x_1^{(k+1)} + a_{n2}x_2^{(k+1)} + \ldots + a_{n,n-1}x_{n-1}^{(k+1)}) + \frac{b_n}{a_{nn}} \tag{3.71}$$

which may be rearranged in the form

$$a_{11}x_1^{(k+1)} = -\sum_{i=2}^{n} a_{1i}x_i^{(k)} + b_1$$

$$a_{21}x_1^{(k+1)} + a_{22}x_2^{(k+1)} = -\sum_{i=3}^{n} a_{2i}x_i^{(k)} + b_2$$

$$\vdots$$

$$a_{n1}x_1^{(k+1)} + \ldots + a_{nn}x_n^{(k+1)} = b_n \tag{3.72}$$

In matrix notation, (3.72) becomes
$$(\mathbf{D} + \mathbf{L})\mathbf{x}^{(k+1)} = -\mathbf{U}\mathbf{x}^{(k)} + \mathbf{b}$$
or
$$\mathbf{x}^{(k+1)} = -(\mathbf{D} + \mathbf{L})^{-1}\mathbf{U}\mathbf{x}^{(k)} + (\mathbf{D} + \mathbf{L})^{-1}\mathbf{b}$$
$$= \mathbf{H}\mathbf{x}^{(k)} + \mathbf{c}, \quad k = 0, 1, 2, \ldots \tag{3.73}$$
where $\mathbf{H} = -(\mathbf{D} + \mathbf{L})^{-1}\mathbf{U}$ and $\mathbf{c} = (\mathbf{D} + \mathbf{L})^{-1}\mathbf{b}$.

Example 3.10 Consider the system of equations
$$2x_1 - x_2 + 0x_3 = 7$$
$$-x_1 + 2x_2 - x_3 = 1$$
$$0x_1 - x_2 + 2x_3 = 1$$

Use the Gauss-Seidel iterative method and perform three iterations.

We have

$$L = \begin{bmatrix} 0 & 0 & 0 \\ -1 & 0 & 0 \\ 0 & -1 & 0 \end{bmatrix}, D = \begin{bmatrix} 2 & 0 & 0 \\ 0 & 2 & 0 \\ 0 & 0 & 2 \end{bmatrix} \text{ and } U = \begin{bmatrix} 0 & -1 & 0 \\ 0 & 0 & -1 \\ 0 & 0 & 0 \end{bmatrix}$$

The Gauss-Seidel method gives

$$x^{(k+1)} = -(D+L)^{-1}Ux^{(k)} + (D+L)^{-1}b = \begin{bmatrix} 0 & \frac{1}{2} & 0 \\ 0 & \frac{1}{4} & \frac{1}{2} \\ 0 & \frac{1}{8} & \frac{1}{4} \end{bmatrix} x^{(k)} + \begin{bmatrix} \frac{7}{2} \\ \frac{9}{4} \\ \frac{13}{8} \end{bmatrix}$$

Starting with the zero initial vector, we get

$$x^{(1)} = \begin{bmatrix} 3.5 \\ 2.25 \\ 1.625 \end{bmatrix}, x^{(2)} = \begin{bmatrix} 4.625 \\ 3.625 \\ 2.3125 \end{bmatrix}, \text{ and } x^{(3)} = \begin{bmatrix} 5.3115 \\ 4.3125 \\ 2.6563 \end{bmatrix}$$

The exact solution is $x = \begin{bmatrix} 6 & 5 & 3 \end{bmatrix}^T$.

3.4.3 Successive Over Relaxation (SOR) Method

This method is a generalization of the Gauss-Seidel method. This method is often used when the coefficient matrix of the system of equations is symmetric and has 'property A'. We define an auxiliary vector $\tilde{x}$

$$\tilde{x}^{(k+1)} = -D^{-1}Lx^{(k+1)} - D^{-1}Ux^{(k)} + D^{-1}b \tag{3.74}$$

The final solution is now written as

$$x^{(k+1)} = x^{(k)} + w(\tilde{x}^{(k+1)} - x^{(k)})$$

or

$$x^{(k+1)} = (1-w)x^{(k)} + w\tilde{x}^{(k+1)} \tag{3.75}$$

Substituting from (3.74) into (3.75) and simplifying we have

$$x^{(k+1)} = (D+wL)^{-1}[(1-w)D - wU]x^{(k)} + w(D+wL)^{-1}b$$
$$= Hx^{(k)} + c, \qquad k = 0, 1, 2, \ldots \tag{3.76}$$

where $H = (D+wL)^{-1}[(1-w)D - wU]$ and $c = w(D+wL)^{-1}b$.

When $w = 1$, the equation (3.76) reduces to the Gauss-Seidel method. The quantity w is called the **relaxation parameter** and $x^{(k+1)}$ is a weighted mean of $\tilde{x}^{(k+1)}$ and $x^{(k)}$. From the equation (3.75) we find that the weights are non-negative for $0 \leq w \leq 1$. If $w > 1$, then the method is called an **over relaxation** method and if $w < 1$ then it is called an **under relaxation** method.

3.4.4 Convergence Analysis

To discuss the convergence of the iteration method (3.64), we study the behaviour of the difference between the exact solution **x** and an approximation $\mathbf{x}^{(k)}$. The exact solution **x** will satisfy

$$\mathbf{x} = \mathbf{H}\mathbf{x} + \mathbf{c} \tag{3.77}$$

Subtracting (3.77) from (3.64) and substituting $\varepsilon^{(k)} = \mathbf{x}^{(k)} - \mathbf{x}$, we get

$$\varepsilon^{(k+1)} = \mathbf{H}\varepsilon^{(k)}, \quad k = 0, 1, 2, \ldots \tag{3.78}$$

from which follows

$$\varepsilon^{(k)} = \mathbf{H}^k \varepsilon^{(0)}, \quad k = 0, 1, 2, \ldots \tag{3.79}$$

where we have assumed that the iteration matrix **H** remains constant for each iteration. We now give few results which we require for proving the convergence of the iterative methods.

THEOREM 3.1 *Let* **A** *be a square matrix. Then*

$$\lim_{m \to \infty} \mathbf{A}^m = \mathbf{0}$$

if $\|\mathbf{A}\| < 1$, *or iff* $\rho(\mathbf{A}) < 1$.

Proof. If $\|\mathbf{A}\| < 1$, we have

$$\|\mathbf{A}^m\| \leqslant \|\mathbf{A}\|^m$$

and

$$\|\lim_{m \to \infty} \mathbf{A}^m\| \leqslant \lim_{m \to \infty} \|\mathbf{A}\|^m = 0$$

For simplicity, assume that all the eigenvalues of **A** are distinct. Then, there exists a similarity transformation **S**, such that

$$\mathbf{A} = \mathbf{S}^{-1}\mathbf{D}\mathbf{S}$$

where **D** is the diagonal matrix having the eigenvalues of **A** on the diagonal. Therefore,

$$\mathbf{A}^m = \mathbf{S}^{-1}\mathbf{D}^m\mathbf{S}$$

where $\mathbf{D}^m = \begin{bmatrix} \lambda_1^m & & & \bigcirc \\ & \lambda_2^m & & \\ & & \ddots & \\ \bigcirc & & & \lambda_n^m \end{bmatrix}$.

Obviously, $\lim_{m \to \infty} \mathbf{A}^m = \mathbf{0}$, iff all the eigenvalues satisfy $|\lambda_i| < 1$, that is, $\rho(\mathbf{A}) < 1$.

THEOREM 3.2 *The infinite series*

$$\mathbf{I} + \mathbf{A} + \mathbf{A}^2 + \ldots$$

converges if $\lim_{m \to \infty} \mathbf{A}^m = \mathbf{0}$. *The series converges to* $(\mathbf{I} - \mathbf{A})^{-1}$.

Proof. Only if, is obvious.

If $\lim_{m \to \infty} \mathbf{A}^m = 0$, then by Theorem 3.1, $\rho(\mathbf{A}) < 1$.
Hence $|\mathbf{I} - \mathbf{A}| \neq 0$ and $(\mathbf{I} - \mathbf{A})^{-1}$ exists.
Consider the identity

$$(\mathbf{I} + \mathbf{A} + \mathbf{A}^2 + \ldots + \mathbf{A}^m)(\mathbf{I} - \mathbf{A}) = \mathbf{I} - \mathbf{A}^{m+1}$$

Post multiplying by $(\mathbf{I} - \mathbf{A})^{-1}$, we have

$$(\mathbf{I} + \mathbf{A} + \mathbf{A}^2 + \ldots + \mathbf{A}^m) = (\mathbf{I} - \mathbf{A}^{m+1})(\mathbf{I} - \mathbf{A})^{-1}$$

As $m \to \infty$, we get

$$\mathbf{I} + \mathbf{A} + \mathbf{A}^2 + \ldots = (\mathbf{I} - \mathbf{A})^{-1}$$

THEOREM 3.3 *No eigenvalue of a matrix* $\mathbf{A}$ *exceeds the norm of a matrix*, *i.e.* $\|\mathbf{A}\| \geqslant \rho(\mathbf{A})$.

Proof. We have

$$\mathbf{A}\mathbf{x} = \lambda \mathbf{x}$$
$$\|\lambda \mathbf{x}\| = \|\mathbf{A}\mathbf{x}\| \leqslant \|\mathbf{A}\| \, \|\mathbf{x}\|$$

or

$$|\lambda| \, \|\mathbf{x}\| \leqslant \|\mathbf{A}\| \, \|\mathbf{x}\|$$
$$|\lambda| \leqslant \|\mathbf{A}\|$$

Hence the result.

THEOREM 3.4 *The iteration method of the form* (3.64) *for the solution of* (3.2) *converges to the exact solution for any initial vector, if* $\|\mathbf{H}\| < 1$.

Proof. Without loss of generality, we take initial vector $\mathbf{x}^{(0)} = 0$. We have

$$\mathbf{x}^{(1)} = \mathbf{c}$$
$$\mathbf{x}^{(2)} = \mathbf{H}\mathbf{x}^{(1)} + \mathbf{c} = (\mathbf{H} + \mathbf{I})\mathbf{c}$$
$$\mathbf{x}^{(3)} = \mathbf{H}\mathbf{x}^{(2)} + \mathbf{c} = (\mathbf{H}^2 + \mathbf{H} + \mathbf{I})\mathbf{c}$$
$$\vdots$$
$$\mathbf{x}^{(k+1)} = (\mathbf{H}^k + \mathbf{H}^{k-1} + \ldots + \mathbf{H} + \mathbf{I})\mathbf{c}$$
$$\lim_{k \to \infty} \mathbf{x}^{(k+1)} = \lim_{k \to \infty} (\mathbf{H}^k + \mathbf{H}^{k-1} + \ldots + \mathbf{H} + \mathbf{I})\mathbf{c}$$
$$= (\mathbf{I} - \mathbf{H})^{-1}\mathbf{c}$$

if $\|\mathbf{H}\| < 1$ or iff $\rho(\mathbf{H}) < 1$ \hfill (3.80)

In the case of the Jacobi method, we have

$$(\mathbf{I} - \mathbf{H})^{-1}\mathbf{c} = [\mathbf{I} + \mathbf{D}^{-1}(\mathbf{L} + \mathbf{U})]^{-1}\mathbf{D}^{-1}\mathbf{b}$$
$$= [\mathbf{D}^{-1}(\mathbf{D} + \mathbf{L} + \mathbf{U})]^{-1}\mathbf{D}^{-1}\mathbf{b} = (\mathbf{D} + \mathbf{L} + \mathbf{U})^{-1}\mathbf{D}\mathbf{D}^{-1}\mathbf{b}$$
$$= \mathbf{A}^{-1}\mathbf{b} = \mathbf{x}$$

Similar result can be proved for the Gauss-Seidel and SOR methods.

THEOREM 3.5 *A necessary and sufficient condition for convergence of an iterative method of the form* (3.64) *is that the eigenvalues of the iteration matrix satisfy*

$$|\lambda_i(\mathbf{H})| < 1, \; i = 1(1)n$$

Proof. We prove the result for the case when the iteration matrix $\mathbf{H}$ has n independent eigenvectors $\mathbf{x}_1, \mathbf{x}_2, \ldots, \mathbf{x}_n$ with eigenvalues $\lambda_1, \lambda_2, \ldots, \lambda_n$ respectively. The error vector $\boldsymbol{\epsilon}^{(0)}$ can be written as

$$\boldsymbol{\epsilon}^{(0)} = c_1 \mathbf{x}_1 + c_2 \mathbf{x}_2 + \ldots + c_n \mathbf{x}_n$$

Using (3.79) we get

$$\boldsymbol{\epsilon}^{(k)} = c_1 \lambda_1^k \mathbf{x}_1 + c_2 \lambda_2^k \mathbf{x}_2 + \ldots + c_n \lambda_n^k \mathbf{x}_n \tag{3.81}$$

(i) *Necessity.* If $\lim_{k \to \infty} \boldsymbol{\epsilon}^{(k)} = \mathbf{0}$ for any arbitrary initial vector $\mathbf{x}^{(0)}$ and thus for any arbitrary error vector $\boldsymbol{\epsilon}^{(0)}$, then by (3.81) the magnitudes of the eigenvalues $|\lambda_i|$, $i = 1(1)n$ must necessarily be less than unity.

(ii) *Sufficiency.* For $|\lambda_i| < 1$, $i = 1(1)n$, the convergence of $\boldsymbol{\epsilon}^{(k)}$ towards the zero vector follows from (3.81).

From (3.81) we notice that to reduce the magnitude of the error component $c_i \mathbf{x}_i$ in $\boldsymbol{\epsilon}^{(0)}$ by a factor of 10^{-m}, we have to perform k iterations, where k is the smallest number such that

$$|\lambda_i|^k \leqslant 10^{-m} \text{ or } k \geqslant \frac{m}{-\log_{10} |\lambda_i|} \tag{3.82}$$

which determines the number of iterations to be performed. Thus the dominant eigenvalue λ_1 of $\mathbf{H}$, that is, the one with the largest magnitude generally governs the rate of convergence of the iteration method.

Definition 3.1 The number

$$\nu = -\log_{10} \rho(\mathbf{H}) \tag{3.83}$$

where $\rho(\mathbf{H})$ is the spectral radius of $\mathbf{H}$, is called the rate of convergence of the iteration method (3.64).

From (3.82) and (3.83) we obtain

$$k \approx \frac{m}{\nu} \tag{3.84}$$

The number of iterations k necessary to reduce the error $\boldsymbol{\epsilon}^{(k)}$ by a factor of 10 is approximately inversely proportional to the convergence rate ν. From (3.83) we deduce that the smaller the spectral radius of $\mathbf{H}$, the higher the order of convergence. The spectral radius of the iteration matrix $\mathbf{H}$ in SOR method satisfies

$$\rho(\mathbf{H}) \geqslant |w - 1| \tag{3.85}$$

with the equality being satisfied only if all eigenvalues of $\mathbf{H}$ are of modulus

$|w - 1|$. Our aim is to find a real value of w, which minimizes $\rho(\mathbf{H})$. From (3.85) we need to consider only values of w lying in the open interval $0 < w < 2$. The analysis of the method (3.75) gives the optimal value of the relaxation parameter w_{opt} as

$$w_{opt} = \frac{2}{\mu^2}(1 - \sqrt{1 - \mu^2}) \tag{3.86}$$

where μ is the spectral radius of the Jacobi iteration matrix.

When $w = w_{opt}$, $\rho(\mathbf{H}) = \frac{\mu^2 w^2}{4} = w - 1$. We also find that

$$\mu = \frac{\lambda + w - 1}{w \lambda^{1/2}} \tag{3.87}$$

where λ is the spectral radius of the SOR method (3.75). When $w = 1$, we get $\lambda = \mu^2$. That is

$$\rho(\text{Gauss-Seidel iteration matrix})$$
$$= \rho^2(\text{Jacobi iteration matrix}) \tag{3.88}$$

Hence, we deduce that the Gauss-Seidel iteration method is at least twice as fast as the Jacobi iteration method.

Example 3.11 Consider the system of equations

$$\begin{bmatrix} 1 & -a \\ -a & 1 \end{bmatrix} \begin{bmatrix} x_1 \\ x_2 \end{bmatrix} = \begin{bmatrix} b_1 \\ b_2 \end{bmatrix}$$

where a is a real constant.

(i) For which values of a, the Jacobi and the Gauss-Seidel methods converge.

(ii) For $a = 0.5$ find the value of w which minimizes the spectral radius of the SOR iteration matrix.

The Jacobi method becomes

$$\mathbf{x}^{(k+1)} = \mathbf{H}\mathbf{x}^{(k)} + \mathbf{c}$$

where $\mathbf{H} = \begin{bmatrix} 0 & a \\ a & 0 \end{bmatrix}$.

The eigenvalues of the Jacobi iteration matrix $\mathbf{H}$ are given by

$$|\mathbf{H} - \lambda \mathbf{I}| = \begin{vmatrix} -\lambda & a \\ a & -\lambda \end{vmatrix} = \lambda^2 - a^2 = 0$$

or

$$\lambda = a, -a$$

The spectral radius of the Jacobi iteration matrix becomes

$$\rho(H_J) = |a|$$

The condition for the convergence of the Jacobi iteration method is

$$|a| < 1$$

The Gauss-Seidel iteration method becomes

$$\begin{bmatrix} 1 & 0 \\ -a & 1 \end{bmatrix} \mathbf{x}^{(k+1)} = \begin{bmatrix} 0 & a \\ 0 & 0 \end{bmatrix} \mathbf{x}^{(k)} + \begin{bmatrix} b_1 \\ b_2 \end{bmatrix}$$

or

$$\mathbf{x}^{(k+1)} = \begin{bmatrix} 0 & a \\ 0 & a^2 \end{bmatrix} \mathbf{x}^{(k)} + \begin{bmatrix} 1 & 0 \\ a & 1 \end{bmatrix} \begin{bmatrix} b_1 \\ b_2 \end{bmatrix}$$

The eigenvalues of the Gauss-Seidel iteration matrix are given by

$$|\mathbf{H} - \lambda \mathbf{I}| = \begin{vmatrix} -\lambda & a \\ 0 & a^2 - \lambda \end{vmatrix} = \lambda(\lambda - a^2) = 0$$

or

$$\lambda = 0, a^2$$

The spectral radius of the Gauss-Seidel iteration matrix becomes

$$\rho(\mathbf{H}_{GS}) = a^2$$

This result is obvious because of (3.88).

For convergence we get $|a| < 1$. The SOR method may be written as

$$\begin{bmatrix} 1 & 0 \\ -wa & 1 \end{bmatrix} \mathbf{x}^{(k+1)} = \begin{bmatrix} 1 - w & wa \\ 0 & 1 - w \end{bmatrix} \mathbf{x}^{(k)} + \begin{bmatrix} b_1 \\ b_2 \end{bmatrix}$$

or

$$\mathbf{x}^{(k+1)} = \mathbf{H}\mathbf{x}^{(k)} + \begin{bmatrix} 1 & 0 \\ wa & 1 \end{bmatrix} \begin{bmatrix} b_1 \\ b_2 \end{bmatrix}$$

where $\mathbf{H} = \begin{bmatrix} 1 - w & wa \\ wa(1 - w) & w^2 a^2 + 1 - w \end{bmatrix}$.

The eigenvalues of the matrix $\mathbf{H}$ are given by

$$\lambda^2 + \lambda(2(w - 1) - w^2 a^2) + (w - 1)^2 = 0$$

We have

$$\lambda = \tfrac{1}{2}[2(1 - w) + w^2 a^2 \pm wa\sqrt{4(1 - w) + w^2 a^2}]$$

For SOR method, $w > 1$ and

$$\rho(\mathbf{H}_S) = \tfrac{1}{2}[2(1 - w) + w^2 a^2 + wa\sqrt{4(1 - w) + w^2 a^2}]$$

For $a = 0.5$, $\rho(\mathbf{H}_S)$ as a function of w is shown in Fig. 3.1. From the Fig. 3.1, we find that $\rho(\mathbf{H}_S)$ is minimum when $w = 1.0718$. For this value of w maximum convergence is achieved.

The optimal relaxation factor can also be determined from

$$w_{opt} = \frac{2(1 - \sqrt{1 - \mu^2})}{\mu^2} = \frac{2(1 - \sqrt{1 - a^2})}{a^2} = \frac{2}{1 + \sqrt{(1 - a^2)}} = 1.0718$$

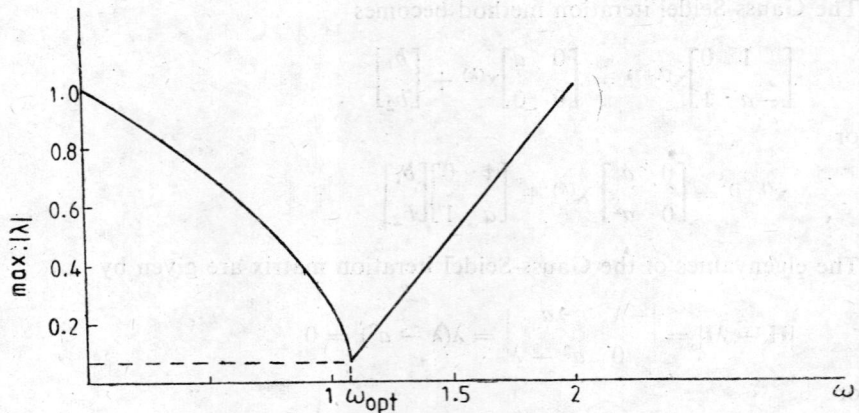

Fig. 3.1 Spectral radius of SOR method

Example 3.12 For the solution of the system of equations

$$\begin{bmatrix} 2 & -1 & 0 \\ -1 & 2 & -1 \\ 0 & -1 & 2 \end{bmatrix} \begin{bmatrix} x_1 \\ x_2 \\ x_3 \end{bmatrix} = \begin{bmatrix} 7 \\ 1 \\ 1 \end{bmatrix}$$

set up the SOR method. Find the optimal relaxation factor and the rate of convergence.

We have

$$\mathbf{L} = \begin{bmatrix} 0 & 0 & 0 \\ -1 & 0 & 0 \\ 0 & -1 & 0 \end{bmatrix}, \quad \mathbf{D} = \begin{bmatrix} 2 & 0 & 0 \\ 0 & 2 & 0 \\ 0 & 0 & 2 \end{bmatrix} \text{ and } \mathbf{U} = \begin{bmatrix} 0 & -1 & 0 \\ 0 & 0 & -1 \\ 0 & 0 & 0 \end{bmatrix}$$

The SOR scheme is obtained as

$$\mathbf{x}^{(k+1)} = \begin{bmatrix} 1-w & \dfrac{w}{2} & 0 \\ \dfrac{w}{2}(1-w) & 1-w+\dfrac{w^2}{4} & \dfrac{w}{2} \\ \dfrac{w^2}{4}(1-w) & (1-w)\dfrac{w}{2}+\dfrac{w^3}{8} & 1-w+\dfrac{w^2}{4} \end{bmatrix} \mathbf{x}^{(k)}$$

$$+ \begin{bmatrix} \dfrac{7w}{2} \\ \dfrac{w}{4}(7w+2) \\ \dfrac{w}{8}(7w^2+2w+4) \end{bmatrix}$$

The spectral radius of the Jacobi iteration matrix is $\mu = 1/\sqrt{2}$. The optimal relaxation factor of the SOR scheme is

$$w_{opt} = \frac{2}{\mu^2}(1 - \sqrt{1-\mu^2}) = 4\left(1 - \frac{1}{\sqrt{2}}\right) = 1.171573$$

When $w = w_{opt}$, $\rho(\mathbf{H}_S) = \frac{\mu^2 w^2}{4} = w - 1 = 0.171573$

The rate of convergence of SOR method is

$$\nu = -\log(0.171573) = 1.7628$$

3.4.5 Iterative Method for A^{-1}

Suppose that we know the approximate inverse $\mathbf{B}$ of a matrix $\mathbf{A}$. Then $\mathbf{AB} \neq \mathbf{I}$. The error matrix is given by

$$\mathbf{E} = \mathbf{AB} - \mathbf{I}$$

or

$$\mathbf{AB} = \mathbf{I} + \mathbf{E} \tag{3.89}$$

From (3.89) we get

$$\mathbf{A}^{-1} = \mathbf{B}(\mathbf{I} + \mathbf{E})^{-1}$$
$$= \mathbf{B}(\mathbf{I} - \mathbf{E} + \mathbf{E}^2 - \dots) \tag{3.90}$$

if $\|\mathbf{E}\| < 1$. Approximating (3.90) we have

$$\mathbf{A}^{-1} \approx \mathbf{B}(\mathbf{I} - \mathbf{E}) = \mathbf{B}(2\mathbf{I} - \mathbf{AB})$$

This gives an iterative method

$$\mathbf{B}^{(k+1)} = \mathbf{B}^{(k)}(2\mathbf{I} - \mathbf{AB}^{(k)}), \; k = 0, 1, 2, \dots \tag{3.91}$$

Premultiplying (3.91) by $\mathbf{A}$ we have

$$\mathbf{AB}^{(k+1)} = 2\mathbf{AB}^{(k)} - (\mathbf{AB}^{(k)})^2$$

or

$$\mathbf{I} - \mathbf{AB}^{(k+1)} = \mathbf{I} - 2\mathbf{AB}^{(k)} + (\mathbf{AB}^{(k)})^2$$
$$= (\mathbf{I} - \mathbf{AB}^{(k)})^2 \tag{3.92}$$

Hence the convergence of iterative method is quadratic.

Example 3.13 Find the inverse of the matrix

$$\mathbf{A} = \begin{bmatrix} 1 & 1 \\ 1 & 2 \end{bmatrix}$$

using the iterative method, given that its approximate inverse is

$$\mathbf{B} = \begin{bmatrix} 1.8 & -0.9 \\ -0.9 & 0.9 \end{bmatrix}$$

Perform two iterations.

We have

$$\mathbf{B}^{(1)} = \begin{bmatrix} 1.8 & -0.9 \\ -0.9 & 0.9 \end{bmatrix} \left\{ \begin{bmatrix} 2 & 0 \\ 0 & 2 \end{bmatrix} - \begin{bmatrix} 0.9 & 0.0 \\ 0.0 & 0.9 \end{bmatrix} \right\}$$

$$= \begin{bmatrix} 1.98 & -0.99 \\ -0.99 & 0.99 \end{bmatrix}$$

$$\mathbf{B}^{(2)} = \begin{bmatrix} 1.98 & -0.99 \\ -0.99 & 0.99 \end{bmatrix} \left\{ \begin{bmatrix} 2 & 0 \\ 0 & 2 \end{bmatrix} - \begin{bmatrix} 0.99 & 0.00 \\ 0.00 & 0.99 \end{bmatrix} \right\}$$

$$= \begin{bmatrix} 1.998 & -0.9999 \\ -0.9999 & 0.9999 \end{bmatrix}$$

The exact inverse is

$$\mathbf{A}^{-1} = \begin{bmatrix} 2 & -1 \\ -1 & 1 \end{bmatrix}$$

3.5 EIGENVALUES AND EIGENVECTORS

The eigenvalues of a matrix $\mathbf{A}$ are given by the roots of the characteristic equation

$$\det(\mathbf{A} - \lambda \mathbf{I}) = 0 \tag{3.93}$$

which when simplified gives the polynomial equation

$$p(\lambda) = (-1)^n \lambda^n + a_1 \lambda^{n-1} + \ldots + a_n = 0 \tag{3.94}$$

where the sign $(-1)^n$ is used to give the terms of the polynomial the same sign that they would have if the polynomial were generated by expanding the determinant. The coefficients of the polynomial (3.94) can be determined with the help of the **Faddeev-Leverrier** method.

We have

$$\begin{aligned}
& \mathbf{B}_1 = \mathbf{A} \text{ and } a_1 = \text{tr } \mathbf{B}_1 \\
& \mathbf{B}_2 = \mathbf{A}(\mathbf{B}_1 - a_1 \mathbf{I}) \text{ and } a_2 = \tfrac{1}{2} \text{tr } \mathbf{B}_2 \\
& \mathbf{B}_3 = \mathbf{A}(\mathbf{B}_2 - a_2 \mathbf{I}) \text{ and } a_3 = \tfrac{1}{3} \text{tr } \mathbf{B}_3 \\
& \vdots \\
& \mathbf{B}_k = \mathbf{A}(\mathbf{B}_{k-1} - a_{k-1} \mathbf{I}) \text{ and } a_k = \tfrac{1}{k} \text{tr } \mathbf{B}_k \\
& \vdots \\
& \mathbf{B}_n = \mathbf{A}(\mathbf{B}_{n-1} - a_{n-1} \mathbf{I}) \text{ and } a_n = \tfrac{1}{n} \text{tr } \mathbf{B}_n
\end{aligned} \tag{3.95}$$

where $\text{tr } \mathbf{A} = a_{11} + a_{22} + \ldots + a_{nn}$.

The roots of the polynomial equation (3.94) may be determined by the methods discussed in Chapter 2. A nonzero vector $\mathbf{x}_i$ such that

$$\mathbf{A}\mathbf{x}_i = \lambda_i \mathbf{x}_i \tag{3.96}$$

is called the eigenvector or characteristic vector corresponding to λ_i. Multiplying (3.96) by an arbitrary constant c and putting $\mathbf{y}_i = c\mathbf{x}_i$, we get

$$\mathbf{A}\mathbf{y}_i = \lambda_i \mathbf{y}_i \tag{3.97}$$

which shows that an eigenvector is determined only to within an arbitrary multiplicative constant. Premultiplying (3.4) m times by $\mathbf{A}$ we may obtain

$$\mathbf{A}^m \mathbf{x} = \lambda^m \mathbf{x} \tag{3.98}$$

which shows that λ^m is an eigenvalue of $\mathbf{A}^m$ and $\mathbf{x}$ is the corresponding eigenvector. Substituting (3.98) into (3.94) we get

$$p(\mathbf{A}) = 0 \tag{3.99}$$

which gives the result that a square matrix $\mathbf{A}$ satisfies its own characteristic equation. This result is known as the **Cayley-Hamilton** theorem. Replacing the matrix $\mathbf{A}$ in (3.93) by the transpose matrix $\mathbf{A}^T$ we find

$$\det(\mathbf{A}^T - \lambda \mathbf{I}) = \det(\mathbf{A} - \lambda \mathbf{I}) = 0 \tag{3.100}$$

Thus $\mathbf{A}$ and $\mathbf{A}^T$ have the same eigenvalues. For distinct eigenvalues, if $\mathbf{u}_1, \mathbf{u}_2, \ldots, \mathbf{u}_n$ are the eigenvectors of $\mathbf{A}$ and $\mathbf{v}_1, \mathbf{v}_2, \ldots, \mathbf{v}_n$ are the eigenvectors of $\mathbf{A}^T$ then we have

$$\mathbf{A}\mathbf{u}_i = \lambda_i \mathbf{u}_i \tag{3.101}$$

$$\mathbf{A}^T \mathbf{v}_j = \lambda_j \mathbf{v}_j \tag{3.102}$$

We obtain

$$\mathbf{v}_j^T \mathbf{A} \mathbf{u}_i = \lambda_i \mathbf{v}_j^T \mathbf{u}_i \tag{3.103}$$

Taking transpose of (3.102) and post-multiplying by $\mathbf{u}_i$, we get

$$\mathbf{v}_j^T \mathbf{A} \mathbf{u}_i = \lambda_j \mathbf{v}_j^T \mathbf{u}_i \tag{3.104}$$

Subtracting (3.104) from (3.103) we get

$$(\lambda_i - \lambda_j) \mathbf{v}_j^T \mathbf{u}_i = 0 \tag{3.105}$$

If $i \neq j$, $\lambda_i \neq \lambda_j$ and we have

$$\mathbf{v}_j^T \mathbf{u}_i = 0 \tag{3.106}$$

If $i = j$, $\mathbf{v}_i^T \mathbf{u}_i \neq 0$ and since the length of eigenvectors is arbitrary we normalize them such that

$$\mathbf{v}_i^T \mathbf{u}_i = 1 \tag{3.107}$$

Thus, we have

$$\mathbf{v}_j^T \mathbf{u}_i = \begin{vmatrix} 0, & i \neq j \\ 1, & i = j \end{vmatrix} \tag{3.108}$$

We conclude that the eigenvectors corresponding to different eigenvalues of a matrix and of its transpose are orthogonal. Such sets of mutually orthogonal vectors are called biorthogonal sets. When $\mathbf{A}$ is a real symmetric matrix, $\mathbf{A} = \mathbf{A}^T$, $\mathbf{u}_i = \mathbf{v}_i$ and the eigenvectors corresponding to different eigenvalues are orthogonal. Premultiplying (3.101) by $\mathbf{u}_i^T$, we get

$$\lambda_i = \frac{\mathbf{u}_i^T \mathbf{A} \mathbf{u}_i}{\mathbf{u}_i^T \mathbf{u}_i} \tag{3.109}$$

which gives an expression for the eigenvalues in terms of the eigenvectors. For arbitrary $\mathbf{u}$, (3.109) is called the **Rayleigh quotient**.

Let $\mathbf{A}$ and $\mathbf{B}$ be two square matrices of same order. If a non-singular matrix $\mathbf{S}$ can be determined such that

$$\mathbf{B} = \mathbf{S}^{-1} \mathbf{A} \mathbf{S} \tag{3.110}$$

then the matrices $\mathbf{A}$ and $\mathbf{B}$ are said to be **similar** and (3.110) is called a **similarity transformation**. The matrix $\mathbf{S}$ is called the **similarity matrix**. From (3.110) we may write

$$\mathbf{A} = \mathbf{S} \mathbf{B} \mathbf{S}^{-1} \tag{3.111}$$

If λ_i is an eigenvalue of $\mathbf{A}$ and $\mathbf{u}_i$ is the corresponding eigenvector then

$$\mathbf{A} \mathbf{u}_i = \lambda_i \mathbf{u}_i$$

or

$$\mathbf{S}^{-1} \mathbf{A} \mathbf{u}_i = \lambda \mathbf{S}^{-1} \mathbf{u}_i \tag{3.112}$$

Substituting $\mathbf{u}_i = \mathbf{S} \mathbf{v}_i$ in (3.112) and using (3.110) we get

$$\mathbf{B} \mathbf{v}_i = \lambda_i \mathbf{v}_i \tag{3.113}$$

Thus $\mathbf{S}^{-1} \mathbf{A} \mathbf{S}$ has the same eigenvalues as $\mathbf{A}$ and its eigenvectors $\mathbf{v}_i$ are obtained from the relation $\mathbf{v}_i = \mathbf{S}^{-1} \mathbf{u}_i$.

Suppose that the matrix $\mathbf{A}$ has eigenvalues λ_i with eigenvectors $\mathbf{u}_i$ and that $\mathbf{A}$ has an inverse $\mathbf{A}^{-1}$. Then

$$\mathbf{A} \mathbf{u}_i = \lambda_i \mathbf{u}_i$$

which may be written as

$$\mathbf{A}^{-1} \mathbf{u}_i = \frac{1}{\lambda_i} \mathbf{u}_i \tag{3.114}$$

The inverse matrix $\mathbf{A}^{-1}$ has the same eigenvectors as $\mathbf{A}$ but has eigenvalues $1/\lambda_i$. A similarity transformation, where $\mathbf{S}$ is the matrix of eigenvectors, reduces a matrix $\mathbf{A}$ to its diagonal form. The eigenvalues of $\mathbf{A}$ are located on the leading diagonal of this diagonal matrix.

The bounds on the eigenvalues of the matrix $\mathbf{A}$ are given by the following theorem.

THEOREM 3.6 (GERSCHGORIN) *The largest eigenvalue in modulus of the square matrix* $\mathbf{A}$ *cannot exceed the largest sum of the moduli of the elements along any row or any column.*

Proof. Let λ_i be an eigenvalue of $\mathbf{A}$ and $\mathbf{x}_i$ be the corresponding eigenvector. Then we have

$$\mathbf{A} \mathbf{x}_i = \lambda_i \mathbf{x}_i$$

which in detail may be written as

$$a_{11}x_{i,1} + a_{12}x_{i,2} + \ldots + a_{1n}x_{i,n} \doteq \lambda_i x_{i,1}$$
$$a_{21}x_{i,1} + a_{22}x_{i,2} + \ldots + a_{2n}x_{i,n} = \lambda_i x_{i,2}$$
$$\vdots$$
$$a_{n1}x_{i,1} + a_{n2}x_{i,2} + \ldots + a_{nn}x_{i,n} = \lambda_i x_{i,n}$$

Let $|x_{i,k}| = \max_r |x_{i,r}|$. Select the kth equation and divide it by $x_{i,k}$. We get

$$\lambda_i = a_{k1}\left(\frac{x_{i,1}}{x_{i,k}}\right) + a_{k2}\left(\frac{x_{i,2}}{x_{i,k}}\right) + \ldots + a_{kk} + \ldots + a_{kn}\left(\frac{x_{i,n}}{x_{i,k}}\right) \quad (3.115)$$

and

$$|\lambda_i| \leq |a_{k1}| + |a_{k2}| + \ldots + |a_{kk}| + \ldots + |a_{kn}| \quad (3.116)$$

because $\left|\frac{x_{i,j}}{x_{i,k}}\right| \leq 1, j = 1(1)n$.

Since the eigenvalues of $\mathbf{A}^T$ are the same as those of $\mathbf{A}$, the theorem is also true for columns.

THEOREM 3.7 (BRAUER) *Let P_k be the sum of the moduli of the elements along the kth row excluding the diagonal element a_{kk}. Then every eigenvalue of $\mathbf{A}$ lies inside or on the boundary of at least one of the circles $|\lambda - a_{kk}| = P_k, k = 1(1)n$.*

Proof. We have from (3.115)

$$\lambda_i - a_{kk} = a_{k1}\left(\frac{x_{i,1}}{x_{i,k}}\right) + a_{k2}\left(\frac{x_{i,2}}{x_{i,k}}\right) + \ldots + a_{kn}\left(\frac{x_{i,n}}{x_{i,k}}\right)$$

$$\therefore \quad |\lambda_i - a_{kk}| \leq \sum_{\substack{i=1 \\ i \neq k}}^{n} |a_{ki}| = P_k \quad (3.117)$$

Therefore, all the eigenvalues of $\mathbf{A}$ lie inside or on the union of the above circles.

Again, since $\mathbf{A}$ and $\mathbf{A}^T$ have the same eigenvalues, we find that all the eigenvalues lie in the union of the n circles

$$|\lambda_i - a_{kk}| \leq \sum_{\substack{j=1 \\ j \neq k}}^{n} |a_{jk}|, \quad k = 1(1)n \quad (3.118)$$

The bounds obtained here are all independent. Hence, all the eigenvalues of $\mathbf{A}$ must lie in the intersection of these bounds. These circles are called the **Gerschgorin circles** and the bounds are called the **Gerschgorin bounds**. If any of the Gerschgorin circles is isolated, then it contains exactly one eigenvalue.

The eigenvalues of the matrix $\mathbf{A}$ are given by the diagonal elements when

it has one of the following three forms:
  (i) $\mathbf{A} = \mathbf{D}$
  (ii) $\mathbf{A} = \mathbf{L}$
  (iii) $\mathbf{A} = \mathbf{U}$ \hfill (3.119)

Thus the methods of finding the eigenvalues of A will generally be based on reducing A to either D or L or U or LU. In addition, we have iteration methods to determine a particular eigenvalue such as the largest or the smallest eigenvalue (in modulus) of a matrix.

Example 3.14 Estimate the eigenvalues of the matrix

$$\mathbf{A} = \begin{bmatrix} 1 & 2 & -1 \\ 1 & 1 & 1 \\ 1 & 3 & -1 \end{bmatrix}$$

using the Gerschgorin bounds.

The eigenvalues lie in the regions

  (i) $|\lambda| \leq 5, \quad |\lambda| \leq 6$

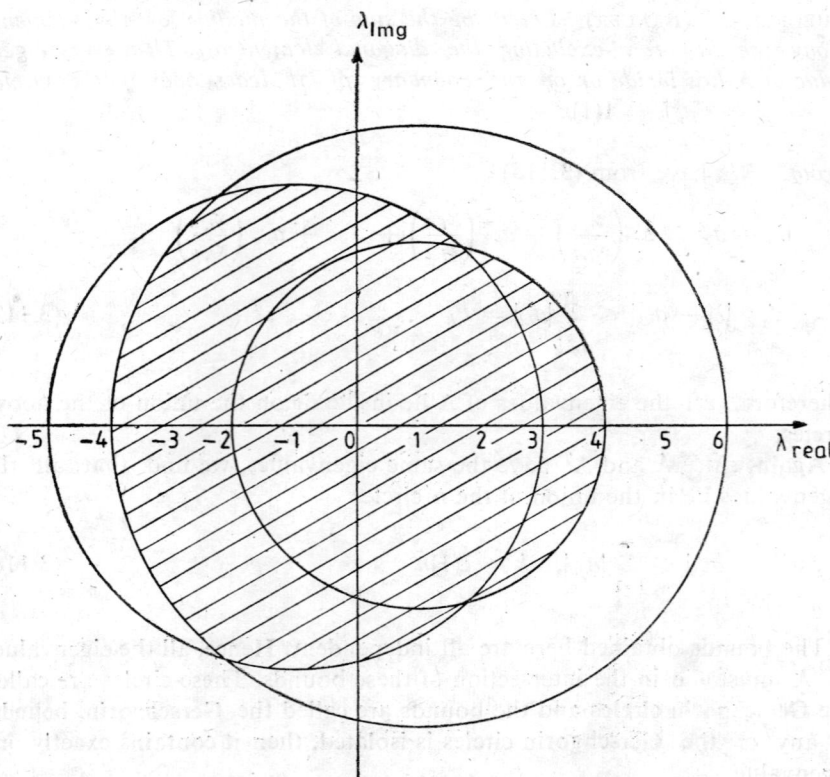

Fig. 3.2 Bounding region for the eigenvalues

(ii) the union of the circles
$$|\lambda - 1| \leqslant 3$$
$$|\lambda - 1| \leqslant 2$$
$$|\lambda + 1| \leqslant 4$$
and
(iii) the union of the circles
$$|\lambda - 1| \leqslant 2$$
$$|\lambda - 1| \leqslant 5$$
$$|\lambda + 1| \leqslant 2$$

The first union gives $|\lambda - 1| \leqslant 3$, $|\lambda + 1| \leqslant 4$ and the second union gives $|\lambda - 1| \leqslant 5$. The intersection of these circles gives the required region as shown in Fig. 3.2.

Example 3.15 Find A^{10} when
$$A = \begin{bmatrix} 2 & 2 \\ 2 & -1 \end{bmatrix}$$

First we reduce A to its diagonal form D by a similarity transformation. Since A is symmetric, there exists an orthogonal matrix S which reduces A to D. We have
$$S^T A S = D \quad \text{or} \quad A = S D S^T$$
Further A^m has eigenvalues λ^m and the same eigenvectors. Hence
$$A^{10} = S D^{10} S^T$$
Let
$$S = \begin{bmatrix} \cos\theta & -\sin\theta \\ \sin\theta & \cos\theta \end{bmatrix}$$
Then
$$S^T A S = \begin{bmatrix} \cos\theta & \sin\theta \\ -\sin\theta & \cos\theta \end{bmatrix} \begin{bmatrix} 2 & 2 \\ 2 & -1 \end{bmatrix} \begin{bmatrix} \cos\theta & -\sin\theta \\ \sin\theta & \cos\theta \end{bmatrix}$$
$$= \begin{bmatrix} 2\cos^2\theta + 2\sin 2\theta - \sin^2\theta & -\tfrac{3}{2}\sin 2\theta + 2\cos 2\theta \\ -\tfrac{3}{2}\sin 2\theta + 2\cos 2\theta & 2\sin^2\theta - 2\sin 2\theta - \cos^2\theta \end{bmatrix}$$

This matrix reduces to a diagonal matrix if we put
$$-\tfrac{3}{2}\sin 2\theta + 2\cos 2\theta = 0$$
or
$$\tan 2\theta = \tfrac{4}{3}$$
Using this value of θ we get
$$S^T A S = D = \begin{bmatrix} 3 & 0 \\ 0 & -2 \end{bmatrix}$$

We obtain

$$\mathbf{A}^{10} = \mathbf{SD}^{10}\mathbf{S}^T = \tfrac{1}{5}\begin{bmatrix} 2 & -1 \\ 1 & 2 \end{bmatrix}\begin{bmatrix} 3^{10} & 0 \\ 0 & 2^{10} \end{bmatrix}\begin{bmatrix} 2 & 1 \\ -1 & 2 \end{bmatrix}$$

$$= \tfrac{1}{5}\begin{bmatrix} 4 \times 3^{10} + 2^{10} & 2 \times 3^{10} - 2^{11} \\ 2 \times 3^{10} - 2^{11} & 3^{10} + 2^{12} \end{bmatrix}$$

3.5.1 Jacobi Method for Symmetric Matrices

Let $\mathbf{A}$ be the given real symmetric matrix. The eigenvalues of $\mathbf{A}$ are real and there exists a real orthogonal matrix $\mathbf{S}$ such that $\mathbf{S}^{-1}\mathbf{A}\mathbf{S}$ is a diagonal matrix $\mathbf{D}$. The diagonalization is done by applying a series of orthogonal transformations $\mathbf{S}_1, \mathbf{S}_2, \ldots, \mathbf{S}_n, \ldots$ as follows.

Among the off-diagonal elements, let $|a_{ik}|$ be the numerically largest element. Then the elements $a_{ii}, a_{ik}, a_{ki}, a_{kk}$ form a 2×2 submatrix $\mathbf{A}_1$ which can be transformed to a diagonal form. We choose

$$\mathbf{S}_1^* = \begin{bmatrix} \cos\theta & -\sin\theta \\ \sin\theta & \cos\theta \end{bmatrix} \tag{3.120}$$

and find θ such that the 2×2 submatrix $\mathbf{A}_1$ is diagonalized. We have

$$\mathbf{S}_1^{*-1}\mathbf{A}_1\mathbf{S} = \begin{bmatrix} \cos\theta & \sin\theta \\ -\sin\theta & \cos\theta \end{bmatrix}\begin{bmatrix} a_{ii} & a_{ik} \\ a_{ki} & a_{kk} \end{bmatrix}\begin{bmatrix} \cos\theta & -\sin\theta \\ \sin\theta & \cos\theta \end{bmatrix}$$

$$= \begin{bmatrix} a_{ii}\cos^2\theta + 2a_{ik}\sin\theta\cos\theta + a_{kk}\sin^2\theta & (a_{kk} - a_{ii})\sin\theta\cos\theta + a_{ik}\cos 2\theta \\ (a_{kk} - a_{ii})\sin\theta\cos\theta + a_{ik}\cos 2\theta & a_{ii}\sin^2\theta - a_{ik}\sin 2\theta + a_{kk}\cos^2\theta \end{bmatrix} \tag{3.121}$$

We now choose θ such that this matrix reduces to a diagonal form. That is, we put

$$\tfrac{1}{2}(a_{kk} - a_{ii})\sin 2\theta + a_{ik}\cos 2\theta = 0$$

or

$$\tan 2\theta = \frac{2a_{ik}}{a_{ii} - a_{kk}} \tag{3.122}$$

This equation produces four values of θ and in order that we may get smallest rotation we require $-\pi/4 \leqslant \theta \leqslant \pi/4$. From (3.122) we get

$$\theta = \frac{1}{2}\tan^{-1}\left(\frac{2a_{ik}}{a_{ii} - a_{kk}}\right) \tag{3.123}$$

if $a_{ii} \neq a_{kk}$. If $a_{ii} = a_{kk}$, then

$$\theta = \begin{vmatrix} \dfrac{\pi}{4}, & a_{ik} > 0 \\ -\dfrac{\pi}{4}, & a_{ik} < 0 \end{vmatrix} \tag{3.124}$$

With the value of θ given in (3.123), the off-diagonal elements in (3.121) vanish and the diagonal elements are simplified. The first step is now completed by performing the rotation $S_1^{-1}AS_1$. In the next step the largest off-diagonal element in magnitude in the new rotated matrix is found and the procedure is repeated. We now perform a series of such two dimensional rotations. After finding θ at each step, the rotation is performed with the corresponding orthogonal matrix. For example, if $|a_{ik}|$ is the largest off-diagonal element then we write S_1 as

$$S_1 = \begin{bmatrix} 1 & & & & & & \\ & 1 & & & & & \\ & & \ddots & & & & \\ & & & \cos\theta & \cdots & -\sin\theta & \\ & & & \vdots & \ddots & \vdots & \\ & & & \sin\theta & \cdots & \cos\theta & \\ & & & & & & \ddots \\ & & & & & & & 1 \end{bmatrix}$$

where $\cos\theta$, $-\sin\theta$, $\sin\theta$, $\cos\theta$ are located in (i, i), (i, k), (k, i) and (k, k) positions respectively. After making r transformations, we get

$$\begin{aligned} B_r &= S_r^{-1} S_{r-1}^{-1} \ldots S_1^{-1} A S_1 \ldots S_{r-1} S_r \\ &= (S_1 S_2 \ldots S_r)^{-1} A (S_1 \ldots S_{r-1} S_r) \\ &= S^{-1} A S \end{aligned} \tag{3.125}$$

where $S = S_1 S_2 \ldots S_r$

As $r \to \infty$, B_r approaches a diagonal matrix with the eigenvalues on the leading diagonal. We then have the eigenvectors as the corresponding columns of S. The minimum number of rotations required to bring A into a diagonal form may be $(n-1)n/2$. This procedure, described to reduce the symmetric matrix A to a diagonal matrix D is called the **Jacobi method**. The method suffers from the following disadvantage. The elements annihilated by a plane rotation may not necessarily remain zero during subsequent transformations. The value of θ must be checked for its accuracy by checking whether $|\sin^2\theta + \cos^2\theta - 1|$, is sufficiently small. The convergence to the eigenvalues takes place even if the pivots are not selected on the basis of their magnitudes, but are selected in the "typewriter fashion." That is annihilate a_{21}, a_{31}, a_{32}, a_{41}, a_{42}, a_{43} etc. This modification is called the **special cyclic Jacobi method**. In this method there is no search for the pivots.

Example 3.16 Find all the eigenvalues and eigenvectors of the matrix

$$\begin{bmatrix} 1 & \sqrt{2} & 2 \\ \sqrt{2} & 3 & \sqrt{2} \\ 2 & \sqrt{2} & 1 \end{bmatrix}$$

The largest off diagonal element is $a_{13} = a_{31} = 2$. The other two elements in this (2, 2) submatrix are $a_{11} = 1$ and $a_{33} = 1$.

$$\theta = \tfrac{1}{2} \tan^{-1}(\tfrac{4}{0}) = \pi/4$$

$$S_1 = \begin{bmatrix} 1/\sqrt{2} & 0 & -1/\sqrt{2} \\ 0 & 1 & 0 \\ 1/\sqrt{2} & 0 & 1/\sqrt{2} \end{bmatrix}$$

The first rotation gives

$$S^{-1}AS = \begin{bmatrix} 1/\sqrt{2} & 0 & 1/\sqrt{2} \\ 0 & 1 & 0 \\ -1/\sqrt{2} & 0 & 1/\sqrt{2} \end{bmatrix} \begin{bmatrix} 1 & \sqrt{2} & 2 \\ \sqrt{2} & 3 & \sqrt{2} \\ 2 & \sqrt{2} & 1 \end{bmatrix} \begin{bmatrix} 1/\sqrt{2} & 0 & -1/\sqrt{2} \\ 0 & 1 & 0 \\ 1/\sqrt{2} & 0 & 1/\sqrt{2} \end{bmatrix}$$

$$= \begin{bmatrix} 3 & 2 & 0 \\ 2 & 3 & 0 \\ 0 & 0 & -1 \end{bmatrix}$$

Again, the largest off diagonal element is $a_{12} = a_{21} = 2$. The other elements are $a_{11} = 3$, $a_{22} = 3$.

$$\theta = \tfrac{1}{2} \tan^{-1}(\tfrac{4}{0}) = \pi/4$$

and

$$S_2 = \begin{bmatrix} 1/\sqrt{2} & -1/\sqrt{2} & 0 \\ 1/\sqrt{2} & 1/\sqrt{2} & 0 \\ 0 & 0 & 1 \end{bmatrix}$$

The second rotation gives

$$S_2^{-1}AS_2 = \begin{bmatrix} 5 & 0 & 0 \\ 0 & 1 & 0 \\ 0 & 0 & -1 \end{bmatrix}$$

We have the matrix of eigenvectors

$$S = S_1 S_2 = \begin{bmatrix} 1/\sqrt{2} & 0 & -1/\sqrt{2} \\ 0 & 1 & 0 \\ 1/\sqrt{2} & 0 & 1/\sqrt{2} \end{bmatrix} \begin{bmatrix} 1/\sqrt{2} & -1/\sqrt{2} & 0 \\ 1/\sqrt{2} & 1/\sqrt{2} & 0 \\ 0 & 0 & 1 \end{bmatrix}$$

$$= \begin{bmatrix} \tfrac{1}{2} & -\tfrac{1}{2} & -1/\sqrt{2} \\ 1/\sqrt{2} & 1/\sqrt{2} & 0 \\ \tfrac{1}{2} & -\tfrac{1}{2} & 1/\sqrt{2} \end{bmatrix}$$

The eigenvalues are 5, 1, −1, and the corresponding eigenvectors are the columns of S.

3.5.2 Givens Method for Symmetric Matrices

We have noted that in the Jacobi's method, the elements which were annihilated by a plane rotation may not remain zero during subsequent rotations. **Givens** proposed an algorithm using plane rotations, which preserves the zeros in the off diagonal elements, once they are created. Let **A** be a real, symmetric matrix. The **Givens method** uses the following steps:
  (a) reduce A to a tridiagonal form using plane rotations
  (b) form a **Sturm sequence**, study the changes in sign in the sequences and find the eigenvalues
  (c) find the eigenvectors

The reduction to a tridiagonal form is achieved by using the orthogonal transformations as in the Jacobi method. However, in this case we start with the subspace containing the elements $a_{22}, a_{23}, a_{32}, a_{33}$. Perform the plane rotation $S_1^{-1} A S_1$ using the orthogonal matrix

$$S_1^* = \begin{bmatrix} \cos\theta & -\sin\theta \\ \sin\theta & \cos\theta \end{bmatrix} \tag{3.126}$$

Let the new matrix obtained be $A' = [a'_{ij}]$.

The angle θ is now obtained by putting $a'_{13} = a'_{31} = 0$ and not by putting $a'_{23} = 0 = a'_{32}$ as in Jacobi method. We find

$$a'_{13} = -a_{12}\sin\theta + a_{13}\cos\theta = 0$$

or

$$\tan\theta = a_{13}/a_{12} \tag{3.127}$$

With this value of θ and performing the plane rotation, we produce zeros in the (3, 1) and (1, 3) positions. Then we perform rotations in (2, 4) space and put $a'_{14} = a'_{41} = 0$. This would not affect zeros that have been obtained earlier. Proceeding in this manner, we put $a'_{15} = a'_{51} = 0$ etc. by performing rotations in (2, 5), ..., (2, n) planes. Then we pass on to the elements $a'_{24}, a'_{25}, \ldots, a'_{2n}$ and make them zero by performing rotations in (3, 4), ..., (3, n) subspaces. Finally, we produce the matrix

$$B = \begin{bmatrix} b_1 & c_1 & & & & \\ c_1 & b_2 & c_2 & & \bigcirc & \\ & c_2 & b_3 & c_3 & & \\ & & & \ddots & & \\ & \bigcirc & & c_{n-2} & b_{n-1} & c_{n-1} \\ & & & & c_{n-1} & b_n \end{bmatrix} \tag{3.128}$$

The number of plane rotations required to bring a matrix of order n to its tridiagonal form is $\frac{1}{2}(n-1)(n-2)$. We already know that **A** and **B** have the same eigenvalues. If $c_i \neq 0$, $i = 1, \ldots, n - 1$, then the eigenvalues are distinct. Now, define

$$f_n = |\lambda I - B|$$

$$= \begin{vmatrix} \lambda - b_1 & -c_1 & & & \\ -c_1 & \lambda - b_2 & -c_2 & & \\ & & \ddots & & \\ & & -c_{n-2} & \lambda - b_{n-1} & -c_{n-1} \\ & & & -c_{n-1} & \lambda - b_n \end{vmatrix}$$

Expanding by minors, the sequence $\{f_n\}$ satisfies

$$f_0 = 1, f_1 = \lambda - b_1$$

and

$$f_r = (\lambda - b_r)f_{r-1} - c_{r-1}^2 f_{r-2}; \ 2 \leqslant r \leqslant n \qquad (3.129)$$

Note that f_n is the characteristic equation. If none of the $c_1, c_2, \ldots, c_{n-1}$ vanish, then $\{f_n\}$ is a Sturm sequence. That is, if $V(x)$ denotes the number of changes in sign in the sequence for a given number x, then the number of zeros of f_n in $[a, b]$ is $V(a) - V(b)$ (provided a or b is not a zero of f_n). In this way one can approximately compute the eigenvalues and by repeated bisections, one can improve these estimates.

The eigenvectors of B are then found. If these are determined then the eigenvectors of A can be determined, since we know that if v and u are the eigenvectors of B and A respectively, then $\mathbf{u} = S\mathbf{v}$; where $S = S_1 S_2 \ldots S_n$ is the product of the orthogonal matrices used in the plane rotations. The eigenvectors of B may be found as follows. Neglect a particular equation (say ith) and then solve the remaining equations. This solution usually satisfies the equation that has been left. Then v is the eigenvector determined from these solutions and by putting a zero in the ith position. An advantage of the Givens method is that it takes only a finite number of plane rotations to reduce A to its tridiagonal form.

Example 3.17 Use the Givens method to find the eigenvalues of the tridiagonal matrix

$$A = \begin{bmatrix} 2 & -1 & 0 \\ -1 & 2 & -1 \\ 0 & -1 & 2 \end{bmatrix}$$

The Sturm sequence is

$$f_0 = 1, f_1 = \lambda - 2; \ f_2 = (\lambda - 2)f_1 - f_0 = (\lambda - 2)^2 - 1;$$
$$f_3 = (\lambda - 2)f_2 - f_1 = (\lambda - 2)^3 - 2(\lambda - 2)$$

We find

λ	f_0	f_1	f_2	f_3	$V(\lambda)$
-1	$+$	$-$	$+$	$-$	3
0	$+$	$-$	$+$	$-$	3
1	$+$	$-$	0	$+$	2
2	$+$	0	$-$	0	$-$
3	$+$	$+$	0	$-$	1
4	$+$	$+$	$+$	$+$	0

Note that $f_3(2) = 0$, so that $\lambda = 2$ is an eigenvalue. There is an eigenvalue in (0, 1) and (3, 4). We now find better estimates of the eigenvalues by repeated bisections. Let us determine the eigenvalue in (0, 1). We have

λ	f_0	f_1	f_2	f_3	$V(\lambda)$
$\tfrac{1}{2}$	$+$	$-$	$+$	$-$	3

The eigenvalue is now located in (0.5, 1). Again

0.75	$+$	$-$	$+$	$+$	2

The eigenvalue is now located in (0.5, 0.75). We have

0.625	$+$	$-$	$+$	$+$	2

The eigenvalue is now located in (0.5, 0.625). Again

0.5625	$+$	$-$	$+$	$-$	3

The eigenvalue is now located in (0.5625, 0.625)

0.59375	$+$	$-$	$+$	$+$	2

The eigenvalue is now located in (0.5625, 0.59375).

We repeat this procedure until the required accuracy is obtained. The exact value of this eigenvalue is $2 - \sqrt{2} \simeq 0.585786$.

3.5.3 Householder's Method for Symmetric Matrices

In the Givens method tridiagonalization is achieved by using $\tfrac{1}{2}(n-1)(n-2)$ plane rotations. However, **Householder** has given a procedure which requires essentially half as much computation as the Givens method for the tridiagonalization. The remaining procedure is same as in the Givens method. In this method **A** is reduced to the tridiagonal form by orthogonal transformations representing reflections. The orthogonal transformations are of the form

$$\mathbf{P} = \mathbf{I} - 2\mathbf{w}\mathbf{w}^T \qquad (3.130)$$

where **w** is a column vector, $\mathbf{w} \in R^n$, such that

$$\mathbf{w}^T\mathbf{w} = w_1^2 + w_2^2 + \ldots + w_n^2 = 1 \tag{3.131}$$

It can be easily shown that $\mathbf{P}$ is symmetric and orthogonal.

$$\mathbf{P}^T = (\mathbf{I} - 2\mathbf{w}\mathbf{w}^T)^T = \mathbf{I} - 2\mathbf{w}\mathbf{w}^T = \mathbf{P}$$

Also

$$\mathbf{P}^T\mathbf{P} = (\mathbf{I} - 2\mathbf{w}\mathbf{w}^T)(\mathbf{I} - 2\mathbf{w}\mathbf{w}^T)$$
$$= \mathbf{I} - 4\mathbf{w}\mathbf{w}^T + 4\mathbf{w}\mathbf{w}^T\mathbf{w}\mathbf{w}^T = \mathbf{I}$$
$$\mathbf{P}^T = \mathbf{P}^{-1}$$

The vectors $\mathbf{w}$ are constructed with the first $(r-1)$ components as zeros, that is,

$$\mathbf{w}_r^T = (0, 0, \ldots, 0, x_r, x_{r+1}, \ldots, x_n) \tag{3.132}$$

Since $\mathbf{w}_r^T\mathbf{w}_r = 1$ we have $x_r^2 + x_{r+1}^2 + \ldots + x_n^2 = 1$

With this choice of $\mathbf{w}_r$, form the matrices

$$\mathbf{P}_r = \mathbf{I} - 2\mathbf{w}_r\mathbf{w}_r^T$$

The similarity transformation is given by

$$\mathbf{P}_r^{-1}\mathbf{A}\mathbf{P}_r = \mathbf{P}_r^T\mathbf{A}\mathbf{P}_r = \mathbf{P}_r\mathbf{A}\mathbf{P}_r \tag{3.133}$$

since $\mathbf{P}_r$ is symmetric and orthogonal. We put $\mathbf{A} = \mathbf{A}_1$ and form successively

$$\mathbf{A}_r = \mathbf{P}_r\mathbf{A}_{r-1}\mathbf{P}_r, \quad r = 2, 3, \ldots, n-1 \tag{3.134}$$

At the first transformation, we find x_r's such that we get zeros in the positions $(1, 3), (1, 4), \ldots, (1, n)$ and in the corresponding positions in the first column. Therefore one rotation brings $n-2$ zeros in the first row and column. In the second rotation, we find x_r's such that we have zeros in $(2, 4), (2, 5), \ldots, (2, n)$ positions. The final matrix is a tridiagonal matrix as in Given's method. The tridiagonalization is completed with exactly $n-2$ **Householder transformations**. Let us illustrate this procedure using a 4×4 matrix

$$\mathbf{A} = \begin{bmatrix} a_{11} & a_{12} & a_{13} & a_{14} \\ a_{21} & a_{22} & a_{23} & a_{24} \\ a_{31} & a_{32} & a_{33} & a_{34} \\ a_{41} & a_{42} & a_{43} & a_{44} \end{bmatrix} \tag{3.135}$$

We note that, since the transformations being used are orthogonal, the sum of squares of the elements in any row is invariant. Choose

$$\mathbf{w}_2^T = [0 \quad x_2 \quad x_3 \quad x_4], \quad x_2^2 + x_3^2 + x_4^2 = 1 \tag{3.136}$$

Now

$$P_2 = I - 2w_2 w_2^T = \begin{bmatrix} 1 & 0 & 0 & 0 \\ 0 & 1 - 2x_2^2 & -2x_2 x_3 & -2x_2 x_4 \\ 0 & -2x_2 x_3 & 1 - 2x_3^2 & -2x_3 x_4 \\ 0 & -2x_2 x_4 & -2x_3 x_4 & 1 - 2x_4^2 \end{bmatrix}. \tag{3.137}$$

Now the (1, 3), (1, 4) elements of $P_2 A P_2$ can become zero only if the corresponding elements in AP_2 are already zero. The first row of AP_2 is given by

$$a_{11},\ a_{12} - 2p_1 x_2,\ a_{13} - 2p_1 x_3,\ a_{14} - 2p_1 x_4$$

where $p_1 = a_{12} x_2 + a_{13} x_3 + a_{14} x_4$. We now require that

$$a_{13} - 2p_1 x_3 = 0 \tag{3.138}$$
$$a_{14} - 2p_1 x_4 = 0 \tag{3.139}$$

so that zeros are obtained in the (1, 3), (1, 4) positions. Since the sum of squares of the elements in any row is invariant under an orthogonal transformation, we have

$$a_{11}^2 + a_{12}^2 + a_{13}^2 + a_{14}^2 = a_{11}^2 + (a_{12} - 2p_1 x_2)^2$$

or

$$a_{12} - 2p_1 x_2 = \pm\sqrt{\{a_{12}^2 + a_{13}^2 + a_{14}^2\}} = \pm s_1 \tag{3.140}$$

Notice that s_1 is a known quantity. Multiply (3.140) by x_2, (3.138) by x_3, (3.139) by x_4 and add

$$a_{12} x_2 + a_{13} x_3 + a_{14} x_4 - 2p_1 = \pm s_1 x_2$$

This gives

$$p_1 = \mp s_1 x_2$$

Substituting in (3.140), we get

$$a_{12} \pm 2s_1 x_2^2 = \pm s_1$$
$$\pm 2s_1 x_2^2 = \pm s_1 - a_{12}$$
$$x_2^2 = \tfrac{1}{2}\left(1 \mp \frac{a_{12}}{s_1}\right) \tag{3.141}$$

This determines x_2. From (3.138) and (3.139) we have

$$x_3 = \mp a_{13}/2s_1 x_2 \quad \text{and} \quad x_4 = \mp a_{14}/2s_1 x_2 \tag{3.142}$$

Usually, we need to find two square roots, one for s_1 and another for x_2. Since x_3 and x_4 contain x_2 in the denominator, we obtain best accuracy if x_2 is large. This can be obtained by taking suitable sign in (3.141). Choose

$$x_2^2 = \tfrac{1}{2}\left(1 + \frac{a_{12}\ \text{sign}\ (a_{12})}{s_1}\right) \tag{3.143}$$

The sign of the square root is irrelevent and taken as plus sign. Hence

$$x_3 = \frac{a_{13} \operatorname{sign}(a_{12})}{2 s_1 x_2}, \quad x_4 = \frac{a_{14} \operatorname{sign}(a_{12})}{2 s_1 x_2}$$

This transformation produces two zeros in the first row and first column. One more transformation produces zeros in the (2, 4) and (4, 2) positions. The resulting matrix is tridiagonal.

3.5.4 Rutishauser Method for Arbitrary Matrices

Rutishauser proposed the LU transformation where **L** is a lower triangular matrix and **U** is an upper triangular matrix. In the limit, we get an upper triangular matrix which displays the eigenvalues of **A** on the leading diagonal. Starting with the matrix $\mathbf{A} = \mathbf{A}_1$ we split it into two triangular matrices

$$\mathbf{A}_1 = \mathbf{L}_1 \mathbf{U}_1 \qquad (3.144)$$

with $l_{ii} = 1$. Then form $\mathbf{A}_2 = \mathbf{U}_1 \mathbf{L}_1$. Since $\mathbf{A}_2 = \mathbf{U}_1 \mathbf{L}_1 = \mathbf{U}_1 \mathbf{A}_1 \mathbf{U}_1^{-1}$, then $\mathbf{A}_1$ and $\mathbf{A}_2$ have the same eigenvalues. We again write

$$\mathbf{A}_2 = \mathbf{L}_2 \mathbf{U}_2 \qquad (3.145)$$

with $l_{ii} = 1$. Form $\mathbf{A}_3 = \mathbf{U}_2 \mathbf{L}_2$ so that $\mathbf{A}_2$ and $\mathbf{A}_3$ have the same eigenvalues. Proceeding in this way, we get a sequence of matrices $\mathbf{A}_1, \mathbf{A}_2, \mathbf{A}_3 \ldots$ which in general reduces to an upper triangular matrix. If the eigenvalues are real, then they all lie on the leading diagonal. However, there are difficulties associated with the practical application of the method. To avoid some of these difficulties, it was proposed that **L** be replaced by a unitary matrix **Q**. If **A** is non-singular then there exists a decomposition $\mathbf{A} = \mathbf{QU}$ where **Q** is unitary and **U** is upper triangular. The **QU** algorithm is also not simple for practical application. It is useful when applied to a matrix **A** in upper Hessenberg form (almost triangular form)

$$\mathbf{A} = \begin{bmatrix} a_{11} & a_{12} & a_{13} & \ldots & a_{1n} \\ a_{21} & a_{22} & a_{23} & \ldots & a_{2n} \\ & a_{32} & a_{33} & \ldots & a_{3n} \\ & & \cdot & \cdot & \\ & & & \cdot & \\ & & & a_{n,n-1} & a_{nn} \end{bmatrix}$$

The number of multiplications and additions in one **QU** transformation is proportional to n^3 for a full matrix whereas it is only n^2 for a Hessenberg matrix. A two step procedure is recommended, first to reduce **A** to upper Hessenberg form and then apply **QU** algorithm to the upper Hessenberg form. Any matrix can be transformed by similarity transformations to the upper Hessenberg form.

Example 3.18 Find all the eigenvalues of the matrix
$$A = \begin{bmatrix} 4 & 3 \\ 1 & 2 \end{bmatrix}$$
using the Rutishauser method.

We write
$$A_1 = \begin{bmatrix} 1 & 0 \\ l_{21} & 1 \end{bmatrix} \begin{bmatrix} u_{11} & u_{12} \\ 0 & u_{22} \end{bmatrix}$$

We find
$$l_{21} = \tfrac{1}{4}, \quad u_{11} = 4, \quad u_{12} = 3, \quad u_{22} = \tfrac{5}{4}$$

Form
$$A_2 = \begin{bmatrix} 4 & 3 \\ 0 & \tfrac{5}{4} \end{bmatrix} \begin{bmatrix} 1 & 0 \\ \tfrac{1}{4} & 1 \end{bmatrix} = \begin{bmatrix} \tfrac{19}{4} & 3 \\ \tfrac{5}{16} & \tfrac{5}{4} \end{bmatrix}$$

Decomposing again, we have
$$A_2 = \begin{bmatrix} 1 & 0 \\ \tfrac{5}{76} & 1 \end{bmatrix} \begin{bmatrix} \tfrac{19}{4} & 3 \\ 0 & \tfrac{20}{19} \end{bmatrix}$$

Form
$$A_3 = \begin{bmatrix} \tfrac{19}{4} & 3 \\ 0 & \tfrac{20}{19} \end{bmatrix} \begin{bmatrix} 1 & 0 \\ \tfrac{5}{76} & 1 \end{bmatrix} = \begin{bmatrix} 4.94737 & 1 \\ 0.06925 & 1.05263 \end{bmatrix}$$

We find
$$A_4 = \begin{bmatrix} 4.98937 & 3 \\ 0.01415 & 1.01063 \end{bmatrix}$$
$$A_5 = \begin{bmatrix} 4.99789 & 3 \\ 0.00285 & 1.00211 \end{bmatrix}$$
$$A_6 = \begin{bmatrix} 4.99960 & 3 \\ 0.00057 & 1.00040 \end{bmatrix}$$

The sequence $\{A_i\}$ converges to an upper triangular matrix and the diagonal elements converge to the eigenvalues of A. The exact eigenvalues are 5 and 1.

3.5.5 Power Method

Power method is used to determine the largest eigenvalue (in magnitude) and the corresponding eigenvector of the system
$$Ax = \lambda x$$

Let $\lambda_1, \lambda_2, \ldots, \lambda_n$ be the distinct eigenvalues such that
$$|\lambda_1| > |\lambda_2| > \ldots > |\lambda_n| \tag{3.146}$$

and $v_1, v_2, \ldots, v_n$ be the corresponding eigenvectors. The procedure is applicable if a complete system of n independent eigenvectors exists, even though some of eigenvalues $\lambda_2, \ldots, \lambda_n$ may not be distinct. Then any eigenvector v in the space of eigenvectors $v_1, v_2, \ldots, v_n$ can be written as

$$v = c_1 v_1 + c_2 v_2 + \ldots + c_n v_n \tag{3.147}$$

Premultiplying by A and substituting $Av_1 = \lambda_1 v_1$, $Av_2 = \lambda_2 v_2$ etc., we obtain

$$Av = c_1 \lambda_1 v_1 + c_2 \lambda_2 v_2 + \ldots + c_n \lambda_n v_n$$

$$= \lambda_1 \left[c_1 v_1 + c_2 \left(\frac{\lambda_2}{\lambda_1}\right) v_2 + \ldots + c_n \left(\frac{\lambda_n}{\lambda_1}\right) v_n \right]$$

Premultiplying by A again and simplifying we get

$$A^2 v = \lambda_1^2 \left[c_1 v_1 + c_2 \left(\frac{\lambda_2}{\lambda_1}\right)^2 v_2 + \ldots + c_n \left(\frac{\lambda_n}{\lambda_1}\right)^2 v_n \right]$$

$$\vdots$$

$$A^k v = \lambda_1^k \left[c_1 v_1 + c_2 \left(\frac{\lambda_2}{\lambda_1}\right)^k v_2 + \ldots + c_n \left(\frac{\lambda_n}{\lambda_1}\right)^k v_n \right] \tag{3.148}$$

$$A^{k+1} v = \lambda_1^{k+1} \left[c_1 v_1 + c_2 \left(\frac{\lambda_2}{\lambda_1}\right)^{k+1} v_2 + \ldots + c_n \left(\frac{\lambda_n}{\lambda_1}\right)^{k+1} v_n \right] \tag{3.149}$$

As $k \to \infty$, the right hand sides of (3.148) and (3.149) tend to $\lambda_1^k c_1 v_1$ and $\lambda_1^{k+1} c_1 v_1$, since $|\lambda_i/\lambda_1| < 1, i = 2, 3, \ldots, n$. The vector $c_1 v_1 + c_2(\lambda_2/\lambda_1)^k v_2 + \ldots + c_n(\lambda_n/\lambda_1)^k v_n$ tends to $c_1 v_1$ which is the eigenvector corresponding to λ_1. The eigenvalue λ_1 is obtained as the ratio of the corresponding components of $A^k v$ and $A^{k+1} v$.

$$\lambda_1 = \lim_{k \to \infty} \frac{(A^{k+1} v)_r}{(A^k v)_r}, \quad r = 1, 2, \ldots, n \tag{3.150}$$

where the suffix r denotes the rth component of the vector.

If $|\lambda_2| \ll |\lambda_1|$, then faster convergence is obtained. In order to keep the roundoff error in control, we normalize (such that the largest element is unity) the vector before premultiplying by A. We can write the method as follows. Let v_0 be an arbitrary initial vector and find

$$y_{k+1} = A v_k \tag{3.151}$$

$$v_{k+1} = y_{k+1}/m_{k+1} \tag{3.152}$$

where m_{k+1} is the largest element in magnitude of y_{k+1}. Then

$$\lambda_1 = \lim_{k \to \infty} \frac{(y_{k+1})_r}{(v_k)_r}, \quad r = 1, 2, \ldots, n \tag{3.153}$$

and v_{k+1} is the required eigenvector. The initial vector is usually chosen as a vector with all components equal to unity.

Shift of Origin

Power method can be used with a shift of origin. We know that A and

$\mathbf{A} - k\mathbf{I}$ have the same set of eigenvectors and for each eigenvalue λ_i of $\mathbf{A}$ we have, for $\mathbf{A} - k\mathbf{I}$, the eigenvalue $\lambda_i - k$.

$$(\mathbf{A} - k\mathbf{I})\mathbf{v} = A\mathbf{v} - k\mathbf{v} = \lambda\mathbf{v} - k\mathbf{v} = (\lambda - k)\mathbf{v}$$

Therefore, if we subtract k from the diagonal elements of $\mathbf{A}$, then each eigenvalue is reduced by the same factor and the eigenvectors are not changed. The power method can now be used as

$$\mathbf{y}_{k+1} = (\mathbf{A} - k\mathbf{I})\mathbf{v}_k \tag{3.154}$$

$$\mathbf{v}_{k+1} = \mathbf{y}_{k+1}/m_{k+1} \tag{3.155}$$

The dominant eigenvalue is then obtained using (3.153). Regardless of the value we choose, the dominant eigenvalue of $\mathbf{A} - k\mathbf{I}$ will always be either $\lambda_1 - k$ or $\lambda_n - k$. For example, if 20, 10, 1 are the eigenvalues of a system, then if we choose, say, $k = 15$, then the new eigenvalues are 5, -5 and -14. Hence, the power method applied on the new system would give the largest eigenvalue of the new system and the corresponding eigenvector, which is the smallest eigenvalue and its eigenvector respectively of the original system. If we choose $k = \frac{1}{2}(\lambda_2 + \lambda_n)$ we have maximum rate of convergence to $\lambda_1 - k$ if we use $\mathbf{A} - k\mathbf{I}$ as the iteration matrix. If we choose $k = \frac{1}{2}(\lambda_1 + \lambda_{n-1})$, then we have maximum rate of convergence to $\lambda_n - k$. In the above example, if we choose $k = \frac{1}{2}(10 + 1) = 5.5$, then the eigenvalues of the new system are 14.5, 4.5, -4.5 and the ratios of $|\lambda_2/\lambda_1|$ and $|\lambda_3/\lambda_1|$ are approximately 0.31 while these ratios for the original system are 0.5, 0.05. All other choices of k would produce larger values for the ratio $|\lambda_2/\lambda_1|$. The convergence is faster for the new system than the original system, both the systems converging to the same eigenvalue. If we choose $k = \frac{1}{2}(20 + 10) = 15$, then the eigenvalues of the new system are 5, -5 and -14. The ratios of $|\lambda_2/\lambda_1|$ and $|\lambda_3/\lambda_1|$ for the new system are approximately 0.357. Convergence to the largest eigenvalue is maximum for this value of k and this eigenvalue corresponds to the smallest eigenvalue of the original system. Of course, the choice of k is difficult unless we know a priori an estimate of the eigenvalues.

Example 3.19 Find the largest eigenvalue in modulus and the corresponding eigenvector of the matrix

$$\mathbf{A} = \begin{bmatrix} -15 & 4 & 3 \\ 10 & -12 & 6 \\ 20 & -4 & 2 \end{bmatrix}$$

using the power method.

We start the iteration using the unit vector as the initial vector

$$\mathbf{v}_0 = [1 \quad 1 \quad 1]^T$$

we find

$$\mathbf{y}_1 = [-8 \quad 4 \quad 18]^T, \qquad \mathbf{v}_1 = [-\tfrac{4}{9} \quad \tfrac{2}{9} \quad 1]^T$$

$$\mathbf{y}_2 = [\tfrac{95}{9} \quad -\tfrac{10}{9} \quad -\tfrac{70}{9}]^T, \quad \mathbf{v}_2 = [1 \quad -\tfrac{2}{19} \quad -\tfrac{14}{19}]^T$$

. . . .

$$\mathbf{y}_7 = [-19.8674 \quad 9.7072 \quad 19.8524]^T, \quad \mathbf{v}_7 = [-1.0 \quad 0.4886 \quad 0.9992]^T$$
$$\mathbf{y}_8 = [19.952 \quad -9.868 \quad -19.956]^T, \quad \mathbf{v}_8 = [0.9998 \quad -0.494 \quad -1]^T$$
$$\mathbf{y}_9 = [-19.973 \quad 9.926 \quad 19.988]^T$$

At this step, the approximations to the largest eigenvalue in modulus are

$$|\lambda| = 19.977, \; 20.093, \; 19.988$$

If we roundoff to 3 digits, we have $|\lambda| = 20$.
The approximate eigenvector is $[0.9998 \quad -0.494 \quad -1]^T$.
The exact eigenvalue is -20 and its eigenvector is $[1 \quad -0.5 \quad -1]^T$.

Inverse Power Method

The inverse power method is usually more powerful than the power method. The method has the advantage that it can give approximation to any eigenvalue rather than only to λ_1 or λ_n. If λ is an eigenvalue of $\mathbf{A}$ and $\mathbf{v}$ is the corresponding eigenvector, then $1/\lambda$ is an eigenvalue of $\mathbf{A}^{-1}$ corresponding to the same eigenvector. Choose any nonzero eigenvector $\mathbf{y}_0 \in R^n$ and express it as a linear combination of $\mathbf{v}_1, \mathbf{v}_2, \ldots, \mathbf{v}_n$. Applying the power method on $\mathbf{A}^{-1}$, we have

$$\mathbf{z}_{k+1} = \mathbf{A}^{-1}\mathbf{y}_k \tag{3.156}$$

$$\mathbf{y}_{k+1} = \mathbf{z}_{k+1}/m_{k+1} \tag{3.157}$$

This gives an approximation to the dominant eigenvalue of $\mathbf{A}^{-1}$, that is, the smallest eigenvalue of $\mathbf{A}$ in modulus. However, one need not find $\mathbf{A}^{-1}$ to find the smallest eigenvalue (in modulus) of $\mathbf{A}$. We write equation (3.156) as

$$\mathbf{A}\mathbf{z}_{k+1} = \mathbf{y}_k \tag{3.158}$$

We find $\mathbf{z}_{k+1}$ by solving the linear system of algebraic equation (3.158). Normalization is done according to (3.157). The coefficient matrix for all iterations is the same.

If we introduce a shift of the origin we have

$$\mathbf{z}_{k+1} = (\mathbf{A} - k\mathbf{I})^{-1}\mathbf{y}_k \tag{3.159}$$

The ratio of the corresponding components tends to $1/(\lambda_i - k)$, where λ_i are the eigenvalues of $\mathbf{A}$.

$$\frac{1}{\lambda_i - k} = \lim_{k \to \infty} \frac{(\mathbf{z}_{k+1})_r}{(\mathbf{y}_k)_r} \tag{3.160}$$

By carefully selecting k, one can find an approximation to any eigenvalue of $\mathbf{A}$. For example, again, let 20, 10, 1 be the eigenvalues of $\mathbf{A}$ and choose, say, $k = 9$. Then the eigenvalues of $\mathbf{A} - k\mathbf{I}$ are 11, 1, -8 and that of $(\mathbf{A} - k\mathbf{I})^{-1}$ are 1/11, 1, $-1/8$. The power method applied to $(\mathbf{A} - k\mathbf{I})^{-1}$ gives the dominant eigenvalue of the new system which is 1. This corresponds to the middle eigenvalue of the original system.

The system (3.159) can be written as

$$(\mathbf{A} - k\mathbf{I})\mathbf{z}_{k+1} = \mathbf{y}_k \qquad (3.161)$$

We find $\mathbf{z}_{k+1}$ by solving the linear system of equations with the right hand side changing at each iteration. We normalize the vectors at each stage of iteration. It is known that this inverse iteration is the most powerful and accurate of all methods for computing eigenvectors.

3.6 Choice of a Method

If we have a system of equations which is to be solved for various right side vectors, then it is preferable to use a direct method. In this case, the inverse of the coefficient matrix can be determined, and the solution in the different cases can be obtained by matrix multiplication. For large systems of equations, iterative methods may be faster than the direct methods. For large n, Gauss elimination requires approximately $n^3/3$ operations while Jordon's method requires approximately $n^3/2$ operations. If the matrix $\mathbf{A}$ is symmetric, then the number of operations is considerably reduced to approximately $n^3/6$ operations. Gauss-Jordan method is useful in finding the inverse of $\mathbf{A}$. In this case, we require approximately n^3 operations. However, triangularization is more often used for the solution of the system of equations as well as finding the inverse of a given matrix.

Gauss-Seidal iterative scheme is twice as fast as the Jacobi method. If the value of the optimal relaxation factor can be determined, then the SOR scheme works many times faster than the Gauss-Seidel scheme. However, for large systems determination of the optimal relaxation factor is difficult. Among the direct methods, triangularization is often used in the software for the computers. When the solution of the partial differential equations is attempted by difference methods, we produce a tridiagonal system, a five band system or a band matrix system for the solution. The tridiagonal and five band systems can be solved easily by the direct methods. The solution of a tridiagonal system requires only $5n - 4$ operations.

While solving the system of equations, we encounter two types of instabilities. One is inherent instability and the other is the induced instability. Inherent instability is a property of the given problem itself. It occurs because the problem is, illconditioned. In most cases, it is possible to determine the condition number and know something about the illconditionedness of the system. A way to avoid inherent instability is a suitable reformulation of the problem. Induced instability occurs because of a wrong choice of the method. It is illustrated well by the following example of Fox.

Consider the eigenvalue problem defined by the equations

$$(10 - \lambda)x_1 + x_2 = 0$$
$$x_1 + (9 - \lambda)x_2 + x_3 = 0$$
$$\vdots$$
$$x_{19} + (-9 - \lambda)x_{20} + x_{21} = 0$$
$$x_{20} + (-10 - \lambda)x_{21} = 0$$

The largest eigenvalue is $\lambda = 10.74619418\ldots$, and it can easily be determined by any well known method. An obvious way of finding the corresponding eigenvector is to substitute λ in the above equations, take $x_1 = 1$ and find $x_2, x_3, \ldots, x_{21}$ from the first, second, ... and the penultimate equation. The last equation is automatically satisfied if all the arithmetic is exact. If we use $\lambda = 10.74619420$, then the eigenvector is produced as

r	1	...	4	...	7	...	12	...	15	...	21
x_r	1	...	10^{-1}	...	10^{-4}	...	0.4	...	10^3	...	10^{10}

while the exact eigenvector is

r	1	...	4	...	7	...	12	...	21
x_r	1	...	10^{-1}	...	10^{-3}	...	10^{-8}	...	10^{-20}

The method used is unstable even though the problem is well-conditioned. In this problem, if we take $x_{21} = 1$ and work backwards leaving the first equation, then the exact solution is obtained. However, this cannot be guaranteed in every example. A successful method may fail in slightly different circumstances.

If we require only one eigenvalue of the system, as often happens in engineering applications, the best method for use is the power or inverse power method. Givens-Householder algorithm finds the eigenvalues one at a time and if we want only one eigenvalue, then this algorithm is also appropriate. Even if we want all the eigenvalues, it requires less computation time than the cyclic Jacobi algorithm. Hence, it is widely used for arbitrary, real, symmetric matrices. The Jacobi algorithm is well suited for finding all the eigenvalues and the complete set of orthonormal eigenvectors, simultaneously. In the Jacobi method, it is better to check whether $|\sin^2 \theta + \cos^2 \theta - 1|$ is small after determining the smallest value of θ which gives the plane rotation. The disadvantage of the Jacobi method is that the elements that are annihilated by plane rotation may not necessarily remain zero during subsequent transformations of the Jacobi algorithm. The Givens' algorithm preserves the zeros in the off-diagonal elements, once they are created. For the reduction to tridiagonal form, Givens' method requires $\frac{1}{2}(n-1)(n-2)$ plane rotations. In the Householder method tridiagonalization can be completed using exactly $(n-2)$ Householder transformations. In this case a single orthogonal similarity transformation produces $n - k - 1$ zeros in the kth row and column. These transformations are more complicated than the plane rotations proposed by Givens.

PROBLEMS

1. $\mathbf{A}$ is a given non-singular $n \times n$ matrix, $\mathbf{u}$ is a given $n \times 1$ vector, $\mathbf{v}^T$ is a given $1 \times n$ vector

   (a) Show that
   $$(\mathbf{A} - \mathbf{u}\mathbf{v}^T)^{-1} = \mathbf{A}^{-1} + \alpha \mathbf{A}^{-1}\mathbf{u}\mathbf{v}^T\mathbf{A}^{-1}$$
   where α is a scalar. Determine α and give conditions for the existence of the inverse on the left hand side.

   (b) Discuss the possibility of a breakdown of the algorithm even though $\mathbf{A}$ is neither singular nor ill-conditioned, and describe how such difficulties may be overcome.

   (Stockholm Univ., Sweden, BIT 4(1964), 61)

2. Show that the matrix
$$\begin{bmatrix} 15 & 4 & -2 & 9 & 0 \\ 4 & 7 & 1 & 1 & 1 \\ -2 & 1 & 18 & 6 & 6 \\ 9 & 1 & 6 & 19 & 3 \\ 0 & 1 & 6 & 3 & 11 \end{bmatrix}$$
is positive definite.

   (Gothenburg Univ., Sweden, BIT 6(1966), 359)

3. Show that the matrix
$$\mathbf{A} = \begin{bmatrix} 2 & 4 & & & & \\ 1 & 2 & 4 & & \bigcirc & \\ & 1 & 2 & 4 & & \\ & & \cdot & \cdot & \cdot & \\ & & & 1 & 2 & 4 \\ & \bigcirc & & & 1 & 2 \end{bmatrix}$$
has real eigenvalues.

   (Lund Univ., Sweden, BIT 12(1972), 435)

4. Compute the spectral radius of the matrix $\mathbf{A}^{-1}$ where
$$\mathbf{A} = \begin{bmatrix} 0 & 1 & 0 & 0 & 0 & 0 \\ 1 & 0 & 1 & 0 & 0 & 0 \\ 0 & 1 & 0 & 1 & 0 & 0 \\ 0 & 0 & 1 & 0 & 1 & 0 \\ 0 & 0 & 0 & 1 & 0 & 1 \\ 0 & 0 & 0 & 0 & 1 & 0 \end{bmatrix}$$

   (Gothenburg Univ., Sweden, BIT 7(1967), 170)

5. Calculate $f(A) = e^A - e^{-A}$ where A is the matrix
$$\begin{bmatrix} 2 & 4 & 0 \\ 6 & 0 & 8 \\ 0 & 3 & -2 \end{bmatrix}$$
(Stockholm Univ., Sweden, BIT 18(1978), 504)

6. Which of the following matrices have spectral radius < 1?

(a) $\begin{bmatrix} 0 & \frac{1}{3} & \frac{1}{4} \\ -\frac{1}{3} & 0 & \frac{1}{2} \\ -\frac{1}{4} & -\frac{1}{2} & 0 \end{bmatrix}$, (b) $\begin{bmatrix} \cos\alpha & 0 & \sin\alpha \\ 0 & 0.5 & 0 \\ -\sin\alpha & 0 & \cos\alpha \end{bmatrix}$, where $\alpha = 5\pi/8$

(c) $\begin{bmatrix} \frac{1}{2} & \frac{1}{4} & -\frac{1}{4} \\ \frac{1}{2} & 0 & -\frac{1}{4} \\ -\frac{1}{4} & \frac{1}{2} & -\frac{1}{4} \end{bmatrix}$, (d) $\begin{bmatrix} 0.5 & -0.25 & 0.75 \\ 0.25 & 0.25 & 0.5 \\ -0.5 & 0.5 & 1.0 \end{bmatrix}$

(Uppsala Univ., Sweden, BIT 12(1972), 272)

7. The following systems of equations are given:
   (a) $4x_1 + x_2 + x_3 = 4$
   $x_1 + 4x_2 - 2x_3 = 4$
   $3x_1 + 2x_2 - 4x_3 = 6$
   (b) $x_1 + x_2 - x_3 = 2$
   $2x_1 + 3x_2 + 5x_3 = -3$
   $3x_1 + 2x_2 - 3x_3 = 6$

   Solve the above systems
   (i) by the Gauss elimination with partial pivoting
   (ii) by the decomposition method.

8. Find the inverse of the matrix
$$\begin{bmatrix} 1 & 2 & 1 \\ 2 & 3 & -1 \\ 2 & -1 & 3 \end{bmatrix}$$
by the Gauss-Jordan method.

9. Find the inverse to the following $n \times n$ matrix:
$$A = \begin{bmatrix} x & & & & & \\ x & 1 & & & & \\ x^2 & x & 1 & & O & \\ x^3 & x^2 & x & 1 & & \\ \vdots & & & & & \\ x^{n-1} & \cdots & & x^2 & x & 1 \end{bmatrix}$$
(Lund Univ., Sweden, BIT 11(1971), 338)

10. Calculate the inverse of the n-rowed square matrix **L**

$$\mathbf{L} = \begin{bmatrix} 1 & & & & \\ -\frac{1}{2} & 1 & & \bigcirc & \\ 0 & -\frac{2}{3} & 1 & & \\ & \bigcirc & 0 & -\frac{n-1}{n} & 1 \end{bmatrix}$$

(Lund Univ., Sweden, BIT 10(1970), 515)

11. Find the Cholesky factorization of

$$\begin{bmatrix} 1 & -1 & 0 & 0 & 0 \\ -1 & 2 & -1 & 0 & 0 \\ 0 & -1 & 2 & -1 & 0 \\ 0 & 0 & -1 & 2 & -1 \\ 0 & 0 & 0 & -1 & 2 \end{bmatrix}$$

(Oslo Univ., Norway, BIT 20(1980), 529)

12. Find the inverse of the matrix

$$\begin{bmatrix} 2 & 3 & -1 \\ 3 & 1 & 2 \\ -1 & 2 & -1 \end{bmatrix}$$

by the Choleski method.

13. Find the inverse of the matrix

$$\begin{bmatrix} 2 & 1 & 0 & 0 \\ 1 & 2 & 1 & 0 \\ 0 & 1 & 2 & 1 \\ 0 & 0 & 1 & 2 \end{bmatrix}$$

by the partition method.

14. Suppose the system of linear equations

Mx = y

is given. Suppose the system can be partitioned in the following way

$$\mathbf{M} = \begin{bmatrix} A_1 & B_1 & 0 & 0 \\ B_1 & A_2 & B_2 & 0 \\ 0 & B_2 & A_3 & B_3 \\ 0 & 0 & B_3 & A_4 \end{bmatrix}, \quad \mathbf{x} = \begin{bmatrix} x_1 \\ x_2 \\ x_3 \\ x_4 \end{bmatrix} \quad \text{and} \quad \mathbf{y} = \begin{bmatrix} y_1 \\ y_2 \\ y_3 \\ y_4 \end{bmatrix}$$

A_i and B_i are $p \times p$ matrices and x_i and y_i are column vectors ($p \times 1$). Suppose that A_i, $i = 1, 2, 3, 4$ are strictly diagonally dominant and tridiagonal. In that case, systems of the type $A_i \mathbf{v} = \mathbf{w}$ are easily solved. For system **Mx = y** we therefore propose the following iterative method

$$A_1 x_1^{(n+1)} = y_1 - B_1 x_2^{(n)}$$

$$A_2 x_2^{(n+1)} = y_2 - B_1 x_1^{(n)} - B_2 x_3^{(n)}$$

$$A_3 x_3^{(n+1)} = y_3 - B_2 x_2^{(n)} - B_3 x_4^{(n)}$$

$$A_4 x_4^{(n+1)} = y_4 - B_3 x_3^{(n)}$$

(i) If $p = 1$, do you recognize the method?

(ii) Show that for $p > 1$ the method converges if $\|A_i^{-1}\| < \tfrac{1}{2}$ and $\|B_i\| \leq 1$.

15. The system of equation $Ax = b$ is to be solved iteratively by
$$x_{n+1} = M x_n + b$$
Suppose $A = \begin{bmatrix} 1 & k \\ 2k & 1 \end{bmatrix}$, $k \neq \sqrt{2}/2$, k real

(a) Find a necessary and sufficient condition on k for convergence of the Jacobi method.

(b) For $k = 0.25$ determine the optimal relaxation factor w, if the system is to be solved with relaxation method.

(Lund Univ., Sweden, BIT 13(1973), 375)

16. (a) Determine the convergence factor for the Jacobi and Gauss-Seidel methods for the system

$$\begin{bmatrix} 4 & 0 & 2 \\ 0 & 5 & 2 \\ 5 & 4 & 10 \end{bmatrix} \begin{bmatrix} x_1 \\ x_2 \\ x_3 \end{bmatrix} = \begin{bmatrix} 4 \\ -3 \\ 2 \end{bmatrix}$$

(b) This system can also be solved by the relaxation method. Determine w_{opt} and write down the iteration formula exactly.

(Lund Univ., Sweden, BIT 13(1973), 493)

17. The following system of equations is given

$$4x + y + 2z = 4$$
$$3x + 5y + z = 7$$
$$x + y + 3z = 3$$

(a) Set up the Jacobi and Gauss-Seidel iterative schemes for the solution and iterate three times starting with the initial vector $x^{(0)} = 0$. Compare with the exact solution.

(b) Find the spectral radii of the iteration matrices and hence find the rate of convergence of these schemes.

18. The following system of equations is given

$$3x + 2y = 4.5$$
$$2x + 3y - z = 5$$
$$-y + 2z = -0.5$$

(a) Set up the SOR iterative scheme for the solution.

(b) Find the optimal relaxation factor and determine the rate of convergence.

(c) Using the optimal relaxation factor, iterate three times with the above scheme. Compare with the exact solution.

19. The system of equations $AX = Y$ where

$$A = \begin{bmatrix} 3 & 2 \\ 1 & 2 \end{bmatrix}, \quad X = \begin{bmatrix} x_1 \\ x_2 \end{bmatrix} \quad \text{and} \quad Y = \begin{bmatrix} 1 \\ 2 \end{bmatrix}$$

can be solved by the following iteration

$$X^{(n+1)} = X^{(n)} + \alpha(AX^{(n)} - Y)$$

$$X^{(0)} = \begin{bmatrix} 1 \\ 1 \end{bmatrix}$$

How should the parameter α be chosen to produce optimal convergence?

(Uppsala Univ., Sweden, BIT 10(1970), 228)

20. Given a system of equations $Ax = B$ where

$$A = \begin{bmatrix} 2 & -1 & 0 & 0 & 0 \\ -1 & 2 & -1 & 0 & 0 \\ 0 & -1 & 2 & -1 & 0 \\ 0 & 0 & -1 & 2 & -1 \\ 0 & 0 & 0 & -1 & 2 \end{bmatrix}, \quad B = \begin{bmatrix} 1 \\ 1 \\ 1 \\ 1 \\ 1 \end{bmatrix}$$

Prove that the matrix A has 'property A' and find the optimum value of the relaxation factor w for the method of successive over-relaxation.

(Gothenburg Univ., Sweden, BIT 8(1968), 138)

21. (a) Let $A = B - C$ where A, B and C are non-singular matrices, and set

$$Bx^{(m)} = Cx^{(m-1)} + y, \quad m = 1, 2, \ldots$$

Give a necessary and sufficient condition so that

$$\lim_{m \to \infty} x^{(m)} = A^{-1}y$$

for every choice of $x^{(0)}$.

(b) Let A be an $n \times n$ matrix with real positive elements a_{ij} fulfilling the condition

$$\sum_{j=1}^{n} a_{ij} = 1, \quad i = 1, 2, \ldots, n$$

Show that $\lambda = 1$ is an eigenvalue of the matrix A, and give the corresponding eigenvector. Then show that the spectral radius $\rho(A) \leqslant 1$.

(Lund Univ., Sweden, BIT 9(1969), 174)

22. The matrix **B** is defined by

$$\mathbf{B} = \mathbf{I} + ir\mathbf{A}^2$$

where **I** is the identity matrix, **A** is a Hermitian matrix and $i^2 = -1$. Show that $\|\mathbf{B}\| > 1$ for all real $r \neq 0$. $\|\,.\,\|$ denotes the Hilbert norm.

(Lund Univ., Sweden, BIT 9(1969), 87)

23. Let R be a $n \times n$ triangular matrix with unit diagonal elements and with the absolute value of non-diagonal elements less than or equal to 1. Determine the maximum possible value of the maximum norm $\|\mathbf{R}^{-1}\|$.

(Stockholm Univ., Sweden, BIT 8(1968), 59)

24. The $n \times n$ matrix A satisfies

$$\mathbf{A}^4 = -1.6\mathbf{A}^2 - 0.64\mathbf{I}$$

Show that $\lim_{m \to \infty} \mathbf{A}^m$ exists and determine this limit.

(Inst. Tech., Gothenburg, Sweden, BIT 11(1971), 455)

25. Consider the matrix

$$\mathbf{A} = \begin{bmatrix} 2 & -1 & -1 & 1 \\ -1 & 2 & 1 & -1 \\ -1 & 1 & 2 & -1 \\ 1 & -1 & -1 & 2 \end{bmatrix}$$

(a) Determine the spectral norm $\rho(\mathbf{A})$.

(b) Determine a vector x with $\|\mathbf{x}\|_2 = 1$ satisfying $\|\mathbf{Ax}\|_2 = \rho(\mathbf{A})$.

(Inst. Tech., Lund, Sweden, BIT 10(1970), 228)

26. Given the system of equation

$$\mathbf{Ax} = \mathbf{b}$$

where $\mathbf{A} = \begin{bmatrix} \frac{1}{2} & \frac{1}{3} & \frac{1}{4} \\ \frac{1}{3} & \frac{1}{4} & \frac{1}{5} \\ \frac{1}{4} & \frac{1}{5} & \frac{1}{6} \end{bmatrix}$

The vector **b** consists of three quantities measured with an error bounded by ϵ. Derive error bounds for

(a) the components of x

(b) the sum of the components $y = x_1 + x_2 + x_3$

(Royal Inst. Tech., Stockholm, Sweden, BIT 8(1968), 343)

27. Let
$$A(\alpha) = \begin{bmatrix} 0.1\alpha & 0.1\alpha \\ 1.0 & 1.5 \end{bmatrix}$$

Determine α such that cond $(A(\alpha))$ is minimized. Use the maximum norm.

(Uppsala Univ., Sweden, BIT 16(1976), 466)

28. Estimate the effect of a disturbance $[\epsilon_1 \quad \epsilon_2]^T$ on the right hand side of the system of equations

$$\begin{bmatrix} 1 & 2 \\ 2 & -1 \end{bmatrix} \begin{bmatrix} x_1 \\ x_2 \end{bmatrix} = \begin{bmatrix} 5 \\ 0 \end{bmatrix}$$

if $|\epsilon_1|, |\epsilon_2| \leqslant 10^{-4}$

(Uppsala Univ., Sweden, BIT 15(1975), 335)

29. (a) Prove that if the perturbed matrix $(A + \Delta A)$ is singular then

cond $(A) \geqslant \|A\|/\|\Delta A\|$

for any matrix norm consistent with some vector norm.

(b) Show that equality holds in the relation (a) for the L_2-norm for

$\Delta A = -yx^T/y^Tx, \; x = A^{-1}y$

where y is a vector for which

$\|A^{-1}\|_2 \|y\|_2 = \|A^{-1}y\|_2$

(c) Use the inequality in (a) to get a lower bound for

cond$_\infty (A) = \|A\|_\infty \|A^{-1}\|_\infty$

for the matrix

$$A = \begin{bmatrix} 1 & -1 & 1 \\ -1 & \epsilon & \epsilon \\ 1 & \epsilon & \epsilon \end{bmatrix}, \quad 0 < \epsilon < 1$$

(Stockholm Univ., Sweden, BIT 8(1968), 246)

30. Let

$$A = \begin{bmatrix} -2 & -1 & 2 \\ 2 & 1 & 0 \\ 0 & 0 & 1 \end{bmatrix} \text{ and } B = \begin{bmatrix} -1 & 1 & -1 \\ 1 & -1 & 1 \\ -1 & 1 & -1 \end{bmatrix}$$

$\lambda_i(\epsilon)$ are the eigenvalues of $A + \epsilon B$, $\epsilon \geqslant 0$. Estimate
$|\lambda_i(\epsilon) - \lambda_i(0)|$, $i = 1, 2, 3$

(Gothenburg Univ., Sweden, BIT 9(1969), 174)

31. The matrix
$$A = \begin{bmatrix} 1 & -2 & 3 \\ 6 & -13 & 18 \\ 4 & -10 & 14 \end{bmatrix}$$
is transformed to diagonal form by the matrix
$$T = \begin{bmatrix} 1 & 0 & 1 \\ 3 & 3 & 4 \\ 2 & 2 & 3 \end{bmatrix}, \quad \text{i.e. } T^{-1}AT$$

Calculate the eigenvalues and the corresponding eigenvectors of A.
(Uppsala Univ., Sweden, BIT 9(1969), 174)

32. (a) A and B are (2×2)-matrices with spectral radii $\rho(A) = 0$ and $\rho(B) = 1$. How big can $\rho(AB)$ be?

(b) Let
$$A = \begin{bmatrix} 1 & 1 \\ 1 & 1 \end{bmatrix} \quad \text{and} \quad B = \begin{bmatrix} \beta_1 & 1 \\ 0 & \beta_2 \end{bmatrix}$$
For which β_1, β_2 does $(AB)^k \to 0$ as $k \to \infty$?
(Gothenburg Univ., Sweden, BIT 9(1969), 294)

33. Show that the symmetric matrix
$$A = \begin{bmatrix} 2 & 0 & 0 & \epsilon & 0 \\ 0 & 2 & -1 & 1 & 0 \\ 0 & -1 & 1 & 2 & 1 \\ \epsilon & 1 & 2 & 1 & -1 \\ 0 & 0 & 1 & -1 & 1 \end{bmatrix}$$
has an eigenvalue λ which satisfies $|\lambda - 2| \leq |\epsilon|$.
(Uppsala Univ., Sweden, BIT 9(1969), 400)

34. Given
$$A = \begin{bmatrix} \frac{3}{2} & \frac{1}{2} \\ \frac{1}{2} & \frac{3}{2} \end{bmatrix}$$
for which values of α does the vector sequence $\{y_n\}_0^\infty$ defined by
$$y_n = (I + \alpha A + \alpha^2 A^2) y_{n-1}, \quad n = 1, 2, \ldots$$
y_0 arbitrary, converge to 0 as $n \to \infty$?
(Uppsala Univ., Sweden, BIT 14(1974), 366)

35. (a) Show that the smallest eigenvalue of the matrix
$$A = \begin{bmatrix} \epsilon^2 & \epsilon^2 & \epsilon^2 \\ \epsilon^2 & \epsilon & \epsilon \\ \epsilon^2 & \epsilon & 1 \end{bmatrix}, \quad |\epsilon| \ll 1$$
is equal to $\lambda_1 = \epsilon^2 + 0(\epsilon^{5/2})$.

(b) Determine realistic upper bound for the relative perturbation in λ_1 when $\mathbf{A}$ is perturbed by $\Delta \mathbf{A} = (\Delta a_{ij})$ where
(i) $|\Delta a_{ij}| \leqslant \delta |a_{ij}|$
(ii) $\|\Delta \mathbf{A}\|_2 \leqslant \delta \|\mathbf{A}\|_2$, $\delta \ll 1$

(Stockholm Univ., Sweden, BIT 8(1968), 246)

36. Show that the eigenvalues of the tridiagonal matrix

$$\mathbf{A} = \begin{bmatrix} a & b_1 & & & & \\ c_1 & a & b_2 & & & \\ & c_2 & a & b_3 & & \\ & & & \ddots & & \\ & & & & c_{n-1} & a \end{bmatrix}$$

satisfy the inequality

$$|\lambda - a| \leqslant 2\sqrt{\max_i |b_i| \max_i |c_i|}$$

(Uppsala Univ., Sweden, BIT 8(1968), 246)

37. Let $P_n(\lambda) = \det(\mathbf{A}_n - \lambda \mathbf{I})$ where

$$\mathbf{A}_n = \begin{bmatrix} a & 0 & \ldots & 0 & a_n \\ 0 & a & \ldots & 0 & a_{n-1} \\ \vdots & \vdots & \ldots & \vdots & \vdots \\ 0 & 0 & \ldots & a & a_2 \\ a_n & a_{n-1} & \ldots & a_2 & a_1 \end{bmatrix}, \quad n \geqslant 0$$

Prove the recurrence relation

$$P_n(\lambda) = (a - \lambda) P_{n-1}(\lambda) - a_n^2 (a - \lambda)^{n-2}, \quad P_1(\lambda) = a_1 - \lambda$$

and determine all eigenvalues of $\mathbf{A}_n$.

(Royal Inst. Tech., Stockholm, Sweden, BIT 8(1968), 343)

38. (a) What is wrong in the following computation?

$$\begin{bmatrix} 1 & 0.01 \\ 1 & 1 \end{bmatrix}^n = \left\{ \begin{bmatrix} 1 & 0 \\ 1 & 1 \end{bmatrix} + 10^{-2} \begin{bmatrix} 0 & 1 \\ 0 & 0 \end{bmatrix} \right\}^n$$

$$= \begin{bmatrix} 1 & 0 \\ 1 & 1 \end{bmatrix}^n + n \times 10^{-2} \begin{bmatrix} 1 & 0 \\ 1 & 1 \end{bmatrix}^{n-1} \begin{bmatrix} 0 & 1 \\ 0 & 0 \end{bmatrix}$$

since $\begin{bmatrix} 0 & 1 \\ 0 & 0 \end{bmatrix}^k = \begin{bmatrix} 0 & 0 \\ 0 & 0 \end{bmatrix}$ for $k \geqslant 2$.

(b) Compute $\begin{bmatrix} 1 & 0.1 \\ 0.1 & 1 \end{bmatrix}^{10}$ exactly.

(Lund Univ., Sweden, BIT 12(1972), 589)

39. Compute A^{10} where

$$A = \tfrac{1}{9}\begin{bmatrix} 4 & 1 & -8 \\ 7 & 4 & 4 \\ 4 & -8 & 1 \end{bmatrix}$$

(Uppsala Univ., Sweden, BIT 14(1974), 254)

40. Give a good upper estimate of the eigenvalues of the matrix **A** in the complex number plane. Also give an upper estimate of the matrix norm of **A**, which corresponds to the Euclidean vector norm

$$A = \begin{bmatrix} -1 & 0 & 1+2i \\ 0 & 2 & 1-i \\ 1-2i & 1+i & 0 \end{bmatrix}$$

(Uppsala Univ., Sweden, BIT 9(1969), 294)

41. Use Gerschgorin's theorem to estimate

$$|\lambda_i - \bar{\lambda}_i|, \quad i = 1, 2, 3$$

where λ_i are the eigenvalues of

$$A = \begin{bmatrix} 2 & \tfrac{3}{2} & 0 \\ \tfrac{1}{2} & 1 & 0 \\ 0 & 0 & -1 \end{bmatrix}$$

and $\bar{\lambda}_i$ are the eigenvalues of

$$\tilde{A} = A + 10^{-2}\begin{bmatrix} 1 & -1 & 1 \\ -1 & 1 & -1 \\ 1 & -1 & 1 \end{bmatrix}$$

(Lund Univ., Sweden, BIT 11(1971), 225)

42. Using Gerschgorin's theorems, find bounds for the eigenvalues λ of the real $n \times n$ matrix **A** ($n \geqslant 3$).

$$A = \begin{bmatrix} a & -1 & & & & \\ -1 & a & -1 & & \bigcirc & \\ 0 & -1 & a & -1 & & \\ & & \ddots & \ddots & \ddots & \\ & \bigcirc & & -1 & a & -1 \\ & & & & -1 & a \end{bmatrix}$$

Show that the components x_i of the eigenvector **x** obey a linear difference equation, and find all the eigenvalues and eigenvectors.

(Bergen Univ., Norway, BIT 5(1965), 214)

43. Find all the eigenvalues and eigenvectors of the matrix
$$\begin{bmatrix} 2 & 3 & 1 \\ 3 & 2 & 2 \\ 1 & 2 & 1 \end{bmatrix}$$
by the (i) Jacobi method, (ii) special cyclic Jacobi method.

44. Transform the matrix
$$\mathbf{A} = \begin{bmatrix} 1 & 2 & 2 \\ 2 & 1 & 2 \\ 2 & 2 & 1 \end{bmatrix}$$
to tri-diagonal form by Givens method. Find the eigenvector corresponding to the largest eigenvalue from the eigenvectors of the tri-diagonal matrix.
        (Uppsala Univ., Sweden, BIT 6(1966), 270)

45. Transform, using Givens method, the symmetric matrix $\mathbf{A}$, by a sequence of orthogonal transformations to tri-diagonal form. Use exact arithmetic
$$\mathbf{A} = \begin{bmatrix} 1 & \sqrt{2} & \sqrt{2} & 2 \\ \sqrt{2} & -\sqrt{2} & -1 & \sqrt{2} \\ \sqrt{2} & -1 & \sqrt{2} & \sqrt{2} \\ 2 & \sqrt{2} & \sqrt{2} & -3 \end{bmatrix}$$
        (Inst. Tech., Lund, Sweden, BIT 4(1964), 261)

46. Find all the eigenvalues of the matrix
$$\begin{bmatrix} 1 & 2 & -1 \\ 2 & 1 & 2 \\ -1 & 2 & 1 \end{bmatrix}$$
using the Householder method.

47. Find all the eigenvalues of the matrices

(i) $\begin{bmatrix} -15 & 4 & 3 \\ 10 & -12 & 6 \\ 20 & -4 & 2 \end{bmatrix}$ (ii) $\begin{bmatrix} 3 & 1 \\ 1 & 1 \end{bmatrix}$

using the Rutishauser method.

48. Find the largest eigenvalue of the matrix
$$\mathbf{A} = \begin{bmatrix} 2 & 1 & 1 & 0 \\ 1 & 1 & 0 & 1 \\ 1 & 0 & 1 & 1 \\ 0 & 1 & 1 & 2 \end{bmatrix}$$
using power method.
        (Stockholm Univ., Sweden, BIT 7(1967), 81)

49. Determine the largest eigenvalue and the corresponding eigenvector to the matrix
$$\begin{bmatrix} 4 & 1 & 0 \\ 1 & 20 & 1 \\ 0 & 1 & 4 \end{bmatrix}$$
to 3 correct decimal places using the power method.

(Royal Inst. Tech., Stockholm, Sweden, BIT 11(1971), 125)

50. Compute with an iterative method the greatest characteristic number λ of the matrix
$$\mathbf{A} = \begin{bmatrix} 0 & 0 & 1 & 1 & 0 \\ 0 & 0 & 1 & 0 & 1 \\ 1 & 1 & 0 & 0 & 1 \\ 1 & 0 & 0 & 0 & 1 \\ 0 & 1 & 1 & 1 & 0 \end{bmatrix}$$
with four correct decimal places.

(Lund Univ., Sweden, BIT 4(1964), 131)

CHAPTER 4

Interpolation and Approximation

4.1 INTRODUCTION

In this chapter we consider the problem of approximating a given function by a class of simpler functions mainly polynomials. There are two main uses of interpolation or interpolating polynomials. The first use is in reconstructing the function $f(x)$ when it is not given explicitly and only the values of $f(x)$ and/or its certain order derivatives at a set of points, called **nodes, tabular points** or **arguments** are known. The second use is to replace the function $f(x)$ by the interpolating polynomial $P(x)$ so that many common operations such as determination of roots, differentiation and integration etc. which are intended for the function $f(x)$ may be performed using $P(x)$. In approximation, we measure the deviation of the given function $f(x)$ from the approximating polynomial $P(x)$ for all values of x over a given interval $[a, b]$. We first discuss the methods to construct the interpolating polynomials $P(x)$ to a given function $f(x)$.

Definition 4.1 A polynomial $P(x)$ is called an interpolating polynomial if the values of $P(x)$ and/or its certain order derivatives coincide with those of $f(x)$ and/or its same order derivatives at one or more tabular points.

For example if the polynomial $P(x)$ is written as the Taylor's expansion, for the function $f(x)$ at a point x_0, $x_0 \in [a, b]$, in the form

$$P(x) = f(x_0) + (x - x_0)f'(x_0) + \frac{1}{2!}(x - x_0)^2 f''(x_0) + \ldots$$
$$+ \frac{1}{n!}(x - x_0)^n f^{(n)}(x_0) \qquad (4.1)$$

then $P(x)$ may be regarded as an interpolating polynomial of degree n, satisfying the conditions

$$P^{(k)}(x_0) = f^{(k)}(x_0), \quad k = 0, 1, \ldots, n \qquad (4.2)$$

The term

$$R_n = \frac{1}{(n+1)!}(x - x_0)^{n+1} f^{(n+1)}(\xi), \quad x_0 < \xi < x \qquad (4.3)$$

which has been neglected in (4.1), is called the **remainder** or the **truncation error**.

The number of terms to be included in (4.1) may be determined by the acceptable error. If this error is $\epsilon > 0$ and the series is truncated at the term $f^{(n)}(x_0)$, then

$$\frac{1}{(n+1)!} |x - x_0|^{n+1} |f^{(n+1)}(\xi)| \leq \epsilon$$

or

$$\frac{1}{(n+1)!} |x - x_0|^{n+1} M_{n+1} \leq \epsilon \qquad (4.4)$$

where $M_{n+1} = \max_{a \leq x \leq b} |f^{(n+1)}(x)|$.

For a given ϵ, (4.4) will determine n, and if n is prescribed, it will determine ϵ. When both n and ϵ are given, it will give an upper bound on $(x - x_0)$.

Example 4.1 Obtain polynomial approximation $P(x)$ to $f(x) = e^{-x}$ using the Taylor's expansion about $x_0 = 0$ and determine

(i) x when the error in $P(x)$ obtained from the first four terms only is to be less than 10^{-6} after rounding

(ii) the number of terms in the approximation to find results correct to 10^{-10} for $0 \leq x \leq 1$.

(i) From $f(x) = e^{-x}$, we have

$$f^{(r)}(x) = (-1)^r e^{-x}$$

and

$$f^{(r)}(0) = (-1)^r, \quad r = 0, 1, \ldots$$

Therefore we get from (4.1)

$$P(x) = 1 - x + \frac{x^2}{2} - \frac{x^3}{6}$$

and from (4.4)

$$x^4 M_4 < 24 \times 5 \times 10^{-7}$$

where $M_4 = \max_{0 \leq x \leq 1} |f^{(4)}(x)| = \max_{0 \leq x \leq 1} |e^{-x}| = 1$

$$x^4 < 120 \times 10^{-7}$$

or

$$x < .06$$

(ii) From (4.4) we obtain

$$\frac{1}{(n+1)!} < 5 \times 10^{-11}$$

which gives

$$n \geqslant 14$$

In general, if there are $n+1$ distinct points $a \leqslant x_0 < x_1 < x_2 \ldots < x_n \leqslant b$, then the problem of interpolation is to obtain $P(x)$ satisfying the conditions

(i) $P(x_i) = f(x_i), \quad i = 0, 1, \ldots, n$ \hfill (4.5)

or

(ii) $P(x_i) = f(x_i)$
$P'(x_i) = f'(x_i), \quad i = 0, 1, \ldots, n$ \hfill (4.6)

The derivative conditions in (4.6) may be replaced by more general derivative conditions involving higher order derivatives. The condition (4.5) gives rise to **Lagrange interpolating polynomial** and (4.6) gives rise to **Hermite** interpolating polynomial.

4.2 LAGRANGE AND NEWTON INTERPOLATIONS

We assume that we are given an interval $[a, b]$ and a function $f(x)$ which is continuous on $[a, b]$. Further we assume that we have $n+1$ distinct points $a \leqslant x_0 < x_1 < x_2 \ldots < x_{n-1} < x_n \leqslant b$ of $[a, b]$ and that the values of a function $f(x)$ are known at these points. We seek to find the polynomial

$$P(x) = a_0 + a_1 x + a_2 x^2 + \ldots + a_n x^n \tag{4.7}$$

satisfying the property (4.5), that is

$$P(x_i) = f(x_i), \quad i = 0, 1, 2, \ldots, n$$

The polynomial $P(x)$ so obtained is unique.

To prove this, we assume that there is another polynomial $P^*(x)$ which also satisfies

$$P^*(x_i) = f(x_i), \quad i = 0, 1, 2, \ldots, n \tag{4.8}$$

Consider the polynomial

$$Q(x) = P(x) - P^*(x) \tag{4.9}$$

Since $P(x)$ and $P^*(x)$ are both polynomials of degree $\leqslant n$, then $Q(x)$ is also a polynomial of degree $\leqslant n$ satisfying the conditions

$$Q(x_i) = P(x_i) - P^*(x_i) = 0, \quad i = 0, 1, 2, \ldots, n \tag{4.10}$$

Therefore $Q(x)$ is a polynomial of degree $\leqslant n$ which has $n+1$ distinct roots $x_0, x_1, x_2, \ldots, x_n$. This implies that $Q(x) \equiv 0$, because a polynomial

$Q(x)$ of degree n has exactly n roots, real or complex. Therefore, $P^*(x) \equiv P(x)$.

Thus the interpolating polynomials obtained in two different ways may be different in form, but are identical otherwise. Depending on its form the polynomial is called either the **Lagrange interpolation** or the **Newton interpolation with divided differences**.

We discuss the interpolations of various degrees.

4.2.1 Linear Interpolation

Here $n = 1$ and we want to determine a polynomial

$$P(x) = a_1 x + a_0 \tag{4.11}$$

with a_0 and a_1 being arbitrary constants, for a function $f(x)$, $x \in [x_0, x_1]$ such that

$$f(x_0) = P(x_0) = a_1 x_0 + a_0$$
$$f(x_1) = P(x_1) = a_1 x_1 + a_0 \tag{4.12}$$

Eliminating a_0 and a_1 from (4.11) and (4.12), the required linear interpolation is

$$\begin{vmatrix} P(x) & x & 1 \\ f(x_0) & x_0 & 1 \\ f(x_1) & x_1 & 1 \end{vmatrix} = 0 \tag{4.13}$$

which is shown graphically in Fig. 4.1.

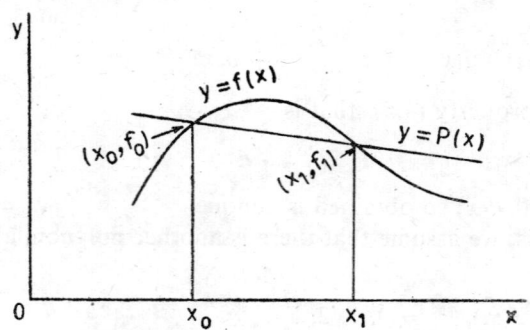

Fig. 4.1 Linear interpolation

Lagrange Interpolation

We simplify (4.13) in terms of the first column and obtain

$$P(x) = \frac{x - x_1}{x_0 - x_1} f(x_0) + \frac{x - x_0}{x_1 - x_0} f(x_1)$$
$$= l_0(x) f(x_0) + l_1(x) f(x_1) \tag{4.14}$$

where $l_0(x) = \dfrac{x - x_1}{x_0 - x_1}$, $l_1(x) = \dfrac{x - x_0}{x_1 - x_0}$.

The functions $l_0(x)$ and $l_1(x)$ are called the *Lagrange fundamental polynomials* and satisfy the conditions

$$l_0(x) + l_1(x) = 1$$
$$l_0(x_0) = 1, \quad l_0(x_1) = 0$$
$$l_1(x_0) = 0, \quad l_1(x_1) = 1$$

or

$$l_i(x_j) = \delta_{ij} = \begin{vmatrix} 1 & \text{if } i = j \\ 0 & \text{if } i \neq j \end{vmatrix} \tag{4.15}$$

The degree of the polynomials $l_0(x)$ and $l_1(x)$ is one.

The equation (4.14) is the linear Lagrange interpolating polynomial. Thus, to obtain the Lagrange interpolating polynomial, we first determine Lagrange fundamental polynomials, multiply by the corresponding function values and add them together.

Iterated Linear Interpolation

We can write (4.14) as

$$P(x) = \frac{1}{x_1 - x_0} [(x_1 - x) f(x_0) - (x_0 - x) f(x_1)]$$

$$= \frac{1}{x_1 - x_0} \begin{vmatrix} I_0(x) & x_0 - x \\ I_1(x) & x_1 - x \end{vmatrix} \tag{4.16}$$

$$= I_{0,1}(x)$$

where $I_0(x) = f(x_0)$ and $I_1(x) = f(x_1)$.

We may regard $I_0(x)$ and $I_1(x)$ as two independent zero degree interpolating polynomials to $f(x)$. It is easily verified that $I_{0,1}(x_0) = f(x_0)$ and $I_{0,1}(x_1) = f(x_1)$. The equation (4.16) is called **Aitken's** or **iterated** linear interpolating polynomial.

Newton's Divided Difference Interpolation

Next, we expand the determinant (4.13) in terms of the first row and get

$$P(x) = f(x_0) + \frac{f(x_1) - f(x_0)}{x_1 - x_0} (x - x_0)$$

$$= f(x_0) + f[x_0, x_1](x - x_0) \tag{4.17}$$

where

$$\frac{f(x_1) - f(x_0)}{x_1 - x_0} = f[x_0, x_1] \tag{4.18}$$

The ratio $f[x_0, x_1]$ is called the **first divided difference** of $f(x)$ relative to x_0 and x_1.

We may write (4.17) as

$$\frac{P(x) - f(x_0)}{x - x_0} = f[x_0, x_1] \tag{4.19}$$

The equation (4.17) or (4.19) is the linear **Newton** interpolating polynomial with divided differences.

Truncation Error Bounds

The polynomial $P(x)$ coincides with the function $f(x)$ at x_0 and x_1, and it deviates at all other points, in the interval (x_0, x_1) as shown in Fig. 4.1. This deviation is called the **truncation error** and may be written as

$$E_1(f; x) = f(x) - P(x) \tag{4.20}$$

We will now derive an expression for $E_1(f; x)$ for $x \in [x_0, x_1]$. We use the result

THEOREM (ROLLE) 4.1 *If $g(x)$ is continuous on the interval $[a, b]$ and if $g(a) = 0$, $g(b) = 0$, then there is at least one point ξ inside (a, b) for which $g'(\xi) = 0$.*

We notice that if $x = x_0$ or $x = x_1$ then $E_1(f; x) = 0$. If $x \in [a, b]$, $x \neq x_0, x_1$ be fixed, then for this x we define a function $g(t)$ as

$$g(t) = f(t) - P(t) - [f(x) - P(x)]\frac{(t - x_0)(t - x_1)}{(x - x_0)(x - x_1)} \tag{4.21}$$

It is easy to verify that $g(t) = 0$ at the three distinct points $t = x_0$, $t = x_1$, $t = x$. Differentiating (4.21) twice with respect to t, we obtain

$$g''(t) = f''(t) - \frac{2(f(x) - (Px))}{(x - x_0)(x - x_1)} \tag{4.22}$$

Now by Theorem 4.1, $g''(\xi) = 0$. Solving (4.22) for $f(x)$ we obtain

$$f(x) = P(x) + \tfrac{1}{2}(x - x_0)(x - x_1)f''(\xi) \tag{4.23}$$

where $\min(x_0, x_1, x) < \xi < \max(x_0, x_1, x)$.

Therefore, the truncation error in linear interpolation is given by

$$E_1(f; x) = \tfrac{1}{2}(x - x_0)(x - x_1)f''(\xi) \tag{4.24}$$

If we can determine a bound for $f''(x)$ in $[x_0, x_1]$, i.e.

$$|f''(x)| \leq M_2, \; x \in [x_0, x_1]$$

then

$$|f(x) - P(x)| = |\tfrac{1}{2}(x - x_0)(x - x_1)f''(\xi)|$$

$$\leq \tfrac{1}{2} \max_{x_0 \leq x \leq x_1} |(x - x_0)(x - x_1)f''(x)|$$

$$\leq \tfrac{1}{2} \max_{x_0 \leq x \leq x_1} |(x - x_0)(x - x_1)|M_2 \tag{4.25}$$

The maximum value of $|(x - x_0)(x - x_1)|$ occurs at $x = \tfrac{1}{2}(x_0 + x_1)$ and (4.25) becomes

$$|f(x) - P(x)| \leq \tfrac{1}{8}(x_1 - x_0)^2 M_2 \tag{4.26}$$

Further equation (4.26) may also be used to construct a table of values for a function $f(x)$ for equally spaced nodal points $x_i = a + ih$, $i = 0, 1, \ldots, n$, $h = (b-a)/n$; so that the maximum truncation error using the linear interpolating polynomial $P(x)$ is less than a given $\epsilon > 0$. We have

$$\frac{h^2}{8} \max_{a \leqslant x \leqslant b} |f''(x)| \leqslant \epsilon \tag{4.27}$$

Example 4.2 Using $\sin(0.1) = 0.09983$ and $\sin(0.2) = 0.19867$, find an approximate value of $\sin(0.15)$ by Lagrange interpolation. Obtain a bound on the truncation error.

We have

$$P_1(.15) = \frac{.15 - .2}{.1 - .2}(0.09983) + \frac{.15 - .1}{.2 - .1}(0.19867)$$

$$= (.5)(0.09983) + (.5)(0.19867) = 0.14925$$

The truncation error is

$$E_1(f; x) = \frac{(x-.1)(x-.2)}{2}(-\sin \xi)$$

where $.1 < \xi < .2$.

The maximum value of $|-\sin \xi|$, $\xi \in [.1, .2]$ is $\sin(0.2) = 0.19867$.
Thus

$$|E_1(f; x)| \leqslant \left|\frac{(.15-.1)(.15-.2)}{2}\right|(0.19867)$$

$$= (0.19867)(0.00125) \approx 0.00025$$

Example 4.3 Determine the stepsize h to be used in the tabulation of $f(x) = \sin x$ in the interval $[1, 3]$ so that linear interpolation will be correct to four decimal places.

$$f(x) = \sin x, \, f'(x) = \cos x, \, f''(x) = -\sin x$$

$$\max_{1 \leqslant x \leqslant 3} |-\sin x| = 1$$

Hence we obtain

$$\frac{h^2}{8} \leqslant 5 \times 10^{-5}$$

This gives $h \leqslant .02$.

4.2.2 Higher Order Interpolation

The Lagrange fundamental polynomials of degree n based on $n+1$ distinct points $a \leqslant x_0 < x_1 < x_2 \ldots < x_n \leqslant b$ and which satisfy (4.15) can be written in the form

$$l_i(x) = \frac{(x-x_0)(x-x_1)\ldots(x-x_{i-1})(x-x_{i+1})\ldots(x-x_n)}{(x_i-x_0)(x_i-x_1)\ldots(x_i-x_{i-1})(x_i-x_{i+1})\ldots(x_i-x_n)} \tag{4.28}$$

$$i = 0, 1, \ldots, n$$

An alternative form of (4.28) is given by

$$l_i(x) = \frac{w(x)}{(x - x_i)w'(x_i)}$$

where

$$w(x) = (x - x_0)(x - x_1) \ldots (x - x_n)$$

and a prime represents differentiation with respect to x.

Thus the polynomial

$$P(x) = \sum_{i=0}^{n} l_i(x) f(x_i) \tag{4.29}$$

where $l_i(x)$ are given by (4.28), is the Lagrange interpolating polynomial of degree n.

The truncation error in the Lagrange interpolation is given by

$$E_n(f; x) = f(x) - P(x)$$

Since $E_n(f; x) = 0$ at $x = x_i$, $i = 0, 1, \ldots, n$, then for $x \in [a, b]$ and $x \neq x_i$, we define a function $g(t)$ as

$$g(t) = f(t) - P(t) - [f(x) - P(x)] \frac{(t - x_0)(t - x_1) \ldots (t - x_n)}{(x - x_0)(x - x_1) \ldots (x - x_n)} \tag{4.30}$$

We observe that $g(t) = 0$ at $t = x$ and $t = x_i$, $i = 0, 1, \ldots, n$.

Differentiating (4.30) $n + 1$ times with respect to t we get

$$g^{(n+1)}(t) = f^{(n+1)}(t) - \frac{(n + 1)! \, [f(x) - P(x)]}{(x - x_0)(x - x_1) \ldots (x - x_n)} \tag{4.31}$$

Now by Theorem (4.1), $g^{(n+1)}(\xi) = 0$. Solving (4.31) for $f(x)$ we get

$$f(x) = P(x) + \frac{w(x)}{(n + 1)!} f^{(n+1)}(\xi)$$

where ξ is some point in $[x_0, x_1, \ldots, x_n, x]$. Hence, the truncation error in Lagrange interpolation is

$$E_n(f; x) = \frac{w(x)}{(n + 1)!} f^{(n+1)}(\xi) \tag{4.32}$$

Iterated Interpolation

The iterated form of the Lagrange interpolation can be written as

$$I_{0, 1, 2, \ldots, n}(x) = \frac{1}{x_n - x_{n-1}} \begin{vmatrix} I_{0, 1, \ldots, n-1}(x) & x_{n-1} - x \\ I_{0, 1, \ldots, n-2, n}(x) & x_n - x \end{vmatrix} \tag{4.33}$$

The interpolating polynomials appearing on the right side of (4.33) are any two independent $(n - 1)$th degree polynomials which could be constructed in a number of ways. In the **Aitken** method, we construct the successive iterated interpolations as given in Table 4.1.

Interpolation and Approximation   143

Table 4.1 Iterated Interpolation

x_0	$x_0 - x$	$I_0(x)$			
x_1	$x_1 - x$	$I_1(x)$	$I_{0,1}(x)$		
x_2	$x_2 - x$	$I_2(x)$	$I_{0,2}(x)$	$I_{0,1,2}(x)$	
$\vdots$	$\vdots$	$\vdots$	$\vdots$	$\vdots$	
x_{n-1}	$x_{n-1} - x$	$I_{n-1}(x)$	$I_{0,n-1}(x)$	$I_{0,1,n-1}(x)$	
x_n	$x_n - x$	$I_n(x)$	$I_{0,n}(x)$	$I_{0,1,n}(x)$	$\ldots$ $I_{0,1,2,\ldots,n}(x)$

where $I_i(x) = f(x_i)$ and $I_{0,1,n}(x) = \dfrac{1}{x_n - x_1} \begin{vmatrix} I_{0,1}(x) & x_1 - x \\ I_{0,n}(x) & x_n - x \end{vmatrix}$.

Newton's Divided Difference Interpolation

The linear Newton divided difference interpolation (4.17) is easy to generalize. We define the higher order divided differences as

$$f[x_0, x_1, x_2] = \frac{f[x_1, x_2] - f[x_0, x_1]}{x_2 - x_0}$$

$$f[x_0, x_1, x_2, \ldots, x_{k-1}, x_k] = \frac{f[x_1, x_2, \ldots, x_k] - f[x_0, x_1, \ldots, x_{k-1}]}{x_k - x_0},$$
$$k = 3, 4, \ldots, n$$

In terms of function values, the nth divided difference can be written as

$$f[x_0, x_1, x_2, \ldots, x_n] = \sum_{i=0}^{n} \frac{f(x_i)}{\prod_{\substack{j=0 \\ j \neq i}}^{n} (x_i - x_j)} \tag{4.34}$$

The divided differences may be calculated with the help of Table 4.2.

Table 4.2 Divided Difference Table

x_0	$f[x_0]$			
x_1	$f[x_1]$	$f[x_0, x_1]$		
x_2	$f[x_2]$	$f[x_1, x_2]$	$f[x_0, x_1, x_2]$	
x_3	$f[x_3]$	$f[x_2, x_3]$	$f[x_1, x_2, x_3]$	$f[x_0, x_1, x_2, x_3]$

Note that
$$f[x_0, x_1] = f[x_1, x_0]$$
$$f[x_0, x_1, x_2] = f[x_2, x_1, x_0] \text{ etc.}$$

The interpolating polynomial $P_n(x)$, interpolating at the $n + 1$ distinct points $x_0, x_1, \ldots, x_n$ can be written as

$$P_n(x) = a_0 + (x - x_0)a_1 + (x - x_0)(x - x_1)a_2 + \cdots$$
$$+ (x - x_0) \cdots (x - x_{n-1})a_n \tag{4.35}$$

Substituting, successively $x_0, x_1, \ldots, x_n$ we have

$$a_0 = f[x_0],\ a_1 = f[x_0, x_1],\ a_2 = f[x_0, x_1, x_2], \ldots$$
$$a_n = f[x_0, x_1, \ldots x_n] \tag{4.36}$$

The divided difference interpolation polynomial becomes

$$P(x) = f[x_0] + (x - x_0) f[x_0, x_1] + \cdots$$
$$+ (x - x_0) \cdots (x - x_{n-1}) f[x_0, x_1, \ldots, x_n] \tag{4.37}$$

Example 4.4 Find the unique polynomial of degree 2 or less, such that $f(0) = 1, f(1) = 3, f(3) = 55$, using
  (i) the Lagrange interpolation
  (ii) the iterated interpolation
  (iii) the Newton divided difference interpolation.

The Lagrange fundamental polynomials are given by

$$l_0(x) = \frac{(x-1)(x-3)}{(0-1)(0-3)} = \tfrac{1}{3}(x-1)(x-3)$$

$$l_1(x) = \frac{(x-0)(x-3)}{(1-0)(1-3)} = -\tfrac{1}{2}x(x-3)$$

$$l_2(x) = \frac{(x-0)(x-1)}{(3-0)(3-1)} = \tfrac{1}{6}x(x-1)$$

The Lagrange interpolating polynomial becomes

$$P_2(x) = \frac{(x-1)(x-3)}{3}(1) - \frac{x(x-3)}{2}(3) + \tfrac{1}{6}x(x-1)(55)$$
$$= 8x^2 - 6x + 1$$

The iterated interpolating polynomial becomes

$$I_{0,1}(x) = \frac{1}{(1-0)} \begin{vmatrix} 1 & 0-x \\ 3 & 1-x \end{vmatrix} = 1 + 2x$$

$$I_{0,2}(x) = \frac{1}{(3-0)} \begin{vmatrix} 1 & 0-x \\ 55 & 3-x \end{vmatrix} = 1 + 18x$$

$$I_{0,1,2}(x) = \frac{1}{3-1} \begin{vmatrix} I_{0,1}(x) & 1-x \\ I_{0,2}(x) & 3-x \end{vmatrix}$$

$$= \tfrac{1}{3}[(1+2x)(3-x) - (1-x)(1+18x)]$$
$$= 8x^2 - 6x + 1$$

The divided differences are given by

$$f[0,1] = \frac{3-1}{1-0} = 2,\quad f[1,3] = \frac{55-3}{3-1} = 26,$$

$$f[0,1,3] = \frac{26-2}{3-0} = 8$$

The Newton divided difference interpolating polynomial becomes
$$P_2(x) = f[0] + (x-0)f[0,1] + (x-0)(x-1)f[0,1,3]$$
$$= 1 + 2x + 8x(x-1)$$
$$= 8x^2 - 6x + 1$$

4.3 FINITE DIFFERENCE OPERATORS

Let the tabular points $x_0, x_1, \ldots, x_n$ be equally spaced, that is, $x_i = x_0 + ih$, $i = 0, 1, \ldots, n$. We now define the following operators:

$Ef(x_i) = f(x_i + h)$	The shift operator
$\Delta f(x_i) = f(x_i + h) - f(x_i)$	The forward-difference operator
$\nabla f(x_i) = f(x_i) - f(x_i - h)$	The backward-difference operator
$\delta f(x_i) = f\left(x_i + \frac{h}{2}\right) + f\left(x_i - \frac{h}{2}\right)$	The central-difference operator
$\mu f(x_i) = \frac{1}{2}\left[f\left(x_i + \frac{h}{2}\right) + f\left(x_i - \frac{h}{2}\right)\right]$	The averaging operator

Repeated application of the difference operators lead to the following higher order difference

$$E^n f(x_i) = f(x_i + nh) \tag{4.38}$$

$$\Delta^n f(x_i) = \Delta^{n-1} f_{i+1} - \Delta^{n-1} f_i = \sum_{k=0}^{n} (-1)^k \frac{n!}{k!(n-k)!} f_{i+n-k} \tag{4.39}$$

$$\nabla^n f(x_i) = \nabla^{n-1} f_i - \nabla^{n-1} f_{i-1} = \sum_{k=0}^{n} (-1)^k \frac{n!}{k!(n-k)!} f_{i-k} \tag{4.40}$$

$$\delta^n f(x_i) = \delta^{n-1} f_{i+1/2} - \delta^{n-1} f_{i-1/2} = \sum_{k=0}^{n} (-1)^k \frac{n!}{k!(n-k)!} f_{i+n/2-k} \tag{4.41}$$

where $f_i = f(x_i)$.

We may use Tables 4.3, 4.4 and 4.5 to calculate respectively the forward, backward and central differences of various orders.

Table 4.3 Forward Difference Table

x	$f(x)$	Δf	$\Delta^2 f$	$\Delta^3 f$
x_0	f_0			
		Δf_0		
x_1	f_1		$\Delta^2 f_0$	
		Δf_1		$\Delta^3 f_0$
x_2	f_2		$\Delta^2 f_1$	
		Δf_2		
x_3	f_3			

Note that the differences $\Delta^k f_0$ lie on a straight line sloping downward to the right.

Table 4.4 Backward Difference Table

x	$f(x)$	∇f	$\nabla^2 f$	$\nabla^3 f$
x_0	f_0			
		∇f_1		
x_1	f_1		$\nabla^2 f_2$	
		∇f_2		$\nabla^3 f_3$
x_2	f_2		$\nabla^2 f_3$	
		∇f_3		
x_3	f_3			

Note that the differences $\nabla^k f_3$ lie on a straight line sloping upward to the right.

Table 4.5 Central Difference Table

x	$f(x)$	δf	$\delta^2 f$	$\delta^3 f$	$\delta^4 f$
x_0	f_0				
		$\delta f_{1/2}$			
x_1	f_1		$\delta^2 f_1$		
		$\delta f_{3/2}$		$\delta^3 f_{3/2}$	
x_2	f_2		$\delta^2 f_2$		$\delta^4 f_2$
		$\delta f_{5/2}$		$\delta^3 f_{5/2}$	
x_3	f_3		$\delta^2 f_3$		
		$\delta f_{7/2}$			
x_4	f_4				

Note that the differences $\delta^{2k} f_2$ lie on a horizontal line. It can be easily verified that

$$\Delta f_i = \nabla f_{i+1} = \delta f_{i+1/2}$$
$$\Delta = E - 1, \quad \nabla = 1 - E^{-1}, \quad \delta = E^{1/2} - E^{-1/2} \text{ and}$$
$$\mu = \tfrac{1}{2}(E^{1/2} + E^{-1/2})$$

In Table 4.6, we list the relations among various operators.

Table 4.6 Relationship between the Operators

	E	Δ	∇	δ
E	E	$\Delta + 1$	$(1 - \nabla)^{-1}$	$1 + \tfrac{1}{2}\delta^2 + \delta\sqrt{1 + \tfrac{1}{4}\delta^2}$
Δ	$E - 1$	Δ	$(1 - \nabla)^{-1} - 1$	$\tfrac{1}{2}\delta^2 + \delta\sqrt{1 + \tfrac{1}{4}\delta^2}$
∇	$1 - E^{-1}$	$1 - (1 + \Delta)^{-1}$	∇	$-\tfrac{1}{2}\delta^2 + \delta\sqrt{1 + \tfrac{1}{4}\delta^2}$
δ	$E^{1/2} - E^{-1/2}$	$\Delta(1 + \Delta)^{-1/2}$	$\nabla(1 - \nabla)^{-1/2}$	δ
μ	$\tfrac{1}{2}(E^{1/2} + E^{-1/2})$	$(1 + \tfrac{1}{2}\Delta)(1 + \Delta)^{1/2}$	$(1 - \tfrac{1}{2}\nabla)(1 - \nabla)^{-1/2}$	$\sqrt{1 + \tfrac{1}{4}\delta^2}$

Example 4.5 Construct the difference table for the sequence of values
$$f(x) = (0, 0, 0, \epsilon, 0, 0, 0)$$
where ϵ is an error. Also show that (i) the error spreads and increases in magnitude as the order of the differences is increased, (ii) the errors in each column have binomial coefficients.

We have

$f(x)$	Δf	$\Delta^2 f$	$\Delta^3 f$	$\Delta^4 f$	$\Delta^5 f$	$\Delta^6 f$
0						
	0					
0		0				
	0		ϵ			
0		ϵ		-4ϵ		
	ϵ		-3ϵ		10ϵ	
ϵ		-2ϵ		6ϵ		-20ϵ
	$-\epsilon$		3ϵ		-10ϵ	
0		ϵ		-4ϵ		
	0		$-\epsilon$			
0		0				
	0					
0						

The above table shows how the error affects the differences. The second difference column has the errors ϵ, -2ϵ, ϵ. In the third difference column, the errors are ϵ, -3ϵ, 3ϵ, $-\epsilon$. The fourth difference column is not sufficient to show all of the expected errors ϵ, -4ϵ, 6ϵ, -4ϵ, ϵ. Thus, we may conclude that the errors in each column have binomial coefficients. Further, the maximum error occurs directly opposite the entry where the function value is in error.

The Newton divided differences in terms of forward, backward and central differences become

$$f[x_0, x_1] = \frac{f(x_1) - f(x_0)}{h} = \frac{1}{h}\Delta f_0$$

$$f[x_0, x_1, x_2] = \frac{f[x_1, x_2] - f[x_0, x_1]}{x_2 - x_0}$$

$$= \frac{\frac{1}{h}\Delta f_1 - \frac{1}{h}\Delta f_0}{2h} = \frac{1}{2!h^2}\Delta^2 f_0$$

By induction, we can show that

$$f[x_0, x_1, \ldots, x_n] = \frac{1}{n!h^n}\Delta^n f_0 \cdot \qquad (4.42)$$

Similarly

$$f[x_0, x_1] = \frac{f(x_1) - f(x_0)}{h} = \frac{1}{h} \nabla f_1$$

$$f[x_0, x_1, x_2] = \frac{f[x_1, x_2] - f[x_0, x_1]}{2h}$$

$$= \frac{1}{2h^2}(\nabla f_2 - \nabla f_1) = \frac{1}{2!h^2} \nabla^2 f_2$$

and

$$f[x_0, x_1, \ldots, x_n] = \frac{f[x_1, x_2, \ldots, x_n] - f[x_0, x_1, \ldots, x_{n-1}]}{nh}$$

$$= \frac{1}{n!h^n} \nabla^n f_n \qquad (4.43)$$

We also have

$$f[x_0, x_1] = \frac{1}{h} \delta f_{1/2}$$

$$f[x_0, x_1, x_2] = \frac{1}{2!h^2} \delta^2 f_1$$

and

$$f[x_0, x_1, \ldots, x_{2m}] = \frac{1}{(2m)!h^{2m}} \delta^{2m} f_m$$

$$f[x_0, x_1, \ldots, x_{2m+1}] = \frac{1}{(2m+1)!h^{2m+1}} \delta^{2m+1} f_{m+1/2} \qquad (4.44)$$

4.4 INTERPOLATING POLYNOMIALS USING FINITE DIFFERENCES

Gregory-Newton Forward Difference Interpolation

Substituting the divided differences from (4.42) in terms of forward differences in the Newton divided difference interpolating polynomial (4.37), we get

$$P(x) = f_0 + \frac{(x - x_0)}{h} \Delta f_0 + \frac{(x - x_0)(x - x_1)}{2!h^2} \Delta^2 f_0 + \cdots$$

$$+ \frac{(x - x_0)(x - x_1) \cdots (x - x_{n-1})}{n!h^n} \Delta^n f_0 \qquad (4.45)$$

This is called the **Gregory-Newton** forward difference interpolation. If we put $(x - x_0)/h = u$, then (4.45) and the truncation error (4.32) become

$$P(x_0 + hu) = f_0 + u\Delta f_0 + \frac{u(u-1)}{2!} \Delta^2 f_0$$

$$+ \cdots + \frac{u(u-1) \cdots (u-n+1)}{n!} \Delta^n f_0$$

Interpolation and Approximation 149

$$= \sum_{i=0}^{n} \binom{u}{i} \Delta^i f_0 \qquad (4.46)$$

and

$$E_n(f; x) = \frac{u(u-1)\ldots(u-n)}{(n+1)!} h^{n+1} f^{(n+1)}(\xi) \qquad (4.47)$$

We now give an alternate derivation of the interpolation polynomial (4.46). We have

$$f(x) = f\left(x_0 + \frac{x - x_0}{h} h\right) = f(x_0 + uh) = E^u f(x_0) \qquad (4.48)$$

which may be written as

$$f(x) = (1 + \Delta)^u f(x_0) \qquad (4.49)$$

Symbolically, expanding the right-hand side of (4.49) in binomial series we obtain

$$f(x) = f_0 + u \Delta f_0 + \frac{u(u-1)}{2!} \Delta^2 f_0$$
$$+ \ldots + \frac{u(u-1)\ldots(u-n+1)}{n!} \Delta^n f_0 + \ldots$$

Neglecting the difference $\Delta^{n+1} f_0$ and higher order differences, we get (4.46).

Gregory-Newton Backward Difference Interpolation

From the backward difference Table 4.4, it is obvious that the Newton interpolation with divided differences in terms of backward differences should be in terms of the differences at the end point x_n. Thus we may write

$$f(x) = f\left(x_n + \frac{x - x_n}{h} h\right)$$
$$= f(x_n + hu) = E^u f(x_n)$$
$$= (1 - \nabla)^{-u} f(x_n)$$
$$= f(x_n) + u \nabla f(x_n) + \frac{u(u+1)}{2!} \nabla^2 f(x_n)$$
$$+ \ldots + \frac{u(u+1)\ldots(u+n-1)}{n!} \nabla^n f(x_n) + \ldots$$
$$(4.50)$$

where $(x - x_n)/h = u$.

Neglecting the difference $\nabla^{n+1} f(x_n)$ and higher order differences, we get the interpolation polynomial

$$P(x_n + hu) = f_n + u \nabla f_n + \frac{u(u+1)}{2!} \nabla^2 f_n + \ldots$$
$$+ \frac{u(u+1)\ldots(u+n-1)}{n!} \nabla^n f_n$$

$$= \sum_{i=0}^{n}(-1)^i \binom{-u}{i} \nabla^i f_n \qquad (4.51)$$

The polynomial (4.51) is called the **Gregory-Newton** backward difference interpolation. The truncation error (4.32) becomes

$$E_n(f; x) = \frac{u(u+1)\ldots(u+n)}{(n+1)!} h^{n+1} f^{(n+1)}(\xi) \qquad (4.52)$$

Stirling and Bessel Interpolations

From the central differences table 4.5, it may be seen that there are a number of ways in which the central differences along the horizontal line can be used to construct the interpolations. Herein, we give two interpolation polynomials with central differences. For n even, we assume that the nodal points are

$$x_{-p}, x_{-(p-1)}, \ldots, x_{-1}, x_0, x_1, x_2, \ldots, x_{p-1}, x_p.$$

The **Stirling** interpolation is given by

$$\begin{aligned}P(x) = &f(x_0) + \frac{u}{2}[\delta f_{1/2} + \delta f_{-1/2}] + \frac{u^2}{2!}\delta^2 f_0 \\ &+ \frac{u(u^2-1^2)}{3!}\frac{1}{2}[\delta^3 f_{1/2} + \delta^3 f_{-1/2}] + \ldots \\ &+ \frac{u(u^2-1^2)(u^2-2^2)\ldots(u^2-(p-1)^2)}{(2p-1)!}\tfrac{1}{2}[\delta^{2p-1}f_{1/2}+\delta^{2p-1}f_{-1/2}] \\ &+ \frac{u^2(u^2-1^2)(u^2-2^2)\ldots(u^2-(p-1)^2)}{(2p)!}\delta^{2p}f_0 \qquad (4.53)\end{aligned}$$

where $u = (x - x_0)/h$.

For n odd, we take the nodal points as $x_{-p}, x_{-(p-1)}, \ldots, x_{-1}, x_0, x_1, x_2, \ldots, x_{p+1}$ and write the **Bessel** interpolation as

$$\begin{aligned}P(x) = &\tfrac{1}{2}[f_0 + f_1] + v\delta f_{1/2} + \frac{v^2 - \tfrac{1}{4}}{2!}\tfrac{1}{2}[\delta^2 f_0 + \delta^2 f_1] \\ &+ \frac{v(v^2 - \tfrac{1}{4})}{3!}\delta^3 f_{1/2} + \ldots \\ &+ \frac{(v^2 - \tfrac{1}{4})(v^2 - \tfrac{9}{4})\ldots(v^2 - (2p-1)^2/4)}{(2p)!}\tfrac{1}{2}[\delta^{2p}f_0 + \delta^{2p}f_1] \\ &+ \frac{v(v^2 - \tfrac{1}{4})(v^2 - \tfrac{9}{4})\ldots(v^2 - (2p-1)^2/4)}{(2p+1)!}\delta^{2p+1}f_{1/2} \qquad (4.54)\end{aligned}$$

where $v = u - \tfrac{1}{2}$.

The Newton-Gregory interpolations with forward and backward differences are used if the interpolation is desired near the beginning and end respectively, of a table of values. The Stirling interpolation is used for calculations between $x_0 - \tfrac{1}{4}h$ and $x_0 + \tfrac{1}{4}h$, and also when the first neglected highest difference is of odd order. The Bessel interpolation is used if the calculations are performed between $x_0 + h/4$ and $x_1 - h/4$, and also when the first neglected highest difference is of even order. It may also be noted that when the same number of tabular points are used, all interpolating polynomials are identical. If the entire data is not to be used, we can choose a suitable initial point, so that u is small and terms in the interpolation formulas reduce fast.

Example 4.6 For linear interpolation, in the case of equispaced tabular data, show that the error does not exceed 1/8 of the second difference.
The error in the case of linear interpolation is given by

$$|E_1| \approx \left|\frac{(x-x_0)(x-x_1)}{2!}\right| \left|\frac{\Delta^2 f_0}{h^2}\right|$$

$$\leqslant \max_{x_0 \leqslant x \leqslant x_0+h} |(x-x_0)(x-x_0-h)| \left|\frac{\Delta^2 f_0}{2h^2}\right|$$

The maximum of $(x-x_0)(x-x_0-h)$ occurs at $x = x_0 + h/2$ and this gives

$$|E| \leqslant \tfrac{1}{8}|\Delta^2 f_0|.$$

Example 4.7 For the following data, calculate the differences and obtain the forward and backward difference polynomials. Interpolate at $x = 0.25$ and $x = 0.35$.

x	0.1	0.2	0.3	0.4	0.5
$f(x)$	1.40	1.56	1.76	2.00	2.28

The difference table is obtained as

0.1	1.40				
		0.16			
0.2	1.56		0.04		
		0.20		0.0	
0.3	1.76		0.04		0.0
		0.24		0.0	
0.4	2.00		0.04		
		0.28			
0.5	2.28				

The forward difference polynomial is given by

$$P(x) = 1.4 + (x-0.1)\frac{0.16}{0.1} + \frac{(x-0.1)(x-0.2)}{2}\frac{0.04}{0.01}$$

$$= 2x^2 + x + 1.28$$

The backward difference polynomial is obtained as

$$P(x) = 2.28 + (x-0.5)\frac{0.28}{0.1} + \frac{(x-0.5)(x-0.4)}{2}\frac{0.04}{0.01}$$

$$= 2x^2 + x + 1.28.$$

Both the polynomials are identical and we obtain $f(0.25) = 1.655$ and $f(0.35) = 1.875$.

If we use only first differences in (4.45) we get with $x_0 = 0.1$

$$P(x) = f(x_0) + (x - x_0)\frac{\Delta f_0}{h}$$

$$P(0.25) = 1.40 + \frac{0.15 \times 0.16}{0.1} = 1.64$$

Similarly using only the first difference in (4.50) we get with $x_n = 0.5$

$$P(x) = f(x_n) + (x - x_n)\frac{\nabla f_n}{h}$$

$$P(0.35) = 2.28 + \frac{(-0.15)(0.38)}{0.1} = 1.86$$

However, if we write

$$x = 0.25 = 0.2 + (0.5)(.1)$$

and use the initial point as $x_0 = 0.2$, $u = 0.5$ and keeping only the first differences in (4.46), we get

$$P(0.25) = 1.56 + (0.5)(0.2) = 1.66$$

Again, we write

$$x = 0.35 = 0.4 - (0.5)(0.1)$$

and use the initial point as $x_n = 0.4$. Keeping only the first differences in (4.50), we obtain

$$P(0.35) = 2.00 + \frac{(-0.05)(0.24)}{0.1} = 1.88$$

Thus when the degree of polynomial is fixed, a judicious choice of the initial point may improve the result considerably.

Example 4.8 Determine the stepsize that can be used in the tabulation of $f(x) = \sin x$ in the interval $[0, \pi/4]$ at equally spaced nodal points so that the truncation error of the quadratic interpolation is less than 5×10^{-8}.

Let x_{i-1}, x_i, x_{i+1} denote three equispaced points with step-size h. Since $|f'''(x)| \leq 1$, $x \in [0, \pi/4]$, the truncation error of the quadratic Lagrange interpolation is bounded by

$$|E_2(f; x)| \leq \tfrac{1}{6} |(x - x_{i-1})(x - x_i)(x - x_{i+1})|$$

where $x_{i-1} \leq x \leq x_{i+1}$.

In order to find the maximum value of this expression, we substitute $t = (x - x_i)/h$ and determine the maximum of $|(t - 1)t(t + 1)|h^3/6$, $-1 \leq t \leq 1$. We obtain

$$\max_{-1 \leq t \leq 1} \frac{|(t-1)t(t+1)|}{6} h^3 = \frac{h^3}{9\sqrt{3}}$$

Thus we have

$$|E_2(f;x)| \leq \frac{h^3}{9\sqrt{3}}$$

The stepsize h is given by

$$\frac{h^3}{9\sqrt{3}} \leq 5 \times 10^{-8}$$

or

$$h \approx .009$$

4.5 HERMITE INTERPOLATION

The Hermite interpolating polynomial interpolates not only the function $f(x)$ but also its (certain order) derivatives at a given set of tabular points. The simple interpolating conditions are given in (4.6). We now give an explicit expression for the interpolating polynomial satisfying (4.6), that is

$$P(x_i) = f(x_i)$$
$$P'(x_i) = f'(x_i), \quad i = 0, 1, \ldots, n$$

Since there are $2n + 2$ conditions to be satisfied, $P(x)$ must be a polynomial of degree $\leq 2n + 1$. The required polynomial may be written as

$$P(x) = \sum_{i=0}^{n} A_i(x) f(x_i) + \sum_{i=0}^{n} B_i(x) f'(x_i) \qquad (4.55)$$

where $A_i(x)$ and $B_i(x)$ are polynomials of degree $\leq 2n + 1$ and satisfy

(i) $A_i(x_j) = \begin{cases} 0, & i \neq j \\ 1, & i = j \end{cases}$

(ii) $A_i'(x_j) = 0$ for all i and j

(iii) $B_i(x_j) = 0$ for all i and j

(iv) $B_i'(x_j) = \begin{cases} 0, & i \neq j \\ 1, & i = j \end{cases}$ \qquad (4.56)

Using the Lagrange fundamental polynomials $l_i(x)$, we write

$$A_i(x) = \gamma_i(x) l_i^2(x)$$
$$B_i(x) = \delta_i(x) l_i^2(x) \qquad (4.57)$$

Since $l_i^2(x)$ is a polynomial of degree $2n$, $\gamma_i(x)$ and $\delta_i(x)$ must be linear polynomials. Let

$$\gamma_i(x) = a_i x + b_i$$
$$\delta_i(x) = c_i x + d_i \qquad (4.58)$$

Using the conditions (4.56), we obtain

$$a_i = -2l_i'(x_i), \quad b_i = 1 + 2x_i l_i'(x_i)$$
$$c_i = 1 \quad \text{and} \quad d_i = -x_i \qquad (4.59)$$

154 *Numerical Methods*

Substituting (4.59) into (4.58) and using (4.57), the equation (4.55) becomes

$$P(x) = \sum_{i=0}^{n} [1 - 2(x - x_i)l_i'(x_i)]l_i^2(x) f(x_i)$$

$$+ \sum_{i=0}^{n} (x - x_i)l_i^2(x) f'(x_i) \tag{4.60}$$

which is called the **Hermite interpolating polynomial**. It is easy to verify that

$$l_i'(x_i) = \frac{w''(x_i)}{2w'(x_i)}$$

The truncation error associated with (4.60) can be written as

$$E_{2n+1}(f\,;\,x) = \frac{w^2(x)}{(2n+2)!} f^{(2n+2)}(\xi), \quad x_0 < \xi < x_n \tag{4.61}$$

Example 4.9 Given the following values of $f(x)$ and $f'(x)$

x	$f(x)$	$f'(x)$
-1	1	-5
0	1	1
1	3	7

estimate the values of $f(-0.5)$ and $f(0.5)$ using the Hermite interpolation. The exact values are $f(-0.5) = 33/64$ and $f(0.5) = 97/64$.
Here $n = 2$; $x_0 = -1$, $x_1 = 0$, $x_2 = 1$.

$$P(x) = \sum_{i=0}^{2} A_i(x) f(x_i) + \sum_{i=0}^{2} B_i(x) f'(x_i)$$

$$A_0(x) = [1 - 2(x - x_0)l_0'(x_0)]l_0^2(x)$$

$$A_1(x) = [1 - 2(x - x_1)l_1'(x_1)]l_1^2(x)$$

$$A_2(x) = [1 - 2(x - x_2)l_2'(x_2)]l_2^2(x)$$

$$B_0(x) = (x - x_0)l_0^2(x)$$

$$B_1(x) = (x - x_1)l_1^2(x)$$

$$B_2(x) = (x - x_2)l_2^2(x)$$

$$l_0(x) = \frac{(x - 0)(x - 1)}{(-1 - 0)(-1 - 1)} = \frac{x(x - 1)}{2};\ l_0'(-1) = -\frac{3}{2}$$

$$l_1(x) = \frac{(x + 1)(x - 1)}{(0 + 1)(0 - 1)} = -(x^2 - 1),\ l_1'(0) = 0$$

$$l_2(x) = \frac{(x + 1)(x - 0)}{(1 + 1)(1 - 0)} = \frac{x(x + 1)}{2};\ l_2'(1) = \frac{3}{2}$$

$$A_0(x) = [1 + 3(x + 1)]\frac{x^2(x - 1)^2}{4} = \tfrac{1}{4}(3x^5 - 2x^4 - 5x^3 + 4x^2)$$

$$A_1(x) = [1 - 2(x - 0)(0)](x^2 - 1)^2 = x^4 - 2x^2 + 1$$

$$A_2(x) = [1 - 3(x - 1)]\frac{x^2(x + 1)^2}{4} = \tfrac{1}{4}(-3x^5 - 2x^4 + 5x^3 + 4x^2)$$

$$B_0(x) = \frac{(x + 1)x^2(x - 1)^2}{4} = \tfrac{1}{4}(x^5 - x^4 - x^3 + x^2)$$

$$B_1(x) = x(x^2 - 1)^2 = x^5 - 2x^3 + x$$

$$B_2(x) = \frac{(x - 1)x^2(x + 1)^2}{4} = \tfrac{1}{4}(x^5 + x^4 - x^3 - x^2)$$

Thus we obtain

$$\begin{aligned}P(x) = &\tfrac{1}{4}(3x^5 - 2x^4 - 5x^3 + 4x^2)(1) + (x^4 - 2x^2 + 1)(1) \\ &+ \tfrac{1}{4}(-3x^5 - 2x^4 + 5x^3 + 4x^2)(3) \\ &+ \tfrac{1}{4}(x^5 - x^4 - x^3 + x^2)(-5) + (x^5 - 2x^3 + x)(1) \\ &+ \tfrac{1}{4}(x^5 + x^4 - x^3 - x^2)(7) \\ = &\, 2x^4 - x^2 + x + 1\end{aligned}$$

Substituting $x = -0.5$ and 0.5, we get

$$f(-0.5) \approx \tfrac{3}{8}, \quad \text{exact value is } \tfrac{33}{64}$$

$$f(0.5) \approx \tfrac{11}{8}, \quad \text{exact value is } \tfrac{97}{64}$$

4.6 PIECEWISE AND SPLINE INTERPOLATION

To obtain reasonably accurate results using interpolation, we may have to use polynomials of high degrees. With polynomials of high degrees, not only the computation becomes costly, but also computed results become unreliable because of roundoff errors. In order to keep the degree of interpolating polynomial small and also to achieve accurate results, we use **piecewise interpolation**. We subdivide the given interval $[a, b]$ into a number of subintervals $[x_{i-1}, x_i]$, $i = 1, 2, \ldots, n$ and approximate the function by some lower degree polynomial in each subinterval.

Piecewise Linear Interpolation

We have $n + 1$ distinct nodal points and we want to determine an interpolation as shown in Fig. 4.2. The interpolation is linear in each subinterval $[x_{i-1}, x_i]$ and it agrees with the function $f(x)$ at $n + 1$ nodal points. The subintervals or the line segments are called **finite elements** in one space dimension and the nodal points are called **knots**. The interpolating polynomial is the piecewise linear polynomial. Using the linear Lagrange interpolation (4.14), we have for $x \in [x_{i-1}, x_i]$ the piecewise linear interpolation

$$P_{i,1}(x) = \frac{x - x_i}{x_{i-1} - x_i} f(x_{i-1}) + \frac{x - x_{i-1}}{x_i - x_{i-1}} f(x_i), \quad i = 1, 2, \ldots, n \quad (4.62)$$

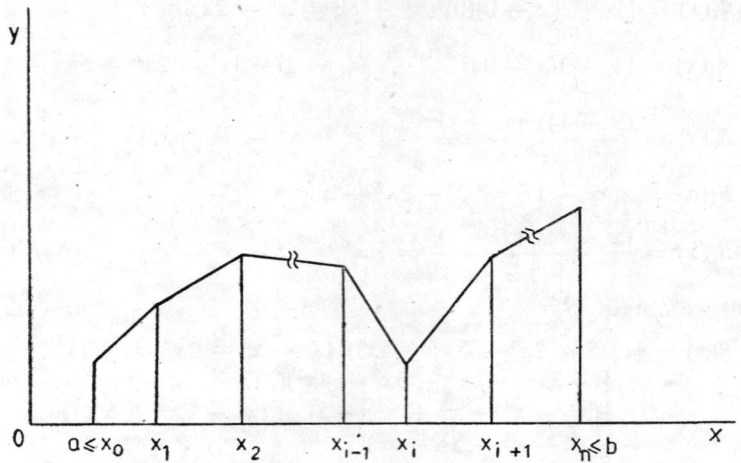

Fig. 4.2 A piecewise linear Lagrange interpolation

The interpolation polynomial

$$P(x) = \sum_{i=1}^{n} P_{i,1}(x) \tag{4.63}$$

which agrees with $f(x)$ at x_i, $i = 0, 1, \ldots, n$ and also linear in each subinterval $[x_{i-1}, x_i]$ can be written as

$$P(x) = \sum_{i=0}^{n} N_i(x) f(x_i) \tag{4.64}$$

where

$$N_i(x) = \begin{vmatrix} 0, & x \leqslant x_{i-1} \\ \dfrac{x - x_{i-1}}{x_i - x_{i-1}}, & x_{i-1} \leqslant x \leqslant x_i \\ \dfrac{x_{i+1} - x}{x_{i+1} - x_i}, & x_i \leqslant x \leqslant x_{i+1} \\ 0, & x \geqslant x_{i+1} \end{vmatrix} \tag{4.65}$$

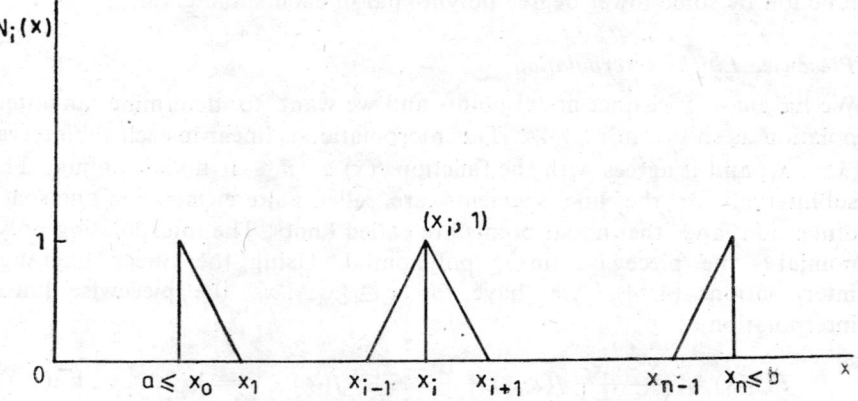

Fig. 4.3 Shape function $N_i(x)$

The function $N_i(x)$ is called a **shape** function and it is shown in Fig. 4.3. The error in piecewise linear interpolation is given by

$$f(x) - P_{i,1}(x) = \frac{1}{2!}(x - x_{i-1})(x - x_i)f''(\xi_i), \quad x_{i-1} < \xi_i < x_i \quad (4.66)$$

Piecewise Cubic Interpolation

If in each subinterval $[x_{i-1}, x_i]$, $i = 1, 2, \ldots, n$, we approximate the function $f(x)$ by a cubic polynomial $P_{i,3}(x)$, we obtain the **piecewise cubic interpolation**. The cubic polynomial on the interval $[x_{i-1}, x_i]$ can be obtained by using the conditions

$$P_{i,3}(x_{i-1}) = f_{i-1}, \quad P_{i,3}(x_i) = f_i$$
$$P'_{i,3}(x_{i-1}) = f'_{i-1}, \quad P'_{i,3}(x_i) = f'_i \quad (4.67)$$

Since we are using Hermite type of conditions, the polynomial thus obtained is also called **piecewise cubic Hermite interpolation**. Using (4.60), we can write this polynomial in the form

$$P_{i,3}(x) = A_{i-1}(x)f_{i-1} + A_i(x)f_i + B_{i-1}(x)f'_{i-1} + B_i(x)f'_i \quad (4.68)$$

where
$$A_{i-1}(x) = \frac{(x - x_i)^2}{(x_{i-1} - x_i)^2}\left[1 + \frac{2(x_{i-1} - x)}{x_{i-1} - x_i}\right]$$

$$A_i(x) = \frac{(x - x_{i-1})^2}{(x_{i-1} - x_i)^2}\left[1 + \frac{2(x - x_i)}{x_{i-1} - x_i}\right]$$

$$B_{i-1}(x) = \frac{(x - x_{i-1})(x - x_i)^2}{(x_{i-1} - x_i)^2}$$

$$B_i(x) = \frac{(x - x_i)(x - x_{i-1})^2}{(x_{i-1} - x_i)^2} \quad (4.69)$$

The interpolation polynomial

$$P_3(x) = \sum_{i=1}^{n} P_{i,3}(x) \quad (4.70)$$

which agrees with $f(x)$ and $f'(x)$ at x_i, $i = 0, 1, \ldots, n$ and is cubic in each subinterval $[x_{i-1}, x_i]$ can be written as

$$P_3(x) = \sum_{i=0}^{n} N_i(x)f(x_i) + \sum_{i=0}^{n} H_i(x)f'(x_i) \quad (4.71)$$

where

$$N_i(x) = \begin{vmatrix} 0 & , & x \leq x_{i-1} \\ \frac{(x - x_{i-1})^2}{(x_i - x_{i-1})^2}\left[1 + \frac{2(x_i - x)}{x_{i-1} - x_i}\right], & x_{i-1} \leq x \leq x_i \\ \frac{(x - x_{i+1})^2}{(x_{i+1} - x_i)^2}\left[1 + \frac{2(x - x_i)}{x_{i-1} - x_i}\right], & x_i \leq x \leq x_{i+1} \\ 0 & , & x \geq x_{i+1} \end{vmatrix} \quad (4.72)$$

and

$$H_i(x) = \begin{vmatrix} 0 & , & x \leqslant x_{i-1} \\ \dfrac{(x - x_{i-1})^2(x - x_i)}{(x_i - x_{i-1})^2}, & x_{i-1} \leqslant x \leqslant x_i \\ \dfrac{(x - x_{i+1})(x - x_i)^2}{(x_{i+1} - x_i)^2}, & x_i \leqslant x \leqslant x_{i+1} \\ 0 & , & x \geqslant x_{i+1} \end{vmatrix} \quad (4.73)$$

It may also be noted that
$$P_{i-1,3}(x_i) = P_{i,3}(x_i) = f_i, \quad i = 1, 2, \ldots, n$$
and
$$P'_{i-1,3}(x_i) = P'_{i,3}(x_i) = f'_i, \quad i = 1, 2, \ldots, n$$

Thus $P_3(x)$ is continuously differentiable on $[a, b]$.

The error in the piecewise cubic Hermite interpolation is given by
$$f(x) - P_{i,3}(x) = \frac{1}{4!}(x - x_{i-1})^2(x - x_i)^2 f^{(4)}(\xi_i), \ x_{i-1} < \xi_i < x_i \quad (4.74)$$

Example 4.10 Using the following values of $f(x)$ and $f'(x)$

x	$f(x)$	$f'(x)$
-1	1	-5
0	1	1
1	3	7

estimate the values of $f(-0.5)$ and $f(0.5)$ using piecewise cubic Hermite interpolation.

Here
$$x_{i-1} = -1, \quad x_i = 0, \quad x_{i+1} = 1$$

Since $x = -0.5 \in [x_{i-1}, x_i]$, piecewise cubic Hermite interpolation becomes

$$\begin{aligned}P_3(x) &= [1 + 2(x + 1)]x^2(1) + [1 - 2(x - 0)](x + 1)^2(1) \\ &\quad + (x + 1)x^2(-5) + x(x + 1)^2(1) \\ &= -(4x^3 + 3x^2 - x - 1)\end{aligned}$$

from which we get
$$f(-0.5) \approx \tfrac{1}{4}$$

Similarly, since $x = 0.5 \in [x_i, x_{i+1}]$, piecewise cubic Hermite interpolation becomes

$$\begin{aligned}P_3(x) &= [1 + 2(x - 0)](x - 1)^2(1) + [1 - 2(x - 1)]x^2(3) \\ &\quad + x(x - 1)^2(1) + (x - 1)x^2(7) \\ &= 4x^3 - 3x^2 + x + 1\end{aligned}$$

we obtain

$$f(0.5) \approx \tfrac{5}{4}$$

Spline Interpolation

The piecewise cubic Hermite interpolation requires prior knowledge of $f'(x_i)$, $i = 0, 1, \ldots, n$. In practical problems, it is often difficult to know these values. If we continue to use $f_i = f(x_i)$, $i = 0, 1, \ldots, n$, then regardless of the particular choice of the numbers $m_i = f'(x_i)$, the resulting piecewise cubic polynomial $P_3(x)$ will interpolate $f(x)$ at $x_0, x_1, \ldots, x_n$. Also $P_3(x)$ is continuously differentiable on $[a, b]$.

The second order derivative of $P_3(x)$ exists but may not be continuous at the knots. It is possible to determine $m_0, m_1, \ldots, m_n$ in such a way that the resulting piecewise cubic interpolation is twice continuously differentiable. Such an interpolation is called **cubic spline** interpolation. In general, we have

Definition 4.2 A spline function of degree n with knots x_i, $i = 0, 1, \ldots, n$ is a function $F(x)$ with the properties

(i) On each subinterval $[x_{i-1}, x_i]$, $1 \leqslant i \leqslant n$, $F(x)$ is a polynomial of degree n.

(ii) $F(x)$ and its first $(n - 1)$ derivatives are continuous on $[a, b]$.

We shall only discuss cubic splines.

In order to satisfy the continuity of the second derivative of $F(x)$ at $x = x_i$, we differentiate the equation (4.71) twice at $x_i \pm \epsilon$, $\epsilon > 0$, $1 \leqslant i \leqslant n$ and get

$$\begin{aligned}F''(x_i + \epsilon) &= N_i''(x_i + \epsilon)f(x_i) + H_i''(x_i + \epsilon)f'(x_i) \\ &\quad + N_{i+1}''(x_i + \epsilon)f(x_{i+1}) + H_{i+1}''(x_i + \epsilon)f'(x_{i+1}) \\ &= \frac{6(f_{i+1} - f_i)}{h_{i+1}^2} - \frac{4f_i'}{h_{i+1}} - \frac{2f_{i+1}'}{h_{i+1}}\end{aligned} \quad (4.75)$$

and

$$\begin{aligned}F''(x_i - \epsilon) &= N_{i-1}''(x_i - \epsilon)f(x_{i-1}) + H_{i-1}''(x_i - \epsilon)f'(x_{i-1}) \\ &\quad + N_i''(x_i - \epsilon)f(x_i) + H_i''(x_i - \epsilon)f'(x_i) \\ &= \frac{6(f_{i-1} - f_i)}{h_i^2} + \frac{2}{h_i}f_{i-1}' + \frac{4}{h_i}f_i'\end{aligned} \quad (4.76)$$

where $h_i = x_i - x_{i-1}$.

Equating the right hand sides of (4.75) and (4.76), we obtain

$$\frac{1}{h_i}f_{i-1}' + \left(\frac{2}{h_i} + \frac{2}{h_{i+1}}\right)f_i' + \frac{1}{h_{i+1}}f_{i+1}' = -\frac{3(f_{i-1} - f_i)}{h_i^2} + \frac{3(f_{i+1} - f_i)}{h_{i+1}^2},$$

$$i = 1, 2, \ldots, n - 1 \quad (4.77)$$

160 *Numerical Methods*

These are $n-1$ equations in $n+1$ unknowns $f'_0, f'_1, \ldots, f'_n$. If f''_0 and f''_n are prescribed, then from (4.75) and (4.76) for $i=0$ and $i=n$ respectively, we obtain

$$\frac{2}{h_1} f'_0 + \frac{1}{h_1} f'_1 = \frac{3(f_1 - f_0)}{h_1^2} - \tfrac{1}{2} f''_0 \tag{4.78}$$

$$\frac{1}{h_n} f'_{n-1} + \frac{2}{h_n} f'_n = \frac{3(f_n - f_{n-1})}{h_n^2} + \tfrac{1}{2} f''_n \tag{4.79}$$

The derivatives f'_i, $i=0, 1, \ldots, n$ are determined by solving the equations (4.77)–(4.79). If f'_0 and f'_n are specified then we determine $f'_1, f'_2, \ldots, f'_{n-1}$ from the equation (4.77).

For equispaced points, the equations (4.75)–(4.77) become, respectively,

$$f'_{i-1} + 4f'_i + f'_{i+1} = \frac{3}{h}(f_{i+1} - f_{i-1}), \quad i = 1, 2, \ldots, n-1$$

$$2f'_0 + f'_1 = \frac{3}{h}(f_1 - f_0) - \frac{h}{2} f''_0$$

$$f'_{n-1} + 2f'_n = \frac{3}{h}(f_n - f_{n-1}) + \frac{h}{2} f''_n$$

where $x_i - x_{i-1} = h$, $i = 1(1)n$.

Alternatively, we may construct the cubic spline function as follows. Since $F(x)$ is to be a piecewise cubic polynomial, $F''(x)$ is a linear function of x in the interval $x_{i-1} \leqslant x \leqslant x_i$ and hence can be written as

$$F''(x) = \frac{x_i - x}{x_i - x_{i-1}} F''(x_{i-1}) + \frac{(x - x_{i-1})}{x_i - x_{i-1}} F''(x_i) \tag{4.80}$$

Integrating (4.80) two times with respect to x, we get

$$F(x) = \frac{(x_i - x)^3}{6h_i} M_{i-1} + \frac{(x - x_{i-1})^3}{6h_i} M_i + c_1 x + c_2 \tag{4.81}$$

where $M_i = F''(x_i)$ and c_1 and c_2 are constants to be determined by using the conditions $F_i(x_{i-1}) = f(x_{i-1})$ and $F_i(x_i) = f(x_i)$, we have

$$c_1 = \frac{(f_i - f_{i-1})}{h_i} - \frac{1}{6}(M_i - M_{i-1}) h_i$$

$$c_2 = \frac{(x_i f_{i-1} - x_{i-1} f_i)}{h_i} - \frac{1}{6}(x_i M_{i-1} - x_{i-1} M_i) h_i \tag{4.82}$$

Substituting the values of c_1 and c_2 into (4.81), we get

$$F(x) = \left(\frac{(x_i - x)((x_i - x)^2 - h_i^2)}{6h_i} \right) M_{i-1}$$
$$+ \left(\frac{(x - x_{i-1})((x - x_{i-1})^2 - h_i^2)}{6h_i} \right) M_i$$
$$+ \frac{1}{h_i}(x_i - x) f_{i-1} + \frac{1}{h_i}(x - x_{i-1}) f_i \tag{4.83}$$

and
$$F'(x) = -\frac{(x_i - x)^2}{2h_i} M_{i-1} + \frac{(x - x_{i-1})^2}{2h_i} M_i - \frac{(M_i - M_{i-1})h_i}{6}$$
$$+ \frac{f_i - f_{i-1}}{h_i} \qquad (4.84)$$

Now we require that the derivative $F'(x)$ be continuous at $x = x_i \pm \epsilon$ as $\epsilon \to 0$. Letting $F'(x_i - \epsilon) = F'(x_i + \epsilon)$ as $\epsilon \to 0$, we get

$$\frac{h_i}{6} M_{i-1} + \frac{h_i}{3} M_i + \frac{1}{h_i}(f_i - f_{i-1})$$
$$= -\frac{h_{i+1}}{3} M_i - \frac{h_{i+1}}{6} M_{i+1} + \frac{1}{h_{i+1}}(f_{i+1} - f_i)$$

which may be written as

$$\frac{h_i}{6} M_{i-1} + \frac{h_i + h_{i+1}}{3} M_i + \frac{h_{i+1}}{6} M_{i+1}$$
$$= \frac{1}{h_{i+1}}(f_{i+1} - f_i) - \frac{1}{h_i}(f_i - f_{i-1}), \quad i = 1, 2, \ldots, n-1 \qquad (4.85)$$

This gives a system of $n - 1$ linear equations in $n + 1$ unknowns $M_0, M_1, \ldots, M_n$. The two additional conditions may be taken in one of the following forms.

(i) $M_0 = M_n = 0$ \hfill (4.86)

(A spline satisfying these conditions is called a **natural spline**)

(ii) $M_0 = M_n, M_1 = M_{n+1}, f_0 = f_n, f_1 = f_{n+1}, h_1 = h_{n+1}$ \hfill (4.87)

(A spline satisfying these conditions is called a **Periodic** spline)

(iii) For a **non-periodic spline**, we use the conditions

$F'(a) = f'(a) = f'_0$ and $F'(b) = f'(b) = f'_n$

which give

$$2M_0 + M_1 = \frac{6}{h_1}\left(\frac{f_1 - f_0}{h_1} - f'_0\right)$$
$$M_{n-1} + 2M_n = \frac{6}{h_n}\left(f'_n - \frac{f_n - f_{n-1}}{h_n}\right) \qquad (4.88)$$

For equispaced knots $h_i = h$ for all i, the equations (4.83) and (4.85) reduce to

$$F(x) = \frac{1}{6h}[(x_i - x)^3 M_{i-1} + (x - x_{i-1})^3 M_i] + \frac{1}{h}(x_i - x)\left(f_{i-1} - \frac{h^2}{6} M_{i-1}\right)$$
$$+ \frac{1}{h}(x - x_{i-1})\left(f_i - \frac{h^2}{6} M_i\right) \qquad (4.89)$$

and

$$M_{i-1} + 4M_i + M_{i+1} = \frac{6}{h^2}(f_{i+1} - 2f_i + f_{i-1}) \qquad (4.90)$$

Example 4.11 Obtain the cubic spline approximation for the function given in the tabular form

x	0	1	2	3
$f(x)$	1	2	33	244

and
$$M(0) = 0, \quad M(3) = 0$$

Since the points are equispaced with $h = 1$, we obtain from (4.90)
$$M_{i-1} + 4M_i + M_{i+1} = 6(f_{i+1} - 2f_i + f_{i-1}), \quad i = 1, 2$$
Therefore,
$$M_0 + 4M_1 + M_2 = 6(f_2 - 2f_1 + f_0)$$
$$M_1 + 4M_2 + M_3 = 6(f_3 - 2f_2 + f_1)$$
Using $M_0 = 0$, $M_3 = 0$ and the given function values we get
$$4M_1 + M_2 = 180$$
$$M_1 + 4M_2 = 1080$$
which gives
$$M_1 = -24, \quad M_2 = 276$$

Thus, the cubic splines in corresponding intervals become:

Interval	Corresponding cubic spline
[0, 1]	$-4x^3 + 5x + 1$
[1, 2]	$50x^3 - 162x^2 + 167x - 53$
[2, 3]	$-46x^3 + 414x^2 - 985x + 715$

4.7 BIVARIATE INTERPOLATION

The problem of polynomial interpolation for functions of several independent variables is quite important. For the sake of simplicity, we shall only consider functions of two variables, the extension to higher dimensions is straightforward.

4.7.1 Lagrange Bivariate Interpolation

Let $f(x, y)$ be defined at $(m + 1)(n + 1)$ distinct points (x_i, y_j), $i = 0, 1, \ldots, m$, $j = 0, 1, \ldots, n$ and denote $f(x_i, y_j)$ by $f_{i, j}$. We want to obtain a polynomial $P(x, y)$ of degree at most m in x and n in y, such that
$$P(x_i, y_j) = f_{i, j}, \quad i = 0(1)m, \quad j = 0(1)n \qquad (4.91)$$

Using the Lagrange fundamental polynomials (4.28) of a single variable, we define
$$X_{m, i}(x) = \frac{w(x)}{(x - x_i) w'(x_i)}, \quad i = 0, 1, \ldots, m$$

$$Y_{n,j}(y) = \frac{w^*(y)}{(y - y_j)w^{*\prime}(y_j)}, \quad j = 0, 1, \ldots, n$$

where

$$w(x) = (x - x_0)(x - x_1) \ldots (x - x_m)$$
$$w^*(y) = (y - y_0)(y - y_1) \ldots (y - y_n)$$

Obviously, $X_{m,i}(x)$ and $Y_{n,j}(y)$ are polynomials of degree m in x, and n in y respectively. These polynomials also satisfy

$$X_{m,i}(x_k) = \delta_{ik}$$
$$Y_{n,j}(y_k) = \delta_{jk}$$

Thus, the polynomial which satisfies (4.91) can be written as

$$P_{m,n}(x, y) = \sum_{i=0}^{m} \sum_{j=0}^{n} X_{m,i}(x) Y_{n,j}(y) f_{i,j} \qquad (4.92)$$

and is called the **Lagrange bivariate interpolation** polynomial. This may also be interpreted as double application of the Lagrange interpolation polynomial in a single variable.

4.7.2 Newton's Bivariate Interpolation for Equispaced Points

With equispaced points, with spacing h in x and k in y, we define

$$\Delta_x f(x, y) = f(x + h, y) - f(x, y) = (E_x - 1)f(x, y)$$
$$\Delta_y f(x, y) = f(x, y + k) - f(x, y) = (E_y - 1)f(x, y)$$
$$\Delta_{xx} f(x, y) = \Delta_x f(x + h, y) - \Delta_x f(x, y)$$
$$= (E_x - 1)^2 f(x, y)$$
$$\Delta_{yy} f(x, y) = \Delta_y f(x, y + k) - \Delta_y f(x, y)$$
$$= (E_y - 1)^2 f(x, y)$$
$$\Delta_{xy} f(x, y) = \Delta_x f(x, y + k) - \Delta_x f(x, y)$$
$$= \Delta_y f(x + h, y) - \Delta_y f(x, y)$$
$$= (E_x - 1)(E_y - 1)f(x, y)$$

and so on.

Now

$$f(x_0 + mh, y_0 + nk) = E_x^m E_y^n f(x_0, y_0)$$
$$= (1 + \Delta_x)^m (1 + \Delta_y)^n f(x_0, y_0)$$
$$= \left[1 + \binom{m}{1}\Delta_x + \binom{m}{2}\Delta_{xx} + \ldots\right]\left[1 + \binom{n}{1}\Delta_y \right.$$
$$\left. + \binom{n}{2}\Delta_{yy} + \ldots\right] f(x_0, y_0)$$
$$= \left[1 + \binom{m}{1}\Delta_x + \binom{n}{1}\Delta_y + \binom{m}{2}\Delta_{xx} \right.$$
$$\left. + \binom{m}{1}\binom{n}{1}\Delta_{xy} + \binom{n}{2}\Delta_{yy} + \ldots\right] f(x_0, y_0)$$

This gives the interpolation polynomial

$$P(x, y) = f(x_0, y_0) + \left[\frac{1}{h}(x - x_0)\Delta_x + \frac{1}{k}(y - y_0)\Delta_y\right]f(x_0, y_0)$$

$$+ \frac{1}{2!}\left[\frac{1}{h^2}(x - x_0)(x - x_1)\Delta_{xx} + \frac{2}{hk}(x - x_0)(y - y_0)\Delta_{xy}\right.$$

$$\left. + \frac{1}{k^2}(y - y_0)(y - y_1)\Delta_{yy}\right]f(x_0, y_0) + \cdots \quad (4.93)$$

and is called the **Newton's bivariate interpolation polynomial**, for equispaced points.

Example 4.12 The following data for a function $f(x, y)$ is given:

y \ x	0	1
0	1	1.414214
1	1.732051	2

Find $f(0.25, 0.75)$, using linear interpolation.
The linear interpolation is given by

$$P(x, y) = f(x_0, y_0) + \frac{1}{h}(x - x_0)\Delta_x f(x_0, y_0) + \frac{1}{k}(y - y_0)\Delta_y f(x_0, y_0)$$

We have

$$\Delta_x f(x_0, y_0) = f(x_0 + h, y_0) - f(x_0, y_0)$$
$$= 1.414214 - 1 = .414214$$
$$\Delta_y f(x_0, y_0) = f(x_0, y_0 + k) - f(x_0, y_0)$$
$$= 1.732051 - 1 = .732051$$

We find with $h = k = 1$

$$P(0.25, 0.75) = 1 + .25(.414214) + .75(.732051)$$
$$= 1.652592$$

4.8 APPROXIMATION

The commonly used classes of approximating functions include polynomials, trigonometric functions, exponential functions and rational functions. However, from application view point, the polynomial functions are mostly used, although in special cases, the trigonometric, rational and other functions are also used. The existence of a polynomial function $P(x)$ which approximates any continuous function $f(x)$ on a finite interval $[a, b]$ is guaranteed by the **Weierstrass** approximation theorem which states:

THEOREM (WEIERSTRASS) 4.2 *If the function $f(x)$ is continuous on a finite interval $[a, b]$, then given any $\epsilon > 0$, there exists a $n = n(\epsilon)$ and a polynomial $P(x)$ of degree n such that*

$$|f(x) - P(x)| < \epsilon \quad \text{for all } x \in [a, b].$$

In order to determine an approximation to $f(x)$, we assume an expression of the form

$$f(x) \approx P(x, c_0, c_1, \ldots, c_n)$$
$$= c_0 \phi_0(x) + c_1 \phi_1(x) + \ldots + c_n \phi_n(x) \tag{4.94}$$

where $\phi_i(x)$, $i = 0, 1, \ldots, n$ are n appropriately chosen linearly independent functions and c_i, $i = 0, 1, \ldots, n$ are parameters to be determined. The functions $\phi_i(x)$ are also called **coordinate functions,** and are usually taken as $\phi_i(x) = x^i$, $i = 0, 1, \ldots, n$, for polynomial approximation. The error of approximation is defined as

$$E(f; c) = \| f(x) - (c_0 \phi_0(x) + \ldots + c_n \phi_n(x)) \| \tag{4.95}$$

where $\|\cdot\|$ is a well defined norm. The problem of approximation is to determine c_i, $i = 0, 1, \ldots, n$, such that this error is as small as possible in some sense. By using different norms, we obtain different types of approximations. Once a criterion (or a particular norm) is fixed, the function (out of a class of given functions) which makes this error smallest according to this criterion is called the **best approximation**. Thus the minimization of error norm (4.95) solves the problem of best approximation.

The most commonly used norms are

For Discrete Data

L^p-norm

$$\|x\| = \left(\sum_{i=1}^{n} |x_i|^p \right)^{1/p}, \quad p \geq 1 \tag{4.96}$$

where $x = \{x_i\}$ is a sequence of real or complex numbers.

Euclidean norm

$$\|x\| = \left(\sum_{i=1}^{n} |x_i|^2 \right)^{1/2} \tag{4.97}$$

which is a particular case of (4.96) for $p = 2$ and is also called *square norm* and written as $\|x\|_2$.

Uniform Norm

$$\|x\| = \max_{1 \leq j \leq n} |x_j| \tag{4.98}$$

which is a particular case of (4.96) for $p = \infty$.

For Continuous Data

If the function $f(x)$ is continuous on $[a, b]$ and $|f(x)|^p$ is integrable on $[a, b]$, then

$$\|f\| = \left(\int_a^b W(x)|f(x)|^p \, dx\right)^{1/p}, \quad p \geq 1 \tag{4.99}$$

where $W(x) > 0$ is the weight function, is called the L^p **norm**. For $p = 2$, we have the **Euclidean** or **square norm**

$$\|f\| = \left(\int_a^b W(x)f^2(x) \, dx\right)^{1/2} \tag{4.100}$$

For $p = \infty$, we have the **uniform norm**

$$\|f\| = \max_{a \leq x \leq b} |f(x)| \tag{4.101}$$

When we use the euclidean norm, we obtain the least square approximation and when uniform norm is used, we get the uniform approximation.

4.9 LEAST SQUARES APPROXIMATION

Least square approximations are most commonly used approximations for approximating a function $f(x)$ which may be given in tabular form or known explicitly over a given interval. In this case we use the euclidean norm (4.97) or (4.100). The best approximation in the least square sense is defined as that for which the constants c_i, $i = 0, 1, \ldots, n$ are determined so that the aggregate of $W(x)E^2$ over a given domain D is as small as possible, where $W(x) > 0$ is the weight function. For functions whose values are given at $N+1$ points $x_0, x_1, \ldots, x_N$, we have

$$I(c_0, c_1, \ldots, c_n) = \sum_{k=0}^{N} W(x_k)\left[f(x_k) - \sum_{i=0}^{n} c_i\phi_i(x_k)\right]^2 = \text{minimum} \tag{4.102}$$

For functions which are continuous on $[a, b]$ and are given explicitly, we have

$$I(c_0, c_1, \ldots, c_n) = \int_a^b W(x)\left[f(x) - \sum_{i=0}^{n} c_i\phi_i(x)\right]^2 dx = \text{minimum} \tag{4.103}$$

The coordinate functions $\phi_i(x)$ are usually chosen as

$$\phi_i(x) = x^i, \quad i = 0, 1, \ldots, n$$

and $W(x) = 1$. The necessary conditions for (4.102) or (4.103) to have a minimum value is that

$$\frac{\partial I}{\partial c_i} = 0, \quad i = 0, 1, \ldots, n \tag{4.104}$$

This gives a system of $n+1$ linear equations in $n+1$ unknowns $c_0, c_1, \ldots, c_n$. These equations are called **normal equations**. The normal equations for (4.102) and (4.103) become, respectively

$$\sum_{k=0}^{N} W(x_k)\left[f(x_k) - \sum_{i=0}^{n} c_i\phi_i(x_k)\right]\phi_j(x_k) = 0, \quad j = 0(1)n \tag{4.105}$$

and
$$\int_a^b W(x)\left[f(x) - \sum_{i=0}^{n} c_i\phi_i(x)\right]\phi_j(x)\,dx = 0, \quad j = 0(1)n \qquad (4.106)$$

Example 4.13 Obtain a linear polynomial approximation to the function $f(x) = x^3$ on the interval [0, 1] using the least square approximation with $W(x) = 1$.

Consider a linear polynomial
$$P(x) = a_0 x + a_1$$
where a_0 and a_1 are arbitrary parameters.

Using (4.103), we get
$$I(a_0, a_1) = \int_0^1 [x^3 - (a_0 x + a_1)]^2\,dx = \text{minimum}$$
or
$$I(a_0, a_1) = \frac{1}{7} - 2\left(\frac{a_0}{5} + \frac{a_1}{4}\right) + \frac{a_0^2}{3} + a_0 a_1 + a_1^2$$
$$= \text{minimum}$$

The necessary conditions for $I(a_0, a_1)$ to be minimum are given by
$$\frac{\partial I}{\partial a_0} = -\frac{2}{5} + \frac{2}{3}a_0 + a_1 = 0$$
$$\frac{\partial I}{\partial a_1} = -\frac{1}{2} + a_0 + 2a_1 = 0$$

whose solution is $a_0 = 9/10$ and $a_1 = -1/5$. The desired linear polynomial approximation is $P(x) = (9x - 2)/10$. The value of $I(a_0, a_1)$ is $9/700$.

If we take the linear polynomial approximation through the origin, then we get $P(x) = 3x/5$ and $I(a_0, a_1) = 16/700$. The approximations are plotted in Fig. 4.4. It may be noted that the approximating polynomial $P(x)$ may or may not have common values with $f(x)$.

Example 4.14 Obtain the least square polynomial approximation of degree one and two for $f(x) = x^{1/2}$ on [0, 1].

For $n = 1$, we have
$$I(c_0, c_1) = \int_0^1 (x^{1/2} - c_0 - c_1 x)^2\,dx = \text{minimum}$$

The normal equations are
$$\frac{\partial I}{\partial c_0} = -\int_0^1 2(x^{1/2} - c_0 - c_1 x)\,dx = -2\left(\frac{2}{3} - c_0 - \frac{c_1}{2}\right) = 0$$
$$\frac{\partial I}{\partial c_1} = -\int_0^1 2(x^{1/2} - c_0 - c_1 x)x\,dx = -2\left(\frac{2}{5} - \frac{c_0}{2} - \frac{c_1}{3}\right) = 0$$

We obtain $c_0 = 4/15$ and $c_1 = 4/5$. Thus, the first degree least squares approximation to $x^{1/2}$ on [0, 1] is
$$P(x) = 4(1 + 3x)/15$$

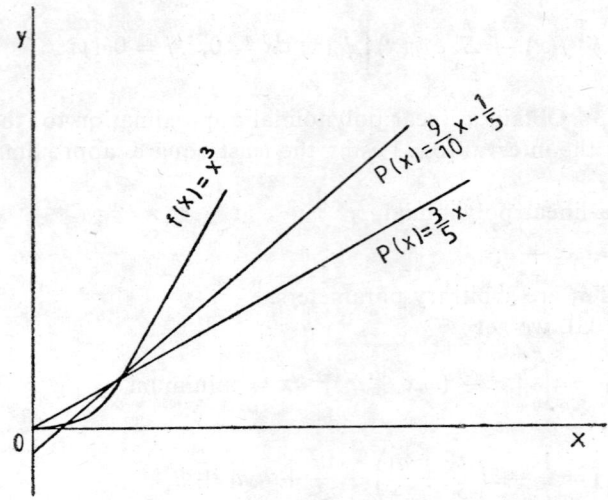

Fig. 4.4 Linear approximation

For $n = 2$, we have

$$I(c_0, c_1, c_2) = \int_0^1 (x^{1/2} - c_0 - c_1 x - c_2 x^2)^2 \, dx = \text{minimum}$$

The normal equations are

$$\frac{\partial I}{\partial c_0} = 2 \int_0^1 (x^{1/2} - c_0 - c_1 x - c_2 x^2) \, dx = 0$$

$$\frac{\partial I}{\partial c_1} = -2 \int_0^1 (x^{1/2} - c_0 - c_1 x - c_2 x^2) x \, dx = 0$$

$$\frac{\partial I}{\partial c_2} = -2 \int_0^1 (x^{1/2} - c_0 - c_1 x - c_2 x^2) x^2 \, dx = 0$$

which give

$$c_0 + \tfrac{1}{2} c_1 + \tfrac{1}{3} c_2 = \tfrac{2}{3}$$
$$\tfrac{1}{2} c_0 + \tfrac{1}{3} c_1 + \tfrac{1}{4} c_2 = \tfrac{2}{5}$$
$$\tfrac{1}{3} c_0 + \tfrac{1}{4} c_1 + \tfrac{1}{5} c_2 = \tfrac{2}{7}$$

The solution of this system is

$$c_0 = \tfrac{6}{35}, \quad c_1 = \tfrac{48}{35}, \quad c_2 = -\tfrac{20}{35}$$

The required approximation is

$$P(x) = \tfrac{1}{35}(6 + 48x - 20x^2)$$

Example 4.15 Derive the least square straight line and quadratic fits for a discrete data (x_i, f_i), $i = 0, 1, \ldots, N$.

Let $P(x) = c_0 + c_1 x$ be a straight line approximation. We have
$$\sum_{i=0}^{N} [f(x_i) - (c_0 + c_1 x_i)]^2 = \text{minimum}$$
The normal equations are
$$\frac{\partial I}{\partial c_0} = -\sum_{i=0}^{N} 2[f(x_i) - (c_0 + c_1 x_i)] = 0$$
$$\frac{\partial I}{\partial c_1} = -\sum_{i=0}^{N} 2[f(x_i) - (c_0 + c_1 x_i)] x_i = 0$$
These equations simplify to
$$c_0(N+1) + c_1 \Sigma x_i = \Sigma f(x_i)$$
$$c_0 \Sigma x_i + c_1 \Sigma x_i^2 = \Sigma x_i f(x_i)$$
For the second degree least squares approximation $P(x) = a + bx + cx^2$, the normal equations simplify to
$$a(N+1) + b\Sigma x_i + c\Sigma x_i^2 = \Sigma f(x_i)$$
$$a\Sigma x_i + b\Sigma x_i^2 + c\Sigma x_i^3 = \Sigma x_i f(x_i)$$
$$a\Sigma x_i^2 + b\Sigma x_i^3 + c\Sigma x_i^4 = \Sigma x_i^2 f(x_i).$$

Example 4.16 Find the least squares approximation of second degree for the discrete data

x	-2	-1	0	1	2
$f(x)$	15	1	1	3	19

The normal equations for fitting a second degree polynomial
$$P_2(x) = c_0 + c_1 x + c_2 x^2$$
are
$$5c_0 + 10c_2 = 39$$
$$10c_1 = 10$$
$$10c_0 + 34c_2 = 140$$
The solution of this system is
$$c_0 = -\frac{37}{35}, \quad c_1 = 1, \quad c_2 = \frac{31}{7}$$
The required approximation is
$$P_2(x) = \frac{1}{35}(-37 + 35x + 155x^2)$$

For large n, the normal equations become ill-conditioned, which cause large errors in the parameters c_i, $i = 0, 1, \ldots, n$. This difficulty can be avoided, if the functions $\phi_i(x)$ are so chosen that they are orthogonal with respect to the weight function $W(x)$ over D.

Definition 4.3 A set of functions $\{\phi_i(x)\}$ is said to be orthogonal over a set of points $\{x_i\}$ with respect to the weight function $W(x)$, if

$$\sum_{i=0}^{N} W(x_i)\phi_j(x_i)\phi_k(x_i) = 0, \quad i \neq k \tag{4.107}$$

Definition 4.4 A set of functions $\{\phi_i(x)\}$ is said to be orthogonal on an interval $[a, b]$, with respect to the weight function $W(x)$, if

$$\int_a^b W(x)\phi_i(x)\phi_j(x)\,dx = 0, \quad i \neq j \tag{4.108}$$

If the functions $\phi_i(x)$, $i = 0, 1, \ldots, n$ are orthogonal, then, we obtain from (4.105)

$$\sum_{k=0}^{N} W(x_k)\phi_i^2(x_k)c_i = \sum_{k=0}^{n} W(x_k)\phi_i(x_k)f(x_k).$$

Hence,

$$c_i = \frac{\sum_{k=0}^{N} W(x_k)\phi_i(x_k)f(x_k)}{\sum_{k=0}^{N} W(x_k)\phi_i^2(x_k)}, \quad i = 0, 1, \ldots, n \tag{4.109}$$

Thus, the use of orthogonal functions as coordinate functions in (4.102), not only avoids the problem of ill-conditioning in normal equations but also determines the constants c_i, $i = 0, 1, \ldots, n$ directly.

It may be noted that not every set of linearly independent polynomials satisfies the condition of orthogonality but, it can be orthogonalized by using the following method.

Gram-Schmidt Orthogonalizing Process

Given the polynomials $\phi_i(x)$ of degree i, the polynomials $\phi_i^*(x)$ of degree i which are orthogonal over $[a, b]$ with respect to the weight function $W(x)$ can be generated recursively from the relation

$$\phi_i^*(x) = x^i - \sum_{r=0}^{i-1} a_{ir}\phi_r^*(x), \quad i = 1, 2, \ldots, n \tag{4.110}$$

where

$$a_{ir} = \frac{\int_a^b W(x)x^i\phi_r^*(x)\,dx}{\int_a^b W(x)\phi_r^{*2}(x)\,dx}$$

and $\phi_0^*(x) = 1$.

On a discrete set of points, the integral is replaced by summation.

Example 4.17 Using the Gram-Schmidt orthogonalization process, compute the first three orthogonal polynomials $P_0(x)$, $P_1(x)$, $P_2(x)$ which are orthogonal on $[0, 1]$ with respect to the weight function $W(x) = 1$. Using these polynomials, obtain the least square approximation of second degree for $f(x) = x^{1/2}$ on $[0, 1]$.

We have
$$P_0(x) = 1 = \phi_0(x)$$
$$P_1(x) = x - a_{10}\phi_0^*(x)$$
where
$$a_{10} = \frac{\int_0^1 x\, dx}{\int_0^1 dx} = \frac{1}{2}$$

Thus, we obtain
$$P_1(x) = x - \tfrac{1}{2} = \phi_1^*(x)$$

Similarly, we have
$$P_2(x) = x^2 - a_{20}\phi_0^*(x) - a_{21}\phi_1^*(x)$$
where
$$a_{20} = \frac{\int_0^1 x^2\, dx}{\int_0^1 dx} = \frac{1}{3}$$

$$a_{21} = \frac{\int_0^1 x^2(x - \tfrac{1}{2})\, dx}{\int_0^1 (x - \tfrac{1}{2})^2\, dx} = 1$$

We find
$$P_2(x) = x^2 - x + \tfrac{1}{6}$$

Using these polynomials, we have for $n = 2$,
$$I(c_0, c_1, c_2) = \int_0^1 [x^{1/2} - c_0 P_0(x) - c_1 P_1(x) - c_2 P_2(x)]^2\, dx = \text{minimum}$$

$$\frac{\partial I}{\partial c_0} = -2\int_0^1 [x^{1/2} - c_0 P_0 - c_1 P_1 - c_2 P_2] P_0\, dx = 0$$

$$\frac{\partial I}{\partial c_1} = -2\int_0^1 [x^{1/2} - c_0 P_0 - c_1 P_1 - c_2 P_2] P_1\, dx = 0$$

and
$$\frac{\partial I}{\partial c_2} = -2\int_0^1 [x^{1/2} - c_0 P_0 - c_1 P_1 - c_2 P_2] P_2\, dx = 0$$

Using the orthogonality conditions, we obtain
$$c_0 = \frac{\int_0^1 x^{1/2} P_0(x)\, dx}{\int_0^1 P_0^2(x)\, dx} = \frac{2}{3}$$

$$c_1 = \frac{\int_0^1 x^{1/2} P_1(x)\, dx}{\int_0^1 P_1^2(x)\, dx} = \frac{4}{5}$$

$$c_2 = \frac{\int_0^1 x^{1/2} P_2(x)\, dx}{\int_0^1 P_2^2(x)\, dx} = -\frac{4}{7}$$

The required least squares approximation is

$$y(x) = \tfrac{2}{3} P_0(x) + \tfrac{4}{5} P_1(x) - \tfrac{4}{7} P_2(x)$$
$$= \tfrac{2}{3} + \tfrac{4}{5}(x - \tfrac{1}{2}) - \tfrac{4}{7}(x^2 - x + \tfrac{1}{6}) = \tfrac{1}{35}(6 + 48x - 20x^2)$$

Even though the Gram-Schmidt orthogonalization process is available to find a set of orthogonal polynomials, often we use the well known orthogonal polynomials, **Legendre** and **Chebyshev** polynomials, as the coordinate functions in (4.102). We now define these polynomials and give their properties.

Legendre Polynomials

The Legendre polynomials $P_n(x)$ defined on $[-1, 1]$ are given by

$$P_n(x) = \sum_{m=0}^{M} (-1)^m \frac{(2n - 2m)! x^{n-2m}}{2^n m!(n - m)!(n - 2m)!} \quad (4.111)$$

where $M = n/2$ or $(n - 1)/2$ whichever is an integer. In particular

$$P_0(x) = 1, \qquad P_1(x) = x$$
$$P_2(x) = \tfrac{1}{2}(3x^2 - 1), \qquad P_3(x) = \tfrac{1}{2}(5x^3 - 3x)$$
$$P_4(x) = \tfrac{1}{8}(35x^4 - 30x^2 + 3), \qquad P_5(x) = \tfrac{1}{8}(63x^5 - 70x^3 + 15x)$$

The Legendre polynomials satisfy the differential equation

$$(1 - x^2)y'' - 2xy' + n(n + 1)y = 0 \quad (4.112)$$

The Legendre polynomials satisfy the recurrence relation

$$(n + 1)P_{n+1}(x) = (2n + 1)x P_n(x) - n P_{n-1}(x) \quad (4.113)$$

and possess the following properties:

(i) $P_n(x)$ is an even polynomial if n is even and an odd polynomial if n is odd.

(ii) $P_n(x)$ are orthogonal polynomials and satisfy

$$\int_{-1}^{1} P_m(x) P_n(x)\, dx = 0, \qquad m \neq n$$
$$= \frac{2}{2n + 1}, \qquad m = n \quad (4.114)$$

(iii) $P_n(-x) = (-1)^n P_n(x)$ \hfill (4.115)

Chebyshev Polynomials

The Chebyshev polynomials of the first kind $T_n(x)$ defined on $[-1, 1]$ are given by

$$T_n(x) = \cos(n \cos^{-1} x) = \cos n\theta \tag{4.116}$$

where $\theta = \cos^{-1} x$ or $x = \cos \theta$.

These polynomials satisfy the differential equation

$$(1 - x^2)y'' - xy' + n^2 y = 0 \tag{4.117}$$

One independent solution of (4.117) gives $T_n(x)$ and the second independent solution gives Chebyshev polynomials of the second kind $U_n(x)$

$$U_n(x) = \sin[(n + 1)\cos^{-1} x] \tag{4.118}$$

The Chebyshev polynomials $T_n(x)$ satisfy the recurrence relation

$$T_{n+1}(x) = 2xT_n(x) - T_{n-1}(x) \tag{4.119}$$
$$T_0(x) = 1, \quad T_1(x) = x$$

Thus, we have

$$T_0(x) = 1 \qquad\qquad 1 = T_0(x)$$
$$T_1(x) = x \qquad\qquad x = T_1(x)$$
$$T_2(x) = 2x^2 - 1 \qquad\qquad x^2 = \tfrac{1}{2}(T_2(x) + T_0(x))$$
$$T_3(x) = 4x^3 - 3x \qquad\qquad x^3 = \frac{1}{2^2}(T_3(x) + 3T_1(x))$$
$$T_4(x) = 8x^4 - 8x^2 + 1 \qquad\qquad x^4 = \frac{1}{2^3}(T_4(x) + 4T_2(x) + 3T_0(x))$$

We also have

$$T_n(x) = \cos n\theta = \text{real part}(e^{in\theta}) = Rl(\cos\theta + i\sin\theta)^n$$
$$= Rl\left[\cos^n \theta + \binom{n}{1}\cos^{n-1}\theta(i\sin\theta) + \binom{n}{2}\cos^{n-2}\theta(i\sin\theta)^2 + \ldots\right]$$
$$= x^n + \binom{n}{2}x^{n-2}(x^2 - 1) + \binom{n}{4}x^{n-4}(x^2 - 1)^2 + \ldots$$
$$= 2^{n-1}x^n + \text{terms of lower degree.}$$

The Chebyshev polynomials $T_n(x)$ possess the following properties:

(i) $T_n(x)$ is a polynomial of degree n. If n is even $T_n(x)$ is an even polynomial and if n is odd, $T_n(x)$ is an odd polynomial.

(ii) $T_n(x)$ has n simple zeros $x_k = \cos\left(\dfrac{2k - 1}{2n}\pi\right)$, $k = 1, 2, \ldots, n$, on the interval $[-1, 1]$.

(iii) $T_n(x)$ assumes extreme values at $n + 1$ points $x_k = \cos(k\pi/n)$, $k = 0, 1, \ldots, n$ and the extreme value at x_k is $(-1)^k$.

(iv) $|T_n(x)| \leq 1$, $x \in [-1, 1]$.

(v) If $p_n(x)$ is any polynomial of degree n with leading coefficient unity (monomial) and $\tilde{T}_n(x) = \frac{1}{2^{n-1}} T_n(x)$ is the monic Chebyshev polynomial then

$$\max_{-1 \leq x \leq 1} |\tilde{T}_n(x)| \leq \max_{-1 \leq x \leq 1} |p_n(x)|$$

This property is called the **minimax property**.

(vi) $T_n(x)$ are orthogonal with respect to the weight function

$$W(x) = \frac{1}{\sqrt{1-x^2}}$$

$$\int_{-1}^{1} \frac{T_m(x) T_n(x)}{\sqrt{1-x^2}} dx = \begin{cases} 0, & m \neq n \\ \frac{\pi}{2}, & m = n \neq 0 \\ \pi, & m = n = 0 \end{cases} \quad (4.120)$$

Example 4.18 Using the Chebyshev polynomials, obtain the least squares approximation of second degree for $f(x) = x^4$ on $[-1, 1]$.

We write $f(x) = c_0 T_0(x) + c_1 T_1(x) + c_2 T_2(x)$.
We have

$$I(c_0, c_1, c_2) = \int_{-1}^{1} \frac{1}{\sqrt{1-x^2}} [x^4 - (c_0 T_0 + c_1 T_1 + c_2 T_2)]^2 \, dx = \text{minimum}$$

$$\frac{\partial I}{\partial c_0} = -2 \int_{-1}^{1} [x^4 - (c_0 T_0 + c_1 T_1 + c_2 T_2)] \frac{T_0 \, dx}{\sqrt{1-x^2}} = 0$$

$$\frac{\partial I}{\partial c_1} = -2 \int_{-1}^{1} [x^4 - (c_0 T_0 + c_1 T_1 + c_2 T_2)] \frac{T_1 \, dx}{\sqrt{1-x^2}} = 0$$

$$\frac{\partial I}{\partial c_2} = -2 \int_{-1}^{1} [x^4 - (c_0 T_0 + c_1 T_1 + c_2 T_2)] \frac{T_2 \, dx}{\sqrt{1-x^2}} = 0$$

$$c_0 = \frac{1}{\pi} \int_{-1}^{1} \frac{x^4 T_0}{\sqrt{1-x^2}} dx = \frac{3}{8}$$

$$c_1 = \frac{2}{\pi} \int_{-1}^{1} \frac{x^4 T_1}{\sqrt{1-x^2}} dx = 0$$

$$c_2 = \frac{2}{\pi} \int_{-1}^{1} \frac{x^4 T_2}{\sqrt{1-x^2}} dx = \frac{1}{2}$$

The required approximation is
$$f(x) = \tfrac{3}{8} T_0 + \tfrac{1}{2} T_2$$

4.10 UNIFORM APPROXIMATION

Weierstrass approximation theorem 4.2 asserts the possibility of a uniform approximation, using uniform norm (4.101) to continuous functions over a closed interval. This theorem uses the **Bernstein polynomials**

$$B_n(f, x) = \sum_{k=0}^{n} f\left(\frac{k}{n}\right)\binom{n}{k} x^k (1-x)^{n-k} \qquad (4.121)$$

defined on [0, 1] and has been shown that

$$\operatorname*{Lt}_{n \to \infty} B_n(f, x) = f(x)$$

uniformly on [0, 1]. However, the convergence of the approximation using the Bernstein polynomials is very slow. The best known polynomials which give a good uniform approximation to a given continuous function $f(x)$ are Chebyshev polynomials.

4.10.1 Uniform (minimax) Polynomial Approximation

Let $f(x)$ be continuous on $[a, b]$. Let $f(x)$ be approximated by the polynomial $P_n(x) = c_0 + c_1 x + \ldots + c_n x^n$. Then, the problem of minimax polynomial approximation is to determine $c_0, c_1, \ldots, c_n$ such that the deviation

$$|E_n(f, c_0, c_1, \ldots, c_n)| = |f(x) - P_n(x)|$$

satisfies

$$\max_{a \leqslant x \leqslant b} |E_n(f, c_0, c_1, \ldots, c_n)| = \min \max_{a \leqslant x \leqslant b} |E_n(f, c_0, c_1, \ldots, c_n)| \qquad (4.122)$$

Consider the case $n = 0$, that is, we want to approximate the function $f(x)$ by a constant c_0. Let

$$M = \max_{a \leqslant x \leqslant b} |f(x)|$$

$$m = \min_{a \leqslant x \leqslant b} |f(x)|$$

Thus we require

$$\max_{a \leqslant x \leqslant b} |f(x) - c_0| = \min_{a \leqslant x \leqslant b} |f(x) - c_0|$$

which gives

$$M - c_0 = -m + c_0$$

or

$$c_0 = \tfrac{1}{2}(m + M)$$

$$E_0(f, c_0) = M - \tfrac{1}{2}(M + m) = \tfrac{1}{2}(M - m)$$

We find from Fig. 4.5, that when the error curve

$$\epsilon(x) = f(x) - c_0$$

is drawn, the value $\pm E_0(f, c_0)$ is assumed at least twice, once with plus sign and once with minus sign, but of equal magnitude.

Consider now the case $n = 1$. In this case we want to approximate $f(x)$

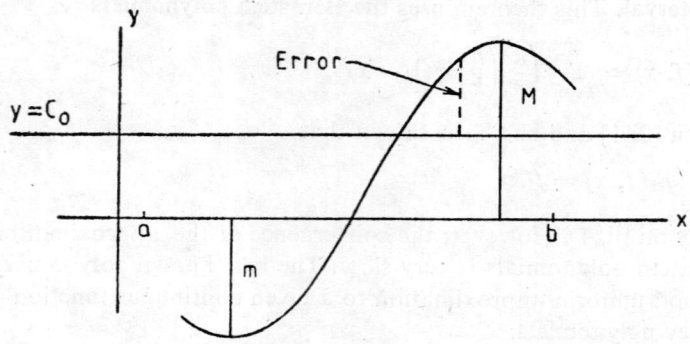

Fig. 4.5 Constant minimax approximation

by a first degree polynomial $P_1(x) = c_0 + c_1 x$. The parameters c_0 and c are to be determined such that

$$E_1(f, c_0, c_1) = f(x) - P_1(x)$$

satisfies

$$\max_{a \leq x \leq b} |E_1(f, c_0, c_1)| = \min_{a \leq x \leq b} |E_1(f, c_0, c_1)| \qquad (4.123)$$

We take three points x_1, x_2, x_3, $a \leq x_1 < x_2 < x_3 \leq b$ such that $\epsilon(x_i) = \pm E_1$ and the error must alternate in signs at these points. Assume that $x_1 = a$, $x_3 = b$ and x_2 is some interior point in the interval (a, b). Hence, $\epsilon'(x_2) = 0 = f'(x_2) - c_1$. Thus, we have the equations

$$f(a) - (c_0 + c_1 a) = -[f(x_2) - (c_0 + c_1 x_2)]$$
$$f(x_2) - (c_0 + c_1 x_2) = -[f(b) - (c_0 + c_1 b)]$$

and

$$f'(x_2) - c_1 = 0$$

These equations determine the values of c_0, c_1 and x_2. The graph of the linear minimax approximation is given in Fig. 4.6. In general, we have

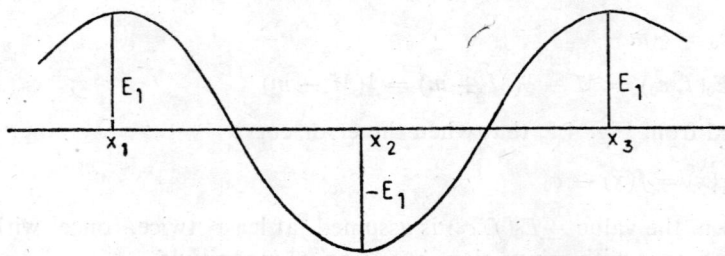

Fig. 4.6 Linear minimax approximation

THEOREM (CHEBYSHEV EQUIOSCILLATION) 4.3 *Let $f(x)$ be continuous on $[a, b]$ and $p_n(x)$ be the best uniform approximation in the sense of (4.122). Denote $E_n(f, x) = \max\limits_{a \leq x \leq b} |f(x) - p_n(x)|$ and $\epsilon_n(x) = f(x) - p_n(x)$. Then there are at least $n + 2$ points $a = x_0 < x_1 < x_2 \ldots < x_n < x_{n+1} = b$ where*

(i) $\epsilon(x_i) = \pm E_n$, $i = 0, 1, 2, \ldots, n+1$

(ii) $\epsilon(x_i) = -\epsilon(x_{i+1})$, $i = 0, 1, 2, \ldots, n$ (4.124)

The best uniform approximation $p_n(x)$ is completely determined and is unique under the above conditions.

It may be observed that (4.124 (ii)) implies

$$\epsilon'(x_i) = 0, \quad i = 0, 1, 2, \ldots, n \qquad (4.125)$$

Example 4.19 Obtain the Chebyshev linear polynomial approximation to the function $f(x) = x^3$ on $[0, 1]$.

Let
$$P_1(x) = c_0 + c_1 x$$
and $x_0 = 0$, $x_1 = \alpha$ and $x_2 = 1$.

We have
$$\epsilon(x) = x^3 - c_0 - c_1 x$$

Using (4.124) we have

$\epsilon(0) = -\epsilon(\alpha)$ or $\epsilon(0) + \epsilon(\alpha) = 0$

$\epsilon(\alpha) = -\epsilon(1)$ or $\epsilon(\alpha) + \epsilon(1) = 0$

and
$$\epsilon'(x) = 3x^2 - c_1 = 0 \quad \text{at } x = \alpha$$

Thus, we have

$\alpha^3 - \alpha c_1 - 2c_0 = 0$

$\alpha^3 - (\alpha + 1)c_1 - 2c_0 + 1 = 0$

and

$3\alpha^2 - c_1 = 0$

which gives

$$c_1 = 1, \quad \alpha = \frac{1}{\sqrt{3}} \quad \text{and} \quad c_0 = -\frac{\sqrt{3}}{9}$$

The Chebyshev linear approximation is given by

$$P_1(x) = x - \frac{\sqrt{3}}{9}$$

Example 4.20 Obtain the Chebyshev polynomial approximation of second degree to the function $f(x) = x^3$ on $[0, 1]$.

We have
$$P_2(x) = c_0 + c_1 x + c_2 x^2$$
$$x_0 = 0, \quad x_1 = \alpha, \quad x_2 = \beta, \quad x_3 = 1$$
and
$$\epsilon(x) = x^3 - c_0 - c_1 x - c_2 x^2$$

The second degree minimax approximation is given in Fig. 4.7. The equations

$$\epsilon(0) + \epsilon(\alpha) = 0$$
$$\epsilon(\alpha) + \epsilon(\beta) = 0$$
$$\epsilon(\beta) + \epsilon(1) = 0$$

give

$$c_2\alpha^2 + c_1\alpha + 2c_0 - \alpha^3 = 0$$
$$c_2(\alpha^2 + \beta^2) + c_1(\alpha + \beta) + 2c_0 - \alpha^3 - \beta^3 = 0$$
$$c_2(\beta^2 + 1) + c_1(1 + \beta) + 2c_0 - 1 - \beta^3 = 0$$

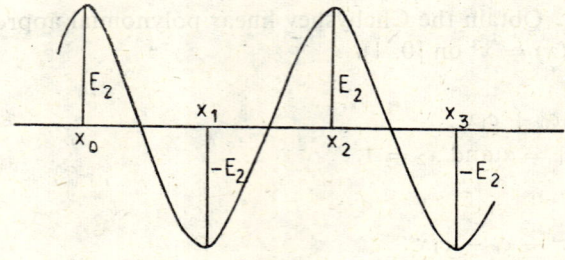

Fig. 4.7 Second degree minimax approximation

The solution of this system is

$$c_2 = \frac{1 + \alpha + \alpha^2 - \beta^2}{1 + \alpha - \beta}$$

$$c_1 = \frac{\beta^2 + \alpha\beta(\beta - \alpha) - \beta(1 + \alpha)}{1 + \alpha - \beta}$$

$$c_0 = \frac{\alpha(1 - \beta)(\beta - \alpha)}{2(1 + \alpha - \beta)}$$

We also have

$$\epsilon'(x) = 0 \quad \text{at } x = \alpha \text{ and } \beta$$

which gives

$$3\alpha^2 - 2c_2\alpha - c_1 = 0$$
$$3\beta^2 - 2c_2\beta - c_1 = 0$$

or

$$c_2 = \frac{3(\alpha + \beta)}{2}, \quad c_1 = -3\alpha\beta$$

Comparing the values of c_2 and c_1 we have

$$\frac{3(\alpha + \beta)}{2} = \frac{1 + \alpha + \alpha^2 - \beta^2}{1 + \alpha - \beta}$$

and
$$-3\alpha\beta = \frac{\beta^2 + \alpha\beta(\beta - \alpha) - \beta(1 + \alpha)}{1 + \alpha - \beta}$$

we obtain
$$\alpha = \frac{1}{4}, \quad \beta = \frac{3}{4}, \quad c_2 = \frac{3}{2},$$
$$c_1 = -\frac{9}{16} \quad \text{and} \quad c_0 = \frac{1}{32}$$

Therefore, the second degree Chebyshev polynomial approximation to x^3 on $[0, 1]$ becomes
$$P_2(x) = \frac{1}{32}(48x^2 - 18x + 1)$$

4.10.2 Chebyshev Polynomial Approximation and Lanczos Economization

Let
$$\frac{a_0}{2} + \sum_{i=1}^{\infty} a_i T_i(x)$$
be the Chebyshev series expansion for a continuous function $f(x)$. Then the partial sum
$$P_n(x) = \frac{a_0}{2} + \sum_{i=1}^{n} a_i T_i(x) \tag{4.126}$$
is very nearly the solution to the problem
$$\max_{-1 \leq x \leq 1} \left| f(x) - \sum_{i=0}^{n} c_i x^i \right| = \text{minimum}$$

This means that the partial sum (4.126) is nearly the best uniform approximation to $f(x)$. The reasoning for this is as follows. Suppose we write
$$f(x) = \frac{a_0}{2} + a_1 T_1(x) + a_2 T_2(x) + \ldots + a_n T_n(x) + a_{n+1} T_{n+1}(x)$$
$$+ \text{remainder} \tag{4.127}$$

Neglecting the remainder, we obtain from (4.127)
$$f(x) - \left(\frac{a_0}{2} + \sum_{i=1}^{n} a_i T_i(x) \right) = a_{n+1} T_{n+1}(x) \tag{4.128}$$

Since $a_{n+1} T_{n+1}(x)$ has $n + 2$ equal maxima and minima, alternating in sign, then by Chebyshev equi-oscillation theorem, the polynomial (4.126) of degree n is the best uniform approximation to $f(x)$. Thus, to determine the best uniform approximation to a given continuous function $f(x)$, it is advantageous to start with the truncated Chebyshev expansion of the function and then improve upon it by the Chebyshev equi-oscillation theorem.

180 Numerical Methods

Example 4.21 Find the best uniform approximant of degree 3 or less to x^4 on $[-1, 1]$.

Expressing x^4 in terms of Chebyshev polynomials we get

$$x^4 = \frac{3}{8} T_0 + \frac{1}{2} T_2 + \frac{1}{8} T_4$$

Since T_4 is a polynomial of degree 4, we approximate $f(x) = x^4$ by $\frac{3}{8}T_0 + \frac{1}{2}T_2$. Therefore we have

$$x^4 - \left(\frac{3}{8} T_0 + \frac{1}{2} T_2\right) = \frac{1}{8} T_4$$

The uniform polynomial approximation of degree three or less to x^4 is

$$\frac{3}{8} T_0 + \frac{1}{2} T_2 = x^2 - \frac{1}{8}$$

and the error of this approximation on $[-1, 1]$ is

$$\max_{-1 \leqslant x \leqslant 1} \left|x^4 - \left(\frac{3}{8} T_0 + \frac{1}{2} T_2\right)\right| = \max_{-1 \leqslant x \leqslant 1} \frac{1}{8}|T_4(x)| = \frac{1}{8}$$

In this process, we have approximated a polynomial by a lower degree polynomial. For a general function, we first express it as a power series in x and then write each term in terms of the Chebyshev polynomials, using the relation

$$x^m = 2^{1-m} \sum_{k=0}^{[m/2]}{}' \binom{m}{k} T_{m-2k}(x) \qquad (4.129)$$

where the prime over the sigma means that the last term is to be halved and $[m/2]$ means the greatest integer $\leqslant m/2$. Thus if we have

$$f(x) = \sum_{i=0}^{\infty} c_i T_i(x)$$

as the Chebyshev series expansion for a continuous function $f(x)$ on $[-1, 1]$, the partial sum

$$P_n(x) = \sum_{i=0}^{n} c_i T_i(x) \qquad (4.130)$$

is a good uniform approximation to $f(x)$ in the sense

$$\max_{-1 \leqslant x \leqslant 1} |f(x) - P_n(x)| \leqslant |c_{n+1}| + |c_{n+2}| + \ldots$$
$$\leqslant \epsilon \qquad (4.131)$$

Given ϵ, it is possible to find the number of terms, that should be retained in (4.130). This process is known as **Lanczos economization**. Replacing each $T_i(x)$ by its polynomial form and rearranging the terms, we get the required economized polynomial approximation.

Example 4.22 Find a uniform polynomial approximation of degree four or less to e^x on $[-1, 1]$ using Lanczos economization with a tolerance of $\epsilon = 0.01$.

The power series expansion of e^x within this tolerance is given by

$$f(x) = e^x = 1 + x + \frac{x^2}{2} + \frac{x^3}{6} + \frac{x^4}{24}$$

$$= T_0 + T_1 + \frac{1}{4}(T_0 + T_2) + \frac{1}{24}(3T_1 + T_3)$$

$$+ \frac{1}{192}(3T_0 + 4T_2 + T_4)$$

$$= \frac{81}{64} T_0 + \frac{9}{8} T_1 + \frac{13}{48} T_2 + \frac{1}{24} T_3 + \frac{T_4}{192}$$

Since the magnitude of the last term on the right hand side is less than 0.01, the economized polynomial approximation becomes

$$f(x) = \frac{81}{64} T_0 + \frac{9}{8} T_1 + \frac{13}{48} T_2 + \frac{1}{24} T_3$$

$$= \frac{x^3}{6} + \frac{13}{24} x^2 + x + \frac{191}{192}$$

4.11 RATIONAL APPROXIMATION

In the earlier sections, we have discussed the polynomial approximation of a continuous function. Implicitly we had assumed that between the nodal points or over the given interval, the function $f(x)$ behaved as a polynomial. However, the polynomial approximation is not always the best approximation when the function behaves as the quotient of two polynomials which is usually called a **rational** function. A rational approximation is written in the form

$$f(x) \approx R_{n,m}(x) = \frac{a_0 + a_1 x + \ldots + a_n x^n}{b_0 + b_1 x + \ldots + b_m x^m} \qquad (4.132)$$

The suffix n, m in $R_{n,m}(x)$ indicates that the numerator and denominator are polynomials of degree n and m respectively. Without any loss of generality we have taken $b_0 = 1$, since we can always divide the numerator and the denominator in (4.132) by b_0. The constant b_0 will, generally, not be zero, else the approximation becomes undefined at $x = 0$. The rational approximation can also be used for functions behaving as polynomials. The same or higher order of accuracy can be achieved by using lower degree polynomials in $R_{n,m}(x)$ than the direct polynomial approximation.

When the function $f(x)$ is such that it retains a finite value when x tends to infinity, polynomial approximations give very poor results. In comparison, rational approximations give much better results. We generally compute functions like $\sin x$, $\cos x$ etc. for large arguments using rational approximations. The approximation (4.132) is also called **Padé** approximation. Using the **Maclaurin** series expansion for $f(x)$ given by

$$f(x) = \sum_{j=0}^{\infty} c_j x^j$$

182 *Numerical Methods*

in which c_j's are known constants, we obtain from (4.132)

$$f(x) - R_{n,m}(x) = \frac{\left(\sum_{j=0}^{\infty} c_j x^j\right)\left(\sum_{j=0}^{m} b_j x^j\right) - \sum_{j=0}^{n} a_j x^j}{\sum_{j=0}^{m} b_j x^j} \tag{4.133}$$

The $n + m + 1$ constants a_i, $i = 0, 1, \ldots, n$; b_j, $j = 1, 2, \ldots, m$ are determined by equating the coefficients of x^j, $j = 0, 1, \ldots, (m+n)$ in the numerator of (4.133) to zero. The first non-zero term in the numerator gives the order of approximation. If $m = 0$, we get Maclaurin series expansion, otherwise we get an approximation of order $N = (m+n)$.

Example 4.23 Obtain the rational approximation of the form

(i) $\dfrac{a_0 + a_1 x}{1 + b_1 x}$

(ii) $\dfrac{a_0 + a_1 x}{1 + b_1 x + b_2 x^2}$

to e^x.

(i) We have for $n = m = 1$, $N = 2$

$$e^x = 1 + x + \frac{x^2}{2} + \ldots$$

Therefore

$$e^x - \frac{a_0 + a_1 x}{1 + b_1 x} = \frac{(1 + x + x^2/2 + \ldots)(1 + b_1 x) - (a_0 + a_1 x)}{1 + b_1 x}$$

Set

$$\left(1 + x + \frac{x^2}{2} + \ldots\right)(1 + b_1 x) - (a_0 + a_1 x) = \sum_{j=0}^{\infty} d_j x^j$$

Putting $d_i = 0$, $i = 0, 1, 2$, we have

$1 - a_0 = 0$

$1 + b_1 - a_1 = 0$

$\frac{1}{2} + b_1 = 0$

which gives $a_0 = 1$, $a_1 = \frac{1}{2}$, $b_1 = -\frac{1}{2}$. Therefore

$$e^x = \frac{1 + \frac{1}{2}x}{1 - \frac{1}{2}x} + 0(x^3)$$

(ii) In this case we have $n = 1$, $m = 2$, $N = 3$.
Set

$$\left(1 + x + \frac{x^2}{2} + \frac{x^3}{6} + \ldots\right)(1 + b_1 x + b_2 x^2) - (a_0 + a_1 x) = \sum_{j=0}^{\infty} d_j x^j$$

Putting $d_i = 0$, $i = 0, 1, 2, 3$, we get

$1 - a_0 = 0$

$1 + b_1 - a_1 = 0$

$\frac{1}{2} + b_1 + b_2 = 0$

$\frac{1}{6} + \frac{1}{2}b_1 + b_2 = 0$

The solution of this system is

$a_0 = 1,\ a_1 = \frac{1}{3},\ b_1 = -\frac{2}{3},\ b_2 = \frac{1}{6}$

The rational approximation is given by

$$e^x = \frac{1 + \frac{1}{3}x}{1 - \frac{2}{3}x + \frac{1}{6}x^2} + O(x^4)$$

Various order rational approximations (Padé approximations) to e^x are given in Table 4.7.

Table 4.7 Rational approximations to e^x

m \ n	0	1	2	3
0	$\frac{1}{1}$	$\frac{1+x}{1}$	$\frac{1+x+\frac{1}{2}x^2}{1}$	$\frac{1+x+\frac{1}{2}x^2+\frac{1}{6}x^3}{1}$
1	$\frac{1}{1-x}$	$\frac{1+\frac{1}{2}x}{1-\frac{1}{2}x}$	$\frac{1+\frac{2}{3}x+\frac{1}{6}x^2}{1-\frac{1}{3}x}$	$\frac{1+\frac{3}{4}x+\frac{1}{4}x^2+\frac{1}{24}x^3}{1-\frac{1}{4}x}$
2	$\frac{1}{1-x+\frac{1}{2}x^2}$	$\frac{1+\frac{1}{3}x}{1-\frac{2}{3}x+\frac{1}{6}x^2}$	$\frac{1+\frac{1}{2}x+\frac{1}{12}x^2}{1-\frac{1}{2}x+\frac{1}{12}x^2}$	$\frac{1+\frac{3}{5}x+\frac{3}{20}x^2+\frac{1}{60}x^3}{1-\frac{2}{5}x+\frac{1}{20}x^2}$
3	$\frac{1}{1-x+\frac{1}{2}x^2-\frac{1}{6}x^3}$	$\frac{1+\frac{1}{4}x}{1-\frac{3}{4}x+\frac{1}{4}x^2-\frac{1}{24}x^3}$	$\frac{1+\frac{2}{5}x+\frac{1}{20}x^2}{1-\frac{3}{5}x+\frac{3}{20}x^2-\frac{1}{60}x^3}$	$\frac{1+\frac{1}{2}x+\frac{1}{20}x^2+\frac{1}{120}x^3}{1-\frac{1}{2}x+\frac{1}{10}x^2-\frac{1}{120}x^3}$

4.12 CHOICE OF THE METHOD

In this chapter we have considered various forms of the approximating polynomials. The interpolating polynomial and the function $f(x)$ agree in value at a set of points x_i, $i = 0, 1, \ldots, n$ belonging to an interval $[a, b]$. We use this interpolating polynomial to determine the value of $f(x)$ at any point $x \neq x_i$. If $x \in (a, b)$, we call it an interpolation problem, otherwise it is called an **extrapolation** problem. Normally interpolation problem produces better results. The same interpolating polynomial can be used for obtaining roots of $f(x)$, certain order derivatives of $f(x)$ at tabular or nontabular points, integral of $f(x)$ between known limits or for any other operation desired on $f(x)$.

For a given set of tabular data, the Lagrange interpolating polynomial can be easily obtained. In many applications, the exact number of tabular points to be used to achieve a certain degree of accuracy is not readily known. It is not uncommon to obtain a sequence of interpolants, each one based on one more tabular point than the previous one. The results thus obtained are compared and computation of successive interpolants is stopped when the required accuracy is achieved. Adding of one more tabular point means getting an interpolating polynomial of degree higher by one than the preceding polynomial. It can be seen from the form of Lagrange interpolating polynomial that if we add one more data item to the tabular data already available, then the Lagrange fundamental polynomials $l_i(x)$ have all to be recalculated. The Lagrange fundamental polynomials of lower degree which had already been obtained are of no use.

When a data item is added at the beginning or at the end of the tabular data and if it is possible to derive an interpolating polynomial by adding one more term to the previously calculated interpolating polynomial, then such an interpolating polynomial is said to possess **permanence property**. Obviously Lagrange interpolating polynomial does not possess this property. Interpolating polynomials based on divided differences or finite differences have the permanence property. If one more data item (x_{n+1}, f_{n+1}) is added to the given data (x_i, f_i), $i = 0, 1, \ldots, n$, then in the case of Newton's divided difference formula, we need to add the term $(x - x_0)(x - x_1) \ldots (x - x_n) f[x_0, x_1, \ldots, x_{n+1}]$ to the previously calculated nth degree interpolating polynomial. In the case of Gregory-Newton's forward and backward difference formulas, we need to add the term $\binom{u}{n+1} \Delta^{n+1} f_0$ and $(-1)^{n+1} \binom{-u}{n+1} \nabla^{n+1} f_{n+1}$ respectively, to obtain the $(n+1)$th degree polynomial.

Therefore, if the problem requires calculation of various degree interpolating polynomials to achieve the desired accuracy, formulas based on divided differences should be used rather than Lagrange interpolating polynomial. If the points are equispaced, formulas based on finite differences may be used. If the value of $f(x)$, at any point $x \in [a, b]$ of an equispaced tabular data, is required, we use interpolation formula, based on forward differences if x is near the beginning of the table, based on backward differences if x is near the end of the table and based on central differences if x is near the centre of the table. The initial point can be suitably chosen to obtain better results.

For a given point x and the tabular points x_i, $i = 0, 1, \ldots, n$, we can obtain an upper bound on the error of interpolation, by using the expression

$$|E_n(f, x)| = |f(x) - P(x)| \leq \max_{a \leq x \leq b} |(x - x_0)(x - x_1) \ldots (x - x_n)| \frac{M_{n+1}}{(n+1)!}$$

where $M_{n+1} = \max_{a \leq x \leq b} |f^{(n+1)}(x)|$.

If the tabular points x_i, $i = 0, 1, \ldots, n$ are at our choice, then the error of interpolation is minimum if x_i are taken as zeros of the Chebyshev polynomial $T_{n+1}(x)$. In this case

$$|(x - x_0)(x - x_1) \ldots (x - x_n)| = |\tilde{T}_{n+1}(x)| \leq \frac{1}{2^n}$$

where
$$\tilde{T}_{n+1}(x) = \frac{1}{2^n} T_{n+1}(x)$$
and the error of interpolation becomes
$$|E_n(f, x)| \leq \frac{1}{2^n(n+1)!} M_{n+1}$$

Interpolating polynomials are easy to derive and easy to implement. To obtain reasonably accurate results, one may be forced to use polynomials of very high degree. With polynomials of high degree, not only the computation becomes costly, but also, the computed results may become unreliable because of round-off errors. There are situations when higher degree interpolating polynomials produce poorer results compared to lower degree polynomials. One such well known situation is the **Runge's example**

$$f(x) = \frac{1}{1+x^2}, \quad x \in [-5, 5]$$

The function $f(x) = \frac{1}{1+x^2}$ is a very well behaved function. The function and its derivatives of all orders are bounded in the given interval. The interpolating polynomial $P(x)$ based on equispaced points belonging to the interval $[-5, 5]$ diverges in the interval $3.63 < |x| < 5.0$, as the degree of the polynomial is increased. The reason for this divergence seems to be that it has two simple poles at $x = \pm i$ in the complex plane, which are too close to real axis. Therefore the use of higher degree interpolating polynomials should be avoided and piecewise interpolation or spline interpolation may be used. Spline interpolation may be preferred because of its versatility.

For approximating a given function or a discrete data, the decisions to be taken are: (i) the choice of the form of the approximating functions, and (ii) the choice of the norm. It is known that the polynomial interpolation based on equispaced points does not work well as an approximation function for $y(x) = 1/(1+x^2)$. However, Chebyshev approximation or rational approximation works well for this function. Further, for large values of x, the rational approximation will always produce bounded results. Usually, the form of approximation is chosen keeping in view the form of the given function and the domain of its applications.

If the least square approximation is used, it is advisable to take the coordinate functions as some orthogonal functions to avoid ill-conditioning of the normal equations. A satisfactory least squares approximation is not expected if the number of data points is not adequate, generally not less than twice the number of arbitrary constants to be determined in the approximation. The minimax approximation is best when one has to represent a mathematical function over a given interval. But, finding points where the minimum or maximum errors occur is a difficult job. By using truncated

Chebyshev approximation, we may get nearly minimax approximation as the coefficients tend to settle down to approximately constant values, as the degree of the approximating polynomial increases. Almost all the commonly used mathematical functions can be approximated by using the Chebyshev approximation combined with Lanczos economization.

PROBLEMS

1. Obtain a second degree polynomial approximation to $f(x) = (1 + x)^{1/2}$ over [0, 1] by means of the Taylor expansion about $x = 0$. Use the first three terms of the expansion to approximate $f(0.05)$ and obtain a bound of the error.

2. Expand $\log(1 + x)$ in a Taylor expansion about $x_0 = 1$, through terms of degree 4. Obtain a bound of the truncation error when approximating $\log 1.2$ using this expansion.

3. Obtain polynomial approximation to $f(x) = (1 - x)^{1/2}$ over [0, 1], by means of Taylor expansion about $x = 0$. Find the number of terms required in the expansion to obtain results correct to 5×10^{-3} for $0 \leqslant x < 1/2$.

4. If we use the somewhat unsuitable method of Taylor expansion around $x = 0$ for computation of $\sin x$ in the interval $[0, 2\pi]$ and if we want 4 accurate decimal places, how many terms are needed. If instead we use the fact that $\sin(\pi + x) = -\sin x$, we only need the expansion in $[0, \pi]$. How many terms do we then need for the same accuracy. (Lund Univ., Sweden, BIT 16 (1976), 228)

5. Let $f(x) = \log(1 + x)$, $x_0 = 1$ and $x_1 = 1.1$. Use linear interpolation to calculate an approximate value for $f(1.04)$, and obtain a bound on the truncation error.

6. Determine an appropriate step size to use, in the construction of a table of $f(x) = (1 + x)^6$ on [0, 1]. The truncation error for linear interpolation is to be bounded by 5×10^{-5}.

7. Determine the maximum step size that can be used in the tabulation of $f(x) = e^x$ in [0, 1], so that the error in the linear interpolation will be less than 5×10^{-4}. Find also the step size if quadratic interpolation is used.

8. By considering the limit of the three point Lagrangian interpolation formula relative to x_0, $x_0 + \epsilon$ and x_1 as $\epsilon \to 0$, obtain the formula
$$f(x) = \frac{(x_1 - x)(x + x_1 - 2x_0)}{(x_1 - x_0)^2} f(x_0) + \frac{(x - x_0)(x_1 - x)}{(x_1 - x_0)} f'(x_0)$$
$$+ \frac{(x - x_0)^2}{(x_1 - x_0)^2} f(x_1) + E(x)$$
where $E(x) = \frac{1}{6}(x - x_0)^2 (x - x_1) f'''(\xi)$.

9. Denoting the interpolant of $f(x)$ on the set of (distinct) points $x_0, x_1, \ldots, x_n$ by $\sum_{k=0}^{n} l_k(x)f(x_k)$, find an expression for $\sum_{k=0}^{n} l_k(0)x_k^{n+1}$.
   (Gothenburg Univ., Sweden, BIT 15 (1975), 224)

10. Find the unique polynomial $P(x)$ of degree 2 or less such that
    $$P(1) = 1, \quad P(3) = 27, \quad P(4) = 64$$
    using each of the following methods: the Lagrange interpolation formula, the Newton-divided difference formula and the Aitken's iterated interpolation formula. Evaluate $P(1.5)$.

11. Suppose that $f(x) = e^x \cos x$ is to be approximated on $[0, 1]$ by an interpolating polynomial on $n + 1$ equally spaced points $0 = x_0 < x_1 < x_2 \ldots < x_n = 1$. Determine n so that the truncation error will be less than .0001 in this interval.

12. If $f(x) = e^{ax}$, show that $\Delta^n f(x) = (e^a - 1)^n e^{ax}$.

13. Calculate the nth divided difference of $f(x) = \dfrac{1}{x}$.

14. If $f(x) = u(x)v(x)$, show that
    $$f[x_0, x_1] = u[x_0]\, v[x_0, x_1] + u[x_0, x_1]\, v[x_1].$$

15. Prove the relations:
    (i) $\nabla - \Delta = \Delta \nabla$
    (ii) $\Delta + \nabla = \Delta/\nabla - \nabla/\Delta$
    (iii) $\sum_{k=0}^{n-1} \Delta^2 f_k = \Delta f_n - \Delta f_0$
    (iv) $\Delta(f_i g_i) = f_i \Delta g_i + g_{i+1} \Delta f_i$
    (v) $\Delta f_i^2 = (f_i + f_{i+1}) \Delta f_i$
    (vi) $\Delta(f_i/g_i) = (g_i \Delta f_i - f_i \Delta g_i)/g_i g_{i+1}$
    (vii) $\Delta(1/f_i) = -\Delta f_i/f_i f_{i+1}$

16. The following are supposed to be the values of a polynomial of degree $n < 10$. Locate any errors in the following table:

x	0	0.2	0.4	0.6	0.8	1.0
$p(x)$	1.500	1.590	1.392	0.954	0.324	−0.540
x	1.2	1.4	1.6	1.8	2.0	
$p(x)$	−1.320	−2.238	−3.156	−4.156	−4.800	

17. Use the Lagrange and the Newton-divided difference formulas to calculate $f(3)$ from the following table:

x	0	1	2	4	5	6
$f(x)$	1	14	15	5	6	19

18. The following data are part of a table for $g(x) = \sin x/x^2$:

x	0.1	0.2	0.3	0.4	0.5
$g(x)$	9.9833	4.9667	3.2836	2.4339	1.9177

Calculate $g(0.25)$ as accurately as possible.
(a) by interpolating directly in this table,
(b) by first tabulating $xg(x)$ and then interpolating in that table.
(c) explain the difference between the results in (a) and (b) respectively. (Umea Univ., Sweden, BIT 19(1979), 285)

19. Determine the constants a, b, c and d such that the interpolation polynomial
$$y_s = y(x_0 + sh) = ay_0 + by_1 + h^2(cy_0'' + dy_1'')$$
becomes correct to the highest possible order.

20. Determine the constants a, b, c and d such that the interpolation polynomial
$$y(x_0 + sh) = ay(x_0 - h) + by(x_0 + h) + h[cy'(x_0 - h) + dy'(x_0 + h)]$$
becomes correct to the highest possible order. Find the error term.

21. Determine the parameters in the formula
$$P(x) = a_0(x-a)^3 + a_1(x-a)^2 + a_2(x-a) + a_3$$
such that
$$P(a) = f(a), \quad P'(a) = f'(a)$$
$$P(b) = f(b), \quad P'(b) = f'(b)$$

22. Obtain the unique polynomial $P(x)$ of degree 5 or less corresponding to a function $f(x)$, where
$$f(x_0) = 1, \quad f'(x_0) = 2, \quad f''(x_0) = 1$$
$$f(x_1) = 3, \quad f'(x_1) = 0, \quad f''(x_1) = -2$$
Also find $P\left(\dfrac{x_0 + x_1}{2}\right)$.

23. Fit the following four points by the cubic splines:

i	0	1	2	3
x_i	1	2	3	4
y_i	1	5	11	8

Use the end conditions
$$y_0'' = y_3'' = 0$$
Hence compute (i) $y(1.5)$, and (ii) $y'(2)$.

24. Suppose $f_i = x_i^{-2}$ and $f_i' = -2x_i^{-3}$ where $x_i = \frac{1}{2}i$, $i = 1(1)4$ are given. Fit these values by the piecewise cubic Hermite polynomial.

25. Determine the piecewise quadratic fit $P(x)$ to $f(x) = (1 + x^2)^{-1/2}$ with knots at $-1, -\frac{1}{2}, 0, \frac{1}{2}, 1$. Estimate the error $|f - P|$ and compare this with full Lagrange polynomial fit.

26. The piecewise interpolating function $S_\Delta(x)$ which satisfies the following relations:

(a) $S_\Delta'' - \alpha S_\Delta = \dfrac{(x - x_{j-1})}{h}(M_j - \alpha f_j)$
$\qquad + \dfrac{(x_j - x)}{h}(M_{j-1} - \alpha f_{j-1}), \quad \alpha > 0$

(b) $S_\Delta(x_{j-1}) = f_{j-1}, \quad S_\Delta(x_j) = f_j$

(c) $S_\Delta'(x_j + h) = S_\Delta'(x_j - h)$

where $S_\Delta''(x_j) = M_j$, α an arbitrary parameter and h is the length of the interval, is called **spline in tension**.

(i) Find the function $S_\Delta(x)$ and show that for $\alpha \gg 1$ the spline in tension becomes the piecewise linear polynomial

$$S_\Delta(x) = \dfrac{(x_j - x)}{h} f_{j-1} + \dfrac{(x - x_{j-1})}{h} f_j + O(\alpha^{-1})$$

(ii) Prove that the relation (c) gives

$$f_{j+1} - 2f_j + f_{j-1} = h^2(\gamma M_{j+1} + 2\delta M_j + \gamma M_{j-1})$$

where

$$\gamma = \dfrac{\sinh w - w}{w^2 \sinh w}, \quad \delta = \dfrac{w \cosh w - \sinh w}{w^2 \sinh w}, \quad w^2 = \alpha h^2$$

27. The piecewise interpolating polynomial function $S_\Delta(x)$ which satisfies the following relations:

(a) $S_\Delta + \alpha S_\Delta = \dfrac{(x - x_{j-1})}{h}(M_j + \alpha f_j)$
$\qquad + \dfrac{(x_i - x)}{h}(M_{j-1} + \alpha f_{j-1}), \quad \alpha > 0$

(b) $S_\Delta(x_{j-1}) = f_{j-1}, \quad S_\Delta(x_j) = f_j$

(c) $S_\Delta'(x_j+) = S_\Delta'(x_j-)$

(d) $\dfrac{w}{2} = \tan \dfrac{w}{2}, \quad w^2 = \alpha h^2$

where $S_\Delta''(x_j) = M_j$, α an arbitrary parameter and h is the length of interval, is called **spline in compression**. Prove that the relations (c) and (d) give the condition

$$f_{j-1} - 2f_j + f_{j+1} = \dfrac{h^2}{4}(M_{j-1} + 2M_j + M_{j+1})$$

28. $S_3(x)$ is the piecewise cubic Hermite interpolating approximant of $f(x) = \sin x \cos x$ in the abscissas 0, 1, 1.5, 2, 3. Estimate the error $\max\limits_{0 \leqslant x \leqslant 3} |f(x) - S_3(x)|$. (Uppsala Univ., Sweden, BIT 19(1979), 425)

29. Determine the piecewise quadratic approximating function of the form

$$S_\Delta(x, y) = \sum_{i=0}^{8} N_i f_i$$

for the following configuration of rectangular network

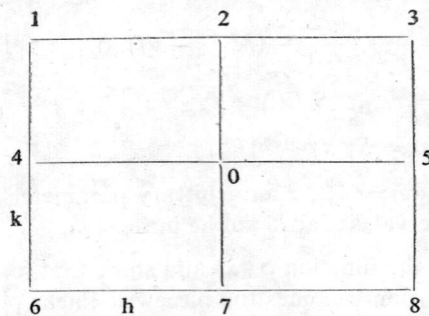

30. Using the following data obtain the Lagrange and Newton's bivariate interpolating polynomials

x \ y	0	1	2
0	1	3	7
1	3	6	11
2	7	11	17

31. The Bernstein polynomial of degree n approximating a function $f(x)$ defined in $(0, 1)$ is given by

$$B_n(f, x) = \sum_{m=0}^{n} f\left(\frac{m}{n}\right) \binom{n}{m} x^m (1-x)^{n-m}$$

Prove that

$$\frac{d^2}{dx^2} B_n(x^3, x) = 3x^2 + \frac{3x(2-3x)}{n} + \frac{(1-6x+6x^2)}{n^2}, \quad n \geq 3$$

32. A person runs the same race track for five consecutive days and is timed as follows:

day (x)	1	2	3	4	5
time (y)	15.30	15.10	15.00	14.50	14.00

Make a least square fit to the above data using a function $a + b/x + c/x^2$. (Uppsala Univ., Sweden, BIT 18(1978), 115)

33. Obtain an approximation in the sense of the principle of least squares in the form of a polynomial of the degree 4 to the function $1/(1+x^2)$ in the range $-1 \leq x \leq 1$.

34. Determine the least-squares approximation of the type $ax^2 + bx + c$, to the function 2^x at the points $x_i = 0, 1, 2, 3, 4$.

 (Royal Inst. Tech., Stockholm, Sweden, BIT 10(1970), 398)

35. We are given the following values of a function f of the variable t:

t	0.1	0.2	0.3	0.4
f	0.76	0.58	0.44	0.35

 Obtain a least squares fit of the form

 $$g = ae^{-3t} + be^{-2t}$$

 (Royal Inst. Tech., Stockholm, Sweden, BIT 17(1977), 115)

36. Let $l(x)$ be a straight line which is the best approximation of $\sin x$ in the sense of the method of least squares over the interval $[-\pi/2, \pi/2]$. Show that the residual $d(x) = \sin x - l(x)$ is orthogonal to any second degree polynomial. The scalar product is given by

 $$(f, g) = \int_{-\pi/2}^{\pi/2} \bar{f}(x) g(x)\, dx$$

 (Uppsala Univ., Sweden, BIT 14(1974), 122)

37. Experiments with a periodic process gave the following data:

t (degrees)	0	50	100	150	200	250	300	350
y	0.754	1.762	2.041	1.412	0.303	−0.484	−0.380	0.520

 Estimate the parameters a and b in the model $y = b + a \sin t$, using the least squares approximation.

 (Lund Univ., Sweden, BIT 21(1981), 242)

38. A physicist wants to approximate the following data:

x	0.0	0.5	1.0	2.0
$f(x)$	0.00	0.57	1.46	5.05

 using a function $ae^{bx} + c$.
 He believes that $b \approx 1$.

 (i) Compute the values of a and c that give the best least squares approximation assuming that $b = 1$.
 (ii) Use these values of a and c to obtain a better value of b.

 (Uppsala Univ., Sweden, BIT 17(1977), 369)

39. The second degree polynomial $f(x) = a + bx + cx^2$ is determined from the condition

 $$d = \sum_{i=m}^{n} [f(x_i) - y_i]^2 = \text{minimum},$$

 where (x_i, y_i), $i = m(1)n$, $m < n$, are given real numbers. Putting $X = x - \xi$, $Y = y - \eta$, $X_i = x_i - \xi$, $Y_i = y_i - \eta$ we determine $F(X) = A + BX + CX^2$ from the condition

$$D = \sum_{i=m}^{n} [F(X_i) - Y_i]^2 = \text{minimum}.$$

Show that $F(X) = f(x) - \eta$. Also derive an explicit formula for $F'(0)$ expressed in Y_i when $X_i = ih$ and $m = -n$.

(Bergen Univ., Sweden, BIT 7(1967), 247)

40. A function is approximated by a piecewise linear function in the sense that

$$\int_0^1 \left[f(x) - \sum_{i=0}^{10} a_i \phi_i(x) \right]^2 dx$$

is minimized, where the shape functions ϕ_i are defined by

$$\phi_0 = \begin{vmatrix} 1 - 10x, & 0 \leqslant x \leqslant 0.1 \\ 0, & \text{otherwise} \end{vmatrix}$$

$$\phi_{10} = \begin{vmatrix} 10x - 9, & 0.9 \leqslant x \leqslant 1 \\ 0, & \text{otherwise} \end{vmatrix}$$

$$\phi_i = \begin{vmatrix} 10(x - x_{i-1}), & x_{i-1} \leqslant x \leqslant x_i \\ 10(x_{i+1} - x), & x_i \leqslant x \leqslant x_{i+1}, \\ 0, & \text{otherwise}, \; x_i = 0.1i \end{vmatrix} \quad i = 1(1)9$$

Write down the coefficient matrix of the normal equations.

(Uppsala Univ., Sweden, BIT 19(1979), 552)

41. Polynomials $p_r(x)$, $r = 0(1)n$, are defined by

$$\sum_{j=0}^{n} p_r(x_j) p_s(x_j) \begin{vmatrix} = 0, & r \neq s \\ \neq 0, & r = s \end{vmatrix} \quad r, s \leqslant n$$

$$x_j = -1 + \frac{2j}{n}, \quad j = 0(1)n$$

subject also to $p_r(x)$ being a polynomial of degree r with leading term x^r.

Derive a recurrence relation for these polynomials and obtain $p_0(x)$, $p_1(x)$, $p_2(x)$, when $n = 4$.

Hence obtain coefficients a_0, a_1, a_2 which minimize

$$\sum_{j=0}^{4} [(1 + x_j^2)^{-1} - (a_0 + a_1 x_j + a_2 x_j^2)]^2$$

42. Find suitable values of $a_0, \ldots, a_4$ so that

$$\sum_{r=0}^{4} a_r T_r^*(x)$$

is a good approximation to $1/(1 + x)$ for $0 \leqslant x \leqslant 1$. Estimate the maximum error of this approximation.

(Note $T_r^*(x) = \cos r\theta$ where $\cos \theta = 2x - 1$.)

43. Determine as accurately as possible a straight line $y = ax + b$, approximating $1/x^2$ in the Chebyshev sense on the interval $[1, 2]$.

What is the maximal error? Calculate a and b to two correct decimals.
(Royal Inst. Tech., Stockholm, Sweden, BIT 9 (1969), 87)

44. Suppose that we want to approximate a continuous function $f(x)$ on $|x| \leq 1$ by a polynomial $P_n(x)$ of degree n. Suppose further that we have found

$$f(x) - P_n(x) = \alpha_{n+1} T_{n+1}(x) + r(x)$$

where $T_{n+1}(x)$ denotes the Chebyshev polynomial of degree $n+1$ with

$$1/2^{n+1} \leq |\alpha_{n+1}| \leq 1/2^n$$

and

$$|r(x)| \leq |\alpha_{n+1}|/10, \quad |x| \leq 1.$$

Show that

$$0.4/2^n \leq \max_{|x| \leq 1} |f(x) - p_n^*(x)| \leq 1.1/2^n$$

where finally $p_n^*(x)$ denotes the optimal polynomial of degree n for $f(x)$ on $|x| \leq 1$.   (Uppsala Univ., Sweden, BIT 13 (1973), 375)

45. Determine the polynomial of second degree, which is the best approximation in maximum norm to $\sqrt{x}$ on the point set $\{0, \frac{1}{9}, \frac{4}{9}, 1\}$.
(Gothenburg Univ., Sweden, BIT 8 (1968), 343)

46. Suppose that we want to approximate the function $f(x) = (3+x)^{-1}$ on the interval $-1 \leq x \leq 1$ with a polynomial $p(x)$ such that

$$\max_{|x| \leq 1} |f(x) - p(x)| \leq 0.021$$

(a) Show that there does not exist a first degree polynomial satisfying this condition.

(b) Show that there exists a second degree polynomial satisfying this condition.   (Stockholm Univ., Sweden, BIT 14 (1974), 366)

47. Two terms of the Taylor expansion of e^x around $x = a$ are used to approximate e^x on the interval $0 \leq x \leq 1$. How should a be chosen so as to minimize the error in maximum norm? Compute a correct to two decimal places.
(Lund Univ., Sweden, BIT 12 (1972), 589)

48. The function $P_3(x) = x^3 - 9x^2 - 20x + 5$ is given. Find a second degree polynomial $P_2(x)$ such that

$$\delta = \max_{0 \leq x \leq 4} |P_3(x) - P_2(x)|$$

becomes as small as possible. The value of δ and the values of x for which $|P_3(x) - P_2(x)| = \delta$ should also be given.
(Inst. Tech., Lund, Sweden, BIT 7 (1967), 81)

49. Determine the values of a, b, c and d in the polynomial

$$P(x) = ax^3 + bx^2 + cx + d$$

which minimizes
$$\max_{-1 \leqslant x \leqslant 1} |P(x) - |x||$$
(Uppsala Univ., Sweden, BIT 8 (1968), 59)

50. Find a polynomial $p(x)$ of degree as low as possible such that
$$\max_{|x| \leqslant 1} |\exp(x^2) - p(x)| \leqslant 0.05$$
(Lund Univ., Sweden, BIT 15 (1975), 224)

51. Determine the best minimax approximation to $e^{|x|}$ with a polynomial of degree 0 and 1 for $|x| \leqslant 1$.
(Uppsala Univ., Sweden, BIT 17 (1977), 369)

52. The curve $y = e^{-x}$ is to be approximated by a straight line $y = b - ax$ such that $|b - ax - e^{-x}| \leqslant 0.005$. The line should be chosen in such a way that the criterion is satisfied over as large an interval $(0, c)$ as possible (where $c > 0$). Calculate a, b and c to 3 decimal accuracy.
(Inst. Tech., Lund, Sweden, BIT 5 (1965), 214)

53. Prove that the polynomial of best approximation of degree not exceeding 2 for $|x|$ in the interval $[-1, 1]$ is $x^2 + \frac{1}{8}$.

54. Find the lowest order polynomial which approximates the function
$$f(x) = \sum_{r=0}^{4} (-x)^r$$
in the range $0 \leqslant x \leqslant 1$, with an error less than 0.1.

55. Calculate $\min_{p} \max_{0 \leqslant x \leqslant 1} |f(x) - p(x)|$ when p is a polynomial of degree at most 2 and
$$f(x) = \int_{x}^{1} \frac{x^2}{y^3} dy$$
(Uppsala Univ., Sweden, BIT 18 (1978), 236)

56. Approximate
$$F(x) = \frac{1}{x} \int_{0}^{x} \frac{e^t - 1}{t} dt$$
by a third degree polynomial $P_3(x)$ so that
$$\max_{-1 \leqslant x \leqslant 1} |F(x) - P_3(x)| \leqslant 3 \times 10^{-4}$$
(Inst. Tech., Lund, Sweden, BIT 6 (1966), 270)

57. The function f is defined by
$$f(x) = \frac{1}{x} \int_{0}^{x} \frac{1 - e^{-t^2}}{t^2} dt$$
Approximate f by a polynomial $P(x) = a + bx + cx^2$, such that
$$\max_{|x| \leqslant 1} |f(x) - P(x)| \leqslant 5 \times 10^{-3}$$
(Lund Univ., Sweden, BIT 10 (1970), 228)

58. The function F is defined by

$$F(x) = \int_0^x \exp(-t^2/2)\, dt$$

Determine the coefficients of a fifth order polynomial $P_5(x)$ for which $|F(x) - P_5(x)| \leq 10^{-4}$ when $|x| \leq 1$
(the coefficients should be accurate to within $\pm 2 \times 10^{-5}$).

(Uppsala Univ., Sweden, BIT 5 (1965), 294)

59. Consider the following approximation problem:
Determine $\min_g \|1 - x - g(x)\|$ where $g(x) = ax + bx^2$ and a and b are real numbers.
Determine a best approximation g if

(a) $\|f\| = \sup_{0 \leq x \leq 1} |f(x)|$

(b) $\|f\|^2 = \int_0^1 f^2(x)\, dx$

Are the approximations unique?

(Uppsala Univ., Sweden, BIT 10 (1970), 515)

60. Find the polynomial $P(x)$ of degree 3 minimizing $\|q(x) - P(x)\|_2$ where the norm is defined by

$$(g, h) = \int_0^\infty g(x)h(x)e^{-x}\, dx \quad \text{and} \quad q(x) = x^5 - 3x^3 + 5$$

(Umea Univ., Sweden, BIT 19 (1979), 425)

CHAPTER 5

Differentiation and Integration

5.1 INTRODUCTION

Several methods are available to find the derivative of a function $f(x)$ or to evaluate the definite integral $\int_a^b f(x)\,dx$, a, b are real finite constants, in the closed form. However, when $f(x)$ is a complicated function or when it is given in tabular form, we use numerical methods. In this chapter, we discuss the numerical methods for approximating the derivative $f^{(r)}(x)$, $r \geqslant 1$ of a given function $f(x)$ and for the evaluation of the integrals $\int_a^b f(x)\,dx$ where a, b may be finite or infinite.

5.2 NUMERICAL DIFFERENTIATION

Numerical differentiation methods are obtained using one of the following three techniques:

   (a)   Methods based on Interpolation
   (b)   Methods based on Finite Difference Operators
   (c)   Methods based on Undetermined Coefficients

We now discuss each of the methods in detail.

5.2.1 Methods Based on Interpolation

Given the values of $f(x)$ at a set of points $x_0, x_1, \ldots, x_n$, the general approach for deriving the numerical differentiation methods is to first obtain the interpolating polynomials $P_n(x)$ and then differentiate this polynomial r times $(n \geqslant r)$ to get $P_n^{(r)}(x)$. The value of $P_n^{(r)}(x_k)$ gives the approximate value of $f^{(r)}(x)$ at the nodal point x_k. It may be noted that though $P_n(x)$ and $f(x)$ have the same values at the nodal points, yet the derivatives may differ considerably at these points as seen in Fig. 5.1.

The situation may be worse at a non-nodal point. The approximation may further deteriorate as the order of derivative increases. The quantity

$$E^{(r)}(x) = f^{(r)}(x) - P_n^{(r)}(x) \qquad (5.1)$$

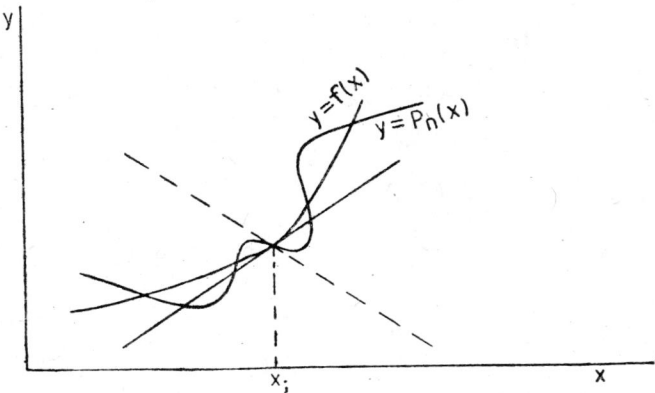

Fig. 5.1 Representation of first derivative

is called the **error of approximation** in the rth order derivative at any point x.

Non-uniform Nodal Points

If (x_i, f_i), $i = 0, 1, \ldots, n$ are $n+1$ distinct tabular points, then the Lagrange interpolating polynomial fitting this data is given by

$$P_n(x) = \sum_{k=0}^{n} l_k(x) f_k \qquad (5.2)$$

where $l_k(x)$ is the Lagrange fundamental polynomial

$$l_k(x) = \frac{\pi(x)}{(x - x_k)\pi'(x_k)}$$

and

$$f_k = f(x_k)$$
$$\pi(x) = (x - x_0)(x - x_1) \ldots (x - x_n)$$

The error of approximation in (5.2) is given by

$$E_n(x) = f(x) - P_n(x) = \frac{\pi(x)}{(n+1)!} f^{(n+1)}(\xi), \quad x_0 < \xi < x_n \qquad (5.3)$$

for any point x. Differentiating (5.2) and (5.3) with respect to x, we obtain

$$P'_n(x) = \sum_{k=0}^{n} l'_k(x) f_k \qquad (5.4)$$

and

$$E'_n(x) = \frac{\pi'(x)}{(n+1)!} f^{(n+1)}(\xi) + \frac{\pi(x)}{(n+1)!} \frac{d}{dx} (f^{(n+1)}(\xi)) \qquad (5.5)$$

Since the function $\xi(x)$ is unknown in the second term on the right hand side of (5.5), we cannot directly evaluate $E'_n(x)$. However at a nodal point x_k, $\pi(x_k) = 0$ and we get

$$E'_n(x_k) = \frac{\pi'(x_k)}{(n+1)!} f^{(n+1)}(\xi), \quad x_0 < \xi < x_n \tag{5.6}$$

provided $\frac{d}{dx}(f^{(n+1)}(\xi))$ remains bounded. For any r, $1 \leq r \leq n$, we obtain from (5.2)

$$f^{(r)}(x) \approx P_n^{(r)}(x) = \sum_{k=0}^{n} l_k^{(r)}(x) f_k \tag{5.7}$$

at any point x. The error term may be obtained by using the relation

$$\frac{1}{(n+1)!} \frac{d^j}{dx^j}(f^{(n+1)}(\xi)) = \frac{j!}{(n+j+1)!} f^{(n+j+1)}(\eta_j), \quad j = 1, 2, \ldots \tag{5.8}$$

where

$$\min(x_0, x_1, \ldots, x_n, x) < \eta_j < \max(x_0, x_1, \ldots, x_n, x)$$

If we use linear interpolation, we have

$$l_0(x) = \frac{x - x_1}{x_0 - x_1}, \quad l_1(x) = \frac{x - x_0}{x_1 - x_0}$$

$$P_1(x) = \frac{x - x_1}{x_0 - x_1} f_0 + \frac{x - x_0}{x_1 - x_0} f_1 \tag{5.9}$$

$$P'_1(x) = \frac{f_1 - f_0}{x_1 - x_0} \tag{5.10}$$

which is constant for all $x \in [x_0, x_1]$. We also have

$$E'_1(x_0) = \frac{x_0 - x_1}{2} f''(\xi) \tag{5.11}$$

$$E'_1(x_1) = \frac{x_1 - x_0}{2} f''(\xi), \quad x_0 < \xi < x_1 \tag{5.12}$$

Similarly for quadratic interpolation, we have

$$l_0(x) = \frac{(x - x_1)(x - x_2)}{(x_0 - x_1)(x_0 - x_2)}, \quad l'_0(x) = \frac{2x - x_1 - x_2}{(x_0 - x_1)(x_0 - x_2)}$$

$$l_1(x) = \frac{(x - x_0)(x - x_2)}{(x_1 - x_0)(x_1 - x_2)}, \quad l'_1(x) = \frac{2x - x_0 - x_2}{(x_1 - x_0)(x_1 - x_2)}$$

$$l_2(x) = \frac{(x - x_0)(x - x_1)}{(x_2 - x_0)(x_2 - x_1)}, \quad l'_2(x) = \frac{2x - x_0 - x_1}{(x_2 - x_0)(x_2 - x_1)}$$

$$P_2(x) = l_0(x) f_0 + l_1(x) f_1 + l_2(x) f_2 \tag{5.13}$$

$$P'_2(x) = l'_0(x) f_0 + l'_1(x) f_1 + l'_2(x) f_2 \tag{5.14}$$

which gives

$$P'_2(x_0) = \frac{2x_0 - x_1 - x_2}{(x_0 - x_1)(x_0 - x_2)} f_0 + \frac{x_0 - x_2}{(x_1 - x_0)(x_1 - x_2)} f_1$$

$$+ \frac{x_0 - x_1}{(x_2 - x_0)(x_2 - x_1)} f_2 \tag{5.15}$$

and
$$E_2'(x_0) = \tfrac{1}{6}(x_0 - x_1)(x_0 - x_2)f'''(\xi), \quad x_0 < \xi < x_2 \tag{5.16}$$

Similarly, we obtain
$$P''(x) = 2\left[\frac{f_0}{(x_0 - x_1)(x_0 - x_2)} + \frac{f_1}{(x_1 - x_0)(x_1 - x_2)}\right.$$
$$\left. + \frac{f_2}{(x_2 - x_0)(x_2 - x_1)}\right] \tag{5.17}$$

which is constant for all $x \in [x_0, x_2]$. The error at the tabular point x_0 is written as
$$\begin{aligned}E''(x_0) &= \tfrac{1}{3}(2x_0 - x_1 - x_2)f'''(\xi) \\ &\quad + \tfrac{1}{24}(x_0 - x_1)(x_0 - x_2)[f^{iv}(\eta_1) + f^{iv}(\eta_2)]\end{aligned} \tag{5.18}$$

where $\xi, \eta_1, \eta_2 \in (x_0, x_2)$.

Similar relations can be obtained at $x = x_1$ and $x = x_2$.

Uniform Nodal Points

When the distinct points $x_0, x_1, \ldots, x_n$ are equispaced with step length h, we have
$$x_i = x_0 + ih, \quad i = 1, 2, \ldots, n$$
$$f_i = f(x_i)$$

Using linear interpolation, we have
$$f'(x_0) \approx P_1'(x_0) = \frac{f_1 - f_0}{h} \tag{5.19}$$

and
$$E_1'(x_0) = f'(x_0) - \frac{f_1 - f_0}{h} \tag{5.20}$$

is the error of approximation, or the local truncation error. Expanding the terms in (5.20) in Taylor series about the point x_0, we get
$$E_1'(x_0) = -\frac{h}{2}f''(\xi), \quad x_0 < \xi < x_1 \tag{5.21}$$

which is same as (5.11).

Similarly, using quadratic interpolation we have
$$f'(x_0) \approx P_2'(x_0) = \frac{-3f_0 + 4f_1 - f_2}{2h} \tag{5.22}$$

with the local truncation error
$$\begin{aligned}E_2'(x_0) &= f'(x_0) - \frac{-3f_0 + 4f_1 - f_2}{2h} \\ &= -\frac{h^2}{3}f'''(\xi), \quad x_0 < \xi < x_2\end{aligned} \tag{5.23}$$

and
$$f''(x_0) \approx P_2''(x_0) = \frac{f_0 - 2f_1 + f_2}{h^2} \tag{5.24}$$

with the local truncation error
$$E_2''(x_0) = hf'''(\xi) + 0(h^2), \quad x_0 < \xi < x_2 \tag{5.25}$$

We now define the order of a numerical differentiation method.

Definition 5.1 A numerical differentiation method is said to be of order p if
$$|f^{(r)}(x) - P^{(r)}(x)| \leqslant ch^p \tag{5.26}$$

where c is a constant independent of h.

Thus the methods (5.19) and (5.24) are of first order, whereas the method (5.22) is of second order.

Example 5.1 Given the following values of $f(x) = \log x$, find the approximate value of $f'(2.0)$ and $f''(2.0)$, using the methods based on linear and quadratic interpolation. Also obtain an upper bound on the error

i	0	1	2
x_i	2.0	2.2	2.6
f_i	0.69315	0.78846	0.95551

(i) Using the method $f'(x_0) = \dfrac{f_1 - f_0}{x_1 - x_0}$ we get

$$f'(2.0) = \frac{0.78846 - 0.69315}{2.2 - 2.0} = 0.47655$$

If we use the method

(ii) $f'(x_0) = \dfrac{2x_0 - x_1 - x_2}{(x_0 - x_1)(x_0 - x_2)} f_0 + \dfrac{x_0 - x_2}{(x_1 - x_0)(x_1 - x_2)} f_1$
$\qquad + \dfrac{x_0 - x_1}{(x_2 - x_0)(x_2 - x_1)} f_2$

we get

$$f'(2.0) = \frac{4 - 2.2 - 2.6}{(2 - 2.2)(2 - 2.6)}(0.69315) + \frac{2 - 2.6}{(2.2 - 2)(2.2 - 2.6)}(0.78846)$$
$$+ \frac{2 - 2.2}{(2.6 - 2)(2.6 - 2.2)}(0.95551)$$
$$= 0.49619$$

The exact value of $f'(2.0) = 0.5$.

Similarly using the method

(iii) $f''(x_0) = 2\left[\dfrac{f_0}{(x_0 - x_1)(x_0 - x_2)} + \dfrac{f_1}{(x_1 - x_0)(x_1 - x_2)}\right.$
$\qquad \left. + \dfrac{f_2}{(x_2 - x_0)(x_2 - x_1)}\right]$

we get

$$f''(2.0) = 2\left[\frac{0.69315}{(2-2.2)(2-2.6)} + \frac{0.78846}{(2.2-2)(2.2-2.6)}\right.$$
$$\left. + \frac{0.95551}{(2.6-2)(2.6-2.2)}\right]$$
$$= -.19642$$

The exact value of $f''(2.0) = -.25$.

The error associated with the methods are given by

$$E_1'(x_0) = \frac{x_0 - x_1}{2} f''(\xi), \quad x_0 < \xi < x_1$$

$$E_2'(x_0) = \frac{1}{6}(x_0 - x_1)(x_0 - x_2) f'''(\xi), \quad x_0 < \xi < x_2$$

$$E_2''(x_0) = \frac{1}{3}(2x_0 - x_1 - x_2) f'''(\xi) + \frac{1}{24}(x_0 - x_1)(x_1 - x_2)$$
$$[f^{iv}(\eta_1) + f^{iv}(\eta_2)], \quad x_0 < \xi, \eta_1, \eta_2 < x_2$$

For $f(x) = \log x$, we can write

$$M_1 = \max_{x_0 < \xi < x_1} |f'(x)| = \max_{2.0 < \xi < 2.2} \left|\frac{1}{x}\right| = 0.5$$

$$M_2 = \max_{x_0 < \xi < x_1} |f''(x)| = \max_{2.0 < \xi < 2.2} \left|-\frac{1}{x^2}\right| = 0.25$$

$$M_3 = \max_{x_0 < \xi < x_2} |f'''(x)| = \max_{2.0 < \xi < 2.6} \left|\frac{2}{x^3}\right| = 0.25$$

$$M_4 = \max_{x_0 < \xi < x_2} |f^{iv}(x)| = \max_{2.0 < \xi < 2.6} \left|-\frac{6}{x^4}\right| = 0.375$$

and obtain

$$|E_1'(2.0)| \leq \left|\frac{2-2.2}{2}\right|(.25) = 0.025$$

$$|E_2'(2.0)| \leq \frac{1}{6}|(2-2.2)(2-2.6)|(.25) = 0.005$$

$$|E_2''(2.0)| \leq \frac{1}{3}|(4-2.2-2.6)|(.25) + \frac{1}{24}|(2-2.2)(2.2-2.6)|(0.75)$$
$$= 0.06917$$

5.2.2 Methods Based on Finite Differences

We consider the relation

$$Ef(x) = f(x+h) = f(x) + hf'(x) + \frac{h^2}{2!} f''(x) + \cdots$$
$$= \left(1 + hD + \frac{h^2 D^2}{2!} + \cdots\right) f(x)$$
$$= e^{hD} f(x) \tag{5.27}$$

in which
$$D = \frac{d}{dx}$$
is called the differential operator.

Symbolically we get from (5.27)
$$e^{hD} = E$$
or
$$hD = \log E$$

Thus we have

$$hD = \log E = \begin{vmatrix} \log(1+\Delta) = \Delta - \tfrac{1}{2}\Delta^2 + \tfrac{1}{3}\Delta^3 - \cdots \\ -\log(1-\nabla) = \nabla + \tfrac{1}{2}\nabla^2 + \tfrac{1}{3}\nabla^3 + \cdots \\ 2\sinh^{-1}\left(\frac{\delta}{2}\right) = \delta - \frac{1^2}{2^2 \cdot 3!}\delta^3 + \cdots \end{vmatrix} \qquad (5.28)$$

We can write

$$hf'(x_k) = hDf(x_k) = \begin{vmatrix} \Delta f_k - \tfrac{1}{2}\Delta^2 f_k + \tfrac{1}{3}\Delta^3 f_k - \cdots \\ \nabla f_k + \tfrac{1}{2}\nabla^2 f_k + \tfrac{1}{3}\nabla^3 f_k + \cdots \\ \delta f_k - \frac{1^2}{2^2 \cdot 3!}\delta^3 f_k + \cdots \end{vmatrix} \qquad (5.29)$$

Since
$$\mu = \sqrt{\left\{1 + \frac{\delta^2}{4}\right\}}$$
we can also write
$$hD = \frac{\mu}{\sqrt{\left\{1 + \frac{\delta^2}{4}\right\}}} \left(2 \sinh^{-1}\frac{\delta}{2}\right)$$
$$= \mu\left(\delta - \frac{1^2}{3!}\delta^3 + \frac{1^2 \cdot 2^2}{5!}\delta^5 - \cdots\right)$$

Thus we get
$$hf'(x_k) = \mu\delta f_k - \frac{1^2}{3!}\mu\delta^3 f_k + \frac{1^2 \cdot 2^2}{5!}\mu\delta^5 f_k - \cdots \qquad (5.30)$$

Similarly, we obtain

$$h^r D^r = \begin{vmatrix} \Delta^r - \frac{1}{2}r\Delta^{r+1} + \frac{r(3r+5)}{24}\Delta^{r+2} - \cdots \\ \nabla^r + \frac{1}{2}r\nabla^{r+1} + \frac{r(3r+5)}{24}\nabla^{r+2} + \cdots \\ \mu\delta^r - \frac{r+3}{24}\mu\delta^{r+2} + \frac{5r^2 + 52r + 135}{5760}\mu\delta^{r+4} - \cdots, \quad r \text{ odd} \\ \delta^r - \frac{r}{24}\delta^{r+2} + \frac{r(5r+22)}{5760}\delta^{r+4} - \cdots, \quad r \text{ even} \end{vmatrix}$$
$$(5.31)$$

In particular, differentiation methods for $r = 1$ and $r = 2$ at $x = x_k$ become

$$hf'(x_k) = \begin{vmatrix} \Delta f_k - \frac{1}{2}\Delta^2 f_k + \frac{1}{3}\Delta^3 f_k - \cdots \\ \nabla f_k + \frac{1}{2}\nabla^2 f_k + \frac{1}{3}\nabla^3 f_k + \cdots \\ \mu\delta f_k - \frac{1}{6}\mu\delta^3 f_k + \frac{1}{30}\mu\delta f_k^5 - \cdots \end{vmatrix} \quad (5.32)$$

$$h^2 f''(x_k) = \begin{vmatrix} \Delta^2 f_k - \Delta^3 f_k + \frac{11}{12}\Delta^4 f_k - \cdots \\ \nabla^2 f_k + \nabla^3 f_k + \frac{11}{12}\nabla^4 f_k + \cdots \\ \delta^2 f_k - \frac{1}{12}\delta^4 f_k + \frac{1}{90}\delta^6 f_k - \cdots \end{vmatrix} \quad (5.33)$$

Keeping only the first term in each of the methods in (5.32), we get

$$f'(x_k) = \begin{vmatrix} \dfrac{f_{k+1} - f_k}{h} & \quad (5.34\text{i}) \\[6pt] \dfrac{f_k - f_{k-1}}{h} & \quad (5.34\text{ii}) \\[6pt] \dfrac{f_{k+1} - f_{k-1}}{2h} & \quad (5.34\text{iii}) \end{vmatrix}$$

It can be verified that the methods (5.34i) and (5.34ii) are of first order whereas the method (5.34iii) is of second order.

Similarly, if we retain only one term in each of the methods in (5.33) then we get

$$f''(x_k) = \begin{vmatrix} \dfrac{f_{k+2} - 2f_{k+1} + f_k}{h^2} & \quad (5.35\text{i}) \\[6pt] \dfrac{f_k - 2f_{k-1} + f_{k-2}}{h^2} & \quad (5.35\text{ii}) \\[6pt] \dfrac{f_{k-1} - 2f_k + f_{k+1}}{h^2} & \quad (5.35\text{iii}) \end{vmatrix}$$

It can be verified that the methods (5.35i) and (5.35ii) are of first order whereas the method (5.35iii) is of second order.

5.2.3 Methods Based on Undetermined Coefficients

Numerical differentiation methods based on interpolating polynomials, express $f^{(r)}(x)$ as a linear combination of the values of $f(x)$ at a set of prechosen tabular points. In the method of undetermined coefficients, we express $f^{(r)}(x)$ as a linear combination of the values of $f(x)$ at an arbitrarily chosen set of tabular points. Assuming that the tabular points are equispaced with the step length h, we write as

$$h^r f^{(r)}(x_k) = \sum_{\nu = -p}^{p} a_\nu f_{k+\nu} \quad (5.36)$$

for symmetric arrangement of tabular points

or

$$h^r f^{(r)}(x_k) = \sum_{\nu=\pm\lambda}^{p} a_\nu f_{k+\nu} \tag{5.37}$$

for non-symmetric arrangement of tabular points. The local truncation error is defined by

$$E^{(r)}(x_k) = \frac{1}{h^r}\left(h^r f^{(r)}(x_k) - \sum_{\nu=-p}^{p} a_\nu f_{k+\nu}\right) \tag{5.38}$$

or

$$E^{(r)}(x_k) = \frac{1}{h^r}\left(h^r f^{(r)}(x_k) - \sum_{\nu=\pm\lambda}^{p} a_\nu f_{k+\nu}\right) \tag{5.39}$$

The coefficients a_ν's in (5.36) or (5.37) are determined by requiring the method to be of a particular order. We expand the right hand side in (5.36) or (5.37) in Taylor's series about the point x_k and on equating the coefficients of various order derivatives on both sides, we obtain the required number of equations to determine these coefficients. The first nonzero term in (5.38) or (5.39) gives the error of approximation. As a particular case we take $r = 2$ and $\rho = 2$ in (5.36) and get

$$h^2 f''(x_k) = a_{-2} f_{k-2} + a_{-1} f_{k-1} + a_0 f_k + a_1 f_{k+1} + a_2 f_{k+2}$$
$$= (a_{-2} + a_{-1} + a_0 + a_1 + a_2) f_k + h(-2a_{-2} - a_{-1} + a_1 + 2a_2) f'_k$$
$$+ \frac{h^2}{2}(4a_{-2} + a_{-1} + a_1 + 4a_2) f''_k + \frac{h^3}{6}(-8a_{-2} - a_{-1} + a_1 + 8a_2) f'''_k$$
$$+ \frac{h^4}{24}(16a_{-2} + a_{-1} + a_1 + 16a_2) f^{iv}_k$$
$$+ \frac{h^5}{120}(-32a_{-2} - a_{-1} + a_1 + 32a_2) f^{v}_k$$
$$+ \frac{h^6}{720}(64a_{-2} + a_{-1} + a_1 + 64a_2) f^{(vi)}(\xi) + \ldots$$

Comparing the coefficients of $f_k^{(i)}$, $i = 0, 1, 2, 3, 4$ on both sides we get the system of equations

$$a_{-2} + a_{-1} + a_0 + a_1 + a_2 = 0$$
$$-2a_{-2} - a_{-1} + a_1 + 2a_2 = 0$$
$$4a_{-2} + a_{-1} + a_1 + 4a_2 = 2$$
$$-8a_{-2} - a_{-1} + a_1 + 8a_2 = 0$$
$$16a_{-2} + a_{-1} + a_1 + 16a_2 = 0$$

Solving the above system of equations, we get

$$a_{-2} = a_2 = -\tfrac{1}{12}$$
$$a_{-1} = a_1 = \tfrac{16}{12}$$
$$a_0 = -\tfrac{30}{12}$$

and the method becomes

$$f''(x_k) = \frac{-f_{k-2} + 16f_{k-1} - 30f_k + 16f_{k+1} - f_{k+2}}{12h^2} \quad (5.40)$$

The first non-zero term in (5.38) gives the error of approximation as

$$\text{Error} = -\frac{h^4}{90} f^{vi}(\xi), \quad x_{k-2} < \xi < x_{k+2}$$

Thus the method (5.40) is of fourth order.

5.3 OPTIMUM CHOICE OF STEP-LENGTH

In numerical differentiation methods, error of approximation or the truncation error is of the form ch^p which tends to zero as $h \to 0$. However the method which approximates $f^{(r)}(x)$ contains h^r in the denominator. As h is successively decreased to smaller values, the truncation error decreases, but the roundoff error in the method may increase as we are dividing by a small number. It may happen that after a certain critical value of h, the roundoff error may become more dominant than the truncation error and the numerical results obtained may start worsening as h is further reduced. When $f(x)$ is given in tabular form, these values may not themselves be exact. These values contain the roundoff errors, that is, $f(x_i) = f_i + \epsilon_i$, where $f(x_i)$ is the exact value and f_i is the tabulated value. To see the effect of this roundoff error in numerical differentiation methods, we consider the method

$$f'(x_0) = \frac{f(x_1) - f(x_0)}{h} - \frac{h}{2} f''(\xi), \quad x_0 < \xi < x_1 \quad (5.41)$$

If the roundoff errors in $f(x_0)$ and $f(x_1)$ are ϵ_0 and ϵ_1 respectively, then we have

$$f'(x_0) = \frac{f_1 - f_0}{h} + \frac{\epsilon_1 - \epsilon_0}{h} - \frac{h}{2} f''(\xi) \quad (5.42)$$

or

$$f'(x_0) = \frac{f_1 - f_0}{h} + \text{RE} + \text{TE} \quad (5.43)$$

where RE and TE denote the roundoff error and truncation error respectively. If we take

$$\epsilon = \max(|\epsilon_1|, |\epsilon_2|)$$

and

$$M_2 = \max_{x_0 \leq x \leq x_1} |f''(x)|$$

then we get

$$|\text{RE}| \leq \frac{2\epsilon}{h}$$

$$|\text{TE}| \leq \frac{h}{2} M_2$$

We may call that value of h as an **optimal value** for which one of the following criteria is satisfied:

(i) $|RE| = |TE|$

(ii) $|RE| + |TE| = $ minimum (5.44)

If we use the criterion (5.44i), then we have

$$\frac{2\epsilon}{h} = \frac{h}{2} M_2$$

which gives

$$h_{\text{opt}} = 2\sqrt{\frac{\epsilon}{M_2}}$$

and

$$|RE| = |TE| = \sqrt{\epsilon M_2}$$

If we use the criterion (5.44ii), then we have

$$\frac{2\epsilon}{h} + \frac{h}{2} M_2 = \text{minimum}$$

Therefore

$$-\frac{2\epsilon}{h^2} + \frac{1}{2} M_2 = 0$$

or

$$h_{\text{opt}} = 2\sqrt{\frac{\epsilon}{M_2}}$$

The minimum total error is $2(\epsilon M_2)^{1/2}$.

This means that if the roundoff error is of the order 10^{-k} (say) and $M_2 \approx 0(1)$, then the accuracy given by the method may be approximately of the order $10^{-k/2}$. Since in any numerical differentiation method, the local truncation error is always proportional to some power of h, whereas the roundoff error is inversely proportional to some power of h, the same technique can be used to determine an optimal value of h for any numerical method which approximates $f^{(r)}(x_k)$, $r \geq 1$.

Example 5.2 For the method

$$f'(x_0) = \frac{-3f(x_0) + 4f(x_1) - f(x_2)}{2h} + \frac{h^2}{3} f'''(\xi), \quad x_0 < \xi < x_2$$

determine the optimal value of h, using the criteria

(i) $|RE| = |TE|$

(ii) $|RE| + |TE| = $ minimum

Using this method and the value of h obtained from the criterion $|RE| = |TE|$, determine an approximate value of $f'(2.0)$ from the following tabulated values of $f(x) = \log x$.

x	2.0	2.01	2.02	2.06	2.12
$f(x)$	0.69315	0.69813	0.70310	0.72271	0.75142

given that the maximum roundoff error in function evaluation is 5×10^{-6}.

If ϵ_0, ϵ_1 and ϵ_2 are the roundoff errors in the given function evaluations f_0, f_1 and f_2 respectively, then we have

$$f'(x_0) = \frac{-3f_0 + 4f_1 - f_2}{2h} + \frac{-3\epsilon_0 + 4\epsilon_1 - \epsilon_2}{2h} + \frac{h^2}{3}f'''(\xi)$$

$$= \frac{-3f_0 + 4f_1 - f_2}{2h} + \text{RE} + \text{TE}$$

Using

$$\epsilon = \max(|\epsilon_0|, |\epsilon_1|, |\epsilon_2|)$$

and

$$M_3 = \max_{x_0 \leq x \leq x_2} |f'''(x)|$$

we obtain

$$|\text{RE}| \leq \frac{8\epsilon}{2h}, \quad |\text{TE}| \leq \frac{h^2 M_3}{3}$$

If we use $|\text{RE}| = |\text{TE}|$, we get

$$\frac{8\epsilon}{2h} = \frac{h^2 M_3}{3}$$

which gives

$$h^3 = \frac{12\epsilon}{M_3}$$

or

$$h_{\text{opt}} = \left(\frac{12\epsilon}{M_3}\right)^{1/3}$$

and

$$|\text{RE}| = |\text{TE}| = \frac{4\epsilon^{2/3} M_3^{1/}}{(\ 2)^{1/3}}$$

If we use $|\text{RE}| + |\text{TE}| = \text{minimum}$, we get

$$\frac{4\epsilon}{h} + \frac{M_3 h^2}{3} = \text{minimum}$$

which gives

$$\frac{-4\epsilon}{h^2} + \frac{2 M_3 h}{3} = 0$$

and

$$h_{\text{opt}} = \left(\frac{6\epsilon}{M_3}\right)^{1/3}$$

Minimum total error $= 6^{2/3}\epsilon^{2/3} M_3^{1/3}$.

When $f(x) = \log(x)$, we have

$$M_3 = \max_{2.0 \leq x \leq 2.12} |f'''(x)| = \tfrac{1}{4}$$

Using the criterion $|RE| = |TE|$ and $\epsilon = 5 \times 10^{-6}$, we get

$$h_{opt} = \left(\frac{12 \times 5 \times 10^{-6}}{\frac{1}{4}}\right)^{1/3} \simeq .06$$

For $h = 0.06$, we get

$$f'(2.0) = \frac{-3(0.69315) + 4(0.72271) - 0.75142}{0.12}$$

$$= 0.49975$$

If we take $h = .01$, we get

$$f'(2.0) = \frac{-3(0.69315) + 4(0.69813) - 0.70310}{0.02}$$

$$= 0.49850$$

The exact value of $f'(2.0) = 0.5$.
This verifies that for $h < h_{opt}$, the results deteriorate.

5.4 EXTRAPOLATION METHODS

To obtain differentiation methods of high order, we require a large number of tabular points and thus a large number of function evaluations. Consequently there is a possibility that the roundoff errors may increase so much that the numerical results may become meaningless. However, it is generally possible to obtain higher order methods by combining the computed values obtained by using a certain method with different step sizes.

Let $g(h)$ denote the approximate value of g, obtained by using a method of order p, with step length h and $g(qh)$ denote the value of g obtained by using the same method of order p, with step length qh. We have

$$g(h) = g + ch^p + O(h^{p+1})$$
$$g(qh) = g + cq^p h^p + O(h^{p+1})$$

Eliminating c from the above equations, we get

$$g = \frac{q^p g(h) - g(qh)}{q^p - 1} + O(h^{p+1})$$

Thus we obtain

$$g^{(1)}(h) = \frac{q^p g(h) - g(qh)}{q^p - 1} = g + O(h^{p+1}) \tag{5.45}$$

which is of order $p + 1$. This technique of combining two computed values, obtained by using the same method with two different step sizes, to obtain a higher order method is called the **Extrapolation method** or **Richardson's extrapolation**.

If the local truncation error associated with the method is known as a power series in h, then by repeating the extrapolation procedure a number of times, we can obtain the methods of any arbitrary order. The application of this procedure becomes simplified when the step lengths form a geometric sequence. For simplicity we generally take $q = \frac{1}{2}$. To illustrate the procedure we consider the method

$$f'(x_0) = \frac{f_1 - f_{-1}}{2h} \tag{5.46}$$

where $f_1 = f(x_0 + h)$ and $f_{-1} = f(x_0 - h)$. The local truncation error associated with the method (5.46) is obtained as

$$E'(x_0) = c_1 h^2 + c_2 h^4 + c_3 h^6 + \ldots \tag{5.47}$$

where $c_1, c_2, c_3, \ldots$ are constants independent of h.

Let $g(x) = f'(x_0)$ be the quantity which is to be obtained and $g(h/2^r)$ denote the approximate value of $g(x)$ obtained by using the method (5.46) with step length $h/2^r$, $r = 0, 1, 2, \ldots$. Thus we have

$$g(h) = g(x) + c_1 h^2 + c_2 h^4 + c_3 h^6 + \ldots$$

$$g\left(\frac{h}{2}\right) = g(x) + \frac{c_1 h^2}{4} + \frac{c_2 h^4}{16} + \frac{c_3 h^6}{64} + \ldots \tag{5.48}$$

$$g\left(\frac{h}{2^2}\right) = g(x) + \frac{c_1 h^2}{16} + \frac{c_2 h^4}{256} + \frac{c_3 h^6}{4096} + \ldots$$

$$\vdots$$

Eliminating c_1 from the above equations we obtain

$$g^{(1)}(h) = \frac{4g\left(\frac{h}{2}\right) - g(h)}{3} = g(x) - \frac{1}{4}c_2 h^4 - \frac{5}{16}c_3 h^6 - \ldots$$

$$g^{(1)}\left(\frac{h}{2}\right) = \frac{4g\left(\frac{h}{2^2}\right) - g\left(\frac{h}{2}\right)}{3} = g(x) - \frac{1}{64}c_2 h^4 - \frac{5}{1024}c_3 h^6 - \ldots \tag{5.49}$$

$$\vdots$$

Thus $g^{(1)}(h), g^{(1)}(h/2), \ldots$ given by (5.49) are $0(h^4)$ approximations to $g(x)$. Eliminating c_2 from (5.49), we obtain

$$g^{(2)}(h) = \frac{4^2 g^{(1)}\left(\frac{h}{2}\right) - g^{(1)}(h)}{4^2 - 1} + \frac{1}{64}c_3 h^6 + \ldots \tag{5.50}$$

which gives an $0(h^6)$ approximation.

Thus the successive higher order methods can be obtained from the formula

$$g^{(m)}(h) = \frac{4^m g^{(m-1)}\left(\frac{h}{2}\right) - g^{(m-1)}(h)}{4^m - 1}, \quad m = 1, 2, \ldots \tag{5.51}$$

where

$$g^{(0)}(h) = g(h)$$

This procedure is called repeated extrapolation to the limit. The successive values of $g^{(m)}(h)$ for various values of m can be calculated as given in Table 5.1.

Table 5.1 Extrapolation Table

Order h	Second	Fourth	Sixth	Eighth
h	$g(h)$			
		$g^{(1)}(h)$		
$\dfrac{h}{2}$	$g\left(\dfrac{h}{2}\right)$		$g^{(2)}(h)$	
		$g^{(1)}\left(\dfrac{h}{2}\right)$		$g^{(3)}(h)$
			$g^{(2)}\left(\dfrac{h}{2}\right)$	
$\dfrac{h}{2^2}$	$g\left(\dfrac{h}{2^2}\right)$			
		$g^{(1)}\left(\dfrac{h}{2^2}\right)$		
$\dfrac{h}{2^3}$	$g\left(\dfrac{h}{2^3}\right)$			

It may be noted, that in Table 5.1, the successive entries in a particular column give better approximations than the preceding entries. Similarly the successive columns give better approximations than the preceding columns. The best results can be obtained from the lower diagonal terms. The extrapolation can be stopped when

$$\left| g^{(k)}(h) - g^{(k-1)}\left(\frac{h}{2}\right) \right| < \epsilon \tag{5.52}$$

for a given error tolerance ϵ.

Example 5.3   The following table of values is given:

x	-1	1	2	3	4	5	7
$f(x)$	1	1	16	81	256	625	2401

Using the formula $f'(x_1) = (f(x_2) - f(x_0))/2h$ and the Richardson extrapolation, find $f'(3)$.

Using the Taylor series expansions, we find

$$\frac{f(x_2) - f(x_0)}{2h} = f'(x_1) + \frac{h^2}{6} f^{(3)}(x_1) + \frac{h^4}{120} f^{(5)}(x_1) + \cdots$$

Hence, for extrapolation, Table 5.1 can be used. We find

h		$f'(3)$	
	$O(h^2)$	$O(h^4)$	$O(h^6)$
4	300		
		103	
2	156		108
		108	
1	120		

Obviously $f'(3) = 108$, must be the exact solution as the exact arithmetic is done and $g^{(1)}(4) = g^{(1)}(2) = g^{(2)}(4)$. This is true, since the data represents $f(x) = x^4$ and the second column must produce the exact solution as the leading error term is $ch^4 f^{(5)}(\xi)$.

5.5 Partial Differentiation

We can use any of the three techniques discussed in the previous sections to obtain numerical partial differentiation methods. We consider only one variable at a time and treat the remaining variables as constants. We consider here a function $f(x, y)$ of two variables only. Let the values of the function $f(x, y)$ be given at a set of points (x_i, y_j) in the (x, y) plane with spacing h and k in x and y directions respectively. We have

$$x_i = x_0 + ih, \quad y_j = y_0 + jk, \quad i, j = 1, 2, \ldots$$

We can now write

$$\left(\frac{\partial f}{\partial x}\right)_{(x_i, y_j)} = \begin{vmatrix} \frac{f_{i+1,j} - f_{i,j}}{h} + O(h) \\ \frac{f_{i,j} - f_{i-1,j}}{h} + O(h) \\ \frac{f_{i+1,j} - f_{i-1,j}}{2h} + O(h^2) \end{vmatrix} \qquad (5.53)$$

where $f_{i,j} = f(x_i, y_j)$.

Similarly we can write

$$\left(\frac{\partial f}{\partial y}\right)_{(x_i, y_j)} = \begin{vmatrix} \frac{f_{i,j+1} - f_{i,j}}{k} + O(k) \\ \frac{f_{i,j} - f_{i,j-1}}{k} + O(k) \\ \frac{f_{i,j+1} - f_{i,j-1}}{2k} + O(k^2) \end{vmatrix} \qquad (5.54)$$

$$\left(\frac{\partial^2 f}{\partial x^2}\right)_{(x_i, y_j)} = \frac{f_{i-1,j} - 2f_{i,j} + f_{i+1,j}}{h^2} + O(h^2) \qquad (5.55)$$

$$\left(\frac{\partial^2 f}{\partial y^2}\right)_{(x_i, y_j)} = \frac{f_{i,j-1} - 2f_{i,j} + f_{i,j+1}}{k^2} + 0(k^2) \tag{5.56}$$

Since

$$\frac{\partial^2 f}{\partial x\, \partial y} = \frac{\partial}{\partial x}\left(\frac{\partial f}{\partial y}\right) = \frac{\partial}{\partial y}\left(\frac{\partial f}{\partial x}\right)$$

we can write

$$\left(\frac{\partial^2 f}{\partial x\, \partial y}\right)_{(x_i, y_j)} = \frac{\partial}{\partial x}\left(\frac{\partial f}{\partial y}\right)(x_i, y_j)$$

$$\approx \frac{\partial}{\partial x}\left(\frac{f_{i,j+1} - f_{i,j-1}}{2k}\right)$$

$$\approx \frac{1}{2k}\left[\frac{f_{i+1,j+1} - f_{i-1,j+1}}{2h} - \frac{f_{i+1,j-1} - f_{i-1,j-1}}{2h}\right]$$

$$\approx \frac{f_{i+1,j+1} - f_{i-1,j+1} - f_{i+1,j-1} + f_{i-1,j-1}}{4hk} \tag{5.57}$$

The method (5.57) is of $0(h^2 + k^2)$. We can also write

$$\left(\frac{\partial^2 f}{\partial x\, \partial y}\right)_{(x_i, y_j)} = \frac{\partial}{\partial y}\left(\frac{\partial f}{\partial x}\right)_{(x_i, y_j)}$$

$$\approx \frac{\partial}{\partial y}\left(\frac{f_{i+1,j} - f_{i-1,j}}{2h}\right)$$

$$\approx \frac{1}{2h}\left[\frac{f_{i+1,j+1} - f_{i+1,j-1}}{2k} - \frac{f_{i-1,j+1} - j_{i-1,j-1}}{2k}\right]$$

$$\approx \frac{f_{i+1,j+1} - f_{i+1,j-1} - f_{i-1,j+1} + f_{i-1,j-1}}{4hk}$$

which is same as (5.57).

Example 5.4 Find the Jacobian matrix for the system of equations

$$f_1(x, y) = x^2 + y^2 - x = 0$$
$$f_2(x, y) = x^2 - y^2 - y = 0$$

at the point (1, 1), using the methods

$$\left(\frac{\partial f}{\partial x}\right)_{(x_i, y_j)} = \frac{f_{i+1,j} - f_{i-1,j}}{2h}$$

$$\left(\frac{\partial f}{\partial y}\right)_{(x_i, y_j)} = \frac{f_{i,j+1} - f_{i,j-1}}{2k}$$

with $h = k = 1$.

The Jacobian matrix is given as

$$J = \begin{bmatrix} \dfrac{\partial f_1}{\partial x} & \dfrac{\partial f_1}{\partial y} \\ \dfrac{\partial f_2}{\partial x} & \dfrac{\partial f_2}{\partial y} \end{bmatrix}$$

We have $x_i = 1$, $y_j = 1$.

$$\left(\frac{\partial f_1}{\partial x}\right)_{(1,1)} = \frac{f_1(1+h, 1) - f_1(1, 1)}{2h} = \frac{f_1(2, 1) - f_1(1, 1)}{2} = 1$$

$$\left(\frac{\partial f_1}{\partial y}\right)_{(1,1)} = \frac{f_1(1, 1+k) - f_1(1, 1)}{2k} = \frac{f_1(1, 2) - f_1(1, 1)}{2} = 2$$

$$\left(\frac{\partial f_2}{\partial x}\right)_{(1,1)} = \frac{f_2(1+h, 1) - f_2(1, 1)}{2h} = \frac{f_2(2, 1) - f_2(1, 1)}{2} = 2$$

$$\left(\frac{\partial f_2}{\partial y}\right)_{(1,1)} = \frac{f_2(1, 1+k) - f_2(1, 1)}{2k} = \frac{f_2(1, 2) - f_2(1, 1)}{2} = -3$$

Hence we get

$$J = \begin{bmatrix} 1 & 2 \\ 2 & -3 \end{bmatrix}$$

5.6 NUMERICAL INTEGRATION

The general problem of numerical integration is to find an approximate value of the integral

$$I = \int_a^b w(x) f(x) \, dx \tag{5.58}$$

where $w(x) > 0$ in $[a, b]$ is the **weight function**. We assume that $w(x)$ and $w(x)f(x)$ are integrable, in the Riemann sense on $[a, b]$. The limits of integration may be finite, semi-infinite or infinite. The integral (5.58) is approximated by a finite linear combination of values of $f(x)$ in the form

$$I = \int_a^b w(x) f(x) \, dx \approx \sum_{k=0}^{n} \lambda_k f_k \tag{5.59}$$

where x_k, $k = 0(1)n$ are called the **abscissas** or **nodes** distributed within the limits of integration $[a, b]$ and λ_k, $k = 0(1)n$ are called the **weights** of the **integration rule** or the **quadrature formula** (5.59). The error of approximation is given as

$$R_n = \int_a^b w(x) f(x) \, dx - \sum_{k=0}^{n} \lambda_k f_k \tag{5.60}$$

We now define the order of the integration method (5.59).

Definition 5.2 An integration method of the form (5.59) is said to be of order p, if it produces exact results ($R_n = 0$) for all polynomials of degree less than or equal to p.

In (5.59) we have $2n + 2$ unknowns ($n + 1$ nodes x_k's and $n + 1$ weights λ_k's) and the method can be made exact for polynomials of degree $\leqslant 2n+1$. Thus the method of the form (5.59) can be of maximum order $2n + 1$. If some of the nodes are prescribed in advance, the order will be reduced. If all the $n + 1$ abscissae are prescribed, then we have to determine only $n + 1$ weights and the corresponding method will be of maximum order n.

5.7 METHODS BASED ON INTERPOLATION

Given the $n+1$ abscissae x_k's and the corresponding values f_k's, the Lagrange interpolating polynomial fitting the data (x_k, f_k), $k = 0(1)n$ is given by

$$f(x) = P_n(x) = \sum_{k=0}^{n} l_k(x) f_k + \frac{\pi(x)}{(n+1)!} f^{(n+1)}(\xi), \quad x_0 < \xi < x_n \quad (5.61)$$

where $l_k(x)$ is the Lagrange fundamental polynomial

$$l_k(x) = \frac{\pi(x)}{(x - x_k)\pi'(x_k)}$$

and

$$\pi(x) = (x - x_0)(x - x_1) \ldots (x - x_n)$$

We replace the function $f(x)$ in (5.58) by the interpolation polynomial (5.61) and integrate within the given limits. We obtain

$$I = \int_a^b w(x) f(x) \, dx = \sum_{k=0}^{n} \left[\int_a^b w(x) l_k(x) \, dx \right] f_k$$
$$+ \int_a^b w(x) \frac{\pi(x)}{(n+1)!} f^{(n+1)}(\xi) \, dx$$
$$= \sum_{k=0}^{n} \lambda_k f_k + R_n \quad (5.62)$$

where

$$\lambda_k = \int_a^b w(x) l_k(x) \, dx \quad (5.63)$$

and

$$R_n = \frac{1}{(n+1)!} \int_a^b w(x) \pi(x) f^{(n+1)}(\xi) \, dx \quad (5.64)$$

If $\pi(x)$ does not change sign in $[a, b]$ and $f^{(n+1)}(x)$ is continuous in $[a, b]$, then, using the Mean Value Theorem of integral calculus, we can write the error of approximation (5.64) in the form

$$R_n = \frac{f^{(n+1)}(\eta)}{(n+1)!} \int_a^b w(x) \pi(x) \, dx, \quad \eta \in (a, b) \quad (5.65)$$

If $\pi(x)$ changes sign in $[a, b]$ we can write (5.64) as

$$|R_n| \leq \frac{1}{(n+1)!} \int_a^b w(x) |\pi(x)| |f^{(n+1)}(\xi)| \, dx$$
$$\leq \frac{M_{n+1}}{(n+1)!} \int_a^b w(x) |\pi(x)| \, dx \quad (5.66)$$

where

$$|f^{(n+1)}(x)| \leq M_{n+1}, \quad x \in [a, b]$$

The error term can also be obtained in the following manner: Since the method is exact for polynomials of degree $\leq n$,

$$R_n = 0 \text{ when } f(x) = x^i, \, i = 0, 1, \ldots, n$$

and
$$R_n \neq 0 \text{ when } f(x) = x^{n+1}$$

Thus we can write the error term in the form

$$R_n = \frac{C}{(n+1)!} f^{(n+1)}(\eta) \tag{5.67}$$

where

$$C = \int_a^b w(x) x^{n+1} \, dx - \sum_{k=0}^{n} \lambda_k x_k^{n+1} \tag{5.68}$$

is called the **error constant**. If C is zero for $f(x) = x^{n+1}$, then we take higher degree polynomials.

Neglecting the error in (5.62) we get the integration method

$$\int_a^b w(x) f(x) \, dx = \sum_{k=0}^{n} \lambda_k f_k \tag{5.69}$$

Newton-Cotes Methods

When $w(x) = 1$ and the nodes x_k's are equispaced with $x_0 = a$, $x_n = b$ with spacing $h = (b-a)/n$, the methods (5.69) are called **Newton-Cotes** integration methods. The weights λ_k's are called **Cotes numbers**. Setting $x = x_0 + sh$, we get

$$\pi(x) = h^{n+1} s(s-1) \ldots (s-n)$$

$$l_k(x) = \frac{(-1)^{n-k}}{k!(n-k)!} s(s-1) \ldots (s-k+1)(s-k-1) \ldots (s-n)$$

and

$$\lambda_k = \frac{(-1)^{n-k}}{k!(n-k)!} h \int_0^n s(s-1) \ldots (s-k+1)(s-k-1) \ldots (s-n) \, ds \tag{5.70}$$

$$R_n = \frac{h^{n+2}}{(n+1)!} \int_0^n s(s-1) \ldots (s-n) f^{(n+1)}(\xi) \, ds \tag{5.71}$$

For $n = 1$, we have $x_0 = a$, $x_1 = b$, $h = b - a$

$$\lambda_0 = -h \int_0^1 (s-1) \, ds = \frac{h}{2}$$

$$\lambda_1 = h \int_0^1 s \, ds = \frac{h}{2}$$

and we get the method

$$\int_a^b f(x) \, dx = \frac{b-a}{2} [f(a) + f(b)] \tag{5.72}$$

which is the **trapezoidal rule**.

Geometrically, it is the area of the trapezoid with width $b - a$ and ordinates $f(a)$, $f(b)$, which is an approximation to the area under the curve $y = f(x)$, above the x-axis and the ordinates $x = a$ and $x = b$. This approximation is shown in Fig. 5.2.

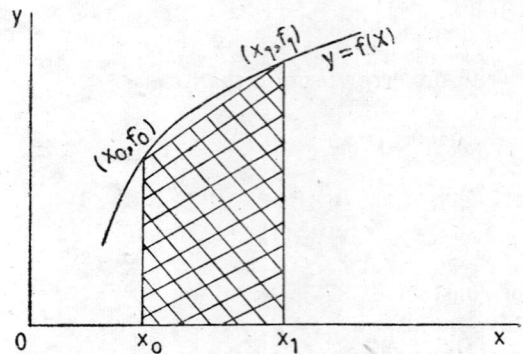

Fig. 5.2 Trapezoidal rule

The error in the trapezoidal rule becomes

$$R_1 = \frac{h^3}{2} \int_0^1 s(s-1) f''(\xi)\, ds$$

Since $s(s-1)$ does not change sign in $[0, 1]$, we get

$$R_1 = \frac{h^3}{2} f''(\eta) \int_0^1 s(s-1)\, ds, \qquad \eta \in (0, 1)$$

$$= -\frac{h^3}{12} f''(\eta)$$

$$= -\frac{(b-a)^3}{12} f''(\eta) \tag{5.73}$$

Thus the trapezoidal rule is exact for polynomials of degree ≤ 1 and hence is of order 1. Alternatively, we get from (5.68)

$$C = \int_a^b x^2\, dx - \tfrac{1}{2}(b-a)(b^2 + a^2)$$

$$= -\tfrac{1}{6}(b-a)^3$$

Using (5.67) we have

$$R_1 = -\frac{(b-a)^3}{12} f''(\eta)$$

which is same as (5.73).

For $n = 2$, we have $h = \dfrac{b-a}{2}$, $x_0 = a$, $x_1 = \dfrac{a+b}{2}$, $x_2 = b$. From (5.70) we get

$$\lambda_0 = \frac{h}{2} \int_0^2 (s-1)(s-2)\, ds = \frac{h}{3}$$

$$\lambda_1 = -h \int_0^2 s(s-2)\, ds = \frac{4h}{3}.$$

$$\lambda_2 = \frac{h}{2}\int_0^2 s(s-1)\,ds = \frac{h}{3}$$

and we get the method

$$I = \int_a^b f(x)\,dx = \frac{b-a}{6}\left[f(a) + 4f\left(\frac{a+b}{2}\right) + f(b)\right] \tag{5.74}$$

which is called the **Simpson's rule**. The error associated with this method is given by

$$R_2 = \frac{h^4}{3!}\int_0^2 s(s-1)(s-2)f'''(\xi)\,d\xi$$

We use (5.67) to obtain the error term. Since the method is exact for polynomials of degree ≤ 2, we have

$$C = \int_a^b x^3\,dx - \frac{b-a}{6}\left[a^3 + 4\left(\frac{a+b}{2}\right)^3 + b^3\right]$$

It is easy to verify that $C = 0$, which shows that the method (5.74) is exact for polynomials of degree three also. Hence the error term becomes

$$R_2 = \frac{C}{4!}f^{iv}(\eta), \quad \eta \in (0, 2)$$

where

$$C = \int_a^b x^4\,dx - \frac{b-a}{6}\left[a^4 + 4\left(\frac{a+b}{2}\right)^4 + b^4\right]$$

$$= -\frac{(b-a)^5}{120}$$

Therefore, the error of approximation in the Simpson's rule becomes

$$R_2 = -\frac{(b-a)^5}{2880}f^{iv}(\eta) \tag{5.75}$$

When $n = 3$, the corresponding integration method is called $\frac{3}{8}$th **Simpson's rule**. The weights λ_k of the integration method (5.69) with $w(x) = 1$ for $n \leq 6$ are given in Table 5.2.

Table 5.2 Weights of Newton-Cotes Integration Rule

n \ λ	λ_0	λ_1	λ_2	λ_3	λ_4	λ_5	λ_6
1	1/2	1/2					
2	1/3	4/3	1/3				
3	3/8	9/8	9/8	3/8			
4	14/45	64/45	24/45	64/45	14/45		
5	95/288	375/288	250/288	250/288	375/288	95/288	
6	41/140	216/140	27/140	272/140	27/140	216/140	41/140

Usually for larger values of n, we get better approximation. However, for large n ($n \geqslant 8$, $n \neq 9$) some of the weights become negative. This may cause loss of significant digits in the result, because of mutual cancellation. For this reason higher order Newton-Cotes formulas are not commonly used.

The methods of the form (5.69) include the end points x_0 and x_n as abscissas. Such methods are also called **closed-type** methods. The methods which do not include the end points as abscissas are often called **open-type** methods.

We replace $f(x)$ in (5.58) by the Lagrange interpolating polynomial fitting the $n-1$ data points (x_k, f_k), $k = 1(1)n - 1$ and integrate between the given limits. Some of the open-type integration methods ($w(x) = 1$) together with the associated errors are listed below. The nodes are equispaced with $h = (b-a)/n$ and $x_0 = a$, $x_n = b$.

(i) Mid-point rule ($n = 2$)

$$\int_a^b f(x)\, dx = 2hf(x_0 + h) + \frac{h^3}{3} f''(\xi_1) \qquad (5.76)$$

(ii) Two-point rule ($n = 3$)

$$\int_a^b f(x)\, dx = \frac{3h}{2} [f(x_0 + h) + f(x_0 + 2h)] + \tfrac{3}{4} h^3 f''(\xi_2) \qquad (5.77)$$

(iii) Three-point rule ($n = 4$)

$$\int_a^b f(x)\, dx = \frac{4h}{3} [2f(x_0+h) - f(x_0+2h) + 2f(x_0+3h)] + \frac{14h^5}{45} f^{\text{iv}}(\xi_3) \qquad (5.78)$$

where

$$a < \xi_1, \xi_2, \xi_3 < b$$

Example 5.5

Find the approximate value of

$$I = \int_0^1 \frac{dx}{1+x}$$

using (i) trapezoidal rule, and (ii) Simpson's rule. Obtain a bound for the errors. The exact value of $I = \log 2 = .693147$ correct to six decimal places.

Using the trapezoidal rule, we have

$I \simeq \tfrac{1}{2}(1 + \tfrac{1}{2}) = 0.75$

Error $= 0.75 - 0.693147 = 0.056853$

The error in the trapezoidal rule is given by

$$|R_1| \leqslant \frac{1}{12} \max_{0 \leqslant x \leqslant 1} \left| \frac{2}{(1+x)^3} \right| \leqslant \frac{1}{6}$$

Using the Simpson's rule, we have

$I \simeq \tfrac{1}{6}(1 + \tfrac{8}{3} + \tfrac{1}{2}) = \tfrac{25}{36} = 0.694444$

Error $= 0.694444 - 0.693147 = 0.001297$

The error in the Simpson's rule is given by

$$|R_2| \leq \frac{1}{2880} \max_{0 \leq x \leq 1} \left|\frac{24}{(1+x)^5}\right| = 0.008333$$

We note that in both cases, the actual error is much smaller than the error bounds obtained from theoretical considerations.

5.8 METHODS BASED ON UNDETERMINED COEFFICIENTS

In the integration method (5.69), the nodes $x_k's$ and the weights $\lambda_k's$, $k = 0(1)n$ can also be obtained by making the formula exact for polynomials of degree upto m. When the nodes are known, that is, $m = n$, the corresponding methods are called Newton-Cotes methods. When the nodes are also to be determined, we have $m = 2n + 1$ and the methods are called **Gaussian integration methods**. Since any finite interval $[a, b]$ can always be transformed to $[-1, 1]$, using the transformation

$$x = \frac{b-a}{2}t + \frac{b+a}{2}$$

we consider the integral in the form

$$\int_{-1}^{1} w(x)f(x)\,dx = \sum_{k=0}^{n} \lambda_k f_k \tag{5.79}$$

5.8.1 Gauss-Legendre Integration Methods

For $w(x) = 1$, the method (5.79) reduces to

$$\int_{-1}^{1} f(x)\,dx = \sum_{k=0}^{n} \lambda_k f_k \tag{5.80}$$

In this case, all the nodes and weights are unknown. For $n = 2$, the method becomes

$$\int_{-1}^{1} f(x)\,dx = \lambda_0 f(x_0) + \lambda_1 f(x_1) + \lambda_2 f(x_2) \tag{5.81}$$

There are six unknowns in the method (5.81) and it can be made exact for polynomials of degree upto five. For $f(x) = x^i$, $i = 0(1)5$, we get the system of equations

$$\lambda_0 + \lambda_1 + \lambda_2 = 2$$

$$\lambda_0 x_0 + \lambda_1 x_1 + \lambda_2 x_2 = 0$$

$$\lambda_0 x_0^2 + \lambda_1 x_1^2 + \lambda_2 x_2^2 = \tfrac{2}{3}$$

$$\lambda_0 x_0^3 + \lambda_1 x_1^3 + \lambda_2 x_2^3 = 0$$

$$\lambda_0 x_0^4 + \lambda_1 x_1^4 + \lambda_2 x_2^4 = \tfrac{2}{5}$$

$$\lambda_0 x_0^5 + \lambda_1 x_1^5 + \lambda_2 x_2^5 = 0$$

Solving the above system of equations, we get

$$x_1 = -\sqrt{\tfrac{3}{5}}, \quad x_2 = 0, \quad x_3 = \sqrt{\tfrac{3}{5}}$$

and
$$\lambda_0 = \tfrac{5}{9}, \quad \lambda_1 = \tfrac{8}{9}, \quad \lambda_2 = \tfrac{5}{9}$$

The method (5.81), therefore becomes

$$\int_{-1}^{1} f(x)\,dx = \tfrac{1}{9}[5f(-\sqrt{\tfrac{3}{5}}) + 8f(0) + 5f(\sqrt{\tfrac{3}{5}})] \tag{5.82}$$

The error term associated with the method (5.82) can be written as

$$R_5 = \frac{C}{6!} f^{(6)}(\xi), \quad -1 < \xi < 1 \tag{5.83}$$

where

$$C = \int_{-1}^{1} x^6\,dx - (\lambda_0 x_0^6 + \lambda_1 x_1^6 + \lambda_2 x_2^6)$$

$$= \frac{2}{7} - \frac{6}{25} = \frac{8}{175}$$

It is found that the nodes $x_k's$ are the roots of the **Legendre polynomial** $P_{n+1}(x) = 0$ where

$$P_{n+1}(x) = \frac{1}{2^{n+1}(n+1)!} \frac{d^{n+1}}{dx^{n+1}}[(x^2-1)^{n+1}], \quad n = 0, 1, \ldots \tag{5.84}$$

The nodes and the corresponding weights for the **Gauss-Legendre integration method** (5.80) for $n = 1(1)5$ are given in Table 5.3.

Table 5.3 Nodes and Weights for Gauss-Legendre integration Method (5.80)

n	nodes x_k	weights λ_k
1	± 0.5773502692	1.0000000000
2	0.0000000000	0.8888888889
	± 0.7745966692	0.5555555556
3	± 0.3399810436	0.6521451549
	± 0.8611363116	0.3478548451
4	0.0000000000	0.5688888889
	± 0.5384693101	0.4786286705
	± 0.9061798459	0.2369268851
5	± 0.2386191861	0.4679139346
	± 0.6612093865	0.3607615730
	± 0.9324695142	0.1713244924

Example 5.6 Evaluate the integral

$$I = \int_0^1 \frac{dx}{1+x}$$

using Gauss-Legendre three point formula.

First we transform the interval $[0, 1]$ to the interval $[-1, 1]$. Let

$$t = ax + b$$

we have

$$-1 = b$$
$$1 = a + b$$

or

$$a = 2, \ b = -1, \text{ and } t = 2x - 1$$

$$I = \int_0^1 \frac{dx}{1+x} = \int_{-1}^1 \frac{dt}{t+3}$$

Using Gauss-Legendre three point rule (corresponding to $n = 2$), we get

$$I = \frac{1}{9}\left[8\left(\frac{1}{0+3}\right) + 5\left(\frac{1}{3+\sqrt{\frac{3}{5}}}\right) + 5\left(\frac{1}{3-\sqrt{\frac{3}{5}}}\right)\right]$$
$$= \tfrac{131}{189} = 0.693122$$

The exact solution is $I = 0.693147$.

5.8.2 Lobatto Integration Method

In this case $w(x) = 1$ and the two end points -1 and 1 are always taken as nodes. The remaining $n - 1$ nodes are to be determined. The integration method of the form

$$\int_{-1}^1 f(x)\, dx = \lambda_0 f(-1) + \lambda_n f(1) + \sum_{k=1}^{n-1} \lambda_k f_k \tag{5.85}$$

is called the **Lobatto integration method**. Since there are $2n$ unknowns ($n-1$ nodes and $n + 1$ weights), this method can be made exact for polynomials of degree upto $2n - 1$. For $n = 2$, the method (5.85) becomes

$$\int_{-1}^1 f(x)\, dx = \lambda_0 f(-1) + \lambda_1 f(x_1) + \lambda_2 f(1) \tag{5.86}$$

The method (5.86) is exact for polynomials of degree upto three. For $f(x) = x^i$, $i = 0, 1, 2, 3$, we get the system of equations

$$\lambda_0 + \lambda_1 + \lambda_2 = 2$$
$$-\lambda_0 + \lambda_1 x_1 + \lambda_2 = 0$$
$$\lambda_0 + \lambda_1 x_1^2 + \lambda_2 = \tfrac{2}{3}$$
$$-\lambda_0 + \lambda_1 x_1^3 + \lambda_2 = 0$$

Solving the above system of equations we get

$$x_1 = 0, \quad \lambda_0 = \lambda_2 = \tfrac{1}{3}, \quad \lambda_1 = \tfrac{4}{3}$$

and the method becomes

$$\int_{-1}^1 f(x)\, dx = \tfrac{1}{3}[f(-1) + 4f(0) + f(1)] \tag{5.87}$$

with the error term

$$R_3 = \frac{C}{4!} f^{iv}(\xi), \quad -1 < \xi < 1$$

where

$$C = \int_{-1}^{1} x^4 \, dx - (\lambda_0 + \lambda_1 x_1^4 + \lambda_2) = -\frac{4}{15}$$

The nodes and the corresponding weights for the Lobatto integration method (5.85) for $n = 2(1)5$ are given in Table 5.4.

Table 5.4 Nodes and Weights for Lobatto Integration Method (5.85)

n	nodes x_k	weights λ_k
2	±1.00000000 0.00000000	0.33333333 1.33333333
3	±1.00000000 ±0.44721360	0.16666667 0.83333333
4	±1.00000000 ±0.65465367 0.00000000	0.10000000 0.54444444 0.71111111
5	±1.00000000 ±0.76505532 ±0.28523152	0.06666667 0.37847496 0.55485837

5.8.3 Radau Integration Method

In this case $w(x) = 1$ and the lower limit -1 is fixed as one node. The remaining n nodes are to be determined. The integration method of the form

$$\int_{-1}^{1} f(x) \, dx = \lambda_0 f(-1) + \sum_{k=1}^{n} \lambda_k f_k \qquad (5.88)$$

is called the **Radau integration method**. Since there are $2n + 1$ unknowns (n nodes and $n + 1$ weights), this method can be made exact for polynomials of degree upto $2n$. For $n = 2$, the method (5.88) becomes

$$\int_{-1}^{1} f(x) \, dx = \lambda_0 f(-1) + \lambda_1 f(x_1) + \lambda_2 f(x_2) \qquad (5.89)$$

For $f(x) = x^i$, $i = 0(1)4$, we get the system of equations

$$\lambda_0 + \lambda_1 + \lambda_2 = 2$$
$$-\lambda_0 + \lambda_1 x_1 + \lambda_2 x_2 = 0$$
$$\lambda_0 + \lambda_1 x_1^2 + \lambda_2 x_2^2 = \tfrac{2}{3}$$
$$-\lambda_0 + \lambda_1 x_1^3 + \lambda_2 x_2^3 = 0$$
$$\lambda_0 + \lambda_1 x_1^4 + \lambda_2 x_2^4 = \tfrac{2}{5}$$

Solving the above system of equations, we get

$$x_1 = \frac{1-\sqrt{6}}{5}, \quad x_2 = \frac{1+\sqrt{6}}{5}$$

$$\lambda_0 = \frac{2}{9}, \quad \lambda_1 = \frac{16+\sqrt{6}}{18}, \quad \lambda_2 = \frac{16-\sqrt{6}}{18}$$

and the method becomes

$$\int_{-1}^{1} f(x)\,dx = \frac{2}{9}f(-1) + \frac{16+\sqrt{6}}{18}f\left(\frac{1-\sqrt{6}}{5}\right) + \frac{16-\sqrt{6}}{18}f\left(\frac{1+\sqrt{6}}{5}\right) \tag{5.90}$$

with the error term

$$R_4 = \frac{C}{5!}f^v(\xi), \quad -1 < \xi < 1$$

where

$$C = \int_{-1}^{1} x^5\,dx - (-\lambda_0 + \lambda_1 x_1^5 + \lambda_2 x_2^5) = -\frac{37}{225}$$

The nodes and the corresponding weights for the Radau integration method (5.88) for $n = 1(1)5$ are given in Table 5.5.

Table 5.5 Nodes and Weights for Radau Integration Method (5.88)

n	nodes x_k	weights λ_k
1	−1.0000000 0.3333333	0.5000000 1.5000000
2	−1.0000000 −0.2898979 0.6898979	0.2222222 1.0249717 0.7528061
3	−1.0000000 −0.5753189 0.1810663 0.8228241	0.1250000 0.6576886 0.7763870 0.4409244
4	−1.0000000 −0.7204803 0.1671809 0.4463140 0.8857916	0.0800000 0.4462078 0.6236530 0.5627120 0.2874271
5	−1.0000000 −0.8029298 −0.3909286 0.1240504 0.6039732 0.9203803	0.0555556 0.3196408 0.4853872 0.5209268 0.4169013 0.2015884

5.8.4 Gauss-Chebyshev Integration Methods

When $w(x) = \dfrac{1}{\sqrt{(1-x^2)}}$, the methods of the form

$$\int_{-1}^{1} \frac{1}{\sqrt{(1-x^2)}} f(x)\, dx = \sum_{k=0}^{n} \lambda_k f_k \tag{5.91}$$

are called **Gauss-Chebyshev integration methods.** These methods are exact for polynomials of degree upto $2n+1$. The nodes x_k's are found to be the roots of the Chebyshev polynomials

$$T_{n+1}(x) = \cos((n+1)\cos^{-1} x) = 0$$

Thus, we get

$$x_k = \cos\left(\frac{(2k+1)\pi}{2n+2}\right), \quad k = 0, 1, \ldots, n \tag{5.92}$$

Taking $n = 2$ in (5.91), we get the method

$$\int_{-1}^{1} \frac{f(x)}{\sqrt{(1-x^2)}}\, dx = \lambda_0 f(x_0) + \lambda_1 f(x_1) + \lambda_2 f(x_2) \tag{5.93}$$

Since the method is exact for $f(x) = x^i$, $i = 0(1)5$, we get the system of equations

$$\begin{aligned}
\lambda_0 + \lambda_1 + \lambda_2 &= \pi \\
\lambda_0 x_0 + \lambda_1 x_1 + \lambda_2 x_2 &= 0 \\
\lambda_0 x_0^2 + \lambda_1 x_1^2 + \lambda_2 x_2^2 &= \frac{\pi}{2} \\
\lambda_0 x_0^3 + \lambda_1 x_1^3 + \lambda_2 x_2^3 &= 0 \\
\lambda_0 x_0^4 + \lambda_1 x_1^4 + \lambda_2 x_2^4 &= \frac{3\pi}{8} \\
\lambda_0 x_0^5 + \lambda_1 x_1^5 + \lambda_2 x_2^5 &= 0
\end{aligned} \tag{5.94}$$

We obtain from (5.92)

$$x_k = \cos(2k+1)\frac{\pi}{6}, \quad k = 0, 1, 2$$

or

$$x_0 = \frac{\sqrt{3}}{2}, \quad x_1 = 0, \quad x_2 = -\frac{\sqrt{3}}{2}$$

Substituting the values of x_0, x_1 and x_2 in (5.94), we get

$$\lambda_0 = \lambda_1 = \lambda_2 = \frac{\pi}{3}$$

Thus we get the method

$$\int_{-1}^{1} \frac{f(x)\, dx}{\sqrt{(x-1^2)}} = \frac{\pi}{3}\left[f\left(\frac{\sqrt{3}}{2}\right) + f(0) + f\left(-\frac{\sqrt{3}}{2}\right)\right] \tag{5.95}$$

with the error term
$$R_5 = \frac{C}{6!} f^{vi}(\xi), \quad -1 < \xi < 1$$
where
$$C = \int_{-1}^{1} \frac{x^6}{\sqrt{(1-x^2)}} dx - (\lambda_0 x_0^6 + \lambda_1 x_1^6 + \lambda_2 x_2^6)$$
$$= \frac{\pi}{32}$$

It may be verified that in the method (5.91), all the weights λ_k's are equal and are given by

$$\lambda_k = \frac{\pi}{n+1}, \quad k = 0, 1, \ldots, n \tag{5.96}$$

Example 5.7 Evaluate the integral
$$I = \int_{-1}^{1} (1-x^2)^{3/2} \cos x \, dx$$
using
(i) Gauss-Legendre three point formula
(ii) Gauss-Chebyshev three point formula.

Gauss-Legendre three point formula is given by
$$\int_{-1}^{1} f(x) \, dx = \tfrac{1}{9}[5f(\sqrt{\tfrac{3}{5}}) + 8f(0) + 5f(-\sqrt{\tfrac{3}{5}})]$$

In this case $f(x) = (1-x^2)^{3/2} \cos x$ and we get
$$I = \int_{-1}^{1} (1-x^2)^{3/2} \cos x \, dx$$
$$= \tfrac{2}{9}[\sqrt{\tfrac{2}{5}} \cos \sqrt{\tfrac{3}{5}} + 4 + \sqrt{\tfrac{2}{5}} \cos \sqrt{\tfrac{3}{5}}]$$
$$= 1.08979$$

Gauss-Chebyshev three point formula is given by
$$\int_{-1}^{1} \frac{f(x)}{\sqrt{(1-x^2)}} dx = \frac{\pi}{3}\left[f\left(\frac{\sqrt{3}}{2}\right) + f(0) + f\left(-\frac{\sqrt{3}}{2}\right)\right]$$

In this case $f(x) = (1-x^2)^2 \cos x$ and we get
$$I = \int_{-1}^{1} (1-x^2)^{3/2} \cos x \, dx$$
$$= \frac{\pi}{3}\left[\frac{1}{16} \cos \frac{\sqrt{3}}{2} + 1 + \frac{1}{16} \cos \frac{\sqrt{3}}{2}\right]$$
$$= 1.13200$$

5.8.5 Gauss-Laguerre Integration Method

Here we consider the integral of the form
$$\int_0^\infty e^{-x} f(x) \, dx = \sum_{k=0}^{n} \lambda_k f_k \tag{5.97}$$

The nodes x_k's are found to be the roots of the **Laguerre polynomial**

$$L_{n+1}(x) = (-1)^{n+1} e^x \frac{d^{n+1}}{dx^{n+1}} (e^{-x} x^{n+1}) \qquad (5.98)$$

The corresponding weights are obtained from the relation

$$\lambda_k = \int_0^\infty \frac{e^{-x} L_{k+1}(x)}{(x - x_k) L'_{k+1}(x_k)} \, dx \qquad (5.99)$$

The method (5.97) produces exact results for polynomials of degree upto $2n + 1$. The nodes and the weights for the method (5.97) for $n = 1(1)5$ are given in Table 5.6. The Laguerre polynomials $L_n(x)$ are orthogonal with respect to the weight function e^{-x} on $(0, \infty)$.

$$\int_0^\infty e^{-x} L_m(x) L_n(x) \, dx = 0, \quad m \neq n$$

Table 5.6 Nodes and Weights for Gauss-Laguerre Integration Method (5.97)

n	nodes x_k	weights λ_k
1	0.5857864376 3.4142135624	0.8535533906 0.1464466094
2	0.4157745568 2.2942803603 6.2899450829	0.7110930099 0.2785177336 0.0103892565
3	0.3225476896 1.7457611012 4.5366202969 9.3950709123	0.6031541043 0.3574186924 0.0388879085 0.0005392947
4	0.2635603197 1.4134030591 3.5964257710 7.0858100059 12.6408008443	0.5217556106 0.3986668111 0.0759424497 0.0036117587 0.0000233700
5	0.2228466042 1.1889321017 2.9927363261 5.7751435691 9.8374674184 15.9828739806	0.4589646740 0.4170008308 0.1133733821 0.0103991975 0.0002610172 0.0000008985

5.8.6 Gauss-Hermite Integration Methods

The methods of the form

$$\int_{-\infty}^{\infty} e^{-x^2} f(x) \, dx = \sum_{k=0}^{n} \lambda_k f_k \qquad (5.100)$$

are called **Gauss-Hermite integration methods**. The nodes x_k's are the roots of the **Hermite polynomial**

$$H_{n+1}(x) = (-1)^{n+1}e^{x^2}\frac{d^{n+1}}{dx^{n+1}}(e^{-x^2}) \tag{5.101}$$

The corresponding weights λ_k's are obtained from the relation

$$\lambda_k = \int_{-\infty}^{\infty} \frac{e^{-x^2}H_{k+1}(x)}{(x-x_k)H'_{k+1}(x_k)} dx \tag{5.102}$$

The method (5.100) produces exact results for polynomials of degree upto $2n+1$. The nodes and the weights for the method (5.100) for $n = 0(1)5$ are given in Table 5.7. The Hermite polynomials are orthogonal with respect to the weight function e^{-x^2} on $(-\infty, \infty)$

$$\int_{-\infty}^{\infty} e^{-x^2}H_m(x)H_n(x)\,dx = 0, \quad m \neq n$$

Table 5.7 Nodes and Weights for Gauss-Hermite Integration Methods (5.100)

n	nodes x_k	weights λ_k
0	0·000000000	1.7724538509
1	±0.7071067812	0.8862269255
2	0.0000000000 ±1.2247448714	1.1816359006 0.2954089752
3	±0.5246476233 ±1.6506801239	0.8049140900 0.0813128354
4	0.0000000000 ±0.9585724646 ±2.0201828705	0.9453087205 0.3936193232 0.0199532421
5	±0.4360774119 ±1.3358490740 ±2.3506049737	0.7264295952 0.1570673203 0.0045300099

5.9 COMPOSITE INTEGRATION METHODS

As the order of the integration method (5.59) is increased the order of the derivative in the error term, associated with the method, also increases. For any method to produce meaningful results, these higher order derivatives must remain continuous in the interval of interest. Also Newton-Cotes type methods of higher order sometimes produce diverging results. An alternative to obtain the accurate results, while using lower order methods is the use of composite integration methods. We subdivide the given interval $[a, b]$ or $[-1, 1]$ into a number of subintervals and evaluate the integral in each subinterval by a particular method.

5.9.1 Trapezoidal Rule

We divide the interval $[a, b]$ into N subintervals, each of length $h=(b-a)/N$. We denote the subintervals as $(x_0, x_1), (x_1, x_2), \ldots, (x_{N-1}, x_N)$ where $x_0 = a, x_N = b$ and $x_i = x_0 + ih, i = 1(1)N - 1$. We write

$$I = \int_a^b f(x)\, dx = \int_{x_0}^{x_1} f(x)\, dx + \int_{x_1}^{x_2} f(x)\, dx + \ldots + \int_{x_{N-1}}^{x_N} f(x)\, dx \tag{5.103}$$

Evaluating each of the integrals on the right hand side of (5.103) by the trapezoidal rule (5.72), we get

$$I = \frac{h}{2}[(f_0 + f_1) + (f_1 + f_2) + \ldots + (f_{N-1} + f_N)]$$

$$= \frac{h}{2}[f_0 + 2(f_1 + f_2 + \ldots + f_{N-1}) + f_N] \tag{5.104}$$

where $f_k = f(x_k), k = 0(1)N$. The formula (5.104) is called **composite trapezoidal rule**. The error in the integration method (5.104) becomes

$$R_1 = -\frac{h^3}{12}[f''(\xi_1) + f''(\xi_2) + \ldots + f''(\xi_N)] \tag{5.105}$$

where

$$x_{i-1} < \xi_i < x_i, \quad i = 1, 2, \ldots, N$$

If $f''(x)$ is constant for all x in $[a, b]$ or if

$$f''(\eta) = \max_{a \leqslant x \leqslant b} |f''(x)|, \quad a < \eta < b$$

we may write (5.105) as

$$R_1 = -\frac{h^3 N}{12} f''(\eta)$$

Since $h = (b - a)/N$, we have

$$R_1 = -\frac{(b-a)^3}{12N^2} f''(\eta) = -\frac{(b-a)}{12} h^2 f''(\eta) \tag{5.106}$$

The factor N in the denominator in the error term (5.106) reduces the error considerably for large N. The number of intervals may be odd or even in this case.

5.9.2 Simpson's Rule

In using the Simpson's rule of integration (5.74), we need three abscissas. We divide the interval $[a, b]$ into an even number of subintervals of equal length giving an odd number of abscissas. If we divide the interval $[a, b]$ into $2N$ subintervals each of length $h = (b - a)/2N$, then we get $2N + 1$ abscissas $x_0, x_1, \ldots, x_{2N}$, $x_0 = a, x_{2N} = b, x_i = x_0 + ih, i = 1, 2, \ldots, 2N - 1$. We write

$$I = \int_a^b f(x)\, dx = \int_{x_0}^{x_2} f(x)\, dx + \int_{x_2}^{x_4} f(x)\, dx + \cdots$$
$$+ \int_{x_{2N-2}}^{x_{2N}} f(x)\, dx \tag{5.107}$$

Evaluating each of the integrals on the right hand side of (5.107) by the Simpson's rule (5.74), we get

$$I = \frac{h}{3}[(f_0 + 4f_1 + f_2) + (f_2 + 4f_3 + f_4) + \cdots + (f_{2N-2} + 4f_{2N-1} + f_{2N})]$$
$$= \frac{h}{3}[f_0 + 4(f_1 + f_3 + \cdots + f_{2N-1}) + 2(f_2 + f_4 + \cdots + f_{2N-2}) + f_{2N}] \tag{5.108}$$

The formula (5.108) is called **Composite Simpson's rule**. The error in the integration method (5.108) becomes

$$R_2 = -\frac{h^5}{90}[f^{iv}(\xi_1) + f^{iv}(\xi_2) + \cdots + f^{iv}(\xi_N)] \tag{5.109}$$

where $x_{2i-2} < \xi_i < x_{2i}$, $i = 1, 2, \ldots, N$.

Using
$$f^{iv}(\eta) = \max_{a \leq x \leq b} |f^{iv}(x)|, \quad a < \eta < b$$

we can write (5.109) in the form

$$R_2 = -\frac{Nh^5}{90} f^{iv}(\eta)$$
$$= -\frac{(b-a)^5}{2880 N^4} f^{iv}(\eta) = -\frac{(b-a)}{180} h^4 f^{iv}(\eta) \tag{5.110}$$

Similarly, composite rules in other cases may be obtained.

Example 5.8 Evaluate the integral

$$I = \int_0^1 \frac{dx}{1 + x}$$

using (i) composite trapezoidal rule, (ii) composite Simpson's rule, with 2, 4 and 8 equal subintervals.

When $N = 2$, we have $h = \frac{1}{2}$ and three nodes 0, $\frac{1}{2}$ and 1. Let I_T and I_s represent the values obtained by using the trapezoidal rule and Simpson's rule respectively. We have two subintervals for trapezoidal rule and only one interval for Simpson's rule. We have

$$I_T = \tfrac{1}{4}[f(0) + 2f(\tfrac{1}{2}) + f(1)] = \tfrac{1}{4}(1 + \tfrac{4}{3} + \tfrac{1}{2}) = \tfrac{17}{24}$$
$$= 0.708333$$
$$I_s = \tfrac{1}{6}[f(0) + 4f(\tfrac{1}{2}) + f(1)] = \tfrac{1}{6}(1 + \tfrac{8}{3} + \tfrac{1}{2}) = \tfrac{25}{36}$$
$$= 0.694444$$

When $N = 4$, we have $h = \frac{1}{4}$ and five nodes $0, \frac{1}{4}, \frac{2}{4}, \frac{3}{4}$ and 1. We have four subintervals for trapezoidal rule and two subintervals for Simpson's rule. We get

$$I_T = \tfrac{1}{8}[f(0) + 2(f(\tfrac{1}{4}) + f(\tfrac{1}{2}) + f(\tfrac{3}{4})) + f(1)]$$
$$= 0.697024$$
$$I_s = \tfrac{1}{12}[f(0) + 4(f(\tfrac{1}{4}) + f(\tfrac{3}{4})) + 2f(\tfrac{1}{2}) + f(1)]$$
$$= 0.693254$$

When $N = 8$ we have $h = \tfrac{1}{8}$ and nine nodes $0, \tfrac{1}{8}, \tfrac{2}{8}, \ldots, 1$. We have eight subintervals for trapezoidal rule and four subintervals for Simpson's rule. We get

$$I_T = \frac{1}{16}\left[f(0) + 2\sum_{i=1}^{7} f\left(\frac{i}{8}\right) + f(1)\right]$$
$$= 0.694122$$
$$I_s = \frac{1}{24}\left[f(0) + 4\sum_{i=1}^{4} f\left(\frac{2i-1}{8}\right) + 2\sum_{i=1}^{3} f\left(\frac{2i}{8}\right) + f(1)\right]$$
$$= 0.693155$$

The exact value of the integral is

$$I = 0.693147$$

Example 5.9 Evaluate the integral

$$I = \int_0^1 \frac{dx}{1+x}$$

by subdividing the interval $[0, 1]$ into two equal parts and then applying the Gauss-Legendre three point formula

$$\int_{-1}^{1} f(x)\, dx = \tfrac{1}{9}\left[5f\left(-\sqrt{\tfrac{3}{5}}\right) + 8f(0) + 5f\left(\sqrt{\tfrac{3}{5}}\right)\right]$$

we write

$$\int_0^1 \frac{dx}{1+x} = \int_0^{1/2} \frac{dx}{1+x} + \int_{1/2}^1 \frac{dx}{1+x}$$
$$= I_1 + I_2$$

The transformations $t = 4x - 1$ and $y = 4x - 3$ in the first and second integrals respectively, change the limits of integration to $[-1, 1]$. Thus we have

$$I = \int_{-1}^{1} \frac{dt}{t+5} + \int_{-1}^{1} \frac{dy}{y+7}$$
$$= I_1 + I_2$$

Evaluating each of the integrals I_1 and I_2 by Gauss-Legendre three point formula, we get

$$I_1 = \frac{1}{9}\left[\frac{5}{5-\sqrt{\frac{3}{5}}} + \frac{8}{5} + \frac{5}{5+\sqrt{\frac{3}{5}}}\right]$$
$$= 0.405464$$
$$I_2 = \frac{1}{9}\left[\frac{5}{7-\sqrt{\frac{3}{5}}} + \frac{8}{7} + \frac{5}{7+\sqrt{\frac{3}{5}}}\right]$$
$$= 0.287682$$

Therefore
$$I = I_1 + I_2$$
$$= 0.405464 + 0.287682$$
$$= 0.693146$$

The exact value of I is 0.693147.

5.10 Romberg Integration

Richardson's extrapolation procedure described in section 5.4, applied to the integration methods, is called **Romberg integration**. First, we find the power series expansion of the error term in the integration method. Then by eliminating the leading terms in the error expression by using the computed results, we obtain the new methods which are of higher order than the previous methods. Consider the integral

$$I = \int_a^b f(x)\, dx \tag{5.111}$$

The errors in the composite trapezoidal rule (5.104) and the composite Simpson's rule (5.108) can be obtained as

$$I = I_T + c_1 h^2 + c_2 h^4 + c_3 h^6 + \ldots \tag{5.112}$$
$$I = I_s + d_1 h^4 + d_2 h^6 + d_3 h^8 + \ldots \tag{5.113}$$

where c's and d's are constants independent of h.

The extrapolation procedure for the trapezoidal rule as given by (5.51) becomes

$$I_T^{(m)}(h) = \frac{4^m I_T^{(m-1)}\left(\frac{h}{2}\right) - I_T^{(m-1)}(h)}{4^m - 1}, \quad m = 1, 2, \ldots \tag{5.114}$$

The extrapolation procedure for the Simpson's rule becomes

$$I_s^{(m)}(h) = \frac{4^{m+1} I_s^{(m-1)}\left(\frac{h}{2}\right) - I_s^{(m-1)}(h)}{4^{m+1} - 1}, \quad m = 1, 2, \ldots \tag{5.115}$$

Example 5.10 Find the approximate value of the integral

$$I = \int_0^1 \frac{dx}{1+x}$$

232 Numerical Methods

using (i) composite trapezoidal rule with 2, 3, 5, 9 nodes and Romberg integration, (ii) composite Simpson's rule with 3, 5, 9 nodes and Romberg integration. Obtain the number of function evaluations required to get an accuracy of 10^{-6}, when the integral is evaluated directly by using the trapezoidal and Simpson's rules.

Using the composite trapezoidal rule

$$I = \int_a^b f(x)\,dx = \frac{h}{2}\left[f_0 + 2\sum_{i=1}^{N-1} f_i + f_N\right]$$

where $x_0 = a$, $x_N = b$, $h = \dfrac{b-a}{N}$, $x_i = x_0 + ih$, we get

$$N = 1,\quad h = 1,\quad I_T = \frac{h}{2}[f_0 + f_1] = 0.750000$$

$$N = 2,\quad h = \tfrac{1}{2},\quad I_T = \frac{h}{2}[f_0 + 2f_1 + f_2] = 0.708333$$

$$N = 4,\quad h = \tfrac{1}{4},\quad I_T = \frac{h}{2}[f_0 + 2f_1 + 2f_2 + 2f_3 + f_4] = 0.697024$$

$$N = 8,\quad h = \tfrac{1}{8},\quad I_T = \frac{h}{2}\left[f_0 + 2\sum_{i=1}^{7} f_i + f_8\right] = 0.694122$$

Using Romberg integration we obtain the results as given in Table 5.8.

Table 5.8   Trapezoidal Rule with Romberg Integration

h	Second order method	Fourth order method	Sixth order method	Eighth order method
1	0.750000			
		0.694444		
$\tfrac{1}{2}$	0.708333		0.693175	
		0.693254		0.693148
$\tfrac{1}{4}$	0.697024		0.693148	
		0.693155		
$\tfrac{1}{8}$	0.694122			

Since the exact solution is 0.693147, we require only nine function evaluations using trapezoidal rule with Romberg integration to achieve the accuracy of 10^{-6}.

The error in the trapezoidal rule is given by

$$R_1 = \frac{h^2}{12} f''(\xi),\quad 0 < \xi < 1$$

Since $f(x) = \dfrac{1}{1+x}$, for $0 \leqslant x \leqslant 1$, $\tfrac{1}{4} \leqslant |f''(x)| \leqslant 2$. Therefore, we have

$$\frac{h^2}{48} \leqslant \frac{h^2}{12}|f''(\xi)| \leqslant \frac{h^2}{6}$$

For achieving accuracy of 10^{-6}, we require

$$\frac{h^2}{48} \leqslant 10^{-6}$$

which gives $h \approx 0.007$.

Hence we require $\frac{1-0}{0.007} \approx 145$ function evaluations to achieve this accuracy if trapezoidal rule is used directly.

Similarly, using the composite Simpson's rule

$$I = \int_a^b f(x)\,dx = \frac{h}{3}\left[f_0 + 4\sum_{i=1}^{N} f_{2i-1} + 2\sum_{i=1}^{N-1} f_{2i} + f_{2N}\right]$$

$x_0 = 0,\ x_{2N} = b,\ h = \dfrac{b-a}{2N}$, we get

$N = 1,\ h = \tfrac{1}{2},\ I_s = \dfrac{h}{3}(f_0 + 4f_1 + f_2) = 0.694444$

$N = 2,\ h = \tfrac{1}{4},\ I_s = \dfrac{h}{3}[f_0 + 4(f_1 + f_3) + 2f_2 + f_4] = 0.693254$

$N = 4,\ h = \tfrac{1}{8},\ I_s = \dfrac{h}{3}[f_0 + 4(f_1 + f_3 + f_5 + f_7)$
$\qquad\qquad\qquad\qquad + 2(f_2 + f_4 + f_6) + f_8] = 0.693155$

Using Romberg integration, we obtain the results as given in Table 5.9.

Table 5.9 Simpson's Rule with Romberg Integration

h	Fourth order method	Sixth order method	Eighth order method
$\tfrac{1}{2}$	0.694444		
		0.693175	
$\tfrac{1}{4}$	0.693254		0.693148
		0.693148	
$\tfrac{1}{8}$	0.693155		

In this case also, we require nine function evaluations to achieve the required accuracy. Also when Simpson's rule is applied directly, we have for $0 \leqslant x \leqslant 1,\ \tfrac{3}{4} \leqslant |f^{\text{iv}}(x)| \leqslant 24$ and therefore

$$\frac{3h^4}{720} \leqslant \frac{h^4}{180}|f^{\text{iv}}(\xi)| \leqslant \frac{24}{180}h^4$$

For getting an accuracy of 10^{-6}, we must have

$$\frac{3h^4}{720} \leqslant 10^{-6}$$

or
$$h \approx 0.1245$$
which also gives $\dfrac{1-0}{0.1245} \approx 9$ function evaluations.

5.11 DOUBLE INTEGRATION

The problem of double integration is to evaluate an integral of the form

$$I = \int_c^d \left(\int_a^b f(x, y)\, dx \right) dy \tag{5.116}$$

over the rectangles $x = a$, $x = b$ and $y = c$, $y = d$. This integral can be evaluated numerically, by two successive integrations in x and y directions respectively, taking into account one variable at one time.

5.11.1 Trapezoidal Method

If we evaluate the inner integral by the trapezoidal rule, we get

$$I = \frac{b-a}{2} \int_c^d [f(a, y) + f(b, y)]\, dy \tag{5.117}$$

Using trapezoidal rule again to evaluate the integral on the right hand side of (5.117), we have

$$I = \frac{(b-a)(d-c)}{4}[f(a, c) + f(a, d) + f(b, c) + f(b, d)] \tag{5.118}$$

The choice of a particular method to evaluate the integral (5.116) depends on the availability of the values of $f(x, y)$ at the nodal points, if the discrete data is given. We can use the composite rules of integrations by dividing the interval $[a, b]$ into N equal subintervals each of length h and the interval $[c, d]$ into M equal subintervals each of length k. We have

$$x_i = x_0 + ih, \quad x_0 = a, \quad x_N = b$$
$$y_j = y_0 + jk, \quad y_0 = c, \quad y_M = d$$

If we use composite trapezoidal rule, in both the directions, we get

$$I = \frac{hk}{4}[f_{00} + 2(f_{01} + f_{02} + \ldots + f_{0,M-1}) + f_{0M}$$
$$+ 2 \sum_{i=1}^{N-1} \{f_{i0} + 2(f_{i1} + f_{i2} + \ldots + f_{i,M-1}) + f_{iM}\}$$
$$+ f_{N0} + 2(f_{N1} + f_{N2} + \ldots + f_{N,M-1}) + f_{NM}] \tag{5.119}$$

where $f_{ij} = f(x_i, y_j)$.

For $M = N = 1$, the method (5.118) is reproduced. This formula is of second order in both h and k.

5.11.2 Simpson's Method

If we apply Simpson's rule to evaluate (5.116) with $h = (b-a)/2$ and

$k = (d - c)/2$ then we obtain

$$I = \frac{hk}{9}[f(a, c) + f(a, d) + f(b, c) + f(b, d) + 4\{f(a, c + k) + f(a + h, c) + f(a + h, d) + f(b, c + k)\} + 16f(a + h, c + k)] \quad (5.120)$$

Example 5.11 Evaluate the integral

$$I = \int_1^2 \int_1^2 \frac{dx\,dy}{x + y}$$

using the trapezoidal rule with $h = k = 0.5$ and $h = k = 0.25$.

With $h = k = 0.5$, we find

$$I = \tfrac{1}{16}[f(1, 1) + f(2, 1) + f(1, 2) + f(2, 2) + 2\{f(\tfrac{3}{2}, 1) + f(1, \tfrac{3}{2}) + f(2, \tfrac{3}{2}) + f(\tfrac{3}{2}, 2)\} + 4f(\tfrac{3}{2}, \tfrac{3}{2})]$$
$$= \tfrac{1}{16}[0.5 + \tfrac{1}{3} + \tfrac{1}{3} + 0.25 + 2(0.4 + 0.4 + \tfrac{2}{7} + \tfrac{2}{7}) + \tfrac{4}{3}]$$
$$= 0.343304$$

The exact solution is ln $(1024/729) = 0.339798$.

With $h = k = 0.25$, we find

$$I = \tfrac{1}{64}[f(1, 1) + f(2, 1) + f(1, 2) + f(2, 2) + 2\{f(\tfrac{5}{4}, 1) + f(\tfrac{3}{2}, 1)$$
$$+ f(\tfrac{7}{4}, 1) + f(1, \tfrac{5}{4}) + f(1, \tfrac{3}{2}) + f(1, \tfrac{7}{4}) + f(2, \tfrac{5}{4}) + f(2, \tfrac{3}{2})$$
$$+ f(2, \tfrac{7}{4}) + f(\tfrac{5}{4}, 2) + f(\tfrac{3}{2}, 2) + f(\tfrac{7}{4}, 2)\} + 4\{f(\tfrac{5}{4}, \tfrac{5}{4})$$
$$+ f(\tfrac{5}{4}, \tfrac{3}{2}) + f(\tfrac{5}{4}, \tfrac{7}{4}) + f(\tfrac{3}{2}, \tfrac{5}{4}) + f(\tfrac{3}{2}, \tfrac{3}{2}) + f(\tfrac{3}{2}, \tfrac{7}{4})$$
$$+ f(\tfrac{7}{4}, \tfrac{5}{4}) + f(\tfrac{7}{4}, \tfrac{3}{2}) + f(\tfrac{7}{4}, \tfrac{7}{4})\}]$$
$$= 0.340668$$

Since the step length in both directions is same, $h = k$, the Romberg integration can also be used. We find that the refined value of the integral is

$$I = \tfrac{1}{3}[4(0.340668) - 0.343304] = 0.339789$$

PROBLEMS

1. A differentiation rule of the form

   $$f'(x_0) = \alpha_0 f_0 + \alpha_1 f_1 + \alpha_2 f_2, \quad (x_k = x_0 + kh)$$

   is given. Find the values of α_0, α_1 and α_2 so that the rule is exact for $f \in P_3$. Find the error term.

2. Using the following data find $f'(6.0)$, error $= 0(h)$, and $f''(6.3)$, error $= 0(h^2)$

x	6.0	6.1	6.2	6.3	6.4
$f(x)$	0.1750	−0.1998	−0.2223	−0.2422	−0.2596

3. Assume that $f(x)$ has a minimum in the interval $x_{n-1} \leq x \leq x_{n+1}$

where $x_k = x_0 + kh$. Show that the interpolation of $f(x)$ by a polynomial of second degree yields the approximation

$$f_n - \frac{1}{8}\left(\frac{(f_{n+1} - f_{n-1})^2}{8f_{n+1} - 2f_n + f_{n-1}}\right), \quad (f_k = f(x_k))$$

for this minimum value of $f(x)$.

(Stockholm Univ., Sweden, BIT 4(1964), 197)

4. Define
$$S(h) = \frac{-y(x+2h) + 4y(x+h) - 3y(x)}{2h}.$$

(a) Show that
$$y'(x) - S(h) = c_1 h^2 + c_2 h^3 + c_3 h^4 + \ldots$$
and state c_1.

(b) Calculate $y'(0.398)$ as accurately as possible using the table below and with the aid of the approximation $S(h)$. Give an error estimate (the values in the table are correctly rounded).

x	0.398	0.399	0.400	0.401	0.402
$y(x)$	0.408591	0.409671	0.410752	0.411834	0.412915

(Royal Inst. Tech., Stockholm, Sweden, BIT 19(1979), 285)

5. Determine α, β, γ and δ such that the relation
$$y'((a+b)/2) = \alpha y(a) + \beta y(b) + \gamma y''(a) + \delta y''(b)$$
is exact for polynomials of as high degree as possible. Give an asymptotically valid expression for the truncation error as $|b - a| \to 0$.

(Royal Inst. Tech., Stockholm, Sweden, BIT 16(1976), 111)

6. Find the coefficients a's in the expansion
$$D = \sum_{s=1}^{\infty} a_s \mu \delta^s$$
($h = 1$, $D =$ differentiation operator, $\mu =$ mean value operator, and $\delta =$ central difference operator).

(Arhus Univ., Denmark, BIT 7(1967), 81)

7. (a) Determine the exponents k_i in the difference formula
$$f''(x_0) = \frac{f(x_0 + h) - 2f(x_0) + f(x_0 - h)}{h^2} + \sum_{i=1}^{\infty} a_i h^{k_i}$$
assuming that $f(x)$ has a convergent Taylor expansion in a sufficiently large interval around x_0.

(b) Compute $f''(0.6)$ from the following table using the formula in (a) with $h = 0.4, 0.2$ and 0.1 and perform repeated Richardson extrapolation.

x	$f(x)$
0.2	1.420072
0.4	1.881243
0.5	2.128147
0.6	2.386761
0.7	2.657971
0.8	2.942897
1.0	3.559753

(Lund Univ., Sweden, BIT 13(1973), 123)

8. (a) Prove that one can use repeated Richardson extrapolation for the formula
$$f''(x) \approx \frac{f(x+h) - 2f(x) + f(x-h)}{h^2}$$
What are the coefficients in the extrapolation scheme?
(b) Apply this to the table given below, and estimate the error in the computed $f''(0.3)$.

x	$f(x)$
0.1	17.60519
0.2	17.68164
0.3	17.75128
0.4	17.81342
0.5	17.86742

(Stockholm Univ., Sweden, BIT 9(1969), 400)

9. By use of repeated Richardson extrapolation find $f'(1)$ from the following values:

x	$f(x)$
0.6	0.707178
0.8	0.859892
0.9	0.925863
1.0	0.984007
1.1	1.033743
1.2	1.074575
1.4	1.127986

Apply the approximate formula
$$f'(x_0) = \frac{f(x_0 + h) - f(x_0 - h)}{2h}$$
with $h = 0.4, \ 0.2, \ 0.1$

(Royal Inst. Tech., Stockholm, Sweden, BIT 6(1966), 270)

10. Compute

$$I_p = \int_0^1 \frac{x^p}{x^3 + 12} \, dx \quad \text{for } p = 0, 1$$

using trapezoidal and Simpson's rules with the number of points 3, 5 and 9. Improve the results using Romberg integration.

11. Determine the coefficients in the formula

$$\int_0^{2h} x^{-1/2} f(x) \, dx = (2h)^{1/2}(A_0 f(0) + A_1 f(h) + A_2 f(2h)) + R$$

and calculate the remainder R, when $f'''(x)$ is constant.
(Gothenburg Univ., Sweden, BIT 4(1964), 61)

12. Determine the coefficients a, b and c in the quadrature formula

$$\int_{x_0}^{x_1} y(x) \, dx = h(ay_0 + by_1 + cy_2) + R$$

where

$$x_i = x_0 + ih, \quad y(x_i) = y_i$$

Prove that the error term R has the form

$$R = ky^{(n)}(\xi), \quad x_0 \leq \xi \leq x_2$$

and determine k and n.
(Bergen Univ., Sweden, BIT 4 (1964), 261)

13. Determine a, b and c such that the formula

$$\int_0^h f(x) \, dx = h\left\{ af(0) + bf\left(\frac{h}{3}\right) + cf(h) \right\}$$

is exact for polynomials of as high order as possible, and determine the order of the truncation error.
(Uppsala Univ., Sweden, BIT 13(1973), 123)

14. Obtain a generalized trapezoidal rule of the form

$$\int_{x_0}^{x_1} f(x) \, dx = \frac{h}{2}(f_0 + f_1) + ph^2(f'_0 - f'_1)$$

Find the constant p and the error term. Deduce the composite rule for integrating

$$\int_a^b f(x) \, dx, \quad a = x_0 < x_1 < x_2 \ldots < x_N = b$$

15. Obtain a generalized Simpson's rule of the form

$$\int_{x_0}^{x_2} f(x) \, dx = \frac{h}{3}(af_0 + bf_1 + af_2) + ph^2(f'_0 - f'_2)$$

where a, b, p are constants and $f_k = f(x_k)$. Find the constants a, b, p

and the error term. Deduce the composite rule for integrating
$$\int_a^b f(x)\,dx; \quad a = x_0 < x_1 < x_2 \ldots < x_{2N} = b$$

16. Determine α and β in the formula
$$\int_a^b f(x)\,dx = h \sum_{i=0}^{n-1} [f(x_i) + \alpha h f'(x_0) + \beta h^2 f''(x_i)] + O(h^p)$$
with the integer p as large as possible.
(Uppsala Univ., Sweden, BIT 11(1971), 225)

17. (a) Compute by using Taylor development
$$\int_{0.1}^{0.2} \frac{x^2}{\cos x}\,dx$$
with an error $< 10^{-6}$.

(b) Using the trapezoidal formula instead, which step length (of form 10^{-k}, 2×10^{-k} or 5×10^{-k}) would be largest giving the accuracy above? How many decimals would be required in the function values?
(Royal Inst. Tech., Stockholm, Sweden, BIT 9(1969), 174)

18. Compute the integral $\int_0^1 y\,dx$ where y is defined through $x = y e^y$, with an error $< 10^{-4}$.
(Uppsala Univ., Sweden, BIT 7 (1967), 170)

19. Find a quadrature formula
$$\int_0^1 f(x) \frac{dx}{\sqrt{x(1-x)}} = \alpha_1 f(0) + \alpha_2 f(\tfrac{1}{2}) + \alpha_3 f(1)$$
which is exact for polynomials of highest possible degree. Then use the formula on
$$\int_0^1 \frac{dx}{\sqrt{x - x^3}}$$
and compare with the exact value.
(Oslo Univ., Norway, BIT 7 (1967), 170)

20. Compute
$$\int_0^{\pi/2} \frac{\cos x \log (\sin x)}{\sin^2 x + 1}\,dx$$
to 2 correct decimal places.
(Uppsala and Umea Univ., Sweden, BIT 11 (1971), 455)

21. Calculate
$$\int_0^{0.8} \left(1 + \frac{\sin x}{x}\right) dx$$

correct to 5 decimal places.

(Umea Univ., Sweden, BIT 20(1980), 261)

22. The following information is given for the function $f(x)$
    (a) $f(0) = f(1) = 0$
    (b) $f'(0) = -f'(1) = -0.45$
    (c) $f''(0) = 0.76$, $f''(0.5) = 0.89$, $f''(1) = 0.93$
    (d) $|f''(x)|$, $|f'''(x)|$ and $|f^{iv}(x)| \leq 1$ for $0 \leq x \leq 1$

    Compute $I = \int_0^1 f(x)e^x \, dx$ and find an error estimate. The function values given above are exact.

    (Uppsala Univ., Sweden, BIT 11(1971), 125)

23. Calculate
    $$\int_0^1 (\cos 2x)(1 - x^2)^{-1/2} \, dx$$
    correct to 4 decimal places.

    (Lund Univ., Sweden, BIT 20(1980), 389)

24. Determine weights and abscissas in the quadrature formula
    $$\int_{-1}^{+1} f(x) \, dx = \sum_{k=1}^{4} A_k f(x_k)$$
    with $x_1 = -1$ and $x_4 = 1$ so that the formula becomes exact for polynomials of highest possible degree.

    (Gothenburg Univ., Sweden, BIT 7(1967), 338)

25. Find the value of the integral
    $$I = \int_2^3 \frac{\cos 2x}{1 + \sin x} \, dx$$
    using Gauss-Legendre two and three point integration rules.

26. In the quadrature formula
    $$\int_{-1}^1 (a - x)f(x) \, dx = A_{-1}f(-x_1) + A_0 f(0) + A_1 f(x_1) + R$$
    the coefficients A_{-1}, A_0, A_1 are functions of the parameter a, x_1 is a constant, and the error R is of the form $cf^{(k)}(\xi)$. Determine A_{-1}, A_0, A_1 and x_1 so that the error R will be of highest possible order. Also investigate if the order of the error is influenced by different values of the parameter a.

    (Inst. Tech., Lund, Sweden, BIT 9(1969), 87)

27. Determine x_i and A_i in the quadrature formula below so that σ, the order of approximation, will be as high as possible
    $$I = \int_{-1}^1 (2x^2 + 1)f(x) \, dx$$

$$= A_1 f(x_1) + A_2 f(x_2) + A_3 f(x_3) + R$$

What is the value of σ? Answer with 4 significant digits.

(Gothenburg Univ., Sweden, BIT 17(1977), 369)

28. Compute by Gaussian quadrature

$$I = \int_0^1 \frac{\log(x+1)}{\sqrt{(x(1-x))}} \, dx$$

The error must not exceed 5×10^{-5}. An error bound is required.

(Uppsala Univ., Sweden, BIT 5(1965), 294)

29. Integrate by Gaussian quadrature ($n = 4$)

$$\int_1^2 \frac{dx}{1+x^3}$$

30. Use Gauss-Laguerre or Gauss-Hermite formulas to evaluate

(i) $\int_0^\infty \frac{e^{-x}}{1+x} \, dx$ (ii) $\int_0^\infty e^{-x} \sin x \, dx$

(iii) $\int_{-\infty}^\infty \frac{e^{-x^2}}{1+x^2} \, dx$ (iv) $\int_{-\infty}^\infty e^{-x^2} \, dx$

Use two point and three point formulas.

31. Obtain an approximate value of

$$I = \int_{-1}^1 (1-x^2)^{1/2} \cos x \, dx$$

using
   (a) Gauss-Legendre integration method for $n = 2, 3$
   (b) Gauss-Chebyshev integration method for $n = 2, 3$

32. The Radau quadrature formula is given by

$$\int_{-1}^1 f(x) \, dx = B_1 f(-1) + \sum_{k=1}^n H_k f(x_k) + R$$

Determine x_k, H_k and R for $n = 1$.

33. The Lobatto quadrature formula is given by

$$\int_{-1}^1 f(x) \, dx = B_1 f(-1) + B_2 f(1) + \sum_{k=1}^{n-1} H_k f(x_k) + R$$

Determine x_k, H_k and R for $n = 3$.

34. Obtain the approximate value of

$$I = \int_{-1}^1 e^{-x^2} \cos x \, dx$$

using
   (a) Gauss-Legendre integration method for $n = 2, 3$
   (b) Radau integration method for $n = 2, 3$
   (c) Lobatto integration method for $n = 2, 3$

35. Evaluate the following integrals:

   (i) $I = \int_0^\infty e^{-5x} \, dx$

   (ii) $I = \int_0^\infty e^{-x} \log(1+x) \, dx$

   correct to two decimal places, using the Gauss-Laguerre's formula.

36. Calculate the weights, abscissas and remainder term in the Gaussian quadrature formula

   $$\frac{1}{\sqrt{\pi}} \int_0^\infty \frac{\exp(-t)f(t)}{\sqrt{t}} \, dt = A_1 f(x_1) + A_2 f(x_2) + c f^{(n)}(\xi)$$

   (Royal Inst. Tech., Stockholm, Sweden, BIT 20(1980), 529)

37. The total emission from an absolutely black body is given by the formula

   $$E = \int_0^\infty E(v) \, dv = \frac{2\pi h}{c^3} \int_0^\infty \frac{v^3 \, dv}{e^{hv/(KT)} - 1}$$

   Defining $x = hv/(KT)$, we get

   $$E = \frac{2\pi h}{c^3} \left(\frac{KT}{h}\right)^4 \int_0^\infty \frac{x^3 \, dx}{e^x - 1}$$

   Calculate the value of the integral correct to 3 decimal places.

   (Royal Inst. Tech., Stockholm, Sweden, BIT 19(1979), 552)

38. A three dimensional Gaussian quadrature formula has the form

   $$\int_{-1}^1 \int_{-1}^1 \int_{-1}^1 f(x, y, z) \, dx \, dy \, dz = f(\alpha, \alpha, \alpha) + f(-\alpha, \alpha, \alpha)$$
   $$+ f(\alpha, -\alpha, \alpha) + f(\alpha, \alpha, -\alpha) + f(-\alpha, -\alpha, \alpha)$$
   $$+ f(-\alpha, \alpha, -\alpha) + f(\alpha, -\alpha, -\alpha) + f(-\alpha, -\alpha, -\alpha) + R$$

   Determine α so that $R = 0$ for every f which is a polynomial of degree 3 in 3 variables, i.e.

   $$f = \sum_{i,j,k=0}^{3} a_{ijk} x^i y^j z^k \qquad \text{(Lund Univ., Sweden, BIT 15 (1975), 111)}$$

39. Evaluate the double integral

   $$\int_0^1 \left(\int_0^2 \frac{2xy}{(1+x^2)(1+y^2)} \, dy \right) dx$$

   using (i) the trapezoidal rule with $h = k = 0.25$, (ii) the Simpson's rule with $h = k = 0.25$. Compare the results obtained with the exact solution.

40. Evaluate the double integral

   $$\int_1^5 \left(\int_1^5 \frac{dx}{(x^2 + y^2)^{1/2}} \right) dy$$

   using the trapezoidal rule with two and four subintervals and extrapolate.

Chapter 6

Ordinary Differential Equations

6.1 INTRODUCTION

An **ordinary differential equation** is a relation between a function, its derivatives, and the variable upon which they depend. The most general form of an ordinary differential equation is given by

$$\phi(t, y, y', y'', \ldots, y^{(m)}) = 0 \tag{6.1}$$

where m represents the highest order derivative, and y and its derivatives are functions of t. The **order** of the differential equation is the order of its highest derivative and its **degree** is the degree of the derivative of the highest order after the equation has been rationalized. If no product of the dependent variable $y(t)$ with itself or any one of its derivatives occur, then the equation is said to be **linear,** otherwise it is **nonlinear.** A linear differential equation of order m can be expressed in the form

$$\sum_{p=0}^{m} \phi_p(t) y^{(p)}(t) = r(t) \tag{6.2}$$

in which $\phi_p(t)$ are known functions. If the general nonlinear differential equation (6.1) of order m can be written as

$$y^{(m)} = F(t, y, y', \ldots, y^{(m-1)}) \tag{6.3}$$

then the equation (6.3) is called a **canonical representation** of the differential equation (6.1). In such a form, the highest order derivative is expressed in terms of the lower order derivatives and the independent variable.

Initial Value Problem

A general **solution** of an ordinary differential equation such as (6.1) is a relation between y, t and m arbitrary constants which satisfies the equation, but which contains no derivatives. The solution may be an implicit relation of the form

$$w(t, y, c_1, c_2, \ldots, c_m) = 0 \tag{6.4}$$

or an explicit function of t of the form

$$y = w(t, c_1, c_2, \ldots, c_m) \tag{6.5}$$

The m arbitrary constants $c_1, c_2, \ldots, c_m$ can be determined by prescribing

m conditions of the form
$$y^{(\nu)}(t_0) = \eta_\nu, \quad \nu = 0, 1, 2, \ldots, m-1 \tag{6.6}$$
at one point $t = t_0$, which are called **initial conditions**. The differential equation (6.1) together with the initial conditions (6.6) is called mth order **initial value problem**.

The mth order differential equation (6.3) with initial conditions (6.6) may be written as an equivalent system of m first order initial value problems:
$$u_1 = y$$
$$u_1' = u_2$$
$$u_2' = u_3$$
$$\vdots$$
$$u_{m-1}' = u_m$$
$$u_m' = F(t, u_1, u_2, \ldots, u_m)$$
$$u_1(t_0) = \eta_0, u_2(t_0) = \eta_1, \ldots, u_m(t_0) = \eta_{m-1}$$
which in vector notation becomes
$$\mathbf{u}' = \mathbf{f}(t, \mathbf{u})$$
$$\mathbf{u}(t_0) = \boldsymbol{\eta} \tag{6.7}$$
where
$$\mathbf{u} = [u_1 \ u_2 \ldots u_m]^T$$
$$\mathbf{f} = [u_2 \ u_3 \ldots u_m \ F]^T$$
$$\boldsymbol{\eta} = [\eta_0 \ \eta_1 \ldots \eta_{m-1}]^T$$
Thus the methods of solution of the first-order initial value problem
$$\frac{du}{dt} = f(t, u), \quad u(t_0) = \eta \tag{6.8}$$
may be used to solve the system of first order initial value problems (6.7) and the mth order initial value problem (6.3).

The existence and uniqueness of the solution of the initial value problem (6.8) is guaranteed by the theorem:

THEOREM 6.1 *We assume that $f(t, u)$ satisfies the following conditions:*
  (i) *$f(t, u)$ is a real function*
  (ii) *$f(t, u)$ is defined and continuous in the strip*
       $t \in [t_0, b], \ u \in (-\infty, \infty)$
  (iii) *there exists a constant L such that for any $t \in [t_0, b]$ and for any u_1 and u_2*
       $$|f(t, u_1) - f(t, u_2)| \leq L|u_1 - u_2|$$

*where L is called the **Lipschitz constant**. Then for any u_0 the initial value problem (6.8) has a unique solution $u(t)$ for $t \in [t_0, b]$.*

Ordinary Differential Equations

We will always assume the existence and uniqueness of the solution and also that $f(t, u)$ has continuous partial derivatives with respect to t and u of as high an order as we desire.

Test Equations

The behaviour of the solution of the intial value problem (6.8) in the neighbourhood of any point $(\bar{t}, \bar{u})$ can be predicted by considering the linearized form of the differential equation

$$u' = f(t, u)$$

The nonlinear function $f(t, u)$ can be linearized by expansion of the function about the point $(\bar{t}, \bar{u})$ in the Taylor series truncated after first order term. The resulting linearized form for (6.8) is given by

$$u' = \lambda u + C \tag{6.9}$$

where

$$\lambda = \left(\frac{\partial f}{\partial u}\right)_{\bar{t}}$$

$$C = f(\bar{t}, \bar{u}) - \bar{u}\left(\frac{\partial f}{\partial u}\right)_{\bar{t}} + \left(\frac{\partial f}{\partial t}\right)_{\bar{t}} (t - \bar{t})$$

Further, the equation (6.9) may be written as

$$w' = \lambda w \tag{6.10}$$

where

$$w = u + C/\lambda$$

The solution $w(t)$ is shown in Fig. 6.1(a).

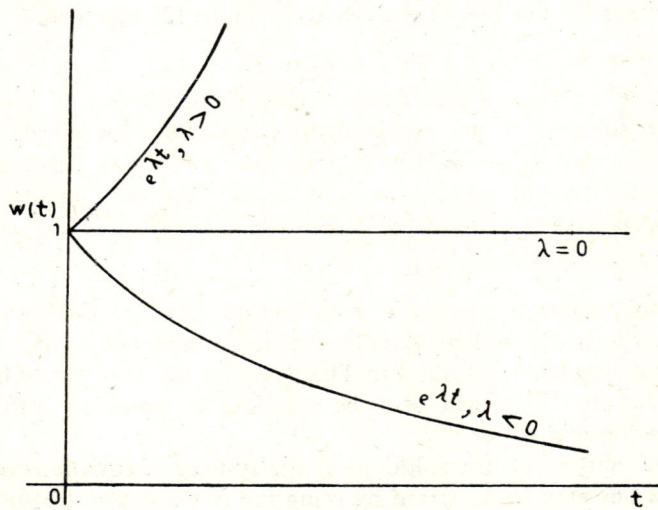

Fig. 6.1(a) Representation of $w = e^{\lambda t}$

Similarly, the test equation for the second order initial value problem

$$y'' = f(t, y, y')$$
$$y(t_0) = \eta_0, \; y'(t_0) = \eta_1 \tag{6.11}$$
may also be obtained in the form
$$w'' = -bw' - cw \tag{6.12}$$
where
$$b = -\frac{\partial f}{\partial y'}, \quad \text{and} \quad c = -\frac{\partial f}{\partial y}$$

The differential equation (6.12) is equivalent to the following system of equations
$$\mathbf{u}' = \mathbf{A}\mathbf{u} \tag{6.13}$$
where
$$\mathbf{u} = [u_1 \; u_2]^T, \quad \mathbf{A} = \begin{bmatrix} 0 & 1 \\ -c & -b \end{bmatrix}$$
$$u_1 = w, \quad u_2 = w'$$

The nature of the solutions of (6.12) or (6.13) depends on the roots ξ_1 and ξ_2 of the characteristic equation of the matrix $\mathbf{A}$,
$$\xi^2 + b\xi + c = 0 \tag{6.14}$$
We now consider the following cases:

(i) $b > 0, c \geqslant 0, b > 2\sqrt{c}$. The solutions are exponentially decreasing. For $c = 0$, the test equation (6.12) becomes
$$w'' + bw' = 0, \; b > 0 \tag{6.15}$$

(ii) $b < 0, c \geqslant 0$ and $|b| > 2\sqrt{c}$. The solutions are exponentially increasing. For $c = 0$, the test equation (6.12) becomes
$$w'' + bw' = 0, \quad b < 0 \tag{6.16}$$

(iii) $c > 0$ and $|b| \leqslant 2\sqrt{c}$. The solutions are oscillating. If $b < 0$ then the solution is an oscillating function whose amplitude becomes unbounded as $t \to \infty$. If $b > 0$ then the solution is a damped oscillating function as $t \to \infty$. For $b = 0$, the test equation (6.12) becomes
$$w'' + cw = 0, \; c > 0 \tag{6.17}$$
whose solution is periodic with period $2\pi/\sqrt{c}$. The solution is shown in Fig. 6.1(b). We observe that ξ_1 and $\xi_2(b = 0, c > 0)$ are pure imaginary numbers. The solution of the test differential equation (6.10) will also be periodic if we allow λ to be pure imaginary number.

Thus the nature of the solutions of the systems of equations or higher order equations may be discussed by using the test equation (6.10) for (i) λ pure real, (ii) λ pure imaginary, and (iii) λ complex.

We shall now develop numerical methods for the solution of the initial value problem (6.8).

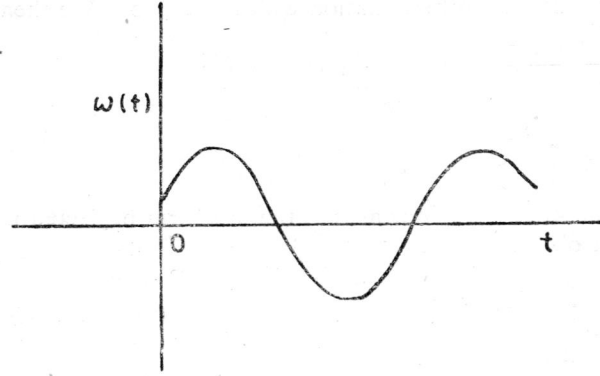

Fig. 6.1(b) Periodic solution

6.2 NUMERICAL METHODS

The first step in obtaining a numerical solution of the differential equation (6.8) is to partition the interval $[t_0, b]$ on which the solution is desired into a finite number of subintervals by the points

$$t_0 < t_1 < t_2 \ldots < t_N = b$$

The points are called the **mesh points** or the **grid points**. The spacing between the points is given by

$$h_j = t_j - t_{j-1}, \quad j = 1, 2, \ldots, N$$

which is called the **mesh spacing** or **step length**. For simplicity we assume that the points are spaced uniformly, i.e.

$$h_j = h = \text{constant}, \quad j = 1, 2, \ldots, N \tag{6.18}$$

The mesh points are given by

$$t_j = t_0 + jh, \quad j = 0, 1, 2, \ldots, N \tag{6.19}$$

In numerical methods we determine a number u_j, which is an approximation to the value of the solution $u(t)$ at the point t_j. The set of numbers $\{u_j\}$, i.e., $u_0, u_1, \ldots, u_N$ is the **numerical solution** of the initial value problem. The numbers $\{u_j\}$ are determined from a set of algebraic equations called the **difference equations**. There are many difference approximations possible for a given differential equation. As an example, let us develop expressions for the first derivative in terms of the forward, backward, and central difference operators. We assume that the function $u(t)$ may be expanded in a Taylor series in the closed interval $t - h \leqslant t \leqslant t + h$. We write as

$$u(t \pm h) = u(t) \pm \frac{h}{1!}u'(t) + \frac{h^2}{2!}u''(t) \pm \ldots + (-1)^p \frac{h^p}{p!}u^{(p)}(t) + \ldots \tag{6.20}$$

where a prime denotes differentiation with respect to t. We then have

$$\frac{u(t+h) - u(t)}{h} = u'(t) + \frac{h}{2} u''(t) + 0(h^2)$$

or

$$\frac{\Delta u(t)}{h} = \frac{du}{dt} + 0(h) \qquad (6.21)$$

where the notation $0(h)$ means that the first term neglected is of order h. Similarly, we obtain

$$\frac{\nabla u(t)}{h} = \frac{du}{dt} + 0(h) \qquad (6.22)$$

and

$$\frac{\mu\delta u(t)}{h} = \frac{du}{dt} + 0(h^2) \qquad (6.23)$$

A difference approximation to $u'(t)$ at $t = t_j$ is obtained by neglecting the error term. We have

$$u'(t_j) \approx \begin{vmatrix} \text{(i)} & (u_{j+1} - u_j)/h \\ \text{(ii)} & (u_j - u_{j-1})/h \\ \text{(iii)} & (u_{j+1} - u_{j-1})/h \end{vmatrix} \qquad (6.24)$$

We use the approximations (6.24) for $u'(t)$ in the differential equation (6.8) at the mesh point t_j. This gives

(i) $\quad \dfrac{u_{j+1} - u_j}{h} = f(t_j, u_j)$

(ii) $\quad \dfrac{u_j - u_{j-1}}{h} = f(t_j, u_j)$

(iii) $\quad \dfrac{u_{j+1} - u_{j-1}}{h} = f(t_j, u_j) \qquad (6.25)$

The equations (6.25) may be considered as a relation between differences of an unknown function u_j and may be called **difference equations**. The **order** of a difference equation is the number of intervals separating the largest and the smallest arguments of the dependent variable. Thus the difference equations (6.25i) and (6.25ii) are of first order and the difference equation (6.25iii) is of second order. The methods (6.25i) and (6.25ii) are called **single step methods** and the method (6.25iii) is called a **two-step** or **multistep method**. The approximate values u_j will contain errors. We must be concerned with the effect of these errors on the solution, and ask what happens as we try to get a more accurate solution, by taking more grid points. A method is **convergent** if, as more grid points are taken or step size is decreased, the numerical solution converges to the exact solution, in the absence of roundoff errors. A method is **stable** if the effect of any single fixed roundoff error is bounded, independent of the number of mesh points.

We now examine each method in turn. We will also discuss the stability and convergence of the methods by applying these on the test equations.

6.2.1 Euler Method

We write (6.25i) as

$$u_{j+1} = u_j + hf_j \qquad (6.26)$$

This is called the **Euler** or the first order **Adams-Bashforth method**. The method is applied at the mesh points t_j, $j = 0(1)N - 1$ to get the numerical solution of the differential equation (6.8). We have

$$u_1 = u_0 + hf_0$$
$$u_2 = u_1 + hf_1$$
$$\vdots$$
$$u_N = u_{N-1} + hf_{N-1} \qquad (6.27)$$

where $f_j = f(t_j, u_j)$.

Choosing h, the value u_1 is determined from the initial condition and the differential equation (6.8), and it is easy to calculate u_2 from u_1 and so on. The method (6.26) is an **explicit method**, since, using u_j, h and f_j we can calculate u_{j+1} from (6.26) directly.

Thus, the Euler method provides a simple procedure for computing approximation u_j to the exact solution $u(t_j)$. The error in the approximation is the difference between the exact solution at $t = t_{j+1}$ and the solution value u_{j+1} determined from (6.26) using the exact arithmetic, and is called the **local truncation error** or **local discretization error**. We have

$$T_{j+1} = u(t_{j+1}) - u_{j+i} \qquad (6.28)$$

Using (6.26), the equation (6.28) becomes

$$T_{j+1} = u(t_{j+1}) - u(t_j) - hf(t_j, u(t_j))$$
$$= \frac{h^2}{2} u''(\xi) \qquad (6.29)$$

where $t_j < \xi < t_{j+1}$. If we denote $\max_{[t_0, b]} |T_{j+1}| = T$ and $\max_{[t_0, b]} |u''(\xi)| = M_2$, then (6.29) becomes

$$T \leqslant \frac{h^2}{2} M_2 \qquad (6.30)$$

The order of the local truncation error is $0(h^2)$ as $h \to 0$. In numerical calculations, we do not deal with the values of u_j, since these values are defined by a procedure which uses exact arithmetic. To include the effect of rounding errors we introduce a new approximation $\bar{u}_j$ which is defined by the same procedure, except that rounding errors are allowed. We have

$$\bar{u}_{j+1} = \bar{u}_j + hf(t_j, \bar{u}_j) - R_{j+1} \qquad (6.31)$$

where the rounding error R_{j+1} is the amount by which the Euler method

(6.26) is not satisfied by $\bar{u}_j$. Applying (6.26) and (6.31) to the test equation $u' = \lambda u$ and subtracting we get

$$u(t_{j+1}) - \bar{u}_{j+1} = u(t_j) - \bar{u}_j + \lambda h(u(t_j) - \bar{u}_j) + T_{j+1} + R_{j+1}$$

We now define $e_j = u(t_j) - u_j$ and substitute in the above equation to get the error equation

$$e_{j+1} = (1 + h\lambda)e_j + T_{j+1} + R_{j+1} \tag{6.32}$$

Let us now introduce the difference equation

$$E_{j+1} = AE_j + B \tag{6.33}$$

where $|1 + h\lambda| \leqslant A$ and $|T_{j+1} + R_{j+1}| \leqslant B$. It is clear that $|e_j| \leqslant E_j$, provided $|e_0| \leqslant E_0$. We obtain

$$E_1 = AE_0 + B$$

$$E_2 = AE_1 + B = A^2 E_0 + (A + 1)B$$

$$\vdots$$

$$E_j = A^j E_0 + \left(\frac{A^j - 1}{A - 1}\right) B$$

provided $A \neq 1$.

We note that

$$(1 + h\lambda)^j < \exp(\lambda(t_j - t_0)), \quad \lambda > 0$$

Putting $E_0 = |e_0|$ we get

$$|e_j| < \exp(\lambda(t_j - t_0))|e_0| + \frac{\exp(\lambda(t_j - t_0)) - 1}{h\lambda}(T + R) \tag{6.34}$$

where $R = \max_{[t_0, b]} |R_{j+1}|$.

We now consider the following cases:

(i) $e_0 = 0$, $R = 0$. If the initial value η_0 is exactly used and the round-off error is negligible then $e_0 = 0$ and $R \approx 0$. The equation (6.34) becomes

$$|e_j| < \frac{T}{h\lambda}(\exp(\lambda(t_j - t_0)) - 1) \tag{6.35}$$

Since $T = 0(h^2)$, it follows that the $|e_j| \to 0$ as $h \to 0$. Thus the Euler method converges.

(ii) $e_0 = 0$. We obtain from (6.34)

$$|e_j| < \frac{\exp(\lambda(t_j - t_0)) - 1}{\lambda}\left(\frac{T}{h} + \frac{R}{h}\right) \tag{6.36}$$

Again, since $T = 0(h^2)$, we see that the bound will decrease as h decreases, until the contribution due to rounding becomes dominant, at which point a further decrease in h causes the bound to increase. This behaviour of the error bound is shown in Fig. 6.2.

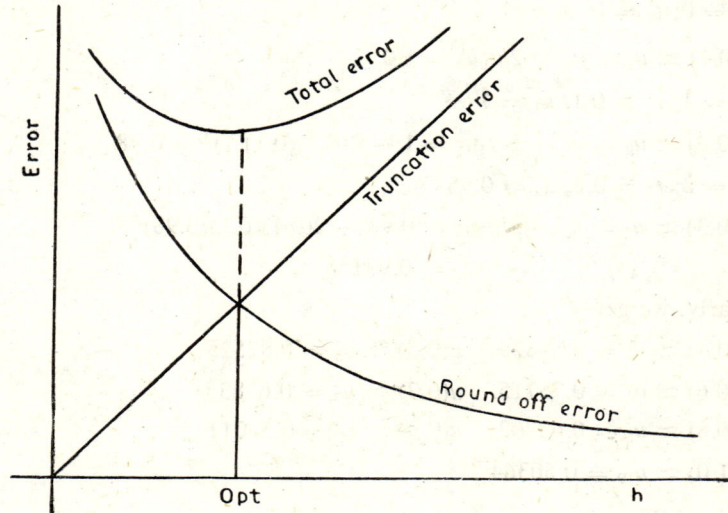

Fig. 6.2 Truncation and round off errors as functions of h

Example 6.1 Use the Euler method to solve numerically the initial value problem

$$u' = -2tu^2, \quad u(0) = 1$$

with $h = 0.2, 0.1$ and 0.05 on the interval $[0, 1]$. Neglecting the roundoff errors, determine the bound for the error. Apply Richardson's extrapolation to improve the computed value $u(1.0)$.

We have

$$u_{j+1} = u_j - 2ht_j u_j^2; \quad j = 0, 1, 2, 3, 4$$

with $h = 0.2$. The initial condition gives $u_0 = 1$.

For $j = 0$; $t_0 = 0$, $u_0 = 1$

$$u(0.2) \approx u_1 = u_0 - 2ht_0 u_0^2 = 1.0$$

For $j = 1$; $t_1 = 0.2$, $u_1 = 1$

$$u(0.4) \approx u_2 = u_1 - 2ht_1 u_1^2$$
$$= 1 - 2(0.2)(0.2)(1)^2 = 0.92$$

For $j = 2$; $t_2 = 0.4$, $u_2 = 0.92$

$$u(0.6) \approx u_3 = u_2 - 2ht_2 u_2^2$$
$$= 0.92 - 2(0.2)(0.4)(0.92)^2 = 0.78458$$

Similarly we get

$$u(0.8) \approx u_4 = 0.63684, \quad u(1) \approx u_5 = 0.50706$$

When $h = 0.1$, we get

For $j = 0$; $t_0 = 0$, $u_0 = 1$

$$u(0.1) \approx u_1 = u_0 - 2ht_0 u_0^2 = 1.0$$

For $j = 1$; $t_1 = 0.1$, $u_1 = 1$

$$u(0.2) \approx u_2 = u_1 - 2ht_1 u_1^2 = 1 - 2(0.1)(0.1)(1)^2 = 0.98$$

For $j = 2$; $t_2 = 0.2$, $u_2 = 0.98$

$$u(0.3) \approx u_3 = u_2 - 2ht_2 u_2^2 = 0.98 - 2(0.1)(0.2)(0.98)^2$$
$$= 0.94158$$

Similarly, we get

$u(0.4) \approx u_4 = 0.88839$, $\quad u(0.5) \approx u_5 = 0.82525$

$u(0.6) \approx u_6 = 0.75715$, $\quad u(0.7) \approx u_7 = 0.68835$

$u(0.8) \approx u_8 = 0.62202$, $\quad u(0.9) \approx u_9 = 0.56011$

$u(1.0) \approx u_{10} = 0.50364$

For $h = 0.05$, we get

$u(0.05) \approx 1.0$, $\quad u(0.1) \approx 0.995$, $\quad u(0.15) \approx 0.9851$

$u(0.2) \approx 0.97054$, $\quad u(0.25) \approx 0.9517$, $\quad u(0.3) \approx 0.92906$

$u(0.35) \approx 0.90316$, $\quad u(0.4) \approx 0.87461$, $\quad u(0.45) \approx 0.84401$

$u(0.5) \approx 0.81195$, $\quad u(0.55) \approx 0.77899$, $\quad u(0.6) \approx 0.74561$

$u(0.65) \approx 0.71225$, $\quad u(0.7) \approx 0.67928$, $\quad u(0.75) \approx 0.64698$

$u(0.8) \approx 0.61559$, $\quad u(0.85) \approx 0.58527$, $\quad u(0.9) \approx 0.55615$

$u(0.95) \approx 0.52831$, $\quad u(1.0) = 0.50179$

The truncation error in the Euler's method is given as

$$TE = \frac{h^2}{2} u''(\xi)$$

$$|TE| = \frac{h^2}{2} |u''(\xi)| \leq \frac{h^2}{2} \max_{0 \leq t \leq 1} |u''(t)|$$

Since the exact solution is $u(t) = \dfrac{1}{(1 + t^2)}$, we get

$$|TE| \leq \frac{h^2}{2} \max_{0 \leq t \leq 1} \left| \frac{2(1 - 3t^2)}{(1 + t^2)^3} \right| \leq h^2$$

The error in Euler's method is of the form

$$u(t_n) - u_n(h) = c_1 h + c_2 h^2 + c_3 h^3 + \ldots$$

Richardson's extrapolation gives

$$u^{(k)}(h) = \frac{2^k u_n^{(k-1)}\left(\dfrac{h}{2}\right) - u_n^{(k-1)}(h)}{2^k - 1}$$

We have the following extrapolated value for $u(1.0)$.

Extrapolated Value for u (1.0)

h	$u^{(0)}(h)$	$u^{(1)}(h)$	$u^{(2)}(h)$	Exact
0.2	0.50706			
		0.50022		
0.1	0.50364		0.49985	0.5
		0.49994		
0.05	0.50179			

6.2.2 Backward Euler Method

The equation (6.25ii) at the mesh point $t = t_{j+1}$ may be written as

$$u_{j+1} = u_j + hf_{j+1} \tag{6.37}$$

where

$$f_{j+1} = f(t_{j+1}, u_{j+1})$$

This is called the **backward Euler** or the first order **Adams-Moulton method**. The solution values $u_1, u_2, \ldots, u_N$ are determined from the following equations:

$$\begin{aligned} u_1 &= u_0 + hf_1 \\ u_2 &= u_1 + hf_2 \\ &\vdots \\ u_N &= u_{N-1} + hf_N \end{aligned} \tag{6.38}$$

The right hand side of (6.37) involves the unknown u_{j+1} which generally must be determined by iteration methods discussed in Chapter II, unless f is linear in u. A numerical method such as (6.37) which involves the unknown value u_{j+1} on both sides, and hence defines u_{j+1} implicitly, is called an **implicit method**. The local truncation error at the point $t = t_{j+1}$ in the method (6.37) is given by

$$\begin{aligned} T_{j+1} &= u(t_{j+1}) - u_{j+1} \\ &= u(t_{j+1}) - (u(t_j) + hf(t_{j+1}, u(t_{j+1}))) \\ &= -\tfrac{1}{2}h^2 u''(\xi), \quad t_j < \xi < t_{j+1} \end{aligned} \tag{6.39}$$

which is equal to the local truncation error of the Euler method but of opposite sign. The error equation (6.33) becomes

$$AE_{j+1} = E_j + B \tag{6.40}$$

where

$$A \leqslant |1 - h\lambda| \tag{6.41}$$

and

$$|T_{j+1} + R_{j+1}| \leqslant B. \tag{6.42}$$

It is easily shown that the backward Euler method converges as $h \to 0$ in the absence of roundoff errors.

Example 6.2 Solve the initial value problem
$$u' = -2tu^2, \quad u(0) = 1$$
with $h = 0.2$ on the interval [0, 1] using the backward Euler method.

The backward Euler method gives
$$u_{j+1} = u_j - 2ht_{j+1}u_{j+1}^2, \quad j = 0, 1, 2, 3, 4$$

This is an implicit equation in u_{j+1} and can be solved by using an iterative method discussed in Chapter II. However, this is a quadratic equation in u_{j+1} and can be solved exactly. One of the solutions will be positive and the other negative. Since we are interested only in the positive solution, we have
$$u_{j+1} = \frac{-1 + \sqrt{1 + 8hu_j t_{j+1}}}{4ht_{j+1}}$$

For $j = 0$; $u_0 = 1$, $t_1 = 0.2$,
$$u(0.2) \approx u_1 = \frac{-1 + \sqrt{1 + 8(.2)(1)(.2)}}{4(.2)(.2)} = 0.9307033$$

For $j = 1$; $u_1 = 0.9307033$, $t_2 = 0.4$
$$u(0.4) \approx u_2 = \frac{-1 + \sqrt{1 + 8(.2)(.9307033)(.4)}}{4(.2)(.4)} = 0.8224701$$

Similarly, we get
$$u(0.6) \approx u_3 = 0.7036429$$
$$u(0.8) \approx u_4 = 0.5916333$$
$$u(1.0) \approx u_5 = 0.4940135$$

6.2.3 Mid-Point Method

The equation (6.25iii) may be written as
$$u_{j+1} = u_{j-1} + 2hf_j \tag{6.43}$$

This is called **mid-point** or the second order **Nyström method**. The solution values are given by
$$u_2 = u_0 + 2hf_1$$
$$u_3 = u_1 + 2hf_2$$
$$\vdots$$
$$u_N = u_{N-2} + 2hf_N \tag{6.44}$$

The value u_0 is known from the initial condition. The value u_1 is unknown and must be determined by some other method after which (6.44) is used to determine successively $u_2, u_3, \ldots, u_N$. The local truncation error is given by

$$T_{j+1} = u(t_{j+1}) - u_{j+1}$$
$$= u(t_{j+1}) - u(t_{j-1}) - 2hf(t_j, u(t_j))$$
$$= \frac{h^3}{3} u'''(\xi), \quad t_{j-1} < \xi < t_{j+1}$$

which may be written as

$$|T_{j+1}| \leq \frac{h^3}{3} M_3 \tag{6.45}$$

where

$$\max_{t \in [t_0, b]} |u'''(\xi)| = M_3$$

Neglecting the roundoff error and applying (6.43) to the test equation, the error equation becomes

$$\epsilon_{j+1} = \epsilon_{j-1} + 2h\lambda\epsilon_j + \frac{h^3}{3} M_3 \tag{6.46}$$

where $u_j - u(t_j) = \epsilon_j$. We wish to determine an explicit bound for the quantity $|\epsilon_j|$. Let $|\epsilon_j| \leq e_j$, $j = 0, 1, 2, \ldots$, where e_j satisfies the difference equation

$$e_{j+1} = e_{j-1} + 2h\lambda e_j + \frac{h^3}{3} M_3 \tag{6.47}$$

This is a second order difference equation with constant coefficients. The solution e_j of (6.47) consists of a solution to the homogeneous equation, say $e_j^{(H)}$, plus a particular solution, say $e_j^{(P)}$, of the inhomogeneous part. The homogeneous difference equation becomes

$$e_{j+1} - 2h\lambda e_j - e_{j-1} = 0 \tag{6.48}$$

To find the solution of (6.48), we use the trial solution

$$e_j = A\xi^j \tag{6.49}$$

where $A \neq 0$ and ξ is a number to be determined. Inserting (6.49) in (6.48), we find that nontrivial solution exists if ξ is a root of the polynomial

$$\xi^2 - 2h\lambda\xi - 1 = 0 \tag{6.50}$$

This equation is called the **characteristic equation** of the difference equation (6.48). It has two roots

$$\xi_1 = h\lambda + \sqrt{h^2\lambda^2 + 1} = e^{h\lambda} - \tfrac{1}{6}h^3\lambda^3 + O(h^4)$$
$$\xi_2 = h\lambda - \sqrt{h^2\lambda^2 + 1} = -e^{-h\lambda} + \tfrac{1}{6}h^3\lambda^3 + O(h^4) \tag{6.51}$$

The nonhomogeneous equation has the particular solution $-h^2 M_3/6\lambda$, so that the general solution is

$$e_j = c_1 \xi_1^j + c_2 \xi_2^j - \frac{h^2 M_3}{6\lambda} \tag{6.52}$$

where c_1 and c_2 are arbitrary constants to be determined. Denoting

$$E_j = e_j + \frac{h^2 M_3}{6\lambda}, \quad j = 0, 1 \tag{6.53}$$

the constants c_1 and c_2 can be found by solving the equations

$$\begin{aligned} E_0 &= c_1 + c_2 \\ E_1 &= c_1 \xi_1 + c_2 \xi_2 \end{aligned} \tag{6.54}$$

If we assume that the initial errors e_0, e_1 are constants and equal to ϵ, then determining c_1 and c_2 from (6.54) and substituting in (6.52) we get

$$e_j = \left(\epsilon + \frac{h^2 M_3}{6\lambda}\right) \frac{1}{\xi_1 - \xi_2} [(1 - \xi_2)\xi_1^j + (\xi_1 - 1)\xi_2^j] - \frac{h^2 M_3}{6\lambda} \tag{6.55}$$

We find

$$\xi_1^j = e^{j\lambda h}\left(1 - \frac{j}{6}h^3\lambda^3 + 0(h^4)\right)$$

$$\xi_2^j = (-1)^j e^{-j\lambda h}\left(1 + \frac{j}{6}h^3\lambda^3 + 0(h^4)\right)$$

$$\xi_1 - \xi_2 = 2 + h^2\lambda^2 + 0(h^4)$$

$$1 - \xi_2 = 2 - h\lambda + \frac{h^2\lambda^2}{2} + 0(h^4) \tag{6.56}$$

$$\xi_1 - 1 = h\lambda + \tfrac{1}{2}h^2\lambda^2 + 0(h^4)$$

Substituting (6.56) into (6.55) we obtain

$$\begin{aligned} e_j = \frac{1}{2 + h^2\lambda^2}\bigg(\epsilon + \frac{h^2 M_3}{6\lambda}\bigg)&\bigg[\bigg(2 - h\lambda + \frac{h^2\lambda^2}{2} + 0(h^4)\bigg) \\ &e^{jh\lambda}\bigg(1 - \frac{jh^3\lambda^3}{6} + 0(h^4)\bigg) + (h\lambda + \tfrac{1}{2}h^2\lambda^2 + 0(h^4)) \\ &(-1)^j e^{-jh\lambda}\bigg(1 + \frac{jh^3\lambda^3}{6} + 0(h^4)\bigg)\bigg] - \frac{h^2 M_3}{6\lambda} \end{aligned} \tag{6.57}$$

Putting $\epsilon = 0$ in (6.57) we find that $e_j \to 0$ as $h \to 0$. But we observe that if λ is positive $\xi_1 > 1$ and if λ is negative $\xi_2 < -1$. Therefore one of the terms in (6.57) increases, for any fixed $h > 0$. Here the extra term in (6.57) is due to the fact that we have approximated the first order differential equation by a second order difference equation. The second term is small when h and $(t_j - t_0)$ are small, but eventually as $(t_j - t_0)$ increases with h fixed, it will dominate over the first term. Thus e_j will oscillate with ever increasing amplitude as $(t_j - t_0)$ increases, for any fixed positive h.

Example 6.3 Solve the initial value problem:

$$u' = -2tu^2, \quad u(0) = 1$$

using the following methods:
  (i) Euler method

(ii) Backward Euler method

(iii) Mid-point method

with $h = 0.2$ over the interval $[0, 1]$. Determine the percentage relative error at $t = 1$.

The mid-point method gives

$$u_{j+1} = u_{j-1} - 4ht_j u_j^2, \quad j = 1, 2, 3, 4$$

We calculate u_1 from the exact solution $u(t) = 1/(1 + t^2)$. We get

$$u_1 = u(0.2) = 1/1.04 = 0.9615385$$

For $j = 1$

$$u_0 = 1, \quad u_1 = 0.9615385, \quad t_1 = 0.2$$

$$u(0.4) \approx u_2 = u_0 - 4ht_1 u_1^2$$

$$= 1 - 4(.2)(.2)(.9615385)^2 = 0.8520710$$

For $j = 2$

$$u_1 = 0.9615385, \quad u_2 = 0.8520710, \quad t_2 = 0.4$$

$$u(0.6) \approx u_3 = u_1 - 4ht_2 u_2^2$$

$$= 0.9615385 - 4(.2)(.4)(.8520710)^2 = 0.7292105$$

Similarly, we get

$$u(0.8) \approx u_4 = 0.5968320$$

$$u(1.0) \approx u_5 = 0.5012371$$

The percentage relative error is given by

$$PRE = \frac{|u - u^*|}{|u|} \times 100$$

where u is the exact solution and u^* is the approximate solution. The exact solution at $t = 1$ is 0.5. The percentage relative error for different methods is obtained as given below. The results for Euler's method and backward Euler's method are taken from Examples 6.1 and 6.2 respectively.

Euler method:

$$\frac{|.5 - .5070602|}{.5} \times 100 = 1.41$$

Backward Euler method:

$$\frac{|.5 - .4940135|}{.5} \times 100 = 1.20$$

Mid-point method:

$$\frac{|.5 - .5012371|}{.5} \times 100 = .25$$

6.3 SINGLESTEP METHODS

A singlestep method for the solution of the initial value problem (6.8)

$$\frac{du}{dt} = f(t, u), \, u(t_0) = \eta_0, \, t \in [t_0, b]$$

is a related first order difference equation. Thus, a general singlestep method may be written as

$$u_{j+1} = u_j + h\phi(t_j, u_j, h) \tag{6.58}$$

where $\phi(t, u, h)$ is a function of the arguments t, h, u and in addition depends on f.

The function $\phi(t, h, u)$ is called the **increment function**. If u_{j+1} can be determined simply by evaluating the right hand side of (6.58), then the singlestep method is called **explicit**, otherwise it is called **implicit**. The local truncation error T_{j+1} is defined by

$$T_{j+1} = u(t_{j+1}) - u(t_j) - h\phi(t_j, u(t_j), h) \tag{6.59}$$

The largest integer p such that

$$|h^{-1}T_{j+1}| = 0(h^p) \tag{6.60}$$

is called the **order** of the singlestep method. We now determine a few specific forms for the increment function $\phi(t, u, h)$.

6.3.1 Taylor Series Method

We assume that the function $u(t)$ can be expanded in a Taylor series about any point t_j.

$$u(t) = u(t_j) + (t - t_j)u'(t_j) + \frac{1}{2!}(t - t_j)^2 u''(t_j) + \cdots$$

$$+ \frac{1}{p!}(t - t_j)^p u^{(p)}(t_j) + \frac{1}{(p+1)!}(t - t_j)^{p+1} u^{(p+1)}(t_j + \theta h) \tag{6.61}$$

This expansion holds good for $t \in [t_0, b]$ and $0 < \theta < 1$.

Substituting $t = t_{j+1}$ in (6.61), we get

$$u(t_{j+1}) = u(t_j) + hu'(t_j) + \frac{h^2}{2!}u''(t_j) + \cdots + \frac{1}{p!}h^p u^{(p)}(t_j)$$

$$+ \frac{1}{(p+1)!}h^{p+1} u^{(p+1)}(t_j + \theta h)$$

We define

$$h\phi(t_j, u(t_j), h) = hu'(t_j) + \frac{h^2}{2!}u''(t_j) + \cdots + \frac{h^p}{p!}u^{(p)}(t_j)$$

and $h\phi(t_j, u_j, h)$ is to be obtained from $\phi(t_j, u(t_j), h)$ by using an approximate value u_j in place of the exact value $u(t_j)$. We compute

$$u_{j+1} = u_j + h\phi(t_j, u_j, h), \quad j = 0, 1, \ldots, N-1 \tag{6.62}$$

to approximate $u(t_{j+1})$. This is called the **Taylor series method** of order p. Substituting $p = 1$, in (6.62) we get

$$u_{j+1} = u_j + hu'_j$$

which is the Euler's method.

To apply (6.62), it is necessary to know $u(t_j), u'(t_j), \ldots, u^{(p)}(t_j)$. If t_j and $u(t_j)$ are known then the derivatives can be calculated as follows:

First the known values t_j and $u(t_j)$ are substituted into the differential equation to give

$$u'(t_j) = f(t_j, u(t_j))$$

Next, the differential equation (6.8) is differentiated to obtain expressions for the higher order derivatives of $u(t)$.

Thus

$$u' = f(t, u)$$
$$u'' = f_t + ff_u$$
$$u''' = f_{tt} + 2ff_{tu} + f^2 f_{uu} + f_u(f_t + ff_u)$$
$$\vdots$$

where $f_t, f_u, \ldots$ represent the partial derivatives of f with respect to t and u and so on. The values $u''(t_j), u'''(t_j), \ldots$ can be computed by substituting $t = t_j$. Therefore, if t_j and $u(t_j)$ are known exactly then (6.62) can be used to compute u_{j+1} with an error

$$\frac{h^{p+1}}{(p+1)!} u^{(p+1)}(t_j + \theta h)$$

The number of terms to be included in (6.62) is fixed by the permissible error. If this error is ϵ and the series is truncated at the term $u^{(p)}(t_j)$ then

$$h^{p+1} |u^{(p+1)}(t_j + \theta h)| < (p+1)! \, \epsilon$$

or

$$h^{p+1} |f^{(p)}(t_j + \theta h)| < (p+1)! \, \epsilon \tag{6.63}$$

For a given h, (6.63) will determine p, and if p is specified then it will give an upper bound on h.

Since $t_j + \theta h$ is not known, $|f^{(p)}(t_j + \theta h)|$ in (6.63) is replaced by its maximum value in $[t_0, b]$. A way of determining this value is as follows. Write one more non-vanishing term in the series than is required and then differentiate this series p times. The maximum value of this quantity in $[t_0, b]$ gives the required bound.

Example 6.4 Given the initial value problem

$$u' = t^2 + u^2, \quad u(0) = 0$$

Determine the first three non-zero terms in the Taylor series for $u(t)$ and hence obtain the value for $u(1)$. Also determine t when the error in $u(t)$

obtained from the first two non-zero terms is to be less than 10^{-6} after rounding.

We have

$$u(0) = 0, \quad u'(0) = 0$$
$$u'' = 2t + 2uu', \quad u''(0) = 0$$
$$u''' = 2 + 2(uu'' + u'^2), \quad u'''(0) = 2$$
$$u^{(4)} = 2(uu''' + 3u'u''), \quad u^{(4)}(0) = 0$$
$$u^{(5)} = 2[uu^{(4)} + 4u'u''' + 3(u'')^2], \quad u^{(5)}(0) = 0$$
$$u^{(6)} = 2(uu^{(5)} + 5u'u^{(4)} + 10u''u'''), \quad u^{(6)}(0) = 0$$
$$u^{(7)} = 2(uu^{(6)} + 6u'u^{(5)} + 15u''u^{(4)} + 10(u''')^2), \quad u^{(7)}(0) = 80$$
$$u^{(8)}(0) = u^{(9)}(0) = u^{(10)}(0) = 0$$
$$u^{(11)} = 2[uu^{(10)} + 10u'u^{(9)} + 45u''u^{(8)} + 120u'''u^{(7)} + 210u^{(4)}u^{(6)} + 126(u^{(5)})^2], \quad u^{(11)}(0) = 38400$$

Thus the Taylor series for $u(t)$ becomes

$$u(t) = \frac{1}{3}t^3 + \frac{1}{63}t^7 + \frac{2}{2079}t^{11}$$

The approximate value of $u(1)$ is given by

$$u(1) \approx 0.3502$$

which is correct to four decimal places.

The value of t is obtained from

$$\left|\frac{2}{2079}t^{11}\right| < 0.5 \times 10^{-7}$$

Solving we get

$$t \approx 0.41$$

Example 6.5 Find the three term Taylor series solution for the third order Blasius equation

$$W''' + WW'' = 0, \quad W(0) = 0, \quad W'(0) = 0, \quad W''(0) = 1$$

Find the bound on the error for $t \in [0, 0.2]$.

We find

$$W''' = -WW'', \quad W'''(0) = 0$$
$$W^{(4)} = -(WW''' + W'W''), \quad W^{(4)}(0) = 0$$
$$W^{(5)} = -[WW^{(4)} + 2W'W''' + (W'')^2], \quad W^{(5)}(0) = -1$$
$$W^{(6)} = 0, \quad W^{(7)}(0) = 0, \quad W^{(8)}(0) = 11$$
$$W^{(9)}(0) = W^{(10)}(0) = 0, \quad W^{(11)}(0) = -375$$

The Taylor series solution is

$$W(t) = \frac{t^2}{2!} - \frac{t^5}{5!} + \frac{11}{8!}t^8 + E_8$$

where
$$|E_8| \leq \max |W^{(9)}(t)| \frac{t^9}{9!}$$

Writing the next term, we have
$$W(t) = \frac{t^2}{2!} - \frac{t^5}{5!} + \frac{11}{8!}t^8 - \frac{375}{11!}t^{11}$$

We find
$$W^{(9)}(t) = -\frac{375}{2}t^2$$
and
$$\max_{0 \leq t \leq 0.2} |W^{(9)}(t)| = 7.5$$

Hence
$$|E_8| \leq \frac{7.5(0.2)^9}{9!} \leq (1.06)10^{-11}$$

6.3.2 Runge-Kutta Methods

We first explain the principle involved in the Runge-Kutta methods. By the Mean Value Theorem any solution of
$$u' = f(t, u), \quad u(t_0) = \eta_0, \quad t \in [t_0, b]$$
satisfies
$$u(t_{j+1}) = u(t_j) + hu'(t_j + \theta h)$$
$$= u(t_j) + hf(t_j + \theta h, u(t_j + \theta h)), \quad 0 < \theta < 1$$

For $\theta = \frac{1}{2}$, we have
$$u(t_{j+1}) = u(t_j) + hf\left(t_j + \frac{h}{2}, u\left(t_j + \frac{h}{2}\right)\right)$$

Euler's method with spacing $h/2$ gives
$$u\left(t_j + \frac{h}{2}\right) \approx u_j + \frac{h}{2}f_j$$

Thus, we have the approximation
$$u_{j+1} = u_j + hf\left(t_j + \frac{h}{2}, u_j + \frac{h}{2}f_j\right)$$

which may be written as
$$K_1 = hf_j$$
$$K_2 = hf\left(t_j + \frac{h}{2}, u_j + \frac{1}{2}K_1\right)$$
$$u_{j+1} = u_j + K_2 \tag{6.64}$$

Alternatively, again using Euler's method, we proceed as follows:

$$u'\left(t_j + \frac{h}{2}\right) \approx \frac{1}{2}(u'(t_j) + u'(t_j + h))$$

$$\approx \tfrac{1}{2}[f(t_j, u_j) + f(t_j + h, u_j + hf_j)]$$

and thus we have the approximation

$$u_{j+1} = u_j + \frac{h}{2}[f(t_j, u_j) + f(t_j + h, u_j + hf_j)] \qquad (6.65)$$

which may be written as

$$K_1 = hf(t_j, u_j)$$
$$K_2 = hf(t_j + h, u_j + K_1)$$
$$u_{j+1} = u_j + \tfrac{1}{2}(K_1 + K_2) \qquad (6.66)$$

This method is also called **Euler-Cauchy method**.

Either (6.64) or (6.65) can be regarded as

$$u_{j+1} = u_j + h \text{ (average slope)} \qquad (6.67)$$

This is the underlying idea of the Runge-Kutta approach. In general, we find the slope at t_j and at several other points, average these slopes, multiply by h and add the result to u_j. Thus the **Runge-Kutta method** with v slopes can be written as

$$K_1 = hf(t_j, u_j)$$
$$K_2 = hf(t_j + c_2 h, u_j + a_{21} K_1)$$
$$K_3 = hf(t_j + c_3 h, u_j + a_{31} K_1 + a_{32} K_2)$$
$$K_4 = hf(t_j + c_4 h, u_j + a_{41} K_1 + a_{42} K_2 + a_{43} K_3)$$
$$\vdots$$
$$K_v = hf(t_j + c_v h, u_j + \sum_{i=1}^{v-1} a_{vi} K_i)$$

and

$$u_{j+1} = u_j + W_1 K_1 + W_2 K_2 + \ldots + W_v K_v \qquad (6.68)$$

From (6.68), we may interpret the increment function as the linear combination of the slopes at t_j and at several other points between t_j and t_{j+1}. Further, knowing the values of the quantities on the right hand side of (6.68), the solution value u_{j+1} may be obtained directly. Thus, (6.68) represents the **explicit Runge-Kutta method** with v slopes. To determine the parameters c's, a's and w's in (6.68), we expand u_{j+1} in powers of h such that it agrees with the Taylor series expansion of the solution of the differential equation upto a certain number of terms.

Consider the following Runge-Kutta method with two slopes

$$K_1 = hf(t_j, u_j)$$
$$K_2 = hf(t_j + c_2 h, u_j + a_{21} K_1)$$
$$u_{j+1} = u_j + W_1 K_1 + W_2 K_2 \qquad (6.69)$$

where the parameters c_2, a_{21}, W_1 and W_2 are chosen to make u_{j+1} closer to $u(t_{j+1})$. Now Taylor's series gives

$$u(t_{j+1}) = u(t_j) + hu'(t_j) + \frac{h^2}{2!}u''(t_j) + \frac{h^3}{3!}u'''(t_j) + \cdots$$

$$= u(t_j) + hf(t_j, u(t_j)) + \frac{h^2}{2!}(f_t + ff_u)_{t_j}$$

$$+ \frac{h^3}{3!}[f_{tt} + 2ff_{tu} + f^2 f_{uu} + f_u(f_t + ff_u)]_{t_j} + \cdots \quad (6.70)$$

We also have

$K_1 = hf_j$

$K_2 = hf(t_j + c_2 h, u_j + a_{21} hf_j)$

$= h[f_j + h(c_2 f_t + a_{21} ff_u)_{t_j}$

$+ \frac{h^2}{2}(c_2^2 f_{tt} + 2c_2 a_{21} ff_{tu} + a_{21}^2 f^2 f_{uu})_{t_j} + \cdots]$

Substituting the values of K_1 and K_2 in (6.69), we get

$$u_{j+1} = u_j + (W_1 + W_2)hf_j + h^2(W_2 c_2 f_t + W_2 a_{21} ff_u)_{t_j}$$

$$+ \frac{h^3}{2} W_2(c_2^2 f_{tt} + 2c_2 a_{21} ff_{tu} + a_{21}^2 f^2 f_{uu})_{t_j} + \cdots \quad (6.71)$$

Comparing the coefficients of various powers of h in (6.70) and (6.71), we obtain

$W_1 + W_2 = 1$

$c_2 W_2 = \frac{1}{2}$

$a_{21} W_2 = \frac{1}{2}$

The solution of this system is

$$a_{21} = c_2, \quad W_2 = \frac{1}{2c_2}, \quad W_1 = 1 - \frac{1}{2c_2} \quad (6.72)$$

where $c_2 \neq 0$, is arbitrary. Substituting (6.72) in (6.71), we get

$$u_{j+1} = u_j + hf_j + \frac{h^2}{2}(f_t + ff_u)_{t_j} + \frac{h^3 c_2}{4}(f_{tt} + 2ff_{tu} + f^2 f_{uu})_{t_j} + \cdots \quad (6.73)$$

The local truncation error is given by

$$T_{j+1} = u(t_{j+1}) - u_{j+1}$$

$$= h^3 \left[\left(\frac{1}{6} - \frac{c_2}{4} \right)(f_{tt} + 2ff_{tu} + f^2 f_{uu})_{t_j} + \frac{1}{6}\{f_u(f_t + ff_u)\}_{t_j} + \cdots \right] \quad (6.74)$$

which shows that the method (6.69) is of second order. The free parameter c_2 is usually taken between 0 and 1. Sometimes c_2 is chosen such that one of the W's in the method (6.69) is zero or the truncation error is minimum. Such a formula is called an **optimal** formula.

It may be noted that every Runge-Kutta method should reduce to a quadrature formula when $f(t, u)$ is independent of u with W's as weights and c's as abscissas.

If $c_2 = \frac{1}{2}$, we get

$$u_{j+1} = u_j + hf\left(t_j + \frac{h}{2}, u_j + \frac{h}{2}f_j\right)$$

which is the Euler's method with spacing $h/2$. It reduces to the mid-point quadrature rule when $f(t, u)$ is independent of u.

For $c_2 = 1$, we get

$$u_{j+1} = u_j + \frac{h}{2}[f(t_j, u_j) + f(t_j + h, u_j + hf_j)]$$

which reduces to the trapezoidal rule when $f(t, u)$ is independent of u.

For $c_2 = \frac{2}{3}$, the truncation error is minimum. We have the optimal method

$$u_{j+1} = u_j + \tfrac{1}{4}hf_j + \tfrac{3}{4}hf(t_j + \tfrac{2}{3}h, u_j + \tfrac{2}{3}hf_j) \tag{6.75}$$

We note that the second order method (6.69) requires two function evaluations for each step of integration. Similarly, we find that the third and fourth order methods require three and four function evaluations respectively for each step of integration. However, for $5 \leqslant v \leqslant 7$, the order of the methods becomes $v - 1$ only. For $v \geqslant 8$, the order of the method reduces further to $v - 2$.

The following two equations occur typically in all Runge-Kutta methods of the form (6.68)

$$c_i = \sum_{j=1}^{i-1} a_{ij}, \quad i = 2, 3, \ldots, v$$

and

$$\sum_{j=1}^{v} w_j = 1$$

The number of unknown parameters are then $v(v + 1)/2$.

We now list the second, third and fourth order Runge-Kutta methods.

Second Order Methods

c_2	a_{21}	
	W_1	W_2
$\frac{1}{2}$	$\frac{1}{2}$	
	0	1

Euler's method with spacing $h/2$

$\frac{2}{3}$	$\frac{2}{3}$	
	$\frac{1}{4}$	$\frac{3}{4}$

Optimal

$$\begin{array}{c|c} 1 & 1 \\ \hline & \frac{1}{2} \quad \frac{1}{2} \end{array}$$

Euler-Cauchy or Heun

Third Order Methods

$$\begin{array}{c|cc} c_2 & a_{21} & \\ c_3 & a_{31} & a_{32} \\ \hline & W_1 & W_2 & W_3 \end{array}$$

$$\begin{array}{c|cc} \frac{2}{3} & \frac{2}{3} & \\ \frac{2}{3} & 0 & \frac{2}{3} \\ \hline & \frac{2}{8} & \frac{3}{8} & \frac{3}{8} \end{array} \qquad \begin{array}{c|cc} \frac{1}{2} & \frac{1}{2} & \\ \frac{3}{4} & 0 & \frac{3}{4} \\ \hline & \frac{2}{9} & \frac{3}{9} & \frac{4}{9} \end{array}$$

Nyström   Nearly optimal

$$\begin{array}{c|cc} \frac{1}{2} & \frac{1}{2} & \\ 1 & -1 & 2 \\ \hline & \frac{1}{6} & \frac{4}{6} & \frac{1}{6} \end{array} \qquad \begin{array}{c|cc} \frac{1}{3} & \frac{1}{3} & \\ \frac{2}{3} & 0 & \frac{2}{3} \\ \hline & \frac{1}{4} & 0 & \frac{3}{4} \end{array}$$

Classical   Heun

Fourth Order Methods

$$\begin{array}{c|ccc} c_2 & a_{21} & & \\ c_3 & a_{31} & a_{32} & \\ c_4 & a_{41} & a_{42} & a_{43} \\ \hline & W_1 & W_2 & W_3 & W_4 \end{array}$$

$$\begin{array}{c|ccc} \frac{1}{2} & \frac{1}{2} & & \\ \frac{1}{2} & 0 & \frac{1}{2} & \\ 1 & 0 & 0 & 1 \\ \hline & \frac{1}{6} & \frac{2}{6} & \frac{2}{6} & \frac{1}{6} \end{array} \qquad \begin{array}{c|ccc} \frac{1}{3} & \frac{1}{3} & & \\ \frac{2}{3} & -\frac{1}{3} & 1 & \\ 1 & 1 & -1 & 1 \\ \hline & \frac{1}{8} & \frac{3}{8} & \frac{3}{8} & \frac{1}{8} \end{array}$$

Classical   Kutta

System of Equations

The fourth order classical Runge-Kutta method for the system of equations

266 Numerical Methods

$$\frac{d\mathbf{u}}{dt} = \mathbf{f}(t, \mathbf{u})$$

$$\mathbf{u}(t_0) = \mathbf{\eta}$$

may be written as

$$\mathbf{u}_{j+1} = \mathbf{u}_j + \tfrac{1}{6}(\mathbf{K}_1 + 2\mathbf{K}_2 + 2\mathbf{K}_3 + \mathbf{K}_4) \tag{6.76}$$

where

$$\mathbf{K}_1 = \begin{bmatrix} K_{11} \\ K_{21} \\ \vdots \\ K_{n1} \end{bmatrix}, \quad \mathbf{K}_2 = \begin{bmatrix} K_{12} \\ K_{22} \\ \vdots \\ K_{n2} \end{bmatrix}, \quad \mathbf{K}_3 = \begin{bmatrix} K_{13} \\ K_{23} \\ \vdots \\ K_{n3} \end{bmatrix}, \quad \mathbf{K}_4 = \begin{bmatrix} K_{14} \\ K_{24} \\ \vdots \\ K_{n4} \end{bmatrix}$$

and

$$K_{i1} = hf_i(t_j, u_{1,j}, u_{2,j}, \ldots, u_{n,j})$$

$$K_{i2} = hf_i\left(t_j + \frac{h}{2}, u_{1,j} + \tfrac{1}{2}K_{11}, u_{2,j} + \tfrac{1}{2}K_{21}, \ldots, u_{n,j} + \tfrac{1}{2}K_{n1}\right)$$

$$K_{i3} = hf_i\left(t_j + \frac{h}{2}, u_{1,j} + \tfrac{1}{2}K_{12}, u_{2,j} + \tfrac{1}{2}K_{22}, \ldots, u_{n,j} + \tfrac{1}{2}K_{n2}\right)$$

$$K_{i4} = hf_i(t_j + h, u_{1,j} + K_{13}, u_{2,j} + K_{23}, \ldots, u_{n,j} + K_{n3}), \quad i = 1(1)n$$

In an explicit form (6.76) becomes

$$\begin{bmatrix} u_{1,j+1} \\ u_{2,j+1} \\ \vdots \\ u_{n,j+1} \end{bmatrix} = \begin{bmatrix} u_{1,j} \\ u_{2,j} \\ \vdots \\ u_{n,j} \end{bmatrix} + \tfrac{1}{6}\left(\begin{bmatrix} K_{11} \\ K_{21} \\ \vdots \\ K_{n1} \end{bmatrix} + 2\begin{bmatrix} K_{12} \\ K_{22} \\ \vdots \\ K_{n2} \end{bmatrix} + 2\begin{bmatrix} K_{13} \\ K_{23} \\ \vdots \\ K_{n3} \end{bmatrix} + \begin{bmatrix} K_{14} \\ K_{24} \\ \vdots \\ K_{n4} \end{bmatrix}\right) \tag{6.77}$$

Example 6.6 Solve the initial value problem

$$u' = -2tu^2, \quad u(0) = 1$$

with $h = 0.2$ on the interval $[0, 1]$. Use the fourth order classical Runge-Kutta method.

For $j = 0$

$$t_0 = 0, \quad u_0 = 1$$

$$K_1 = hf(t_0, u_0) = -2(.2)(0)(1)^2 = 0$$

$$K_2 = hf\left(t_0 + \frac{h}{2}, u_0 + \tfrac{1}{2}K_1\right) = -2(.2)\left(\frac{.2}{2}\right)(1)^2 = -.04$$

$$K_3 = hf\left(t_0 + \frac{h}{2}, u_0 + \tfrac{1}{2}K_2\right) = -2(.2)\left(\frac{.2}{2}\right)(.98)^2 = -.038416$$

$$K_4 = hf(t_0 + h, u_0 + K_3)$$
$$= -2(.2)(.2)(.961584)^2 = -.0739715$$
$$u(0.2) \approx u_1 = 1 + \tfrac{1}{6}[0 - .08 - .076832 - .0739715] = .9615328$$

For $j = 1$

$$t_1 = .2, \quad u_1 = .9615328$$
$$K_1 = hf(t_1, u_1)$$
$$= -2(.2)(.2)(.9615328)^2 = -.0739636$$
$$K_2 = hf\left(t_1 + \frac{h}{2}, u_1 + \frac{K_1}{2}\right)$$
$$= -2(.2)(.3)(.924551)^2 = -.1025754$$
$$K_3 = hf\left(t_1 + \frac{h}{2}, u_1 + \frac{K_2}{2}\right)$$
$$= -2(.2)(.3)(.9102451)^2 = -.0994255$$
$$K_4 = hf(t_1 + h, u_1 + K_3)$$
$$= -2(.2)(.4)(.8621073)^2 = -.1189166$$
$$u(0.4) \approx u_2 = .9615328 + \tfrac{1}{6}(-.0739636 - .2051508$$
$$- .1988510 - .1189166) = .8620525$$

Similarly, we get

$$u(0.6) \approx u_3 = .7352784$$
$$u(0.8) \approx u_4 = .6097519$$
$$u(1.0) \approx u_5 = .5000073$$

Example 6.7 Solve the system of equations

$$u' = -3u + 2v, \quad u(0) = 0$$
$$v' = 3u - 4v, \quad v(0) = \tfrac{1}{2}$$

with $h = 0.2$ on the interval $[0, 1]$. Use the Euler-Cauchy method.

For $j = 0$

$$t_0 = 0, \quad u_0 = 0, \quad v_0 = .5$$
$$K_{11} = hf_1(t_0, u_0, v_0)$$
$$= .2(-3 \times 0 + 2 \times .5) = .2$$
$$K_{21} = hf_2(t_0, u_0, v_0)$$
$$= .2(3 \times 0 - 4 \times .5) = -.4$$
$$K_{12} = hf_1(t_0 + h, u_0 + K_{11}, v_0 + K_{21})$$
$$= .2[-3(0 + .2) + 2(.5 - .4)] = -.08$$

$$K_{22} = hf_2(t_0 + h, u_0 + K_{11}, v_0 + K_{21})$$
$$= .2[3(0 + .2) - 4(.5 - .4)] = .04$$
$$u(0.2) \approx u_1 = u_0 + \tfrac{1}{2}(K_{11} + K_{12}) = .06$$
$$v(0.2) \approx v_1 = v_0 + \tfrac{1}{2}(K_{21} + K_{22}) = .32$$

For $j = 1$

$$t_1 = .2, \quad u_1 = .06, \quad v_1 = .32$$
$$K_{11} = hf_1(t_1, u_1, v_1)$$
$$= .2(-3 \times .06 + 2 \times .32) = .092$$
$$K_{21} = hf_2(t_1, u_1, v_1)$$
$$= .2(3 \times .06 - 4 \times .32) = -.22$$
$$K_{12} = hf_1(t_1 + h, u_1 + K_{11}, v_1 + K_{21})$$
$$= .2[-3(.06 + .092) + 2(.32 - .22)] = -.0512$$
$$K_{22} = hf_2(t_1 + h, u_1 + K_{11}, v_1 + K_{21})$$
$$= .2[3(.06 + .092) - 4(.32 - .22)] = 0.0112$$
$$u(0.4) \approx u_2 = u_1 + \tfrac{1}{2}(K_{11} + K_{12}) = .0804$$
$$v(0.4) \approx v_2 = v_1 + \tfrac{1}{2}(K_{21} + K_{22}) = .2156$$

Similarly, we get

$$u(0.6) \approx u_3 = .082152, \quad v(0.6) \approx v_3 = .152456$$
$$u(0.8) \approx u_4 = .079309, \quad v(0.8) \approx v_4 = .112359$$
$$u(1.0) \approx u_5 = .069000, \quad v(1.0) \approx v_5 = .086190$$

6.3.3 Implicit Runge-Kutta Methods

The implicit Runge-Kutta method using v slopes is defined as

$$K_i = hf\left(t_j + c_i h, u_j + \sum_{m=1}^{v} a_{im} K_m\right)$$
$$u_{j+1} = u_j + \sum_{m=1}^{v} W_m K_m \tag{6.78}$$

where

$$c_i = \sum_{j=1}^{v} a_{ij}, \quad i = 1, 2, \ldots, v$$

and a_{ij}, $1 \leqslant i, j \leqslant v$, $W_1, W_2, \ldots, W_v$ are arbitrary parameters. The slopes K_m are defined implicitly. The number of unknown parameters are $v(v + 1)$. We now give the derivation for the case $v = 1$. We have

$$K_1 = hf(t_j + c_1 h, u_j + a_{11} K_1)$$
$$u_{j+1} = u_j + W_1 K_1 \tag{6.79}$$

The Taylor series gives

$$u(t_{j+1}) = u(t_j) + hu'(t_j) + \frac{h^2}{2}u''(t_j) + \cdots$$

$$= u(t_j) + hf(t_j, u(t_j)) + \frac{h^2}{2}(f_t + ff_u)_{t_j} + \cdots$$

and

$$K_1 = h(f(t_j, u_j) + c_1 h f_t + a_{11} K_1 f_u + \cdots)$$
$$= (hf + c_1 h^2 f_t + h a_{11} f_u K_1)_{t_j} + 0(h^3)$$
$$= hf_j + h^2(c_1 f_t + a_{11} f f_u)_{t_j} + 0(h^3) \qquad (6.80)$$

Substituting (6.80) into (6.79) and comparing the coefficients of h and h^2, we get

$$c_1 = a_{11}$$
$$W_1 = 1$$
$$W_1 c_1 = \tfrac{1}{2}$$

We obtain

$$W_1 = 1, \; c_1 = a_{11} = \tfrac{1}{2}$$

The second order implicit Runge-Kutta method becomes

$$K_1 = hf(t_j + \tfrac{1}{2}h, u_j + \tfrac{1}{2}K_1)$$
$$u_{j+1} = u_j + K_1 \qquad (6.81)$$

For $v = 2$, the implicit Runge-Kutta method (6.78) becomes

$$K_1 = hf(t_j + c_1 h, u_j + a_{11} K_1 + a_{12} K_2)$$
$$K_2 = hf(t_j + c_2 h, u_j + a_{21} K_1 + a_{22} K_2)$$
$$u_{j+1} = u_j + W_1 K_1 + W_2 K_2 \qquad (6.82)$$

where the parameter values

$$W_1 = \tfrac{1}{2}, \; W_2 = \tfrac{1}{2}$$

$$c_1 = \frac{3 + \sqrt{3}}{6}, \; c_2 = \frac{3 - \sqrt{3}}{6}$$

$$a_{11} = \tfrac{1}{4}, \; a_{12} = \frac{1}{4} + \frac{\sqrt{3}}{6}$$

$$a_{21} = \frac{1}{4} - \frac{\sqrt{3}}{6}, \; a_{22} = \frac{1}{4} \qquad (6.83)$$

lead to a fourth order method.

Example 6.8 Solve the initial value problem

$$u' = -2tu^2, \; u(0) = 1$$

with $h = 0.2$ on the interval $[0, 1]$. Use the second order implicit Runge-Kutta method.

The second order implicit Runge-Kutta method is given by
$$u_{j+1} = u_j + K_1, \quad j = 0, 1, 2, 3, 4$$
$$K_1 = hf\left(t_j + \frac{h}{2}, u_j + \tfrac{1}{2}K_1\right)$$
which gives
$$K_1 = -h(2t_j + h)(u_j + \tfrac{1}{2}K_1)^2$$
or
$$h(2t_j + h)K_1^2 + 4(1 + hu_j(2t_j + h))K_1 + 4h(2t_j + h)u_j^2 = 0$$
Solving it as a quadratic in K_1, we get
$$K_1 = \frac{-4(1 + hu_j(2t_j + h)) + \sqrt{16 + 32hu_j(2t_j + h)}}{2h(2t_j + h)}$$
For $j = 0$, $t_0 = 0$, $u_0 = 1$
$$K_1 = \frac{-4(1 + .2(.2)) + \sqrt{16 + 32(.2)(.2)}}{2(.2)(.2)} = -.0384759$$
$$u(0.2) \approx u_1 = u_0 + K_1$$
$$= 1 - .0384759 = .9615241$$
For $j = 1$, $t_1 = .2$, $u_1 = .9615241$
$$K_1 = \frac{-4(1 + .12 \times .9615241) + \sqrt{16 + 3.84 \times .9615241}}{.24}$$
$$= -.0997344$$
$$u(0.4) \approx u_2 = u_1 + K_1$$
$$= .9915241 - .0997344 = .8617897$$
Similarly, we get
$$u(0.6) \approx u_3 = .7343987$$
$$u(0.8) \approx u_4 = .6082158$$
$$u(1.0) \approx u_5 = .4980681$$

6.4 MULTISTEP METHODS

The general multistep method may be written as
$$u_{j+1} = a_1 u_j + a_2 u_{j-1} + \ldots + a_k u_{j-k+1}$$
$$+ h(b_0 u'_{j+1} + b_1 u'_j + \ldots + b_k u'_{j-k+1}) \tag{6.84}$$
or
$$u_{j+1} = \sum_{i=1}^{k} a_i u_{j-i+1} + h \sum_{i=0}^{k} b_i u'_{j-i+1}$$
Symbolically, we can write (6.84) as
$$\rho(E)u_{j-k+1} - h\sigma(E)u'_{j-k+1} = 0$$

where ρ and σ are polynomials defined by

$$\rho(\xi) = \xi^k - a_1\xi^{k-1} - a_2\xi^{k-2} - \ldots - a_k$$
$$\sigma(\xi) = b_0\xi^k + b_1\xi^{k-1} + \ldots + b_k$$

The formula (6.84) can only be used if we know the values of the solution $u(t)$ and $u'(t)$ at k successive points.

These k values will be assumed to be given. Further, if $b_0 = 0$ the method (6.84) is called an explicit or **predictor method**. When $b_0 \neq 0$, it is called an implicit or **corrector method**. We also assume that the polynomials $\rho(\xi)$ and $\sigma(\xi)$ have no common factor, since, otherwise (6.84) can be reduced to a difference equation of lower order. In order to determine the coefficients a's and b's we write the local truncation error as

$$T_{j+1} = u(t_{j+1}) - \sum_{i=1}^{k} a_i u(t_{j-i+1}) - h \sum_{i=0}^{k} b_i u'(t_{j-i+1}) \qquad (6.85)$$

We assume that the function $u(t)$ has continuous derivatives of sufficiently high order. Expanding $u(t_{j-i+1})$ and $u'(t_{j-i+1})$ in Taylor's series, we have

$$u(t_{j-i+1}) = u(t_j) + (1-i)hu'(t_j) + \frac{(1-i)^2}{2!}h^2 u''(t_j) + \ldots$$
$$+ \frac{(1-i)^p}{p!}h^p u^{(p)}(t_j) + \frac{1}{p!}\int_{t_j}^{t_{j-i+1}} (t_{j-i+1} - s)^p u^{(p+1)}(s)\, ds$$

$$u'(t_{j-i+1}) = u'(t_j) + (1-i)hu''(t_j) + \frac{(1-i)^2}{2!}h^2 u'''(t_j) + \ldots$$
$$+ \frac{(1-i)^{p-1}}{(p-1)!}h^{p-1}u^{(p)}(t_j)$$
$$+ \frac{1}{(p-1)!}\int_{t_j}^{t_{j-i+1}} (t_{j-i+1} - s)^{p-1} u^{(p+1)}(s)\, ds \qquad (6.86)$$

Substituting (6.86) into (6.85), we get

$$T_{j+1} = c_0 u(t_j) + c_1 h u'(t_j) + c_2 h^2 u''(t_j) + \ldots + c_p h^p u^{(p)}(t_j) + R_{p+1} \qquad (6.87)$$

where

$$c_0 = 1 - \sum_{i=1}^{k} a_i$$

$$c_q = \frac{1}{q!}\left[1 - \sum_{i=1}^{k} a_i(1-i)^q\right] - \frac{1}{(q-1)!}\sum_{i=0}^{k} b_i(1-i)^{q-1}, \quad q = 1(1)p \qquad (6.88)$$

$$R_{p+1} = \frac{1}{p!}\left[\int_{t_j}^{t_{j+1}} (t_{j+1} - s)^p u^{(p+1)}(s)\, ds\right.$$
$$- \sum_{i=1}^{k} a_i \int_{t_j}^{t_{j-i+1}} (t_{j-i+1} - s)^p u^{(p+1)}(s)\, ds$$
$$- hp \int_{t_j}^{t_{j+1}} b_0(t_{j+1} - s)^{p-1} u^{(p+1)}(s)\, ds$$
$$\left. - hp \sum_{i=1}^{k} b_i \int_{t_j}^{t_{j-i+1}} (t_{j-i+1} - s)^{p-1} u^{(p+1)}(s)\, ds\right] \qquad (6.89)$$

Definition 6.1 The linear multistep method (6.84) is said to be of order p if, in (6.87)

$$c_0 = c_1 = c_2 = \ldots = c_p = 0 \quad \text{and} \quad c_{p+1} \neq 0 \tag{6.90}$$

Thus, for any function $u(t) \in c^{(p+2)}$ and for some nonzero constant c_{p+1}, the pth order difference scheme is given by (6.84). The truncation error becomes

$$T_{j+1} = c_{p+1} h^{p+1} u^{(p+1)}(t_j) + O(h^{p+2}) \tag{6.91}$$

where $c_{p+1}/\sigma(1)$ is called the error constant.

Definition 6.2 The linear multistep method (6.84) is said to be consistent if it has order $p \geqslant 1$.

From the definition 6.2, we get $c_0 = 0 = c_1$, which implies

$$\rho(1) = 0, \quad \rho'(1) = \sigma(1) \tag{6.92}$$

Definition 6.3 The linear multistep method is said to satisfy the **root condition** if the roots of the equation $\rho(\xi) = 0$ lie inside the unit circle in the complex plane, and are simple if they lie on the unit circle.

We now discuss the general method to determine the multistep methods satisfying the consistency and the root condition.

6.4.1 Determination of a_i and b_i

It may be noted from (6.91) that the local truncation error T_{j+1} vanishes identically when $u(t)$ is a polynomial whose degree is less than or equal to p. The constants c_i and p are independent of $u(t)$. These constants can thus be determined by choosing $u(t) = e^t$. Substituting $u_j = e^{t_j}$ in (6.84) and simplifying we get

$$e^{kh} - a_1 e^{(k-1)h} - \ldots - a_k - h(b_0 e^{kh} + b_1 e^{(k-1)h} + \ldots + b_k) = 0$$

Putting $e^h = \xi$, we obtain

$$\rho(\xi) - \log \xi \sigma(\xi) = 0 \tag{6.93}$$

As $h \to 0$, $\xi \to 1$ and we may use (6.93) for determining $\rho(\xi)$ or $\sigma(\xi)$ of maximum order if $\sigma(\xi)$ or $\rho(\xi)$ is given.

If $\sigma(\xi)$ is specified then the equation (6.93) can be used to determine $\rho(\xi)$ of degree k. The term $\log \xi \sigma(\xi)$ can be expanded as a power series in $(\xi - 1)$ and the terms upto $(\xi - 1)^k$ can be used to give $\rho(\xi)$. If, on the other hand, we are given $\rho(\xi)$, we can use the equation

$$\sigma(\xi) - \frac{\rho(\xi)}{\log \xi} = 0 \tag{6.94}$$

to determine $\sigma(\xi)$ of degree $\leqslant k$. The term $\rho(\xi)/\log \xi$ is expanded as a power series in $(\xi - 1)$, and terms upto $(\xi - 1)^k$ for implicit methods and $(\xi - 1)^{k-1}$ for explicit methods are used to get $\sigma(\xi)$. A few well known methods which can be derived using the above technique are listed below.

Adams-Bashforth Methods

$\rho(\xi) = \xi^{k-1}(\xi - 1)$ and $\sigma(\xi)$ of degree $k - 1$

$$\sigma(\xi) = \xi^{k-1} \sum_{m=0}^{k-1} \gamma_m (1 - \xi^{-1})^m \qquad (6.95)$$

where

$$\gamma_m + \tfrac{1}{2}\gamma_{m-1} + \ldots + \frac{1}{m+1}\gamma_0 = 1, \quad m = 0, 1, 2, \ldots$$

Table 6.1 Adams-Bashforth Methods

b_i \ k	b_1	b_2	b_3	b_4	b_5
1	1				
2	3/2	−1/2			
3	23/12	−16/12	5/12		
4	55/24	−59/24	37/24	−9/24	
5	1901/720	−2774/720	2616/720	−1274/720	251/720

Nyström Methods

$\rho(\xi) = \xi^{k-2}(\xi^2 - 1)$ and $\sigma(\xi)$ of degree $k - 1$

$$\sigma(\xi) = \xi^{k-1} \sum_{m=0}^{k-1} \gamma_m (1 - \xi^{-1})^m \qquad (6.96)$$

where

$$\gamma_m + \tfrac{1}{2}\gamma_{m-1} + \ldots + \frac{1}{m+1}\gamma_0 = \begin{cases} 1, & m = 0 \\ 0, & m = 1, 2, \ldots \end{cases}$$

Table 6.2 Nyström Methods

b_i \ k	b_1	b_2	b_3	b_4	b_5
1	2				
2	2	0			
3	7/3	−2/3	1/3		
4	8/3	−5/3	4/3	−1/3	
5	269/90	−266/90	294/90	−146/90	29/90

Adams-Moulton Methods

$\rho(\xi) = \xi^{k-1}(\xi - 1)$ and $\sigma(\xi)$ of degree k

$$\sigma(\xi) = \xi^k \sum_{m=0}^{k} \gamma_m (1 - \xi^{-1})^m \qquad (6.97)$$

where
$$\gamma_m + \tfrac{1}{2}\gamma_{m-1} + \ldots + \frac{1}{m+1}\gamma_0 = \begin{cases} 1, & m = 0 \\ 0, & m = 1, 2, \ldots \end{cases}$$

Table 6.3 Adams-Moulton Methods

k \ b_i	b_0	b_1	b_2	b_3	b_4
1	1/2	1/2			
2	5/12	8/12	−1/12		
3	9/24	19/24	−5/24	1/24	
4	251/720	646/720	−264/720	106/720	−19/720

Milne-Simpson Methods

$$\rho(\xi) = \xi^{k-2}(\xi^2 - 1) \quad \text{and} \quad \sigma(\xi) \text{ of degree } k$$
$$\sigma(\xi) = \xi^k \sum_{m=0}^{k} \gamma_m (1 - \xi^{-1})^m \tag{6.98}$$

where
$$\gamma_m + \tfrac{1}{2}\gamma_{m-1} + \ldots + \frac{1}{m+1}\gamma_0 = \begin{cases} 2, & m = 0 \\ -1, & m = 1 \\ 0, & m = 2, 3, \ldots \end{cases}$$

Table 6.4 Milne-Simpson Methods

k \ b_i	b_0	b_1	b_2	b_3	b_4
1	0	2			
2	1/3	4/3	1/3		
3	1/3	4/3	1/3	0	
4	29/90	124/90	24/90	4/90	−1/90

Numerical Differentiation Methods

$$\sigma(\xi) = \xi^k \quad \text{and} \quad \rho(\xi) \text{ of degree } k \tag{6.99}$$

Table 6.5 Numerical Differentiation Methods

k \ a_i	a_0	a_1	a_2	a_3	a_4
1	1	−1			
2	3/2	−2	1/2		
3	11/6	−3	3/2	−1/3	
4	25/12	−4	3	−4/3	1/4

Even though the number of arbitrary coefficients in (6.84) is $2k + 1$, a method of order $2k$ which satisfies the root condition cannot be obtained. It can be shown that the order of k step method of the type (6.84) which satisfies the root condition cannot exceed $k + 2$ and if k is odd it cannot exceed $k + 1$.

Example 6.9 Given $\sigma(\xi) = (23\xi^2 - 16\xi + 5)/12$, find $\rho(\xi)$ and write down the explicit linear multistep method.

We have

$$\rho(\xi) - \log \xi \sigma(\xi) = \rho(\xi) - \tfrac{1}{12}[(\xi - 1) - \tfrac{1}{2}(\xi - 1)^2 + \tfrac{1}{3}(\xi - 1)^2 + \ldots]$$
$$[23(\xi - 1 + 1)^2 - 16(\xi - 1 + 1) + 5]$$
$$= \rho(\xi) - \tfrac{1}{12}[(\xi - 1) - \tfrac{1}{2}(\xi - 1)^2 + \tfrac{1}{3}(\xi - 1)^3 + \ldots]$$
$$[12 + 30(\xi - 1) + 22(\xi - 1)^2]$$
$$= \rho(\xi) - \tfrac{1}{12}[12(\xi-1) + 24(\xi-1)^2 + 12(\xi-1)^3 + \ldots]$$

Neglecting the higher order terms, we get

$$\rho(\xi) = (\xi - 1) + 2(\xi - 1)^2 + (\xi - 1)^3 = \xi^3 - \xi^2$$

The required explicit linear multistep method is

$$(E^3 - E^2)y_{n-2} - \frac{h}{12}(23E^2 - 16E + 5)y'_{n-2} = 0$$

or

$$y_{n+1} = y_n + \frac{h}{12}(23y'_n - 16y'_{n-1} + 5y'_{n-2})$$

which is the Adams-Bashforth method of order 3.

6.4.2 Convergence of Multistep Methods

Since (6.84) is a k-step difference scheme, we must supply k initial values $u_0, u_1, \ldots, u_{k-1}$ in order to find the solution u_k. We may calculate $u_1, u_2, \ldots, u_{k-1}$ using Taylor series, Runge-Kutta or some other method and u_0 is known from the initial condition. We define convergence of the linear multistep method (6.84) as follows:

Definition 6.4 The linear multistep method of the form (6.84) is said to be convergent if

$$\lim_{h \to 0} u_j = u(t_j), \quad 0 \leqslant j \leqslant N$$

provided the rounding errors arising from all the initial conditions tend to zero.

The necessary and sufficient condition for the convergence of the linear multistep method is given in the Theorem 6.2.

THEOREM 6.2 *The linear multistep method of the form (6.84) is convergent if and only if the method is consistent and satisfies the root condition.*

We illustrate through an example that a method will not converge when the root condition is not satisfied. We consider the second order method

$$u_{j+1} - 3u_j + 2u_{j-1} = \frac{h}{2}(f_j - 3f_{j-1}) \tag{6.100}$$

to solve the test equation

$$u' = \lambda u, \quad u(0) = 1$$

which has the solution $u(t) = e^{\lambda t}$.

Applying (6.100) to the test equation, we get

$$2u_{j+1} - (6 + \bar{h})u_j + (4 + 3\bar{h})u_{j-1} = 0 \tag{6.101}$$

where $\bar{h} = \lambda h$. The difference equation (6.101) is of second order and the characteristic equation is given by

$$2\xi^2 - (6 + \bar{h})\xi + (4 + 3\bar{h}) = 0 \tag{6.102}$$

We find $\rho(\xi) = 2\xi^2 - 6\xi + 4 = 0$ has roots $\xi = 1, 2$. Hence, the root condition is not satisfied. The difference equation (6.102) has the roots

$$\xi = \tfrac{1}{4}[(6 + \bar{h}) \pm \sqrt{\{(6 + \bar{h})^2 - 8(4 + 3\bar{h})\}}]$$

which gives

$$\xi_{1h} = \tfrac{1}{4}[(6 + \bar{h}) - \sqrt{(4 - 12\bar{h} + \bar{h}^2)}]$$

$$= 1 + \bar{h} + \frac{\bar{h}^2}{2} + 0(\bar{h}^3) = e^{\bar{h}} + 0(\bar{h}^3)$$

$$\xi_{2h} = \tfrac{1}{4}[(6 + \bar{h}) + \sqrt{(4 - 12\bar{h} + \bar{h}^2)}]$$

$$= 2 - \frac{\bar{h}}{2} + 0(\bar{h}^2) = 2e^{-\bar{h}/4} + 0(\bar{h}^2)$$

The general solution of the difference equation (6.101) is given as

$$u_j = c_1 \xi_{1h}^j + c_2 \xi_{2h}^j \tag{6.103}$$

To determine the parameters c_1 and c_2 we take

$$u_0 = 1, \quad u_1 = \alpha$$

where α is still arbitrary. We choose $\alpha = e^{\bar{h}}$ which is the exact solution at $t = h$. Using these conditions in (6.103), we get

$$c_1 = (\xi_{2h} - e^{\bar{h}})/(\xi_{2h} - \xi_{1h})$$
$$= (2 - 3\bar{h} + 0(\bar{h}^2))/\sqrt{(4 - 12\bar{h} + \bar{h}^2)}$$
$$c_2 = (e^{\bar{h}} - \xi_{1h})/(\xi_{2h} - \xi_{1h})$$
$$= (-\tfrac{7}{12}\bar{h}^3 + 0(\bar{h}^4))/\sqrt{(4 - 12\bar{h} + \bar{h}^2)}$$

Hence, we have

$$u_j = \frac{1}{\sqrt{(4 - 12\bar{h} + \bar{h}^2)}}[\{2 - 3\bar{h} + 0(\bar{h}^2)\}\{e^{\lambda t} + 0(\bar{h}^3)\}$$
$$+ 2^j\{-\tfrac{7}{12}\bar{h}^3 + 0(\bar{h}^4)\}\{e^{-\lambda t/4} + 0(\bar{h}^2)\}] \tag{6.104}$$

where $jh = t$. As $h \to 0$, the first term converges to the exact solution $e^{\lambda t}$. The second term behaves asymptotically like

$$-\frac{7\lambda^3 t^3 e^{-\lambda t/4}}{12\sqrt{(4 - 12\bar{h} + \bar{h}^2)}} \left(\frac{2^j}{j^3}\right)$$

which tends to $-\infty$ as $j \to \infty$. Hence $u_j \to -\infty$ as $j \to \infty$. The method (6.100) is not convergent for the initial value problem $u' = \lambda u$, $u(0) = 1$.

6.4.3 Predictor-Corrector Methods

We now discuss the application of the explicit and implicit multistep methods for the solution of the initial value problems. We use the explicit (predictor) method for predicting a value $u_{n+1}^{(0)}$ and then use the implicit (corrector) method iteratively until the convergence is obtained. We consider the predictor-corrector set

$$P: u_{j+1}^{(0)} = \sum_{i=1}^{k} a_i^{(0)} u_{j-i+1} + h \sum_{i=1}^{k-1} b_i^{(0)} f_{j-i+1} \tag{6.105}$$

$$C: u_{j+1}^{(p+1)} = \sum_{i=1}^{k} a_i u_{j-i+1} + h b_0 f(t_{n+1}, u_{n+1}^{(p)}) + h \sum_{i=1}^{k} b_i f_{j-i+1},$$

$$p = 0, 1, 2, \ldots \tag{6.106}$$

To solve the initial value problem

$$u' = f(t, u), \quad u(t_0) = \eta$$

the predictor-corrector method may be written as

P : Predict some value $u_{j+1}^{(0)}$

E : Evaluate $f(t_{j+1}, u_{j+1}^{(0)})$

C : Correct $u_{j+1}^{(1)} = \sum_{i=1}^{k} (a_i u_{j-i+1} + h b_i f_{j-i+1}) + h b_0 f(t_{j+1}, u_{j+1}^{(0)})$

E : Evaluate $f(t_{j+1}, u_{j+1}^{(1)})$

C : Correct $u_{j+1}^{(2)} = \sum_{i=1}^{k} (a_i u_{j-i+1} + h b_i f_{j-i+1}) + h b_0 f(t_{j+1}, u_{j+1}^{(1)})$

The sequence of operations

PECECE ...

is denoted by $P(EC)^m E$ and is called a **predictor-corrector** method.

From the application view point we usually take $m = 2$. For $m = 1$, the predictor-corrector procedure becomes *PECE*. We can use the estimate of the truncation error to modify the predicted and corrected values. Thus, we may write this procedure as $PM_p CM_c$. This is called the **modified predictor-corrector method**. The estimates of the local truncation error may be obtained as follows. Let the predictor (6.105) and the corrector (6.106) both have the order p. Thus we have

278 *Numerical Methods*

$$u(t_{j+1}) - u_{j+1}^{(p)} = c_{j+1}^* h^{p+1} u^{(p+1)}(t_j) + 0(h^{p+2}) \qquad (6.107)$$

$$u(t_{j+1}) - u_{j+1}^{(c)} = c_{j+1} h^{p+1} u^{(p+1)}(t_j) + 0(h^{p+2}) \qquad (6.108)$$

where $u_{j+1}^{(p)}$ and $u_{j+1}^{(c)}$ represent the solution value obtained by using the predictor and corrector respectively.

Subtracting (6.107) from (6.108) we get

$$u_{j+1}^{(p)} - u_{j+1}^{(c)} = (c_{j+1} - c_{j+1}^*) h^{p+1} u^{(p+1)}(t_j) + 0(h^{p+2}) \qquad (6.109)$$

Substituting the value of $h^{p+1} u^{(p+1)}(t_j)$ from (6.109) into (6.107) and (6.108) we obtain the modified predicted and corrected values m_{j+1} and u_{j+1} respectively as

$$m_{j+1} = p_{j+1} + c_{j+1}^*(c_{j+1} - c_{j+1}^*)^{-1}(p_{j+1} - C_{j+1})$$

$$u_{j+1} = C_{j+1} + c_{j+1}(c_{j+1} - c_{j+1}^*)^{-1}(p_{j+1} - C_{j+1})$$

where p_{j+1} and C_{j+1} are the predicted and the corrected values respectively. Thus the modified P-C method becomes

Predicted value $\quad p_{j+1} = \sum\limits_{i=1}^{k} (a_i^{(0)} u_{j-i+1} + h b_i^{(0)} f_{j-i+1})$

Modified value $\quad m_{j+1} = p_{j+1} + c_{j+1}^*(c_{j+1} - c_{j+1}^*)^{-1}(p_j - C_j)$

Corrected value $\quad C_{j+1} = \sum\limits_{i=1}^{k} (a_i u_{j-i+1} + h b_i u'_{j-i+1}) + h b_0 m'_{j+1}$ $\qquad (6.110)$

Final value $\quad u_{j+1} = C_{j+1} + c_{j+1}(c_{j+1} - c_{j+1}^*)^{-1}(p_{j+1} - C_{j+1})$

The quantity $p_j - C_j$ required for the modification of the first step is generally put

$$p_j - C_j = 0 \qquad (6.111)$$

Example 6.10 Solve the initial value problem

$$u' = -2tu^2, \; u(0) = 1$$

with $h = 0.2$ on the interval [0, 0.4], using the P-C method

$$P : u_{j+1} = u_j + \frac{h}{2}(3u'_j - u'_{j-1})$$

$$C : u_{j+1} = u_j + \frac{h}{2}(u'_{j+1} + u'_j)$$

as (i) $P(EC)^m E$, $m = 2$, (ii) $PM_p CM_c$.

To use the predictor, we need the values of $u(t)$ and $u'(t)$ at $t = 0.2$. The exact values obtained from the exact solution $u(t) = 1.0/(1 + t^2)$ are

$$u(0.2) = u_1 = .9615385$$

$$u'(0.2) = u'_1 = -.3698225$$

For $j = 1$; $t_0 = 0$, $t_1 = .2$, $t_2 = .4$

$P: u_2^{(0)} = u_1 + \dfrac{h}{2}(3u_1' - u_0')$

$\qquad = .9615385 + .1(-3 \times .3698225 - 0) = .8505918$

$E: f(t, u_2^{(0)}) = u_2'^{(0)} = -.5788051$

$C: u_2^{(1)} = u_1 + \dfrac{h}{2}(u_2'^{(0)} + u_1')$

$\qquad = .9615385 + .1(-.5788051 - .3698225) = .8666757$

$E: f(t_2, u_2^{(1)}) = -.6009015$

$C: u_2^{(2)} = u_1 + \dfrac{h}{2}(u_2'^{(1)} + u_1')$

$\qquad = .9615385 + .1(-.6009015 - .3698225) = .8644661$

$u(0.4) \approx u_2 = u_2^{(2)} = .8644661$

For PM_pCM_c method, we have

$p_{j+1} = u_j + \dfrac{h}{2}(3u_j' - u_{j-1}')$

$m_{j+1} = p_{j+1} - \tfrac{5}{6}(p_j - C_j)$

$C_{j+1} = u_j + \dfrac{h}{2}(m_{j+1}' + u_j')$

$u_{j+1} = C_{j+1} + \tfrac{1}{6}(p_{j+1} - C_{j+1}), \quad j = 1, 2, \ldots$

To start the method, we need the values of $u(t)$ and $u'(t)$ at $t = 0.2$. The exact values are

$t_0 = 0, \; u_0 = 1, \; u_0' = 0$

$t_1 = 0.2, \; u_1 = .9615385, \; u_1' = -.3698225$

For $j = 1$

$p_2 = u_1 + \dfrac{h}{2}(3u_1' - u_0')$

$\qquad = .9615385 + .1(-3 \times .3698225 - 0) = .8505918$

$m_2 = p_2 - \tfrac{5}{6}(p_1 - C_1)$

Taking $p_1 - C_1 = 0$, we obtain

$m_2 = .8505918$

$m_2' = -2t_2 m_2^2 = -2(.4)(.8505918)^2 = -.5788051$

$C_2 = u_1 + \dfrac{h}{2}(m_2' + u_1')$

$\qquad = .9615385 + .1(-.5788051 - .3698225) = .8666757$

$p_2 - C_2 = .8505918 - .8666757 = -.0160839$

$$u(0.4) \approx u_2 = C_2 + \tfrac{1}{6}(p_2 - C_2)$$
$$= .8666757 + \tfrac{1}{6}(-.0160839) = .8639951$$

6.5 STABILITY ANALYSIS

The analytical solution $u(t_j)$ of the differential equation, the difference solution u_j of the difference equation and the numerical solution $\bar{u}_j$ can be related by a relation of the form

$$|u(t_j) - \bar{u}_j| \leq |u(t_j) - u_j| + |u_j - \bar{u}_j| \qquad (6.112)$$

In practice, we would like the difference between the analytical and numerical solution to be small. From (6.112) we find that this difference depends on the values $|u(t_j) - u_j|$ and $|u_j - \bar{u}_j|$. The value $|u(t_j) - u_j|$ is the truncation error which arises because the differential equation is replaced by the difference equation. For a consistent difference method the truncation error tends to zero as h approaches zero. The **numerical error** $|u_j - \bar{u}_j|$ arises because in actual computation we cannot compute the difference solution exactly as we are faced with the roundoff errors. In fact in some cases the numerical solution may differ considerably from the difference solution. If the effect of the roundoff error remains bounded as $j \to \infty$ with fixed step size then the difference method is said to be **stable**, otherwise **unstable**. We now study the stability of the singlestep and multistep methods when applied to the test equation (6.10),

$$u' = \lambda u, \quad u(t_0) = u_0$$

where λ may be a complex number. The analytic solution for this equation, at $t = t_j$, is given by

$$u(t_j) = e^{\lambda j h} u(t_0) = (e^{\lambda h})^j u(t_0) \qquad (6.113)$$

6.5.1 Singlestep Methods

If we use the singlestep method (6.58) to solve the test equation then we get a first order difference equation of the form

$$u_{j+1} = E(\lambda h) u_j, \quad j = 0, 1, 2, \ldots \qquad (6.114)$$

where $E(\lambda h)$ is some approximation to $e^{\lambda h}$.

Substituting $\epsilon_j + u_j = u(t_j)$ in (6.114) we get the error equation

$$\epsilon_{j+1} + u(t_{j+1}) = E(\lambda h)(u(t_j) + \epsilon_j)$$

Using (6.113) we may write this equation as

$$\epsilon_{j+1} = (E(\lambda h) - e^{\lambda h}) u(t_j) + E(\lambda h) \epsilon_j \qquad (6.115)$$

It is obvious from (6.115) that the error at t_{j+1} consists of two parts. The first part is the local truncation error which can be made as small as we like by choosing high order methods. The second part $E(\lambda h)\epsilon_j$ is the propagation of the error from the previous step t_j to t_{j+1} (inherited error) and will not grow if $|E(\lambda h)| \leq 1$. We call a singlestep method

absolutely stable if $|E(\lambda h)| \leq 1$

relatively stable if $|E(\lambda h)| \leq e^{\lambda h}$ (6.116)

If $\lambda < 0$, the exact solution decreases as t_j increases and the important condition is the absolute stability, since the numerical solution must also decrease with t_j. If λ is pure imaginary and $|E(\lambda h)| = 1$, then the absolute stability is called the **Periodic stability** (*P*-stability).

If the Euler method is used, we obtain

$$u_{j+1} = u_j + hf_j = (1 + \lambda h)u_j = E(\lambda h)u_j$$

where

$$E(\lambda h) = 1 + \lambda h$$

From the condition (6.116) we find that the Euler method is absolutely stable if and only if

(i) real λ : $-2 < \lambda h < 0$
(ii) complex λ : λh lies inside the unit circle (Fig. 6.3)

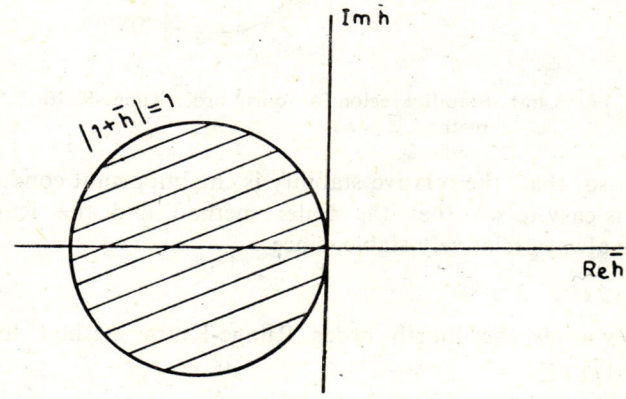

Fig. 6.3 Stability region for Euler's method, $\bar{h} = h\lambda$

Similarly, applying the fourth order Runge-Kutta method to the test equation we get

$$E(\lambda h) = 1 + \lambda h + \frac{(\lambda h)^2}{2!} + \frac{(\lambda h)^3}{3!} + \frac{(\lambda h)^4}{4!}$$

We again find that the fourth order Runge-Kutta method is absolutely stable if and only if

(i) real λ : $-2.785 < \lambda h < 0$
(ii) pure imaginary λ : $0 < |\lambda h| < 2\sqrt{2}$
(iii) complex λ : λh lies inside the region given in Fig. 6.4(a)

If $\lambda > 0$, the exact solution increases with t_j and we do not want

282 Numerical Methods

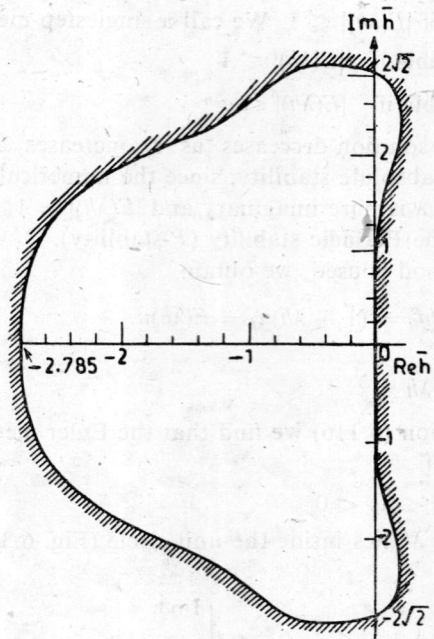

Fig. 6.4(a) Stability region for fourth order Runge-Kutta method, $\bar{h} = h\lambda$

$|E(\lambda h)| \leqslant 1$, so that the relative stability is the important condition to be satisfied. It is easy to see that the Euler method and the Runge-Kutta methods are always relatively stable, since

$$E(\lambda h) \leqslant e^{\lambda h}, \quad \lambda > 0$$

Let us now apply the fourth order Runge-Kutta method to the test equation (6.17)

$$\begin{aligned} w' &= z \\ z' &= -cw, \quad c > 0 \end{aligned} \tag{6.117}$$

which have periodic solutions. We obtain

$$\begin{bmatrix} w_{n+1} \\ z_{n+1} \end{bmatrix} = \begin{bmatrix} a & b \\ -bc & a \end{bmatrix} \begin{bmatrix} w_n \\ z_n \end{bmatrix}$$

where $a = 1 - \frac{1}{2}h_1 + \frac{1}{24}h_1^2$

$$b = h\left(1 - \frac{h_1}{6}\right), \quad h_1 = h^2 c$$

The characteristic equation is given by

$$(a - \mu)^2 + b^2 c = 0 \tag{6.118}$$

We have $\mu_{1,2}$ as a complex pair and

$$|\mu|^2 = a^2 + b^2 c, \quad c > 0$$

We find that the fourth order Runge-Kutta method is P-stable for $h^2 c \leqslant 8$. The region of P-stability is obtained by setting $\mu = \exp(i\theta)$, $0 \leqslant \theta < 2\pi$, in (6.118). The characteristic equation may be written as

$$h_1^4 - 8h_1^3 - 48(\cos\theta + i\sin\theta)h_1^2 + 576(\cos\theta + i\sin\theta)h_1$$
$$+ 576(\cos\theta - 1 + i\sin\theta)^2 = 0$$

The region of periodicity is plotted in Fig. 6.4(b).

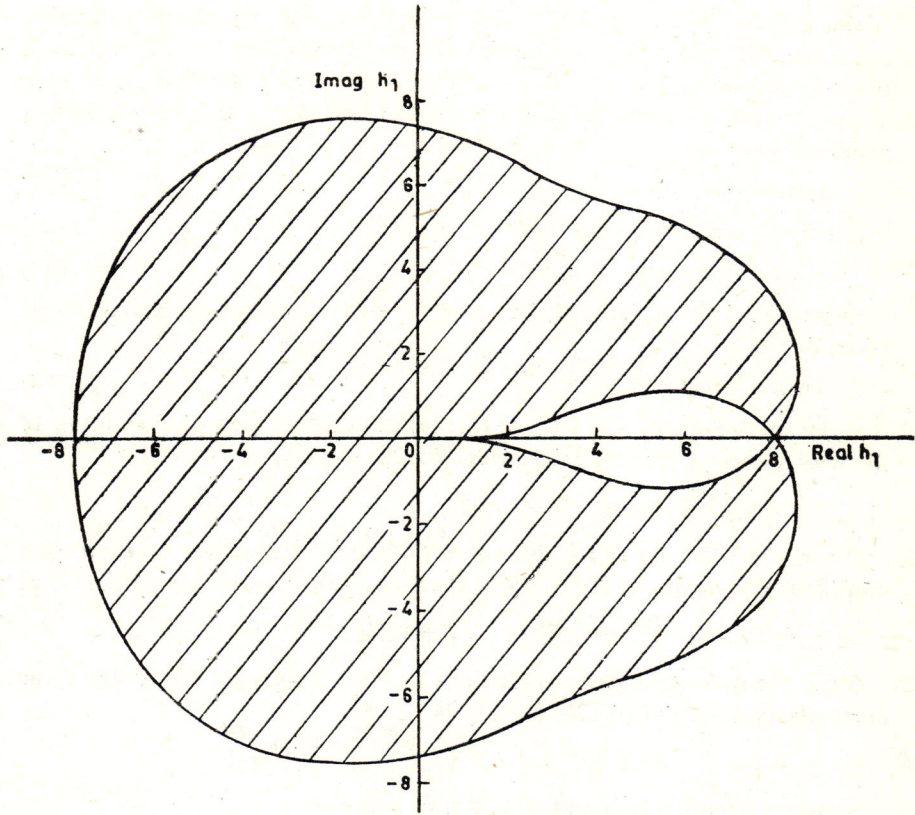

Fig. 6.4(b) Region of periodicity for fourth order Runge-Kutta method for system of equations

6.5.2 Multistep Methods

Ignoring the roundoff errors and applying (6.84) to the test equation (6.10) the computed solution satisfies

$$u_{j+1} = \sum_{i=1}^{k} a_i u_{j-i+1} + h\lambda \sum_{i=0}^{k} b_i u_{j-i+1} \tag{6.119}$$

The true solution satisfies

$$u(t_{j+1}) = \sum_{i=1}^{k} a_i u(t_{j-i+1}) + h\lambda \sum_{i=0}^{k} b_i u(t_{j-i+1}) + T_{j+1} \tag{6.120}$$

Subtracting (6.120) from (6.119) and substituting $\epsilon_j = u_j - u(t_j)$, we get

$$\epsilon_{j+1} = \sum_{i=1}^{k} a_i \epsilon_{j-i+1} + \bar{h} \sum_{i=0}^{k} b_i \epsilon_{j-i+1} - T_{j+1}$$

which may be written as

$$(\rho(E) - \bar{h}\sigma(E))\epsilon_{j-k+1} + T_{j+1} = 0 \tag{6.121}$$

where $\bar{h} = h\lambda$. This is a kth order, linear, non-homogeneous difference equation with constant coefficients. To find its solution for all j, we need the estimates of $\epsilon_0, \epsilon_1, \ldots, \epsilon_{k-1}$ and T_{j+1}. Let us assume that T_{j+1} is a constant and is equal to T. We shall first find the solution of the homogeneous equation

$$(\rho(E) - \bar{h}\sigma(E))\epsilon_{j-k+1} = 0 \tag{6.122}$$

Assume a solution of the form

$$\epsilon_j = \alpha \xi^j, \quad \alpha \neq 0 \tag{6.123}$$

Substituting (6.123) into (6.122) and simplifying we get the characteristic equation

$$\rho(\xi) - \bar{h}\sigma(\xi) = 0 \tag{6.124}$$

If the roots $\xi_{1h}, \xi_{2h}, \ldots, \xi_{kh}$ of (6.124) are distinct then the solution of the homogeneous difference equation (6.122) is given as

$$\epsilon_j = c_1 \xi_{1h}^j + c_2 \xi_{2h}^j + \ldots + c_k \xi_{kh}^j$$

If some of the roots of (6.124) are not distinct, then this solution gets modified. For example, if $\xi_{1h} = \xi_{2h}$, then we have the solution of (6.122) as

$$\epsilon_j = (c_1 + c_2 j)\xi_{1h}^j + c_3 \xi_{3h}^j + \ldots + c_k \xi_{kh}^j$$

Since the non-homogeneous part of (6.121) is a constant we assume the particular solution of (6.121) as

$$\epsilon_j = \beta, \quad \beta \neq 0$$

Substituting in (6.121) and simplifying we get

$$[\rho(1) - \bar{h}\sigma(1)]\beta = -T$$

However, from consistency of the difference scheme we know that $\rho(1) = 0$ and $\sigma(1) = \rho'(1)$. Hence, we get

$$\beta = \frac{T}{\bar{h}\rho'(1)}$$

The general solution of (6.121), when the roots of (6.124) are distinct, is given by

$$\epsilon_j = c_1 \xi_{1h}^j + c_2 \xi_{2h}^j + \ldots + c_k \xi_{kh}^j + \frac{T}{\bar{h}\rho'(1)} \tag{6.125}$$

The coefficients $c_1, c_2, \ldots, c_k$ can be determined from the initial errors $\epsilon_0, \epsilon_1, \ldots, \epsilon_{k-1}$. For $h \to 0$, the roots of the characteristic equation (6.124) tend to the roots of the equation $\rho(\xi) = 0$. This equation is called the **reduced characteristic equation.** Let $\xi_1, \xi_2, \ldots, \xi_k$ be the roots of $\rho(\xi) = 0$. Then the roots of (6.124) may be written as

$$\xi_{ih} = \xi_i(1 + \bar{h}k_i + 0(|\bar{h}|^2)), \quad i = 1, 2, \ldots, k \tag{6.126}$$

for sufficiently small $\bar{h}$. The coefficient k_i is called the **growth parameter.** Substituting (6.126) in (6.124), we have

$$\rho[\xi_i + \bar{h}k_i\xi_i + 0(|\bar{h}|^2)] - \bar{h}\sigma[\xi_i + \bar{h}k_i\xi_i + 0(|\bar{h}|^2)] = 0$$

Expanding in Taylor series, we get

$$\rho(\xi_i) + \bar{h}k_i\xi_i\rho'(\xi_i) - \bar{h}\sigma(\xi_i) + 0(|\bar{h}|^2) = 0$$

Since $\rho(\xi_i) = 0$, we get

$$k_i = \frac{\sigma(\xi_i)}{\xi_i \rho'(\xi_i)} \tag{6.127}$$

We also have from (6.124)

$$\xi_{ih}^j = \xi_i^j (1 + \bar{h}k_i + 0(|\bar{h}|^2))^j$$
$$= \xi_i^j \exp(\lambda h k_i j), \quad i = 1, 2, \ldots, k \tag{6.128}$$

Since the method is consistent, i.e. $\rho(1) = 0, \rho'(1) = \sigma(1)$, we have $\xi_1 = 1$. Substituting in (6.127), we get $k_1 = 1$. Hence from (6.128), we have

$$\xi_{1h}^j = (1 + \bar{h} + 0(|\bar{h}|^2))^j = e^{j\bar{h}} + 0(|\bar{h}|^2) \tag{6.129}$$

Therefore, the root ξ_{1h} approximates the solution of the differential equation $u' = \lambda u$. This is called the **principal root** and the remaining $k - 1$ roots are called the **extraneous roots** which arise because a first order differential equation is replaced by a kth order difference equation.

From (6.125), we find that if any of the roots $\xi_{ih}, i = 1, 2, \ldots, k$ satisfy $|\xi_{ih}| > 1$, then error $|\epsilon_i|$ grows unboundedly. Further, if there is a multiple root of magnitude unity, then again $|\epsilon_i|$ grows unboundedly. However, if $|\xi_{ih}| < 1$, then the error $|\epsilon_i|$ decays as $j \to \infty$. If the roots ξ_{ih} are simple and some of them have magnitude unity, then a fixed amount of error is retained in the numerical solution.

Definition 6.5 The linear multistep method (6.84) is said to be
   (i)   **stable** if $|\xi_j| < 1$, for $j \neq 1$
  (ii)   **unstable** if $|\xi_j| > 1$ for some j or if there is a multiple root of $\rho(\xi) = 0$ of modulus unity.
 (iii)   **conditionally stable** if ξ_j's are simple and if more than one of these roots have modulus unity.

Definition 6.6 The linear multistep method (6.84) is said to be **absolutely stable** if

$$|\xi_{ih}| \leq 1, \quad i = 1, 2, \ldots, k \tag{6.130}$$

The region of absolute stability is the set of points in the $h\lambda$ plane for which the method is absolutely stable. The linear multistep methods having the interval of absolute stability as $(-\infty, 0)$ are called *A-stable* methods. Here, we have $|\xi_{jh}| < 1, j = 1, 2, \ldots, k$ and

$$\lim_{n \to \infty} u_n = 0$$

If all the roots ξ_{jh} are complex and $|\xi_{jh}| = 1, j = 1, 2, \ldots, k$ then the absolute stability is called the Periodic stability (*P*-stability). In practice, it is difficult to test the condition (6.130) if the difference equation is of high order. Further, if the difference equation contains some parameters to be determined, then it is difficult to obtain all the roots $\xi_{ih}, i = 1, 2, \ldots, k$. In such cases, the interval of absolute stability or the values of the parameters for which absolute stability can be achieved, can be determined by using the transformation

$$\xi = \frac{1+z}{1-z} \tag{6.131}$$

which maps the interior of the unit circle $|\xi| = 1$ onto the left half plane $z \leq 0$, the unit circle $|\xi| = 1$ onto the imaginary axis, and the point $\xi = 1$ onto $z = 0$ as given in Fig. 6.5.

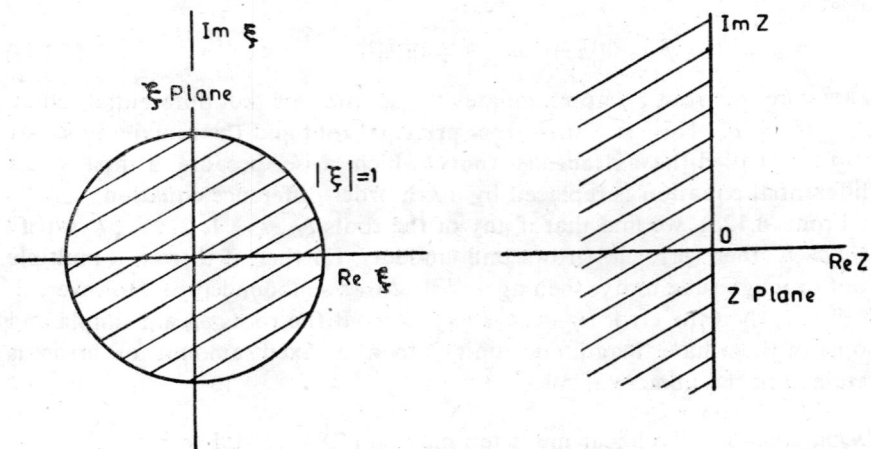

Fig. 6.5 Mapping of the interior of the unit circle onto the left half plane

We can then apply the **Routh-Hurwitz criterion** which gives the necessary and sufficient conditions for the roots of characteristic equation to have negative real parts.

THEOREM (HURWITZ) 6.3 *Let*
$$p(z) = a_0 z^k + a_1 z^{k-1} + \ldots + a_k$$
and
$$D = \begin{vmatrix} a_1 & a_3 & a_5 & \ldots & a_{2k-1} \\ a_0 & a_2 & a_4 & \ldots & a_{2k-2} \\ 0 & a_1 & a_3 & \ldots & a_{2k-3} \\ 0 & a_0 & a_2 & \ldots & a_{2k-4} \\ \vdots & \vdots & \vdots & & \vdots \\ 0 & 0 & 0 & \ldots & a_k \end{vmatrix}$$

where $a_j \geq 0$ *for all j. Then, the real parts of the roots of* $p(z) = 0$ *are negative if and only if the leading principal minors of D are positive.*

We have

$k = 1, \quad a_0 > 0, \quad a_1 > 0$

$k = 2, \quad a_0 > 0, \quad a_1 > 0, \quad a_2 > 0$

$k = 3, \quad a_0 > 0, \quad a_1 > 0, \quad a_2 > 0, \quad a_3 > 0, \quad a_1 a_2 - a_3 a_0 > 0$

as the necessary and sufficient conditions for the real parts of the roots to be negative.

The interval of absolute stability for the Adams methods are given in Table 6.6.

Table 6.6 Stability Interval for Adams Methods

Order	Adams-Bashforth	Adams-Moulton
1	$(-2, 0)$	$(-\infty, 0)$
2	$(-1, 0)$	$(-\infty, 0)$
3	$(-0.5, 0)$	$(-6, 0)$
4	$(-0.3, 0)$	$(-3, 0)$
5	$(-0.2, 0)$	$(-1.9, 0)$

If we apply the second order Nyström method

$$u_{n+1} = u_{n-1} + 2h f_n$$

to the test equation (6.117), we get

$$w_{n+1} = w_{n-1} + 2h z_n$$
$$z_{n+1} = z_{n-1} - 2hc w_n$$

The characteristic equation is obtained as

$$(\mu^2 - 1)^2 = -4h_1 \mu^2$$

or
$$\left(\mu - \frac{1}{\mu}\right)^2 = -4h_1$$

Setting $\mu = \exp(i\theta)$, we get
$$h_1 = \sin^2 \theta$$

Hence, the interval of periodicity becomes $0 < h_1 \leq 1$.

Next, we apply the Milne-Simpson method
$$u_{n+1} = u_{n-1} + \frac{h}{3}(u'_{n+1} + 4u'_n + u'_{n-1})$$

to the test equations (6.117). The characteristic equation is given by
$$9(\mu^2 - 1)^2 = -h_1(\mu^2 + 4\mu + 1)^2$$
or
$$h_1 = -9\left(\frac{\mu^2 - 1}{\mu^2 + 4\mu + 1}\right)^2 \tag{6.132}$$

Setting $\mu = \exp(i\theta)$ in (6.132), we obtain
$$h_1 = \frac{9 \sin^2 \theta}{(2 + \cos \theta)^2}$$

Hence, the interval of periodicity is $0 < h_1 \leq 3$.

Definition 6.7 The linear multistep method (6.84) is said to be relatively stable if
$$|\xi_{ih}| \leq |\xi_{1h}|, \quad i = 2, 3, \ldots, k \tag{6.133}$$

The region of relative stability is defined to be the set of points in the λh-plane for which the method is relatively stable.

To illustrate the difference between absolute and relative stabilities let us apply the second order Adams-Bashforth method
$$u_{j+1} = u_j + \frac{h}{2}(3u'_j - u'_{j-1}) \tag{6.134}$$

to the test equation $u' = \lambda u$. The equation (6.134) becomes
$$u_{j+1} - (1 + \tfrac{3}{2}\bar{h})u_j + \frac{\bar{h}}{2}u_{j-1} = 0 \tag{6.135}$$

where $\bar{h} = \lambda h$. The equation (6.135) is a second order difference equation which will give one extraneous solution.

The roots of the characteristic equation are
$$\xi = \tfrac{1}{4}[2 + 3\bar{h} \pm \sqrt{\{4 + 4\bar{h} + 9\bar{h}^2\}}]$$

The root
$$\xi_{1h} = \tfrac{1}{4}[2 + 3\bar{h} + \sqrt{\{4 + 4\bar{h} + 9\bar{h}^2\}}]$$
$$= 1 + \bar{h} + \frac{\bar{h}^2}{2} - \frac{\bar{h}^3}{4} + O(\bar{h}^4) = e^{\bar{h}} + O(\bar{h}^3)$$

is the principal root approximating $e^{\bar{h}}$ and

$$\xi_{2h} = \tfrac{1}{4}[2 + 3\bar{h} - \sqrt{\{4 + 4\bar{h} + 9\bar{h}^2\}}]$$

$$= \frac{\bar{h}}{2} - \frac{\bar{h}^2}{2} + \frac{\bar{h}^3}{4} + 0(\bar{h}^4) = \frac{\bar{h}}{2} e^{-\bar{h}} + 0(\bar{h}^4)$$

is the extraneous root. The solution of (6.135) is given as

$$u_j = c_1 \xi_{1h}^j + c_2 \xi_{2h}^j \tag{6.136}$$

The roots ξ_{1h} and ξ_{2h} are plotted in Fig. 6.6. We find $|\xi_{2h}| > 1$ for $\bar{h} < -1.0$ so that the method is unstable for $\bar{h} < -1.0$. The method is absolutely stable for $-1.0 < \bar{h} < 0$. For $-\tfrac{2}{3} < \bar{h} < 0$, we find that $|\xi_{2h}| < |\xi_{1h}|$, so that the method is relatively stable in this interval. In this common interval $-\tfrac{2}{3} < \bar{h} < 0$, the method is both absolutely and relatively stable.

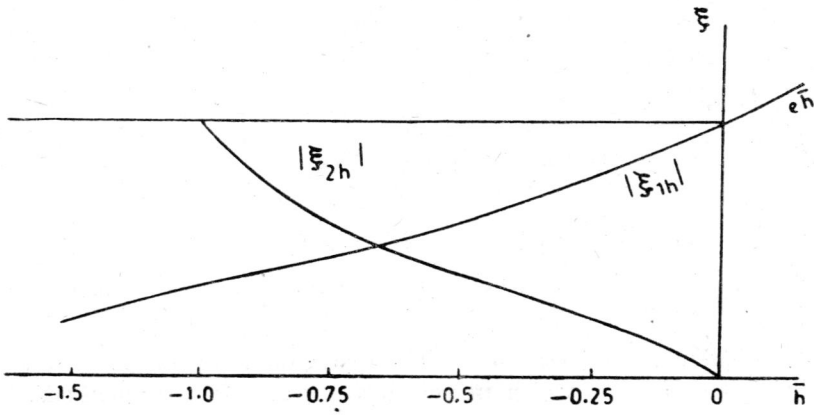

Fig. 6.6 Roots of second order Adams-Bashforth method

6.6 BOUNDARY VALUE PROBLEMS

Consider the two point boundary value problem

$$u'' = f(x, u, u'), \quad x \in [a, b] \tag{6.137}$$

where a prime denotes differentiation with respect to x, with one of the following three boundary conditions.

The boundary conditions of the first kind are:

(i) $u(a) = r_1, \quad u(b) = r_2$ \hfill (6.138)

The boundary conditions of the second kind are:

(ii) $u'(a) = r_1, \quad u'(b) = r_2$ \hfill (6.139)

The boundary conditions of the third kind are:

(iii) $\quad a_0 u'(a) - a_1 u(a) = r_1$
$\qquad b_0 u'(b) + b_1 u(b) = r_2$ \hfill (6.140)

where a_0, b_0, a_1 and b_1 are positive constants.

In (6.137) if all non zero terms involve the dependent variable u then the differential equation is called homogeneous; otherwise it is inhomogeneous. Similarly, the boundary conditions are called homogeneous when r_1 and r_2 are zero; otherwise they are called inhomogeneous. The boundary value problem is called homogeneous if the differential equation and the boundary conditions are homogeneous. A homogeneous boundary value problem possesses only a trivial solution $u(x) = 0$. We, therefore, consider those boundary value problems in which a parameter λ occurs either in the differential equation or in the boundary conditions, and we determine values of λ, called eigenvalues for which the boundary value problem has a nontrivial solution. Such a solution is called an eigenfunction and the entire problem is called an **eigenvalue** or **characteristic value problem**.

The solution of the boundary value problem (6.137) exists and is unique if the following conditions are satisfied: Let $y' = z$, and $-\infty < y, z < \infty$

(i) $f(x, y, z)$ is continuous

(ii) $\dfrac{\partial f}{\partial y}$ and $\dfrac{\partial f}{\partial z}$ exist and are continuous

(iii) $\dfrac{\partial f}{\partial y} > 0$ and $\left|\dfrac{\partial f}{\partial z}\right| \leq w$, $w > 0$.

The numerical methods for solving the boundary value problems may broadly be classified into the following two types:

(i) *Difference methods.* The differential equation is replaced by a set of difference equations which are solved by direct or indirect methods.

(ii) *Shooting methods.* These are initial value problem methods. Here, we add sufficient number of conditions at one point and adjust these conditions until the required conditions are satisfied at the other end.

In practice, the shooting method is quite slow. Therefore we ordinarily use finite difference methods for solving boundary value problems.

6.6.1 Difference Methods

These are the explicit or implicit relations between the derivatives and the function values at the adjacent nodal points. We introduce a finite set of grid points

$$x_j = a + jh, \quad j = 0, 1, 2, \ldots, N+1$$

where $x_0 = a$, $x_{N+1} = b$ and $h = (b - a)/(N + 1)$. We write the difference scheme in the form

$$\sum_{\nu=-m}^{m} (a_\nu u_{j+\nu} + A_\nu h^k u_{j+\nu}^{(k)}) = 0 \hfill (6.141)$$

where $u_{j+\nu}^{(k)}$ represents the kth order derivative of $u(x)$ at $x_{j+\nu}$. The weighting factors a_ν and A_ν are determined by requiring that the method (6.141) satisfies certain accuracy conditions. The local truncation error of the difference method (6.141) may be written as

$$T_j = \sum_{\nu=-m}^{m} (a_\nu u(x_{j+\nu}) + A_\nu h^k u^{(k)}(x_{j+\nu})) \tag{6.142}$$

The largest value of p for which the relation

$$T_j = 0(h^{p+k}) \tag{6.143}$$

holds for all sufficiently differentiable functions $u(x)$ is the order of the difference scheme (6.141). For $m = 1$, $k = 2$, $A_{-1} = A_1 = 0$ and $A_0 = 1$, we get the second order method (see Chapter 5)

$$u_{j-1} - 2u_j + u_{j+1} = h^2 u_j'' \tag{6.144}$$

with the truncation error $(h^4/12) u^{iv}(\xi)$, $x_{j-1} < \xi < x_{j+1}$.

The expressions (6.141) and (6.142), for $k = 2$, $m = 1$ become

$$a_{-1} u_{j-1} + a_0 u_j + a_1 u_{j+1} + h^2 (A_{-1} u_{j-1}'' + A_0 u_j'' + A_1 u_{j+1}'') = 0 \tag{6.145}$$

$$T_j = a_{-1} u(x_{j-1}) + a_0 u(x_j) + a_1 u(x_{j+1}) + h^2 (A_{-1} u''(x_{j-1})$$
$$+ A_0 u''(x_j) + A_1 u''(x_{j+1})) \tag{6.146}$$

Expanding each term on the right hand side of (6.146) in Taylor series about x_j and equating the coefficients of $h^\nu u^{(\nu)}(x_j)/\nu!$, $\nu = 0(1)5$ to zero, we get

$$a_{-1} + a_0 + a_1 = 0$$
$$-a_{-1} + a_1 = 0$$
$$a_{-1} + a_1 + 2(A_{-1} + A_0 + A_1) = 0$$
$$-a_{-1} + a_1 + 6(-A_{-1} + A_1) = 0$$
$$a_{-1} + a_1 + 12(A_{-1} + A_1) = 0$$
$$-a_{-1} + a_1 + 20(-A_{-1} + A_1) = 0 \tag{6.147}$$

and

$$T_j = \frac{1}{5!}\left[a_{-1} \int_{x_j}^{x_{j-1}} (x_{j-1} - t)^5 u^{(6)}(t)\, dt + a_1 \int_{x_j}^{x_{j+1}} (x_{j+1} - t)^5 u^{(6)}(t)\, dt \right.$$
$$+ 20h^2 \left(A_{-1} \int_{x_j}^{x_{j-1}} (x_{j-1} - t)^3 u^{(6)}(t)\, dt \right.$$
$$\left.\left. + A_1 \int_{x_j}^{x_{j+1}} (x_{j+1} - t)^3 u^{(6)}(t)\, dt \right)\right] \tag{6.148}$$

The last equation in (6.147) is automatically satisfied in view of the second and fourth equations. We now have five equations in six unknowns. We can choose one of the unknowns arbitrarily, say $a_1 = 1$, and determine the remaining unknowns. We find

$$a_{-1} = 1,\ a_0 = -2,\ A_{-1} = A_1 = -\tfrac{1}{12},\ A_0 = -\tfrac{10}{12} \tag{6.149}$$

Substituting (6.149) into (6.148) and simplifying we obtain

$$T_j = \frac{h^6}{360}\int_{-1}^{1}(1-|s|^3)(3s^2-6|s|-2)u^{(6)}(x_j+hs)\,ds$$

$$= -\frac{h^6}{240}u^{(6)}(\xi), \quad -1 < \xi < 1 \tag{6.150}$$

where $x_j + hs = t$. Equation (6.145) becomes

$$u_{j-1} - 2u_j + u_{j+1} = \frac{h^2}{12}(u''_{j-1} + 10u''_j + u''_{j+1}) \tag{6.151}$$

This is called the **Numerov** method and the order of the method is four.

The difference expressions for the higher order derivatives may be obtained in a similar manner.

Difference expression for $u^{iv}(x)$

It is easy to show that

$$\delta^4 u_j = \frac{h^4}{720}[474u_j^{(4)} + 124(u_{j-1}^{(4)} + u_{j+1}^{(4)}) - (u_{j-2}^{(4)} + u_{j+2}^{(4)})] \tag{6.152}$$

is a sixth order method with local truncation error of the form

$$T_j = \frac{1}{3024}h^{10}u^{(10)}(t_j) + \cdots \tag{6.153}$$

6.6.2 Boundary Value Problem $u'' = f(x, u)$

Let us consider the numerical solution of the nonlinear differential equation

$$u'' = f(x, u) \tag{6.154}$$

subject to the boundary conditions

$$u(a) = \gamma_1, \quad u(b) = \gamma_2 \tag{6.155}$$

We assume that the solution of the boundary value problem (6.154) exists and is unique, i.e., $f_u \geq 0$ for $x \in [a, b]$.

We approximate (6.154) by the difference scheme

$$-u_{j-1} + 2u_j - u_{j+1} + h^2(\beta_0 f_{j-1} + \beta_1 f_j + \beta_2 f_{j+1}) = 0 \quad 1 \leq j \leq N \tag{6.156}$$

where

$$\beta_0 + \beta_1 + \beta_2 = 1, \quad \beta_0 = \beta_2, \quad u_0 = \gamma_1, \quad u_{N+1} = \gamma_2$$

$$\beta_0 = \begin{vmatrix} 0, & \text{second order method} \\ \frac{1}{12}, & \text{fourth order method} \end{vmatrix}$$

The difference scheme (6.156) represents a system of nonlinear equations in the unknowns u_j, $1 \leq j \leq N$, which in matrix form can be written as

$$\mathbf{Ju} + h^2\mathbf{Bf(u)} + \mathbf{c} = 0 \tag{6.157}$$

where

$$J = \begin{bmatrix} 2 & -1 & & & \\ -1 & 2 & -1 & & \bigcirc \\ & \ddots & \ddots & \ddots & \\ \bigcirc & -1 & 2 & -1 \\ & & & -1 & 2 \end{bmatrix}$$

$$B = \begin{bmatrix} \beta_1 & \beta_2 & & & \bigcirc \\ \beta_0 & \beta_1 & \beta_2 & & \\ & \beta_0 & \beta_1 & \beta_2 & \\ & & \ddots & \ddots & \\ \bigcirc & & & \beta_0 & \beta_1 \end{bmatrix}$$

$$\mathbf{f}(\mathbf{u}) = \begin{bmatrix} f(x_1, u_1) & & & \bigcirc \\ & f(x_2, u_2) & & \\ \bigcirc & & \ddots & \\ & & & f(x_N, u_N) \end{bmatrix}$$

$$\mathbf{c} = \begin{bmatrix} -\gamma_1 + \beta_0 h^2 f(x_0, \gamma_1) \\ 0 \\ \vdots \\ 0 \\ -\gamma_2 + \beta_2 h^2 f(x_{N+1}, \gamma_2) \end{bmatrix}$$

Next, we discuss the numerical solution of the differential equation (6.154) subject to the third boundary conditions

$$a_0 u'(a) - a_1 u(a) = \gamma_1$$
$$b_0 u'(b) + b_1 u(b) = \gamma_2 \qquad (6.158)$$

The system (6.156) contains N equations in $N+2$ unknowns u_j, $j = 0(1)N+1$. We need to find two more equations corresponding to the boundary conditions (6.158). For example, for the fourth order method (6.151), we write a difference scheme for the boundary condition at x_0 as

$$T_0 = u(x_1) - (u(x_0) + hu'(x_0) + h^2(A_0 u''(x_0) + A_1 u''(x_1))) \qquad (6.159)$$

where A_0 and A_1 are arbitrary parameters to be determined and T_0 is the local truncation error.

Expanding $u(x_1)$ and $u''(x_1)$ in the Taylor series and equating the coefficients of $h^r u^{(r)}$, $r = 0(1)3$ to zero and solving we find

$$A_0 = \tfrac{1}{3}, \quad A_1 = \tfrac{1}{6}$$

Thus, neglecting T_0, we obtain

$$u_1 = u_0 + hu'_0 + h^2(\tfrac{1}{3} f_0 + \tfrac{1}{6} f_1) \qquad (6.160)$$

Similarly we may obtain the difference approximation for the boundary condition at $x = x_{N+1}$ as

$$u_N = u_{N+1} - hu'_{N+1} + h^2(\tfrac{1}{6}f_N + \tfrac{1}{3}f_{N+1}) \qquad (6.161)$$

Substituting for u'_0 and u'_{N+1} from (6.158) into (6.160) and (6.161) the third boundary value problem may be replaced by the following equations

$$\left(1 + h\frac{a_1}{a_0}\right)u_0 - u_1 + h^2(\tfrac{1}{3}f_0 + \tfrac{1}{6}f_1) + \frac{h\gamma_1}{a_0} = 0$$

$$-u_{j-1} + 2u_j - u_{j+1} + h^2(\beta_0 u''_{j-1} + \beta_1 u''_j + \beta_2 u''_{j+1}) = 0, \quad 1 \leqslant j \leqslant N$$

$$-u_N + \left(1 + \frac{hb_1}{b_0}\right)u_{N+1} + h^2(\tfrac{1}{6}f_N + \tfrac{1}{3}f_{N+1}) - \frac{h\gamma_2}{b_0} = 0 \qquad (6.162)$$

The system of nonlinear equations (6.162) is generally solved by the Newton-Raphson method discussed in Chapter 2.

6.6.3 Convergence of Difference Schemes

We shall now use some properties of the matrices given in Chapter 3 for establishing the convergence of the difference schemes for the numerical solution of the boundary value problem (6.137–6.138). The exact solution $u(x)$ of (6.156) satisfies

$$-u(x_{j-1}) + 2u(x_j) - u(x_{j+1}) + h^2(\beta_0 f(x_{j-1}, u(x_{j-1})) + \beta_1 f(x_j, u(x_j))$$
$$+ \beta_2 f(x_{j+1}, u(x_{j+1}))) + T_j = 0, \quad 1 \leqslant j \leqslant n \qquad (6.163)$$

where T_j is the truncation error.

Subtracting (6.163) from (6.156) and applying the Mean-Value Theorem, and substituting $\epsilon_j = u_j - u(x_j)$ we get the error equation

$$\mathbf{ME} = (\mathbf{J} + \mathbf{Q})\mathbf{E} = \mathbf{T} \qquad (6.164)$$

where

$$\mathbf{E} = [\epsilon_1 \, \epsilon_2 \ldots \epsilon_N]^T, \quad \mathbf{T} = [T_1 \, T_2 \ldots T_N]^T$$

$$\mathbf{Q} = h^2 \begin{bmatrix} f_{u_1}\beta_1 & f_{u_2}\beta_1 & \mathbf{O} \\ f_{u_1}\beta_0 & f_{u_2}\beta_1 & f_{u_3}\beta_2 \\ & \ddots & \ddots \\ \mathbf{O} & f_{u_{N-1}}\beta_0 & f_{u_N}\beta_1 \end{bmatrix}$$

Thus we note from (6.164) that the convergence of the difference schemes depends on the properties of the matrix $\mathbf{M}$. We now show that the matrix $\mathbf{M} = \mathbf{J} + \mathbf{Q}$ is an irreducible, monotone matrix such that $\mathbf{M} \geqslant \mathbf{J}$ and $\mathbf{Q} \geqslant 0$. Since $\beta_\nu > 0$, $\nu = 0, 1, 2$ and $f_{u_j} > 0, j = 1(1)N$, we have $\mathbf{Q} > 0$ and hence

$$\mathbf{M} = \mathbf{J} + \mathbf{Q} > \mathbf{J}$$

It follows that

$$0 < \mathbf{M}^{-1} < \mathbf{J}^{-1} \qquad (6.165)$$

From (6.164), we have
$$\|\mathbf{E}\| \leqslant \|\mathbf{M}^{-1}\|\,\|\mathbf{T}\| \leqslant \|\mathbf{J}^{-1}\|\,\|\mathbf{T}\| \tag{6.166}$$
In order to simplify (6.166) further we determine $\mathbf{J}^{-1} = (j_{i,\,j})$ explicitly. On multiplying the rows of $\mathbf{J}$ by the jth column of $\mathbf{J}^{-1}$, we have the following equations:

(i) $2j_{1,\,j} - j_{2,\,j} = 0$
(ii) $-j_{i-1,\,j} + 2j_{i,\,j} - j_{i+1,\,j} = 0, \quad 2 \leqslant i \leqslant j - 1$
(iii) $-j_{j-1,\,j} + 2j_{j,\,j} - j_{j+1,\,j} = 1$ (6.167)
(iv) $-j_{i-1,\,j} + 2j_{i,\,j} - j_{i+1,\,j} = 0, \quad j + 1 \leqslant i \leqslant N - 1$
(v) $-j_{N-1,\,j} + 2j_{N,\,j} = 0$

The solution of (6.167ii), using (6.167i), is given by
$$j_{i,\,j} = c_2 i, \quad 2 \leqslant i \leqslant j - 1 \tag{6.168}$$
where c_2 is independent of i, but may depend on j. Similarly the solution of (6.167iv), using (6.167v), is given by
$$j_{i,\,j} = c_1 \left(1 - \frac{i}{N+1}\right), \quad j + 1 \leqslant i \leqslant N - 1 \tag{6.169}$$
The constant c_1 depends only on j. On equating the expression for $j_{i,\,j}$ obtained from (6.168) and (6.169) for $i = j$, we get
$$c_2 j = c_1 \left(1 - \frac{j}{N+1}\right) \tag{6.170}$$
Also, on substituting the values of $j_{i,\,j}\,(i = j - 1, j + 1)$ obtained from (6.168) and (6.169) in (6.167iii), we have
$$c_2 + \frac{c_1}{N+1} = 1 \tag{6.171}$$
Finally from (6.171) and (6.170), we get
$$c_1 = j, \quad c_2 = \frac{N - j + 1}{N + 1} \tag{6.172}$$
On substituting the values of c_1 and c_2, we have
$$j_{i,\,j} = \begin{cases} \dfrac{i(N - j + 1)}{(N + 1)}, & i \leqslant j \\ \dfrac{j(N - i + 1)}{N + 1}, & i \geqslant j \end{cases} \tag{6.173}$$
From (6.173) we see that $\mathbf{J}^{-1}$ is symmetric.
The row sum of $\mathbf{J}^{-1}$ is given as
$$\sum_{j=1}^{N} j_{i,\,j} = \frac{i(N - i + 1)}{2} = \frac{(x_i - a)(b - x_i)}{2h^2}$$

Hence we obtain

$$\|\mathbf{J}^{-1}\| = \max_{1 \leq i \leq N} \sum_{j=1}^{N} |j_{i,j}| \leq \frac{(b-a)^2}{8h^2}$$

The equation (6.166) becomes

$$\|\mathbf{E}\| \leq \frac{(b-a)^2}{8h^2} \|\mathbf{T}\| \qquad (6.174)$$

Substituting for $\|\mathbf{T}\|$ from (6.144) and (6.150), we obtain

$$\|\mathbf{E}\| \leq \frac{1}{96}(b-a)^2 h^2 M_4 = 0(h^2) \qquad (6.175)$$

and

$$\|\mathbf{E}\| \leq \frac{1}{1920}(b-a)^2 h^4 M_6 = 0(h^4) \qquad (6.176)$$

where

$$M_i = \max_{a \leq x \leq b} |u^{(i)}(x)|$$

From equations (6.175) and (6.176) it follows that $\|\mathbf{E}\| \to 0$ or $u_j \to u(x_j)$ as $h \to 0$. This establishes the convergence of the second and fourth order methods.

Example 6.11 Solve the boundary value problem

$$u'' = u + x$$
$$u(0) = 0, \ u(1) = 0$$

with $h = \frac{1}{4}$. Use the Numerov method.

We divide the interval $[0, 1]$ into four subintervals. The nodal points are $x_j = jh$, $0 \leq j \leq 4$ and $h = \frac{1}{4}$. The Numerov method gives the following system of equations

$$u_{j-1} - 2u_j + u_{j+1} = \frac{1}{192}[(u_{j-1} + x_{j-1}) + 10(u_j + x_j)$$
$$+ (u_{j+1} + x_{j+1})], \quad 1 \leq j \leq 3$$

or

$$191u_{j-1} - 394u_j + 191u_{j+1} = x_{j-1} + 10x_j + x_{j+1}$$

For $j = 1$, $191u_0 - 394u_1 + 191u_2 = 3$
For $j = 2$, $191u_1 - 394u_2 + 191u_3 = 6$
For $j = 3$, $191u_2 - 394u_3 + 191u_4 = 9$

Using the boundary conditions $u_0 = u_4 = 0$, we get the system of equations

$$-394u_1 + 191u_2 \qquad\qquad = 3$$
$$191u_1 - 394u_2 + 191u_3 = 6$$
$$191u_2 - 394u_3 = 9$$

which gives
$$u(0.25) \approx u_1 = -\frac{1136118}{32415956} = -.0350481$$
$$u(0.50) \approx u_2 = -\frac{4656}{82274} = -.0565914$$
$$u(0.75) \approx u_3 = -\frac{1629762}{32415956} = -.0502765$$

The exact solution is $u(x) = \frac{\sinh x}{\sinh 1} - x$ and the exact values are
$$u(0.25) = -.0350476, \quad u(0.50) = -.0565906, \quad u(0.75) = -.0502758$$

Example 6.12 Solve the boundary value problem
$$u'' = xu$$
$$u(0) + u'(0) = 1, \quad u(1) = 1$$
with $h = \tfrac{1}{3}$. Use the second order method
$$u_{j-1} - 2u_j + u_{j+1} = h^2 f_j$$
With $h = \tfrac{1}{3}$, we have four nodal points $x_j = jh$, $0 \leqslant j \leqslant 3$.
The second order method gives the following system of equations
$$u_{j-1} - 2u_j + u_{j+1} = \tfrac{1}{9} x_j u_j, \quad 0 \leqslant j \leqslant 2$$
For $j = 0$, $\quad u_{-1} - 2u_0 + u_1 = 0$
For $j = 1$, $\quad u_0 - 2u_1 + u_2 = \tfrac{1}{27} u_1$
For $j = 2$, $\quad u_1 - 2u_2 + u_3 = \tfrac{2}{27} u_2$

Since the method is of second order, we can replace $u'(0)$ in the boundary condition by the relation
$$u'(0) = \frac{u_1 - u_{-1}}{2h}$$
which is also of second order. Thus the boundary conditions become
$$u_0 + \tfrac{3}{2}(u_1 - u_{-1}) = 1$$
and
$$u_3 = 1$$
Thus we get the equations
$$-2u_0 + 3u_1 = 1$$
$$u_0 - \tfrac{55}{27} u_1 + u_2 = 0$$
$$u_1 - \tfrac{56}{27} u_2 = -1$$
Solving this system of equations we get
$$u(0) \approx u_0 = -\tfrac{82}{83} = -.9879518$$
$$u(\tfrac{1}{3}) \approx u_1 = -\tfrac{27}{83} = -.3253012$$
$$u(\tfrac{2}{3}) \approx u_2 = \tfrac{27}{83} = .3253012$$

Example 6.13 Solve the boundary value problem
$$u'' = \tfrac{3}{2}u^2$$
$$u(0) = 4, \quad u(1) = 1$$
with $h = \tfrac{1}{3}$. Use the second order method.

For $h = \tfrac{1}{3}$, we have four nodal points $x_j = jh$, $0 \leqslant j \leqslant 3$. The values at the end points x_0 and x_3 are given from boundary conditions. The second order method gives the following system of equations

$$u_{j-1} - 2u_j + u_{j+1} = \tfrac{1}{6}(u_j^2), \quad j = 1, 2$$

For $j = 1$, $\quad u_0 - 2u_1 + u_2 = \tfrac{1}{6}u_1^2$

For $j = 2$, $\quad u_1 - 2u_2 + u_3 = \tfrac{1}{6}u_2^2$

Using the boundary conditions $u_0 = 4$, $u_3 = 1$ we get the equations

$$f_1(u_1, u_2) = u_1^2 + 12u_1 - 6u_2 - 24 = 0$$
$$f_2(u_1, u_2) = u_2^2 - 6u_1 + 12u_2 - 6 = 0$$

This system of equations can be solved by any iterative method. We use the Newton-Raphson method to solve this system, which is given as

$$\begin{bmatrix} u_1^{(k+1)} \\ u_2^{(k+1)} \end{bmatrix} = \begin{bmatrix} u_1^{(k)} \\ u_2^{(k)} \end{bmatrix} - \begin{bmatrix} 2u_1^{(k)} + 12 & -6 \\ -6 & 2u_2^{(k)} + 12 \end{bmatrix}^{-1} \begin{bmatrix} f_1(u_1^{(k)}, u_2^{(k)}) \\ f_2(u_1^{(k)}, u_2^{(k)}) \end{bmatrix}$$

$$= \begin{bmatrix} u_1^{(k)} \\ u_2^{(k)} \end{bmatrix} - \frac{1}{D} \begin{bmatrix} 2u_2^{(k)} + 12 & 6 \\ 6 & 2u_1^{(k)} + 12 \end{bmatrix} \begin{bmatrix} f_1(u_1^{(k)}, u_2^{(k)}) \\ f_2(u_1^{(k)}, u_2^{(k)}) \end{bmatrix},$$

$$k = 0, 1, \ldots$$

where
$$D = (2u_1^{(k)} + 12)(2u_2^{(k)} + 12) - 36$$

Taking $u_1^{(0)} = 2$ and $u_2^{(0)} = 1.5$, we get

$u_1^{(1)} = 2.3014706$, $\quad u_2^{(1)} = 1.4705882$

$u_1^{(2)} = 2.2950429$, $\quad u_2^{(2)} = 1.4679491$

$u_1^{(3)} = 2.2950399$, $\quad u_2^{(3)} = 1.4679474$

6.6.3 Shooting Method

Consider the solution of the boundary value problem (6.137).

In the **shooting method**, ws solve the initial value problem

$$u'' = f(x, u, u')$$
$$u(a) = r_1, \; u'(a) = \alpha \qquad (6.177)$$

where α is some approximation of the initial slope. Using any of the methods for solving the initial value problems, the approximation $u^{(1)}(b)$ to the solution $u(b)$ is determined. This value is either smaller or larger than the required solution $u(b) = r_2$. Let us denote $g(\alpha_0) = u^{(1)}(b) - u(b)$, where α_0 is the first approximation of α. If $g(\alpha_0) = 0$, then the condition at $x = b$ is satisfied. If this condition is not satisfied, then we repeat the above procedure using $u'(a) = \alpha_1$ to find another estimate $u^{(2)}(b)$ for $u(b)$. The process is usually repeated until the computed value at $x = b$ agrees with the boundary condition $u(b)$. Depending on the choice of $u'(a)$, the computed solution may **overshoot** or **undershoot** the required solution as shown in Fig. 6.7. Shooting method may be described as a procedure which defines a functional relationship $g(\alpha) = 0$, between $u(b)$ and the initial slope $u'(a)$ as given in Fig. 6.8. The problem is then to find the root of this equation. This root cannot be determined by the Newton-Raphson method and the secant method is often used. Secant method gives

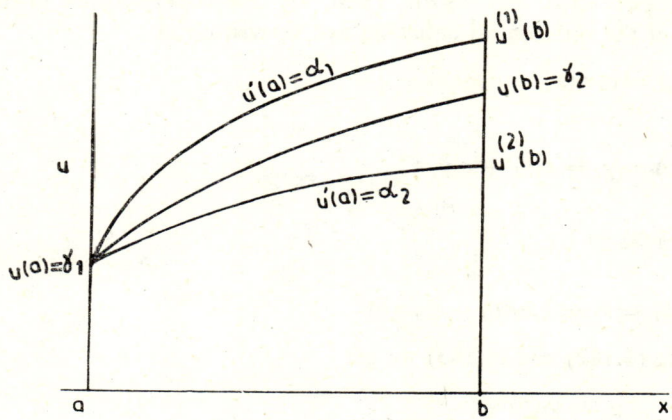

Fig. 6.7 Solution by shooting method

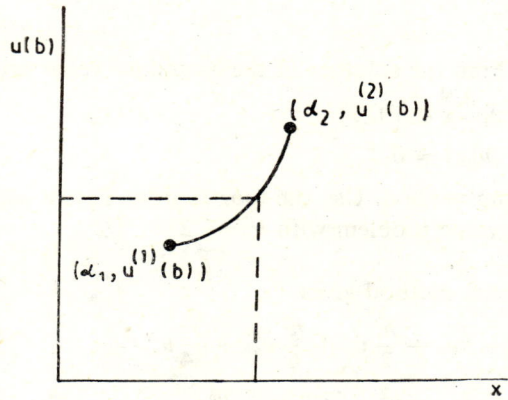

Fig. 6.8 Functional relationship between the final value and initial slope

$$\alpha_{n+1} = \alpha_n - \left[\frac{\alpha_n - \alpha_{n-1}}{g(\alpha_n) - g(\alpha_{n-1})}\right] g(\alpha_n), \quad n = 1, 2 \ldots \qquad (6.178)$$

We have

$$\alpha_2 = \alpha_1 - \left[\frac{\alpha_1 - \alpha_0}{u^{(2)}(b) - u^{(1)}(b)}\right] g(\alpha_1) \tag{6.179}$$

$$\alpha_3 = \alpha_2 - \left[\frac{\alpha_2 - \alpha_1}{u^{(3)}(b) - u^{(2)}(b)}\right] g(\alpha_2) \tag{6.180}$$

. . . .

If the differential equation is linear, then the shooting method becomes very simple. It can be shown that the function relationship $g(\alpha) = 0$ between $u'(a)$ and $u(b)$ is also linear. We suppose that we have computed two solutions $u_1(x)$ and $u_2(x)$ of the differential equation. Both the solutions are obtained using the same initial value $u_1(a) = r_1 = u_2(a)$, but different initial slopes $u_1'(a)$ and $u_2'(a)$. Then, by the superposition principle, the solution of the differential equation can be written as

$$u(x) = c_1 u_1(x) + c_2 u_2(x) \tag{6.181}$$

we have

$$u(a) = \gamma_1 = c_1 \gamma_1 + c_2 \gamma_1 \tag{6.182}$$

or

$$c_1 + c_2 = 1$$

and

$$u(b) = r_2 = c_1 u_1(b) + c_2 u_2(b) \tag{6.183}$$

Solving (6.182) and (6.183) we get

$$c_2 = \frac{\gamma_2 - u_1(b)}{u_2(b) - u_1(b)}, \quad c_1 = 1 - c_2 \tag{6.184}$$

Substituting (6.184) in (6.181), we get the solution of the differential equation.

Example 6.14 Find the solution of the boundary value problem

$$u'' = u + x, \quad x \in [0, 1]$$
$$u(0) = 0, \quad u(1) = 0$$

using the shooting method. Use the fourth order Taylor series method to solve the initial value problem with $h = 0.2$.

We take $u'(0) = \frac{1}{2}$.

The Taylor series method gives

$$u_{n+1} = u_n + h u_n' + \frac{h^2}{2} u_n'' + \frac{h^3}{6} u_n''' + \frac{h^4}{24} u_n^{iv}$$

$$= u_n + h u_n' + \frac{h^2}{2}(u_n + x_n) + \frac{h^3}{6}(u_n' + 1) + \frac{h^4}{24}(u_n + x_n)$$

$$= \left(1 + \frac{h^2}{2} + \frac{h^4}{24}\right) u_n + h u_n' \left(1 + \frac{h^2}{6}\right) + \left(\frac{h^2}{2} + \frac{h^4}{24}\right) x_n + \frac{h^3}{6}$$

$$u'_{n+1} = u'_n + hu''_n + \frac{h^2}{2}u'''_n + \frac{h^3}{6}u_n^{iv}$$

$$= \left(1 + \frac{h^2}{2}\right)u'_n + \left(h + \frac{h^3}{6}\right)u_n + \left(h + \frac{h^3}{6}\right)x_n + \frac{h^2}{2}$$

With $h = 0.2$, we have

$$u_{n+1} = 0.00133 + 0.02007x_n + 1.02007u_n + 0.20133u'_n$$
$$u'_{n+1} = 0.02 + 0.20133x_n + 0.20133u_n + 1.02u'_n$$

We get

$u(0.2) \approx u_2 = 0.10200, \quad u'(0.2) \approx u'_2 = 0.53000$
$u(0.4) \approx u_3 = 0.21610, \quad u'(0.4) \approx u'_3 = 0.62140$
$u(0.6) \approx u_4 = 0.35490, \quad u'(0.6) \approx u'_4 = 0.77787$
$u(0.8) \approx u_5 = 0.53200, \quad u'(0.8) \approx u'_5 = 1.00568$
$u(1.0) \approx u_6 = 0.76254, \quad u'(1.0) \approx u'_6 = 1.31397$

Let the second choice of the initial slope be $u'(0) = -\frac{1}{2}$. We get

$u(0.2) \approx u_2 = -0.09934, \quad u'(0.2) \approx u'_2 = -0.49000$
$u(0.4) \approx u_3 = -0.19464, \quad u'(0.4) \approx u'_3 = -0.45953$
$u(0.6) \approx u_4 = -0.28171, \quad u'(0.6) \approx u'_4 = -0.40738$
$u(0.8) \approx u_5 = -0.35601, \quad u'(0.8) \approx u'_5 = -0.33145$
$u(1.0) \approx u_6 = -0.41250, \quad u'(1.0) \approx u'_6 = -0.22869$

We have

$$c_2 = \frac{0 - 0.76254}{-1.17504} = 0.64895, \quad c_1 = 1 - c_2 = 0.35105$$

Hence

$$u(x) = 0.35105u_1(x) + 0.64895u_2(x)$$

We find

$u(0.2) = 0.35105u_1(0.2) + 0.64895u_2(0.2)$
$\approx 0.35105(0.10200) + 0.64895(-0.09934) = -0.02866$
$u(0.4) = 0.35105u_1(0.4) + 0.64895u_2(0.4)$
$\approx 0.35105(0.21610) + 0.64895(-0.19464) = -0.05045$

Similarly, we get

$u(0.6) \approx -0.05823, u(0.8) \approx -0.04427, u(1.0) \approx -0.000002$

The exact solution is

$u(0.2) = -0.02868, u(0.4) = -0.05048, u(0.6) = -0.05826,$
$u(0.8) = -0.04429, u(1) = 0.$

PROBLEMS

1. Solve the difference equation
   $$\Delta^2 y_n + 3\Delta y_n - 4y_n = n^2$$
   with the initial conditions $y_0 = 0$, $y_2 = 2$.
   (Stockholm Univ., Sweden, BIT 7(1967), 247)

2. Find y_k from the difference equation
   $$\Delta^2 y_{k+1} + \tfrac{1}{2}\Delta^2 y_k = 0, \ k = 0, 1, 2, \ldots$$
   when $y_0 = 0$, $y_1 = \tfrac{1}{2}$, $y_2 = \tfrac{1}{4}$. Is this method numerically stable?
   (Gothenburg Univ., Sweden, BIT 7(1967), 81)

3. Consider the recursion formula
   $$y_{n+1} = y_{n-1} + 2hy_n$$
   $$y_0 = 1$$
   $$y_1 = 1 + h + h^2\left(\frac{1}{2} + \frac{h}{6} + \frac{h^2}{24}\right)$$
   Show that $y_n - e^{nh} = O(h^2)$ as $h \to 0$ for $nh =$ constant.
   (Uppsala Univ., Sweden, BIT 14(1974), 482)

4. Solve the difference equation
   $$u_{n+1} - 2(\sin x)u_n + u_{n-1} = 0$$
   when $u_0 = 0$ and $u_1 = \cos x$.
   (Lund Univ., Sweden, BIT 9(1969), 294)

5. Show that all solutions of the difference equation
   $$y_{n+1} - 2\lambda y_n + y_{n-1} = 0$$
   are bounded, when $n \to \infty$ if $-1 < \lambda < 1$, while for all other complex values of λ there is at least one unbounded solution.
   (Stockholm Univ., Sweden, BIT 4(1964), 261)

6. Apply the Taylor's series second order method to integrate
   $$y' = 2t + 3y, \quad y(0) = 1, \quad t \in (0, 0.4)$$
   with $h = 0.1$.
   (a) If the error in $y(t)$ obtained from the first four terms of the series only, is to be less than 5×10^{-5} after rounding, find t.
   (b) Determine the number of terms in the Taylor's series required to obtain results correct to 5×10^{-6} for $t \leqslant 0.4$.

7. Use the classical Runge-Kutta formula of fourth order to find the numerical solution at $x = 0.8$ for
   $$\frac{dy}{dx} = \sqrt{x+y}, \ y(0.4) = 0.41$$
   Assume the step length $h = 0.2$.

8. Find numerical methods for the differential equation $y' = f(t, y)$, $y(t_0) = y_0$ which produces exact results for
   (i) $y(t) = a + be^{-t}$
   (ii) $y(t) = a + b \cos t + c \sin t$

9. Given the equation
   $$y' = x + \sin y \text{ with } y(0) = 1$$
   show that it is sufficient to use Euler's method with the step $h = 0.2$ to compute $y(0.2)$ with an error less than 0.05.
   (Uppsala Univ., Sweden, BIT 11(1971), 125)

10. (a) Show that if the trapezoidal rule is applied to the equation $y' = ky$, where k is an arbitrary complex constant with negative real part, then for all $h > 0$, $|y_n| < |y_0|$, $n = 1, 2, 3, \ldots$.
    (b) Show that if A is a negative definite symmetric matrix, then a similar conclusion holds for the application of the trapezoidal rule to the system $y' = Ay$, $y(0)$ given, $h > 0$.
    (Stockholm Univ., Sweden, BIT 5(1965), 68)

11. The system
    $$y' = z$$
    $$z' = -by - az$$
    where $0 < a < 2\sqrt{b}$, $b > 0$, is to be integrated by Euler's method with known values. What is the largest step length h for which all solutions of the corresponding difference equation are bounded?
    (Royal Inst. Tech., Stockholm, Sweden, BIT 7(1967), 247)

12. The Runge-Kutta method is used for solving the system
    $$y' = -kz, \quad y(x_0) = y_0$$
    $$z' = ky, \quad z(x_0) = z_0$$
    where $k > 0$ and x, x_0, y, y_0, z and z_0 are real. The step length h is supposed to be > 0. Putting $y_n = y(x_0 + nh)$, $z_n = z(x_0 + nh)$ prove that
    $$\begin{bmatrix} y_{n+1} \\ z_{n+1} \end{bmatrix} = A \begin{bmatrix} y_n \\ z_n \end{bmatrix}$$
    where A is a real 2×2 matrix. Find under what conditions the solutions do not grow exponentially for increasing values of n.
    (Bergen Univ., Norway, BIT 6(1966), 359)

13. The solution of the system of equations
    $$y' = u$$
    $$u' = -4y - 2u$$
    $$y(0) = 1, \quad u(0) = 1$$

is to be obtained by (i) Euler's method and (ii) Runge-Kutta fourth order method. Can a step length $h = 0.1$ be used for integration? If so find $y(0.2)$ and $u(0.2)$.

14. In order to determine the smallest value of λ for which the differential equation

$$y'' = \frac{1}{3+x} y' - \lambda(3+x)y$$

$$y(-1) = y(1) = 0$$

has non-trivial solutions, Runge-Kutta method was used to integrate two first order differential equations equivalent to this equation, but with starting values $y(-1) = 0$, $y'(-1) = 1$. Three step lengths h, and three values λ were tried, with the following results for $y(1)$

h \ λ	0.84500	0.84625	0.84750
2/10	0.0032252	0.0010348	−0.0011504
4/30	0.0030792	0.0008882	−0.0012980
1/10	0.0030522	0.0008608	−0.0013254

(a) Use the above table to calculate λ, with an error less than 10^{-5}.

(b) Rewrite the differential equation so that classical Runge-Kutta method can be used.

(Inst. Tech., Stockholm, Sweden, BIT 5(1965), 214)

15. $y = \log x$ satisfies the differential equation

$$y' = \frac{1}{x}, \quad x \geq 1$$

$$y(1) = 0$$

(a) Use the trapezoidal rule to show that $\log x \approx \dfrac{x^2 - 1}{2x}$.

(b) Estimate

$$\epsilon = \max_{1 \leq x \leq 1.1} \left| \log x - \frac{x^2 - 1}{2x} \right|$$

(Lund Univ., Sweden, BIT 13(1973), 493)

16. (a) Show that Euler's method applied to $y' = \alpha y$, $y(0) = 1$, $\alpha < 0$ is stable for step-sizes satisfying $-2 < \alpha h < 0$ (stability means that $y_n \to 0$ as $n \to \infty$).

(b) Consider the following modified Euler method for $y' = f(x, y)$

$$y_{n+1} = y_n + p_1 h f(x_n, y_n)$$
$$y_{n+2} = y_{n+1} + p_2 h f(x_{n+1}, y_{n+1}) \quad \bigg| \quad n = 0, 2, 4, \ldots$$

where $p_1, p_2 > 0$ and $p_1 + p_2 = 2$.
Apply this method to the problem given in (a) and show that this method is stable for

$$-\frac{2}{p_1 p_2} < \alpha h < 0, \quad \text{if } 1 - \frac{1}{\sqrt{2}} < p_1, p_2 < 1 + \frac{1}{\sqrt{2}}$$

(Linkoping Univ., Sweden, BIT 14(974), 366)

17. Consider the recursion formula for vectors

$$\mathbf{T} \mathbf{y}^{(j+1)} = \mathbf{y}^{(j)} + \mathbf{c}, \quad \mathbf{y}(0) = \mathbf{a}$$

where

$$\mathbf{T} = \begin{bmatrix} 1+2s & -s & & & \bigcirc \\ -s & 1+2s & -s & & \\ & \ddots & \ddots & \ddots & \\ \bigcirc & & -s & 1+2s & -s \\ & & & -s & 1+2s \end{bmatrix}$$

Is the formula stable, i.e. is there any constant k such that $\|\mathbf{y}^{(n)}\| \leqslant k$ for all $n \geqslant 0$?

(Royal Inst. Tech., Stockholm, Sweden, BIT 19(1979), 425)

18. Consider the problem

$$\mathbf{y}' = \mathbf{A}\mathbf{y}$$

$$y(0) = \begin{bmatrix} 1 \\ 0 \end{bmatrix}, \quad \mathbf{A} = \begin{bmatrix} -2 & 1 \\ 1 & -20 \end{bmatrix}$$

(a) Show that the system is asymptotically stable.

(b) Examine the method

$$y_{i+1} = y_i + \frac{h}{2}(3F_{i+1} - F_i)$$

for the equation $y' = F(x, y)$. What is its order of approximation? Is it stable? Is it A-stable?

(c) Choose stepsizes $h = 0.2$ and $h = 0.1$ and compute approximations to $y(0.2)$ using the method in (b). Finally make a suitable extrapolation to $h = 0$. The exact solution is $y(0.2) = [0.68 \quad 0.036]^T$ with 2 significant digits.

(Gothenburg Univ., Sweden, BIT 15(1975), 335)

19. Find the implicit Runge-Kutta method of the form

    $$y_{n+1} = y_n + w_1 k_1 + w_2 k_2$$
    $$k_1 = hf(y_n)$$
    $$k_2 = hf(y_n + a(k_1 + k_2))$$

    for the initial value problem $y' = f(y)$, $y(t_0) = y_0$. Obtain the interval of absolute stability for $y' = \lambda y$, $\lambda < 0$.

20. Find the order of the implicit Runge-Kutta method

    $$y_{n+1} = y_n + \tfrac{3}{4} k_1 + \tfrac{1}{4} k_2$$
    $$k_1 = hf\left(t_n + \frac{h}{3}, y_n + \tfrac{1}{3} k_1\right)$$
    $$k_2 = hf(t_n + h, y_n + k_1)$$

    Determine the interval of absolute stability when applied to the test equation $y' = \lambda y$, $\lambda < 0$.

21. Find the solution at $t = 0.3$ for the differential equation
    $$y' = t - y^2, \quad y(0) = 1$$
    by the Adams-Bashforth method of order two with $h = 0.1$. Determine the starting values using a second order Runge-Kutta method.

22. Find a, b, c and d such that the multistep method
    $$y_{n+1} = (1 - a)y_n + ay_{n-1} + h(by'_{n+1} + cy'_n + dy'_{n-1})$$
    is of highest possible order for the solution of $y' = f(x, y)$. Investigate the stability properties of this method when applied to
    $$y' = \lambda y, \quad \lambda < 0$$

23. Derive a fourth order method of the form
    $$y_{n+1} = ay_{n-2} + h(by'_n + cy'_{n-1} + dy'_{n-2} + ey'_{n-3})$$
    for the solution of $y' = f(x, y)$. Find the truncation error.

24. The formula
    $$y_{n+3} = y_n + \frac{3h}{8}(y'_n + 3y'_{n+1} + 3y'_{n+2} + y'_{n+3})$$
    with a small steplength h is used for solving the equation $y' = -y$. Investigate the convergence properties of the method.
    (Lund Univ., Sweden, BIT 7(1967), 247)

25. Determine constants α, β and γ so that the difference approximation
    $$y_{n+2} - y_{n-2} + \alpha(y_{n+1} - y_{n-1}) = h[\beta(f_{n+1} + f_{n-1}) + \gamma f_n]$$
    for $y' = f(x, y)$ will have the order of approximation 6. Is the difference equation stable for $h = 0$?
    (Uppsala Univ., Sweden, BIT 9(1969), 87)

26. The difference equation

$$\frac{1}{(1+a)}(y_{n+1} - y_n) + \frac{a}{1+a}(y_n - y_{n-1}) = -hy_n, \quad h > 0, a > 0$$

which approximates the differential equation $y' = -y$, is called strongly stable, if for sufficiently small values of h $\lim_{n\to\infty} y_n = 0$ for all solutions y_n. Find the values of a for which strong stability holds.

(Royal Inst. Tech., Stockholm, Sweden, BIT 8(1968), 138)

27. Let a linear multistep method for the initial value problem

$$y'(x) = f(x, y), \quad y(0) = y_0$$

be applied to the test equation $y'(x) = -y(x)$. If the resulting difference equation has at least one characteristic root $\alpha(h)$ such that $|\alpha(h)| > 1$ for arbitrarily small values of h, then the method is called weakly stable. Which of the following methods are weakly stable?

(a) $y_{k+1} = y_{k-1} + 2hf(x_k, y_k)$

(b) $\bar{y}_k = -y_k + 2y_{k-1} + 2hf(x_k, y_k)$
$y_{k+1} = y_{k-1} + 2hf(x_k, \bar{y}_k)$

(c) $\bar{y}_{k+1} = -4y_k + 5y_{k-1} + 2h(2f_k + f_{k-1})$
$y_{k+1} = y_{k-1} + \frac{1}{3}h\{f(x_{k+1}, \bar{y}_{k+1}) + 4f_k + f_{k-1}\}$
$f_i = f(x_i, y_i)$

(Gothenburg Univ., Sweden, BIT 8(1968), 343)

28. Use the two-step formula

$$y_{n+1} = y_{n-1} + \frac{h}{3}(y'_{n+1} + 4y'_n + y'_{n-1})$$

to solve the test problem

$$y' = \alpha y, \quad y(0) = y_0$$

where $\alpha < 0$.
Determine $\lim_{n\to\infty} |y_n|$ and $\lim_{n\to\infty} y(x_n)$ where $x_n = nh$, h fixed, and $y(x)$ is the exact solution of the test problem.

(Uppsala Univ., Sweden, BIT 12(1972), 272)

29. The general solution of the differential equation $y' = 1 + a(1+x+y)$ is $y = 1 + x + c \exp(-ax)$. We attempt to calculate the solution using $y(0) = 1$ numerically. What happens to stability when

(a) $a < 0$; any method
(b) $a > 0$; Runge-Kutta's method
(c) $a > 0$; the midpoint method

$$\frac{y_{n+1} - y_{n-1}}{2h} = 1 + a(1 + x_n + y_n)$$

(Lund Univ., Sweden, BIT 21(1981), 136)

30. To solve the differential equation
$$y' = f(x, y), \quad y(0) = y_0$$
the method
$$y_{n+1} = \tfrac{18}{19}(y_n - y_{n-2}) + y_{n-3} + \frac{6h}{19}(f_{n+1} + 4f_n + 4f_{n-2} + f_{n-3})$$
is suggested.

(a) What is the local truncation error of the method?

(b) Is the method stable?

(Lund Univ., Sweden, BIT 20(1980), 261)

31. A diffusion-transport problem is described by the differential equation for $x > 0$, $py'' + Vy' = 0$, $p > 0$, $V > 0$, $p/V \ll 1$ (and starting conditions at $x = 0$).

We wish to solve the problem numerically by a difference method with stepsize h.

(a) Show that the difference equation which arises when central differences are used for y'' and y' is stable for any $h > 0$ but that when p/h is too small the numerical solution contains slowly damped oscillations with no physical meaning.

(b) Show that when forward-difference approximation is used for y' then there are no oscillations. (This technique is called upstream differencing and is very much in use in the solution of streaming problems by difference methods).

(c) Give the order of accuracy of the method in (b).

(Stockholm Univ., Sweden, BIT 19(1979), 552)

32. Find the characteristic equations for the *PECE* and *PE(CE)²* methods of the *P-C* set
$$y_{n+1}^{*} = y_n + \frac{h}{2}(3y_n' - y_{n-1}')$$
$$y_{n+1} = y_n + \frac{h}{2}(y_{n+1}^{*\prime} + y_n')$$

33. The formulas
$$y_{n+1}^{*} = y_n + \frac{h}{24}(55y_n' - 59y_{n-1}' + 37y_{n-2}' - 9y_{n-3}') + T_1$$
$$y_{n+1} = y_n + \frac{h}{24}(9y_{n+1}^{*\prime} + 19y_n' - 5y_{n-1}' + y_{n-2}') + T_2$$
may be used as a *P-C* set to solve $y' = f(x, y)$. Find T_1 and T_2 and an estimate of the truncation error of the *P-C* set. Construct the corresponding modified *P-C* set.

34. For the corrector formula
$$y_{n+1} - \alpha y_{n-1} = Ay_n + By_{n-2} + h(Cy_{n+1}' + Dy_n' + Ey_{n-1}') + T$$

we have $T = O(h^5)$.

(a) Show that $A = \frac{9}{8}(1 - \alpha)$, $B = -\frac{1}{8}(1 - \alpha)$ and determine C, D and E.

(b) Find the stability conditions.

(Inst. Tech., Lund, Sweden, BIT 13(1973), 375)

35. Which of the following difference methods are applicable for solving the initial value problem

$$y' + ky = 0, \quad y(0) = 1, \quad k > 0$$

For what values of k are the methods stable?

(a) $y'_{n+1} = \frac{1}{2}y_n - \frac{1}{4}y_{n-1} + \frac{h}{3}(2y'_n + y'_{n-1})$

(b) $\begin{cases} y_{n+1} = y_n + h(2y'_n - y'_{n-1}) & \text{(predictor)} \\ y_{n+1} = y_n + \frac{h}{2}(y'_{n+1} + y'_n) & \text{(corrector)} \end{cases}$

with complete iterations of the corrector.

(c) same as (b) but using the corrector just once.

(Gothenburg Univ., Sweden, BIT 6(1966), 83)

36. To integrate a system of differential equations

$$\mathbf{y}' = \mathbf{f}(x, \mathbf{y}), \quad \mathbf{y}_0 \text{ is given}$$

one can use Euler's method as predictor and apply the trapezoidal rule once as corrector, i.e.

$$\mathbf{y}^*_{n+1} = \mathbf{y}_n + h\mathbf{f}(x_n, \mathbf{y}_n)$$

$$\mathbf{y}_{n+1} = \mathbf{y}_n + \tfrac{1}{2}h(\mathbf{f}(x_n, \mathbf{y}_n) + \mathbf{f}(x_{n+1}, \mathbf{y}^*_{n+1}))$$

(also known as **Heun's method**).

(a) If this method is used on $\mathbf{y}' = \mathbf{A}\mathbf{y}$, where $\mathbf{A}$ is a constant matrix, then $\mathbf{y}_{n+1} = \mathbf{B}(h)\mathbf{y}_n$. Find the matrix $\mathbf{B}(h)$.

(b) Assume that $\mathbf{A}$ has real eigenvalues λ satisfying $\lambda_i \in [a, b]$, $a < b < 0$. For what values of h is it true that $\lim_{n \to \infty} \mathbf{y}_n = 0$.

(c) If the scalar equation $y' = qy$ is integrated as above, which is the largest value of p for which

$$\lim_{h \to 0} \frac{y_n - e^{qx}y_0}{h^p}, \quad x = nh$$

x fixed, has a finite limit?

(Royal Inst. Tech., Stockholm, Sweden, BIT 8(1968), 138)

37. The linear tridiagonal system of equations

$$\begin{bmatrix} 2 & -1 & & & & \\ -1 & 2 & -1 & & & \\ & -1 & 2 & -1 & & \\ & & \cdot & \cdot & \cdot & \\ & & & -1 & 2 & -1 \\ & & & & -1 & 2 \end{bmatrix} \mathbf{x} = \begin{bmatrix} 0.001 \\ 0.002 \\ 0.003 \\ \vdots \\ 0.998 \\ 0.999 \end{bmatrix}$$

can be associated with the difference equation

$$-x_{n-1} + 2x_n - x_{n+1} = \frac{n}{1000}, \quad n = 1, 2, \ldots, 999$$

$$x_0 = 0, \quad x_{1000} = 0$$

Solve the system by solving the difference equation.

(Lund Univ., Sweden, BIT 20(1980), 529)

38. Given the boundary value problem

$$(1 + x^2)y'' + \left(5x + \frac{3}{x}\right)y' + \frac{4y}{3} + 1 = 0$$

$$y(-2) = y(2) = 0.6$$

(a) Show that the solution is symmetric, assuming that it is unique.

(b) Show that when $x = 0$ the differential equation is replaced by a central condition $4y'' + \frac{4y}{3} + 1 = 0$.

(c) Discretize the differential equation and the central condition at $x_i = ih$, $i = -N, \ldots, -1, 0, 1, \ldots, N$, $h = \frac{2}{N}$, and formulate the resulting three point numerical problem. Choose $h = 1$ and find approximate values of $y(0)$, $y(1)$ and $y(-1)$.

(Royal Inst. Tech., Stockholm, Sweden, BIT 18(1978), 236)

39. Use a second order method for the solution of the boundary value problem

$$y'' = xy + 1, \quad x \in [0, 1]$$

$$y(0) + y'(0) = 1, \quad y(1) = 1$$

with the step length $h = 0.25$.

40. Solve the following boundary value problems with step lengths $h = 0.5, 0.25$ and extrapolate

(a) $y'' + (1 + x^2)y + 1 = 0$, $y(\pm 1) = 0$.
(b) $y'' + y + 1 = 0$, $y(\pm 1) = 0$.

Use second order and fourth order methods. Note the symmetry in the solutions.

41. (a) Find the coefficients a and b in the operator formula
$$\delta^2 + a\delta^4 = h^2 D^2 (1 + b\delta^2) + O(h^8)$$

(b) Show that this formula defines an explicit multistep method for the integration of the special second order differential equation
$y'' = f(x, y)$
Prove by considering the case $f(x, y) = 0$ that the proposed method is unstable.

(Stockholm Univ., Sweden, BIT 8(1968), 138)

42. A table of the function $y = f(x)$ is given

x	4	5	6	7	8	9	10
y	0.15024	0.56563	1.54068	3.25434	5.51438	7.56171	8.22108

It is known that f satisfies the differential equation
$$y'' + \left(1 - \frac{4}{x} - \frac{n(n+1)}{x^2}\right)y = 0$$
where n is a positive integer.

(a) Find n

(b) Compute $f(12)$ using Numerov's method with step size 1.

(Uppsala Univ., Sweden, BIT 8(1968), 343)

43. The differential equation $y'' = x^3(y + y')$ and the boundary conditions $y(0) = 1$ and $y(0.5) = 1.3$ are given.

Put $y'(0) = \alpha$ and determine by a series expansion a polynomial in x of degree 5 which approximates $y(x)$.
Determine α so that the boundary condition $y(0.5) = 1.3$ is satisfied.

(Lund Univ., Sweden, BIT 14(1974), 122)

44. Determine a difference approximation of the problem
$$\frac{d}{dx}\left[(1 + x^2)\frac{dy}{dx}\right] - y = x^2 + 1;$$
$y(-1) = y(1) = 0$

Find approximate values of $y(0)$ using the steps $h = 1$ and $h = 0.5$, and perform Richardson extrapolation.

(Royal Inst. Tech., Stockholm, Sweden, BIT 7(1967), 338)

45. (a) Determine the constants in the following relations:
$$h^{-4}\delta^4 = D^4(1 + a\delta^2 + b\delta^4) + O(h^6)$$
$$hD = \mu\delta + a_1\Delta^3 E^{-1} + (hD)^4(a_2 + a_3\mu\delta + a_4\delta^2) + O(h^7)$$

(b) Use the relations in (a) to construct a difference method for the boundary value problem
$y^{iv}(x) = p(x)y(x) + q(x)$

$y(0)$, $y(1)$, $y'(0)$ and $y'(1)$ are given. The stepsize is $h = (1/N)$, where N is a natural number. The boundary conditions should not be approximated with substantially lower accuracy than the difference equation. Show that the number of equations and the number of unknowns agree.

(Uppsala Univ., Sweden, BIT 8(1968), 59)

46. In order to illustrate the significance of the fact that even the boundary conditions for a differential equation are to be accurately approximated when difference methods are used, we examine the differential equation

$$y'' = y$$

with boundary conditions

$$y'(0) = 0, \quad y(1) = 1$$

which has the solution

$$y(x) = \frac{\cosh x}{\cosh 1}$$

We put $x_n = nh$, assume that $1/h$ is an integer and use the difference approximation

$$y_n'' \approx (y_{n+1} - 2y_n + y_{n-1})/h^2$$

Two different representations for the boundary conditions are

1. symmetric case:

$$y_{-1} = y_1; \quad y_N = 1, \quad N = \frac{1}{h}$$

2. non-symmetric case

$$y_0 = y_1; \quad y_N = 1$$

(a) Show that the error $y(0) - y_0$ asymptotically approaches ah^2 in the first case, and bh in the second, where a and b are constants to be determined.

(b) Show that the truncation error in the first case is $O(h^2)$ in the closed interval $[0, 1]$.

(Stockholm Univ., Sweden, BIT 5(1965), 294)

47. Find difference approximations of the solution $y(x)$ of the boundary value problem

$$y'' + 8(\sin^2 \pi x)y = 0, \quad 0 \leq x \leq 1, \quad y(0) = y(1) = 1$$

taking step-lengths $h = \frac{1}{4}$ and $h = \frac{1}{6}$. Also find an approximate value for $y'(0)$.

(Chalmer's Inst. Tech., Gothenburg, Sweden, BIT 8(1968), 246)

48. Find the solution of the boundary value problem
$$u'' = u + x, \quad x \in [0, 1]$$
$$u(0) = 0, \quad u(1) = 0$$
using the shooting method. Use the Runge-Kutta method of second order to solve the initial value problem with $h = 0.2$.

49. Find the solution of the boundary value problem
$$x^2 u'' - 2u + x = 0, \quad x \in [2, 3]$$
$$u(2) = u(3) = 0$$
using the shooting method. Use (i) the Taylor series method of order four and (ii) Runge-Kutta method of order four with $h = 0.25$. Compare with the exact solution $u(x) = (19x^2 - 5x^3 - 36)/38x$.

50. Use the shooting method to find the solution of the boundary value problem
$$u'' = 6u^2$$
$$u(0) = 1, \quad u(0.5) = 4/9$$
Assume the initial approximations
$$u'(0) = \alpha_0 = -1.8, \quad u'(0) = \alpha_1 = -1.9$$
and find the solution of the initial value problem using the Taylor series method of order four with $h = 0.1$. Improve the value of $u'(0)$ using the secant method once. Compare with the exact solution $u(x) = 1/(1 + x)^2$.

Chapter 7
Partial Differential Equations

7.1 INTRODUCTION

We consider the general second order partial differential equation of the form

$$A\frac{\partial^2 u}{\partial x^2} + 2B\frac{\partial^2 u}{\partial x\,\partial y} + C\frac{\partial^2 u}{\partial y^2} - F\left(x, y, u, \frac{\partial u}{\partial x}, \frac{\partial u}{\partial y}\right) = 0 \qquad (7.1)$$

If A, B and C are functions of x, y, u, $\partial u/\partial x$ and $\partial u/\partial y$, then (7.1) is called a **quasilinear** partial differential equation. When A, B and C are functions of x, y and F is a linear function of u, $\partial u/\partial x$ and $\partial u/\partial y$ then (7.1) is called **linear**. A linear partial differential equation may be written as

$$A\frac{\partial^2 u}{\partial x^2} + 2B\frac{\partial^2 u}{\partial x\,\partial y} + C\frac{\partial^2 u}{\partial y^2} + D\frac{\partial u}{\partial x} + E\frac{\partial u}{\partial y} + Fu + G = 0 \qquad (7.2)$$

where $A, B, C, \ldots, G$ are constants or functions of x and y only. The equation (7.2) is **homogeneous** if $G \equiv 0$, otherwise **inhomogeneous**. A solution of (7.2) is of the form

$$\phi(x, y, u) = 0 \qquad (7.3)$$

This represents a surface, called an **integral surface** in the (x, y, u) space. If on the integral surface, there exist curves across which the partial derivatives $\partial^2 u/\partial x^2$, $\partial^2 u/\partial x\,\partial y$, $\partial^2 u/\partial y^2$ are discontinuous or indeterminate then these curves are called the **characteristics** of the partial differential equation.

Classification

The classification of the second order linear partial differential equations is very important, as it governs the number and nature of conditions which must be associated with the differential equation in order that a unique solution may exist, in the sense that the solution changes continuously with the change in the prescribed conditions. Using the transformation

$$\xi = \xi(x, y), \quad \eta = \eta(x, y) \qquad (7.4)$$

where ξ, η are twice differentiable functions, in (7.2) and simplifying, we get

$$\left(A\left(\frac{\partial \xi}{\partial x}\right)^2 + 2B\frac{\partial \xi}{\partial x}\frac{\partial \xi}{\partial y} + C\left(\frac{\partial \xi}{\partial y}\right)^2\right)\frac{\partial^2 u}{\partial \xi^2}$$

$$+ 2\left(A\frac{\partial \xi}{\partial x}\frac{\partial \eta}{\partial x} + B\frac{\partial \xi}{\partial x}\frac{\partial \eta}{\partial y} + B\frac{\partial \xi}{\partial y}\frac{\partial \eta}{\partial x} + C\frac{\partial \xi}{\partial y}\frac{\partial \eta}{\partial y}\right)\frac{\partial^2 u}{\partial \xi\, \partial \eta}$$

$$+ \left(A\left(\frac{\partial \eta}{\partial x}\right)^2 + 2B\frac{\partial \eta}{\partial x}\frac{\partial \eta}{\partial y} + C\left(\frac{\partial \eta}{\partial y}\right)^2\right)\frac{\partial^2 u}{\partial \eta^2} + \ldots = 0 \qquad (7.5)$$

Setting the coefficients of $\partial^2 u/\partial \xi^2$ and $\partial^2 u/\partial \eta^2$ to zero and simplifying, we get

$$A\left(\frac{\partial \xi}{\partial x}\bigg/\frac{\partial \xi}{\partial y}\right)^2 + 2B\left(\frac{\partial \xi}{\partial x}\bigg/\frac{\partial \xi}{\partial y}\right) + C = 0 \qquad (7.6)$$

$$A\left(\frac{\partial \eta}{\partial x}\bigg/\frac{\partial \eta}{\partial y}\right)^2 + 2B\left(\frac{\partial \eta}{\partial x}\bigg/\frac{\partial \eta}{\partial y}\right) + C = 0 \qquad (7.7)$$

Since

$$\left(\frac{\partial \xi}{\partial x}\bigg/\frac{\partial \xi}{\partial y}\right) = -\frac{dy}{dx}$$

and

$$\left(\frac{\partial \eta}{\partial x}\bigg/\frac{\partial \eta}{\partial y}\right) = -\frac{dy}{dx}$$

both the equations (7.6) and (7.7) reduce to the same ordinary differential equation

$$A\left(\frac{dy}{dx}\right)^2 - 2B\frac{dy}{dx} + C = 0 \qquad (7.8)$$

Hence, $\xi = $ constant and $\eta = $ constant are respectively the solutions of the ordinary differential equations

$$\frac{dy}{dx} = \frac{B - \sqrt{(B^2 - AC)}}{A} \qquad (7.9)$$

$$\frac{dy}{dx} = \frac{B + \sqrt{(B^2 - AC)}}{A} \qquad (7.10)$$

The curves $\xi = $ const. and $\eta = $ const. are called the characteristics. If $B^2 - AC > 0$ at every point (x, y) then ξ and η are real and (7.2) has two real characteristics. The equation (7.2) is then called an **hyperbolic** partial differential equation. If $B^2 - AC = 0$, then (7.9) and (7.10) coincide and the resulting solution $\xi = $ const. is the only characteristic of the equation. The equation (7.2) is then called a **parabolic** partial differential equation. If $B^2 - AC < 0$ then the characteristics are imaginary. The equation (7.2) is then called an **elliptic** partial differential equation. The trasformation (7.4) reduces the equation (7.2) to a standard form. We obtain

Hyperbolic equation

$$\frac{\partial^2 u}{\partial \xi\, \partial \eta} = F_1\left(\xi, \eta, u, \frac{\partial u}{\partial \xi}, \frac{\partial u}{\partial \eta}\right) \qquad (7.11)$$

or

$$\frac{\partial^2 u}{\partial \xi^2} - \frac{\partial^2 u}{\partial \eta^2} = F_2\left(\xi, \eta, u, \frac{\partial u}{\partial \xi}, \frac{\partial u}{\partial \eta}\right) \quad (7.12)$$

Parabolic equation

$$\frac{\partial^2 u}{\partial \xi^2} = F_3\left(\xi, \eta, u, \frac{\partial u}{\partial \xi}, \frac{\partial u}{\partial \eta}\right)$$

or

$$\frac{\partial^2 u}{\partial \eta^2} = F_4\left(\xi, \eta, u, \frac{\partial u}{\partial \xi}, \frac{\partial u}{\partial \eta}\right) \quad (7.13)$$

Elliptic equation

$$\frac{\partial^2 u}{\partial \xi^2} + \frac{\partial^2 u}{\partial \eta^2} = H\left(\xi, \eta, u, \frac{\partial u}{\partial \xi}, \frac{\partial u}{\partial \eta}\right) \quad (7.14)$$

The well known examples of the three types are

(i) **Wave equation**

$$\frac{\partial^2 u}{\partial t^2} = \frac{\partial^2 u}{\partial x^2} \quad (7.15)$$

which is of the hyperbolic type.

(ii) **Heat flow equation**

$$\frac{\partial u}{\partial t} = \frac{\partial^2 u}{\partial x^2} \quad (7.16)$$

which is of the parabolic type.

(iii) **Laplace equation**

$$\frac{\partial^2 u}{\partial x^2} + \frac{\partial^2 u}{\partial y^2} = 0 \quad (7.17)$$

which is of the elliptic type.

We have used x and t as the independent variables in the case of hyperbolic and parabolic equations and x and y as independent variables in the elliptic equation.

Example 7.1 Classify the partial differential equation

$$\frac{\partial^2 u}{\partial t^2} + 4\frac{\partial^2 u}{\partial t \partial x} + 4\frac{\partial^2 u}{\partial x^2} + 2\frac{\partial u}{\partial x} - \frac{\partial u}{\partial t} = 0$$

and find its characteristics. Reduce the equation to its standard form.

We have $A = 1$, $B = 2$, $C = 4$ and $B^2 - AC = 0$. Hence, the given equation is of parabolic type. The characteristic is the solution of

$$\frac{dx}{dt} = \frac{B}{A} = 2$$

which gives $x = 2t$. Hence $\xi = x - 2t = $ constant, is the characteristic.

Using the transformations

$$\xi = x - 2t$$
$$\eta = t$$

We obtain

$$\frac{\partial u}{\partial x} = \frac{\partial u}{\partial \xi}, \quad \frac{\partial u}{\partial t} = -2\frac{\partial u}{\partial \xi} + \frac{\partial u}{\partial \eta}$$

$$\frac{\partial^2 u}{\partial t^2} = 4\frac{\partial^2 u}{\partial \xi^2} - 4\frac{\partial^2 u}{\partial \xi \, \partial \eta} + \frac{\partial^2 u}{\partial \eta^2}$$

$$\frac{\partial^2 u}{\partial x \, \partial t} = -2\frac{\partial^2 u}{\partial \xi^2} + \frac{\partial^2 u}{\partial \xi \, \partial \eta}$$

and

$$\frac{\partial^2 u}{\partial x^2} = \frac{\partial^2 u}{\partial \xi^2}$$

Substituting in the given differential equation, the standard form is obtained as

$$\frac{\partial^2 u}{\partial \eta^2} + 4\frac{\partial u}{\partial \xi} - \frac{\partial u}{\partial \eta} = 0$$

Example 7.2 Classify the partial differential equation

$$\frac{\partial^2 u}{\partial x^2} + 2x\frac{\partial^2 u}{\partial x \, \partial y} + (1 - y^2)\frac{\partial^2 u}{\partial y^2} = 0$$

We find $B^2 - AC = x^2 + y^2 - 1$. If $x^2 + y^2 - 1 > 0$, that is outside the unit circle, the equation is hyperbolic. If $x^2 + y^2 - 1 = 0$ that is on the unit circle, the equation is parabolic. If $x^2 + y^2 - 1 < 0$, that is inside the unit circle, the equation is elliptic.

Example 7.3 Classify the partial differential equation and find the characteristics of

$$\frac{\partial^2 u}{\partial t^2} + (5 + 2x^2)\frac{\partial^2 u}{\partial x \, \partial t} + (1 + x^2)(4 + x^2)\frac{\partial^2 u}{\partial x^2} = 0$$

We find $B^2 - AC = \frac{1}{4}[(5 + 2x^2)^2 - 4(1 + x^2)(4 + x^2)] = \frac{9}{4} > 0$, so that the equation is of hyperbolic type everywhere. The equation governing the characteristics is

$$\left(\frac{dx}{dt}\right)^2 - (5 + 2x^2)\frac{dx}{dt} + (1 + x^2)(4 + x^2) = 0$$

We obtain

$$\frac{dx}{dt} = \tfrac{1}{2}[(5 + 2x^2) \pm 3] = 4 + x^2;\ 1 + x^2$$

Integrating, we write the characteristics as

$$\xi = \tan^{-1} x - t = \text{constant}$$

$$\eta = \tan^{-1} \frac{x}{2} - 2t = \text{constant}$$

7.2 · DIFFERENCE METHODS

We assume throughout our discussion that our mathematical problem is **well posed**, that is, if its solution exists then it is unique and depends continuously on the given data. The parabolic and the hyperbolic type of equations are either initial value problems or initial boundary value problems whereas the elliptic type equation is always a boundary value problem.

In the finite difference method, we superimpose on the region R of interest a network or a mesh, by lines

$$x_m = a + mh, \quad m = 0, 1, 2, \ldots$$
$$y_n = b + nk, \quad n = 0, 1, 2, \ldots \qquad (7.18)$$

where the quantities h and k are mesh sizes in x and y directions respectively. The points of intersection of the network are called **nodes**. The network and nodes are shown in Fig. 7.1.

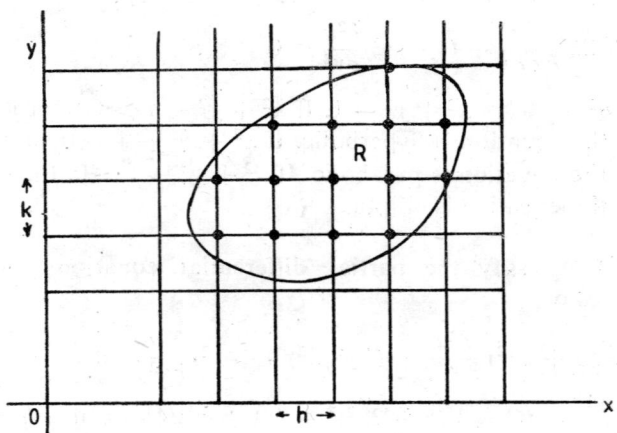

Fig. 7.1 The region R and nodes

The node (x_m, y_n) will be denoted by (m, n). The partial derivatives in the differential equation are replaced by suitable difference quotients, converting the differential equation to a difference equation at each nodal point. The given data is used to modify the difference equation at the nodes near or on the boundary. The solution of this system of equations gives the numerical solution of the partial differential equation.

The difference method is to satisfy the basic requirements such as consistency, stability and convergence. We have already defined these concepts in Chapter 6. The relation between these for initial value problems is given by the **equivalence** theorem.

THEOREM (LAX) 7.1 *For a properly posed initial value problem and a finite difference equation to it that satisfies the consistency condition, stability is the necessary and sufficient condition for convergence.*

We now obtain a few well known difference methods for the three types of equations and also discuss the stability and convergence of these methods.

7.3 PARABOLIC EQUATIONS

The parabolic partial differential equation (7.16) has only one real characteristic. This means that the solution is valid only on one side of this characteristic. If the parabolic equation (7.16) is solved with the initial condition

$$u(x, 0) = f(x), \quad -\infty < x < \infty$$

then this problem is called a **pure initial value problem.** If the parabolic equation (7.16) is solved with any one of the following conditions

(i) **Initial condition**

$$u(x, 0) = f(x), \quad 0 \leqslant x < \infty$$

Boundary conditions

$$a_0(0, t)u + a_1(0, t)\frac{\partial u}{\partial x} = a_2(0, t)$$

where $a_0(0, t) \geqslant 0$, $a_1(0, t) \leqslant 0$ and $a_0 - a_1 > 0$, and also a condition at $x = \infty$, $t \geqslant 0$.

(ii) **Initial condition**

$$u(x, 0) = f(x), \quad a \leqslant x \leqslant b$$

Boundary conditions

$$a_0(a, t)u + a_1(a, t)\frac{\partial u}{\partial x} = a_2(a, t)$$

$$b_0(b, t)u + b_1(b, t)\frac{\partial u}{\partial x} = b_2(b, t)$$

where

$a_0(a, t) \geqslant 0, \quad a_1(a, t) \leqslant 0, \quad a_0 - a_1 > 0$
$b_0(b, t) \geqslant 0, \quad b_1(b, t) \leqslant 0, \quad b_0 - b_1 > 0$

then this problem is called an **initial boundary value problem.**

7.3.1 One Space Dimension

We consider the parabolic equation (7.16)

$$\frac{\partial u}{\partial t} = \frac{\partial^2 u}{\partial x^2}$$

in an arbitrary region $R \times [0, T]$ with suitable boundary conditions, where $R = (a \leqslant x \leqslant b)$ and $0 \leqslant t \leqslant T$. We superimpose on $R \times [0, T]$ a rectilinear grid with grid lines parallel to the coordinate axes with spacing h and k in space and time directions, respectively. The grid points (7.18) on $R \times [0, T]$ are given by

$$t_n = nk, \quad n = 0, 1, 2, \ldots, N$$
$$x_m = a + mh, \quad m = 0, 1, 2, \ldots, M$$

where $x_0 = a$, $x_M = b$, $M = (b - a)/h$. The space nodes on the nth time grid usually constitute the nth **layer** or **level**. Let the solution value $u(x_m, t_n)$ be denoted by U_m^n and its approximate value by u_m^n. The differential equation at the node (m, n) becomes

$$\left[\frac{\partial u}{\partial t}\right]_{(x_m,\, t_n)} = \left[\frac{\partial^2 u}{\partial x^2}\right]_{(x_m,\, t_n)} \tag{7.19}$$

The construction of the difference schemes for the differential equation (7.16) will be based on the approximations to the partial derivatives in (7.19).

The simple approximations to the first derivative in the time direction are given by

$$\left[\frac{\partial u}{\partial t}\right]_{(x_m,\, t_n)} = \begin{cases} \text{(i)} & \dfrac{U_m^{n+1} - U_m^n}{k} + 0(k) \\[6pt] \text{(ii)} & \dfrac{U_m^n - U_m^{n-1}}{k} + 0(k) \\[6pt] \text{(iii)} & \dfrac{U_m^{n+1} - U_m^{n-1}}{2k} + 0(k^2) \end{cases} \tag{7.20}$$

where $U_m^{n+1} = u(x_m, t_{n+1})$. Next, approximation to the second derivative $\partial^2 u / \partial x^2$ in the space direction can be written as

$$\left[\frac{\partial^2 u}{\partial x^2}\right]_{(x_m,\, t_n)} = \frac{U_{m-1}^n - 2U_m^n + U_{m+1}^n}{h^2} + 0(h^2) \tag{7.21}$$

Substituting (7.20) and (7.21) into (7.19), we obtain

(i) $\quad \dfrac{U_m^{n+1} - U_m^n}{k} = \dfrac{U_{m-1}^n - 2U_m^n + U_{m+1}^n}{h^2} + 0(k + h^2)$

(ii) $\quad \dfrac{U_m^n - U_m^{n-1}}{k} = \dfrac{U_{m-1}^n - 2U_m^n + U_{m+1}^n}{h^2} + 0(k + h^2)$

(iii) $\quad \dfrac{U_m^{n+1} - U_m^{n-1}}{2k} = \dfrac{U_{m-1}^n - 2U_m^n + U_{m+1}^n}{h^2} + 0(k^2 + h^2)$ $\tag{7.22}$

The terms $O(k + h^2)$ and $O(k^2 + h^2)$ in (7.22) denote the order of the **local truncation error** and is also known as the **order** of the method. Neglecting the truncation errors in (7.22) and simplifying we obtain the difference methods

$$u_m^{n+1} = (1 - 2\lambda)u_m^n + \lambda(u_{m-1}^n + u_{m+1}^n) \tag{7.23}$$

$$-\lambda u_{m-1}^n + (1 + 2\lambda)u_m^n - \lambda u_{m+1}^n = u_m^{n-1} \tag{7.24}$$

$$u_m^{n+1} = u_m^{n-1} + 2\lambda(u_{m-1}^n - 2u_m^n + u_{m+1}^n) \tag{7.25}$$

where $\lambda = (k/h^2)$ is called the **mesh ratio parameter**.

Two Level Difference Methods

The method (7.23) is called the **Schmidt method**. This is a relation between the function values at the two levels $n + 1$ and n and may be called a **two level formula**. In schematic form (7.23) is shown as in Fig. 7.2. The solution value at any point $(m, n + 1)$ on the $(n + 1)$th level is expressed in terms of the solution values at the points $(m - 1, n)$, (m, n) and $(m + 1, n)$. Such a method is called an **explicit** method. The error T_m^n at the node $(m, n + 1)$ is given by

$$T_m^n = U_m^{n+1} - U_m^n - \lambda(U_{m+1}^n - 2U_m^n + U_{m-1}^n) \tag{7.26}$$

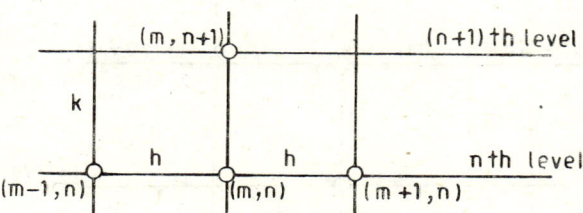

Fig. 7.2 Schematic representation of Schmidt method

Substituting Taylor series expansion of each term about (m, n) on the right side of (7.26) and simplifying we get

$$T_m^n = k\frac{\partial U_m^n}{\partial t} + \frac{k^2}{2}\frac{\partial^2 U_m^n}{\partial t^2} + \frac{k^3}{6}\frac{\partial^3 U_m^n}{\partial t^3} + \cdots$$

$$- \lambda\left(h^2 \frac{\partial^2 U_m^n}{\partial x^2} + \frac{h^4}{12}\frac{\partial^4 U_m^n}{\partial x^4} + \frac{h^6}{360}\frac{\partial^6 U_m^n}{\partial x^6} + \cdots\right)$$

$$= k\left(\frac{\partial U_m^n}{\partial t} - \frac{\partial^2 U_m^n}{\partial x^2}\right) + \frac{k^2}{2}\frac{\partial^2 U_m^n}{\partial t^2} + \frac{k^3}{6}\frac{\partial^3 U_m^n}{\partial t^3}$$

$$- \frac{kh^2}{12}\frac{\partial^4 U_m^n}{\partial x^4} - \frac{kh^4}{360}\frac{\partial^6 U_m^n}{\partial x^6} - \cdots \tag{7.27}$$

If we use the relation (7.19) then (7.27) becomes

$$T_m^n = \frac{k}{2} h^2 \left(\lambda - \frac{1}{6}\right) \frac{\partial^4 U_m^n}{\partial x^4} + \frac{k h^4}{6}\left(\lambda^2 - \frac{1}{60}\right)\frac{\partial^6 U_m^n}{\partial x^6} + \cdots$$

We have the local truncation error as

$$\frac{1}{k} T_m^n = \frac{h^2}{2}\left(\lambda - \frac{1}{6}\right)\frac{\partial^4 U_m^n}{\partial x^4} + \frac{h^4}{6}\left(\lambda^2 - \frac{1}{60}\right)\frac{\partial^6 U_m^n}{\partial x^6} + \cdots \tag{7.28}$$

The method (7.23) is therefore of the order $k + h^2$. When $\lambda = \frac{1}{6}$, the method is of the order $k^2 + h^4$.

The method (7.24) in terms of the function values at the $(n+1)$th and nth levels may be written as

$$-\lambda u_{m-1}^{n+1} + (1+2\lambda) u_m^{n+1} - \lambda u_{m+1}^{n+1} = u_m^n \tag{7.29}$$

This method is called the **Laasonen** method. In schematic form (7.29) is shown in Fig. 7.3. The solution value at any point $(m, n+1)$ on the $(n+1)$th level is dependent on the solution values at the neighbouring points on the same level and on one point on the nth level. Since, the solution values at the $(n+1)$th level are obtained implicitly, the method (7.29) is called an **implicit** method. It is also a two level method. The error T_m^n at the node (m, n) is given by

$$T_m^n = -\lambda U_{m-1}^{n+1} + (1+2\lambda) U_m^{n+1} - \lambda U_{m+1}^{n+1} - U_m^n \tag{7.30}$$

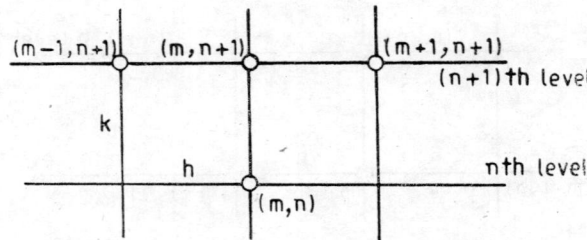

Fig. 7.3 Schematic representation of Laasonen method

Expanding each term on the right hand side of (7.30) about (m, n) and simplifying we get

$$T_m^n = k\left(\frac{\partial U_m^n}{\partial t} - \frac{\partial^2 U_m^n}{\partial x^2}\right) + \tfrac{1}{2} k^2 \frac{\partial^2 U_m^n}{\partial t^2} - k^2 \frac{\partial^3 U_m^n}{\partial x^2 \partial t}$$
$$- \frac{h^2 k}{12}\frac{\partial^4 U_m^n}{\partial x^4} - \tfrac{1}{2} k^3 \frac{\partial^4 U_m^n}{\partial x^2 \partial t^2} - \cdots \tag{7.31}$$

If we use the relation (7.19) then (7.31) becomes

$$k^{-1} T_m^n = \frac{k}{2}\left(\frac{\partial^2 U_m^n}{\partial t^2} - 2 \frac{\partial^3 U_m^n}{\partial x^2 \partial t}\right) - \frac{h^2}{12}\frac{\partial^4 U_m^n}{\partial x^4} - \cdots \tag{7.32}$$

Thus, the method (7.24) is of $0(k + h^2)$.

Averaging (7.26) and (7.30) we get

$$T_m^n = U_m^{n+1} - U_m^n - \frac{\lambda}{2}(U_{m+1}^n - 2U_m^n + U_{m-1}^n)$$

$$+ U_{m-1}^{n+1} - 2U_m^{n+1} + U_{m+1}^{n+1})$$

$$= k\left(\frac{\partial U_m^n}{\partial t} - \frac{\partial^2 U_m^n}{\partial x^2}\right) - \tfrac{1}{2}k^2 \frac{\partial}{\partial t}\left(\frac{\partial U_m^n}{\partial t} - \frac{\partial^2 U_m^n}{\partial x^2}\right)$$

$$- \frac{kh^2}{12}\frac{\partial^4 U_m^n}{\partial x^4} - \frac{k^3}{4}\frac{\partial^4 U_m^n}{\partial x^2 \partial t^2} - \cdots \tag{7.33}$$

Using (7.19), we find

$$T_m^n = 0(k^3 + kh^2) \tag{7.34}$$

or

$$k^{-1}T_m^n = 0(k^2 + h^2)$$

Thus, a method of order $k^2 + h^2$ is given by

$$-\frac{\lambda}{2} u_{m-1}^{n+1} + (1+\lambda)u_m^{n+1} - \frac{\lambda}{2} u_{m+1}^{n+1} = \frac{\lambda}{2} u_{m-1}^n + (1-\lambda)u_m^n + \frac{\lambda}{2} u_{m+1}^n \tag{7.35}$$

which may be written as

$$\left(1 - \frac{\lambda}{2}\delta_x^2\right)u_m^{n+1} = \left(1 + \frac{\lambda}{2}\delta_x^2\right)u_m^n \tag{7.36}$$

The method (7.35) or (7.36) is called the **Crank-Nicolson** method. The schematic representation is shown in Fig. 7.4.

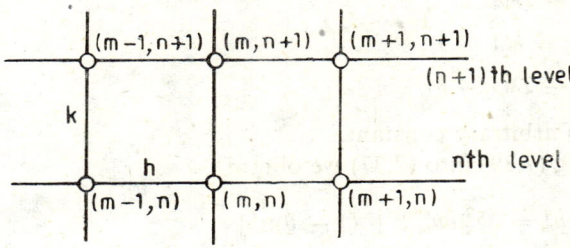

Fig. 7.4 Schematic representation of Crank-Nicolson method

Alternatively, we may define a general two level difference method in the form

$$a_2 u_{m-1}^{n+1} + a_1 u_m^{n+1} + a_0 u_{m+1}^{n+1} - b_2 u_{m-1}^n - b_1 u_m^n - b_0 u_{m+1}^n = 0 \tag{7.37}$$

where the coefficients a_i's and b_i's are arbitrary. The truncation error at the node (m, n) in approximating the differential equation (7.16) by (7.37) may be written as

$$k^{-1}T_m^n = k^{-1}[a_2 U_{m-1}^{n+1} + a_1 U_m^{n+1} + a_0 U_{m+1}^{n+1}$$

$$- b_2 U_{m-1}^n - b_1 U_m^n - b_0 U_{m+1}^n]$$

$$= k^{-1}[(a_2 + a_1 + a_0) - (b_2 + b_1 + b_0)]U_m^n$$
$$+ (a_2+a_1+a_0)\left(\frac{\partial U}{\partial t}\right)_m^n + k^{-1}h[(-a_2+a_0)+(b_2-b_0)]\left(\frac{\partial U}{\partial x}\right)_m^n$$
$$+ \tfrac{1}{2}k^{-1}h^2(a_2+a_0-b_2-b_0)\left(\frac{\partial^2 U}{\partial x^2}\right)_m^n + h(a_0-a_2)\left(\frac{\partial^2 U}{\partial t\, \partial x}\right)_m^n$$
$$+ \tfrac{1}{2}k(a_2 + a_1 + a_0)\left(\frac{\partial^2 U}{\partial t^2}\right)_m^n + \tfrac{1}{2}h^2(a_0 + a_2)\left(\frac{\partial^3 U}{\partial t\, \partial x^2}\right)_m^n$$
$$+ \tfrac{1}{6}k^{-1}h^3[(-a_2 + a_0) + (b_2 - b_0)]\left(\frac{\partial^3 U}{\partial x^3}\right)_m^n$$
$$+ \frac{1}{24}k^{-1}h^4(a_2 + a_0 - b_2 - b_0)\left(\frac{\partial^4 U}{\partial x^4}\right)_m^n + \cdots$$

The necessary conditions for determining the parameters, so that the difference equation is consistent with the differential equation, are given by

$$a_2 + a_1 + a_0 = b_2 + b_1 + b_0$$
$$a_2 + a_1 + a_0 = 1$$
$$a_2 - a_0 = b_2 - b_0$$
$$-(a_2 + a_0) + (b_2 + b_0) = 2\lambda$$
$$a_2 - a_0 = 0 \tag{7.38}$$

The general solution of (7.38) can be expressed in the form

$$a_2 = a_0 = -\lambda\theta$$
$$a_1 = 1 + 2\lambda\theta$$
$$b_2 = b_0 = \lambda(1 - \theta)$$
$$b_1 = 1 - 2\lambda(1 - \theta) \tag{7.39}$$

where θ is an arbitrary constant.

Substituting (7.39) into (7.37) we obtain

$$u_m^{n+1} - u_m^n = \lambda\delta_x^2[\theta u_m^{n+1} + (1 - \theta)u_m^n] \tag{7.40}$$

The values $\theta = 0$ and $\theta = 1$ give the Schmidt and the Laasonen methods respectively. The value $\theta = \tfrac{1}{2}$ gives the Crank-Nicolson method. Using (7.19) and equating the coefficient of $\partial^4 U_m^n/\partial x^4$ to zero, we get

$$\theta = \frac{1}{2} - \frac{1}{12\lambda} \tag{7.41}$$

Substituting (7.41) into (7.40), we get the **Crandall** method which may be written as

$$(1 + \tfrac{1}{12}\delta_x^2)\nabla_t u_m^{n+1} = \frac{\lambda}{2}\delta_x^2(u_m^{n+1} + u_m^n) \tag{7.42}$$

The order of the method (7.42) is $k^2 + h^4$.

Three Level Difference Methods

The difference method (7.25) is called the **Richardson** method. The Richardson method is a three level and an explicit method.

If we replace u_m^n by the mean of the values u_m^{n+1} and u_m^{n-1}, i.e.

$$u_m^n \approx \tfrac{1}{2}(u_m^{n+1} + u_m^{n-1})$$

in (7.25), then we get

$$u_m^{n+1} = u_m^{n-1} + 2\lambda(u_{m-1}^n - u_m^{n+1} - u_m^{n-1} + u_{m+1}^n)$$

which on simplification can be written as

$$u_m^{n+1} = \frac{(1-2\lambda)}{1+2\lambda} u_m^{n-1} + \frac{2\lambda}{1+2\lambda}(u_{m-1}^n + u_{m+1}^n) \tag{7.43}$$

This difference scheme is called the **DuFort and Frankel** method. It is an explicit, three level method. The computational model is given in Fig. 7.5.

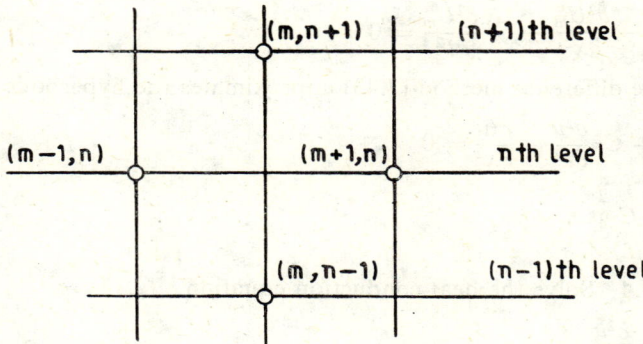

Fig. 7.5 Schematic representation of DuFort-Frankel method

The error of the method (7.43) is given by

$$T_m^n = (1+2\lambda)U_m^{n+1} - (1-2\lambda)U_m^{n-1} - 2\lambda(U_{m-1}^n + U_{m+1}^n) \tag{7.44}$$

Expanding in Taylor's series the terms in (7.44) about the point (x_m, t_n) we have

$$T_m^n = (1+2\lambda)\left(U_m^n + k\frac{\partial U_m^n}{\partial t} + \frac{k^2}{2}\frac{\partial^2 U_m^n}{\partial t^2} + \cdots\right)$$

$$- (1-2\lambda)\left(U_m^n - k\frac{\partial U_m^n}{\partial t} + \frac{k^2}{2}\frac{\partial^2 U_m^n}{\partial t^2} - \cdots\right)$$

$$- 2\lambda\left(2U_m^n + h^2\frac{\partial^2 U_m^n}{\partial x^2} + \frac{h^4}{12}\frac{\partial^4 U_m^n}{\partial x^4} + \cdots\right)$$

$$= 2k\frac{\partial U_m^n}{\partial t} + 2\lambda k^2\frac{\partial^2 U_m^n}{\partial t^2} - 2\lambda h^2\frac{\partial^2 U_m^n}{\partial x^2} - \frac{\lambda h^4}{6}\frac{\partial^4 U_m^n}{\partial x^4} + \cdots$$

The local truncation error is given by

$$k^{-1}T_m^n = 2\left(\frac{\partial U_m^n}{\partial t} - \frac{\partial^2 U_m^n}{\partial x^2}\right) + 2\left(\frac{k}{h}\right)^2 \frac{\partial^2 U_m^n}{\partial t^2} - \frac{h^2}{6}\frac{\partial^4 U_m^n}{\partial x^4} + \frac{k^2}{3}\frac{\partial^3 U_m^n}{\partial t^3} + \cdots \quad (7.45)$$

Thus we have the following cases:

(i) $k^{-1}T_m^n \to 0$ if $\frac{k}{h} \to 0$, as $h \to 0$ and

$$\frac{\partial U_m^n}{\partial t} = \frac{\partial^2 U_m^n}{\partial x^2}$$

Thus, the difference scheme (7.43) is compatible or consistent with the differential equation (7.19) only when $(k/h) \to 0$ as $h \to 0$. In this case the order of method is $k^2 + h^2 + (k/h)^2$.

(ii) $k^{-1}T_m^n \to 0$ if $\frac{k}{h} \to C$ as $h \to 0$ and

$$\frac{\partial U_m^n}{\partial t} - \frac{\partial^2 U_m^n}{\partial x^2} + C^2 \frac{\partial^2 U_m^n}{\partial t^2} = 0$$

Thus, the difference method (7.43) approximates the hyperbolic equation

$$\frac{\partial U}{\partial t} + C^2 \frac{\partial^2 u}{\partial t^2} - \frac{\partial^2 u}{\partial x^2} = 0$$

when $\frac{k}{h} \to C$ as $h \to 0$.

Example 7.4 Solve the heat conduction equation

$$\frac{\partial u}{\partial t} = \frac{\partial^2 u}{\partial x^2}$$

subject to the initial and boundary conditions

$$u(x, 0) = \sin \pi x, \quad 0 \leqslant x \leqslant 1$$
$$u(0, t) = u(1, t) = 0$$

using the following methods
  (i) the Schmidt method
  (ii) the Laasonen method
  (iii) the Crank-Nicolson method
  (iv) the DuFort-Frankel method
for $h = \frac{1}{3}$ and $k = \frac{1}{36}$. Integrate upto two time levels.
The theoretical solution is

$$u(x, t) = \exp(-\pi^2 t) \sin \pi x$$

The nodal points are given by
$$x_m = mh, \quad m = 0, 1, 2, 3$$
$$t_n = nk, \quad n = 0, 1, 2, \ldots$$
which are shown in Fig. 7.6.

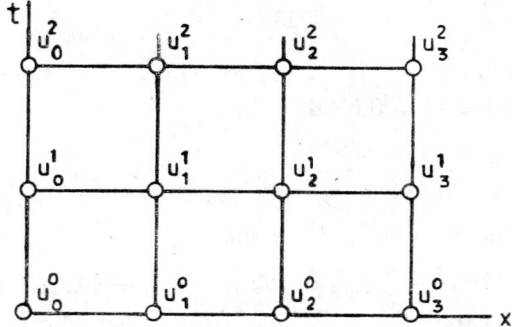

Fig. 7.6 Representation of nodal points

The boundary conditions are given by

$$u_0^n = 0 = u_3^n, \quad n = 0, 1, 2, \ldots$$

(i) The Schmidt method is given by

$$u_m^{n+1} = (1 - 2\lambda)u_m^n + \lambda(u_{m-1}^n + u_{m+1}^n)$$

which becomes

$$u_m^{n+1} = \tfrac{1}{4}(u_{m-1}^n + 2u_m^n + u_{m+1}^n)$$

We obtain

for $n = 0$, $m = 1, 2$

$$u_1^1 = \tfrac{1}{4}(u_0^0 + 2u_1^0 + u_2^0) = \tfrac{1}{4}\left(2\sin\tfrac{\pi}{3} + \sin\tfrac{2\pi}{3}\right) = 0.6495$$

$$u_2^1 = \tfrac{1}{4}(u_1^0 + 2u_2^0 + u_3^0) = \tfrac{1}{4}\left(\sin\tfrac{\pi}{3} + 2\sin\tfrac{2\pi}{3}\right) = 0.6495$$

for $n = 1$, $m = 1, 2$

$$u_1^2 = \tfrac{1}{4}(u_0^1 + 2u_1^1 + u_2^1) = \tfrac{1}{4}(2 \times 0.6495 + 0.6495) = 0.4871$$

$$u_2^2 = \tfrac{1}{4}(u_1^1 + 2u_2^1 + u_3^1) = \tfrac{1}{4}(0.6495 + 2 \times 0.6495) = 0.4871$$

The exact solutions at the second time level are

$$u(\tfrac{1}{3}, \tfrac{1}{18}) = u(\tfrac{2}{3}, \tfrac{1}{18}) = \exp(-\pi^2/18)\sin\pi/3 = 0.5005$$

(ii) Laasonen method becomes

$$-\tfrac{1}{4}u_{m-1}^{n+1} + \tfrac{3}{2}u_m^{n+1} - \tfrac{1}{4}u_{m+1}^{n+1} = u_m^n, \quad m = 1, 2$$

For $n = 0$, we have since $u_0^0 = 0 = u_3^0$

$$\tfrac{3}{2}u_1^1 - \tfrac{1}{4}u_2^1 = u_1^0 = \sin\tfrac{\pi}{3} = \tfrac{\sqrt{3}}{2}$$

$$-\tfrac{1}{4}u_1^1 + \tfrac{3}{2}u_2^1 = u_2^0 = \sin\tfrac{2\pi}{3} = \tfrac{\sqrt{3}}{2}$$

We find $u_1^1 = u_2^1 = \dfrac{2\sqrt{3}}{5} = 0.6928$.

For $n = 1$, we have

$$\tfrac{3}{2}u_1^2 - \tfrac{1}{4}u_2^2 = u_1^1 = 0.6928$$
$$-\tfrac{1}{4}u_1^2 + \tfrac{3}{2}u_2^2 = u_2^1 = 0.6928$$

We get $u_1^2 = u_2^2 = 0.5542$.

(iii) Crank-Nicolson method becomes

$$-\tfrac{1}{8}u_{m-1}^{n+1} + \tfrac{5}{4}u_m^{n+1} - \tfrac{1}{8}u_{m+1}^{n+1} = \tfrac{1}{8}u_{m-1}^n + \tfrac{3}{4}u_m^n + \tfrac{1}{8}u_{m+1}^n, \quad m = 1, 2$$

For $n = 0$, we get

$$\tfrac{5}{4}u_1^1 - \tfrac{1}{8}u_2^1 = \tfrac{3}{4}u_1^0 + \tfrac{1}{8}u_2^0 = \tfrac{7}{16}\sqrt{3}$$
$$-\tfrac{1}{8}u_1^1 + \tfrac{5}{4}u_2^1 = \tfrac{1}{8}u_1^0 + \tfrac{3}{4}u_2^0 = \tfrac{7}{16}\sqrt{3}$$

We obtain

$$u_1^1 = u_2^1 = 0.6736$$

For $n = 1$, we get

$$\tfrac{5}{4}u_1^2 - \tfrac{1}{8}u_2^2 = \tfrac{3}{4}u_1^1 + \tfrac{1}{8}u_2^1 = 0.5894$$
$$-\tfrac{1}{8}u_1^2 + \tfrac{5}{4}u_2^2 = \tfrac{1}{8}u_1^1 + \tfrac{3}{4}u_2^1 = 0.5894$$

We find

$$u_1^2 = u_2^2 = 0.5239$$

(iv) DuFort-Frankel method becomes

$$u_m^{n+1} = \tfrac{1}{3}[u_m^{n-1} + u_{m+1}^n + u_{m-1}^n], \quad m = 1, 2$$

We need the solution on the first level to start the computation. We may use the results on the first level from the Schmidt method. We obtain

$$u_1^2 = \tfrac{2}{3}[0.5u_1^0 + 0.5(u_0^1 + u_2^1)] = 0.5052$$
$$u_2^2 = \tfrac{2}{3}[0.5u_2^0 + 0.5(u_1^1 + u_3^1)] = 0.5052$$

7.3.2 Convergence and Stability Analysis

The analytical solution $u(x_m, t_n)$ of the differential equation, the difference solution u_m^n of the difference equation and the numerical solution $\bar{u}_m^n$ can be related by a relation of the form

$$|u(x_m, t_n) - \bar{u}_m^n| \leqslant |u(x_m, t_n) - u_m^n| + |u_m^n - \bar{u}_m^n| \tag{7.46}$$

In practice, we would like the difference between the analytical and the numerical solution to be small. From (7.46), we find that this difference depends on the values $|u(x_m, t_n) - u_m^n|$ and $|u_m^n - \bar{u}_m^n|$. The value $|u(x_m, t_n) - u_m^n|$ is the local truncation error which arises because the differential equation is replaced by the difference equation. For a convergent difference scheme the truncation error converges to zero as h and k both approach zero. The numerical error $|u_m^n - \bar{u}_m^n|$ arises because in actual computation we cannot construct the difference solution exactly because of the roundoff errors. By stability, we mean that the errors made at one stage of calculations do not cause increasingly large errors as the computations are continued, but rather will eventually damp out. If the difference scheme is stable, then the second term in (7.46) practically equals zero. Thus the results of the convergent and stable methods are very close to the analytical values. We now discuss the convergence and stability analysis of some of the difference schemes developed in the Section 7.3.1.

Convergence Analysis

Definition 7.1 A difference scheme is said to be **convergent** if

$$\lim_{\substack{k \to 0 \\ nk \leqslant T}} u^n(x) = u(x, t) \tag{7.47}$$

for all x and t in the region of interest.

We assume that u_m^n is free from roundoff errors, so that the only difference between $u(x_m, t_n)$ and u_m^n is the error made by replacing (7.16) by the difference equation, that is the local truncation error.

Subtracting (7.22i) from (7.23) we get

$$\epsilon_m^{n+1} = (1 - 2\lambda)\epsilon_m^n + \lambda(\epsilon_{m+1}^n + \epsilon_{m-1}^n) + 0(k^2 + kh^2) \tag{7.48}$$

where $\epsilon_m^n = u_m^n - u(x_m, t_n)$. The equation (7.48) is called the **error equation.**

Since the initial and boundary values are known exactly, the initial and boundary values for the errors are zero.

Let E^n denote the maximal error on the nth level, that is

$$E^n = \max_m |\epsilon_m^n|$$

From (7.48) we have

$$\max_m |\epsilon_m^{n+1}| \leqslant \max_m |\lambda \epsilon_{m+1}^n + (1 - 2\lambda)\epsilon_m^n + \lambda \epsilon_{m-1}^n| + M^*(k^2 + kh^2) \tag{7.49}$$

where M^* is a constant independent of k and h. If all the coefficients in the first term on the right hand side are nonnegative, that is, $\lambda \leqslant \frac{1}{2}$, then we have

$$E^{n+1} \leqslant (\lambda + 1 - 2\lambda + \lambda)E^n + M^*(k^2 + kh^2)$$

Using this recurrence relation we obtain

$$E^{n+1} \leqslant E^n + M^*(k^2 + kh^2)$$
$$\leqslant E^{n-1} + 2M^*(k^2 + kh^2) \leqslant \ldots$$
$$\leqslant E^0 + M^*(n + 1)k(k + h^2)$$
$$= M^* t_{n+1}(k + h^2)$$

because there is no initial error, $E^0 = 0$ and $t_{n+1} = (n + 1)k$. As $h \to 0$,

330 Numerical Methods

$k \to 0$; $E^{n+1} \to 0$. Therefore, we have proved that the method (7.23) is convergent for $\lambda \leqslant \frac{1}{2}$.

Next, we discuss the convergence of the difference solutions given by (7.35). Substituting $\epsilon_m^n = u_m^n - u(x_m, t_n)$ into (7.35) we get the error equation

$$\left(\mathbf{I} - \frac{\lambda}{2}\mathbf{C}\right)\boldsymbol{\epsilon}^{n+1} = \left(\mathbf{I} + \frac{\lambda}{2}\mathbf{C}\right)\boldsymbol{\epsilon}^n + \mathbf{T}^n, \quad n = 0, 1, 2, \ldots \tag{7.50}$$

where

$$\boldsymbol{\epsilon}^n = [\epsilon_1^n \ \epsilon_2^n \ \ldots \ \epsilon_{M-1}^n]^T$$

$$\mathbf{T}^n = [T_1^n \ T_2^n \ \ldots \ T_{M-1}^n]^T$$

$$C = \begin{bmatrix} -2 & 1 & & & \mathbf{O} \\ 1 & -2 & 1 & & \\ & \ddots & \ddots & \ddots & \\ & & 1 & -2 & 1 \\ \mathbf{O} & & & 1 & -2 \end{bmatrix}$$

and

$$T_m^n = 0(k^3 + kh^2)$$

The initial condition gives

$$\boldsymbol{\epsilon}^0 = \mathbf{0}$$

We write (7.50) as

$$\boldsymbol{\epsilon}^{n+1} = \mathbf{G}\boldsymbol{\epsilon}^n + \boldsymbol{\sigma}^n, \quad n = 0, 1, 2, \ldots \tag{7.51}$$

where

$$\mathbf{G} = \left(\mathbf{I} - \frac{\lambda}{2}\mathbf{C}\right)^{-1}\left(\mathbf{I} + \frac{\lambda}{2}\mathbf{C}\right)$$

$$\boldsymbol{\sigma}^n = \left(\mathbf{I} - \frac{\lambda}{2}\mathbf{C}\right)^{-1}\mathbf{T}^n$$

Applying (7.51) recursively, we find

$$\boldsymbol{\epsilon}^n = \mathbf{G}^n \boldsymbol{\epsilon}^0 + \sum_{\nu=1}^{n} \mathbf{G}^{n-\nu} \boldsymbol{\sigma}^{\nu-1} \tag{7.52}$$

or

$$\|\boldsymbol{\epsilon}^n\| \leqslant \|\mathbf{G}\|^n \|\boldsymbol{\epsilon}^0\| + \sum_{\nu=1}^{n} \|\mathbf{G}^{n-\nu}\| \|\boldsymbol{\sigma}^{\nu-1}\|$$

$$\leqslant \|\mathbf{G}\|^n \|\boldsymbol{\epsilon}^0\| + \frac{1 - \|\mathbf{G}\|^n}{1 - \|\mathbf{G}\|} \left(\max_{0 \leqslant \nu \leqslant n-1} \|\boldsymbol{\sigma}^\nu\|\right)$$

The matrix $\mathbf{C}$ is symmetric, and so $\mathbf{G}$ is symmetric. Using the spectral norm, we have

$$\|\mathbf{G}\| = \max_j |\mu_j| \tag{7.53}$$

where μ_j, $1 \leqslant j \leqslant M-1$ are the eigenvalues of $\mathbf{G}$ and these are found to be

$$\mu_j = \left(1 - \frac{\lambda}{2}\lambda_j\right)^{-1}\left(1 + \frac{\lambda}{2}\lambda_j\right) \tag{7.54}$$

where λ_j, $1 \leq j \leq M - 1$ are the eigenvalues of $\mathbf{C}$. The eigenvalues λ_j are given by

$$\lambda_j = -4 \sin^2 \frac{j\pi}{2M}, \qquad 1 \leq j \leq M - 1$$

From (7.53) and (7.54), we find that

$$\|\mathbf{G}\| = \max_{1 \leq j \leq M-1} |\mu_j| < 1, \qquad \lambda > 0$$

The equation (7.52) becomes

$$\|\boldsymbol{\epsilon}^n\| \leq \|\boldsymbol{\epsilon}^0\| + \frac{1}{1 - \|\mathbf{G}\|}\left(\max_{0 \leq \nu \leq n-1} \|\mathbf{T}^\nu\|\right) \tag{7.55}$$

where

$$\max_{0 \leq \nu \leq n-1} \|\mathbf{T}^\nu\| = M^*(k^2 + h^2)$$

and M^* is a constant. Hence, we get

$$\|\boldsymbol{\epsilon}^n\| \leq \|\boldsymbol{\epsilon}^0\| + M^*(k^2 + h^2) \tag{7.56}$$

From this result we deduce that there is unconditional convergence as $k \to 0$ and $h \to 0$.

Fourier Series (Von Neumann) Stability Analysis

The numerical solution will contain errors, especially roundoff error. Let $\bar{u}_m^n$ be the numerical solution of the difference equation (7.23). Then we may write

$$\bar{u}_m^n = u_m^n + \bar{\epsilon}_m^n \tag{7.57}$$

where $\bar{\epsilon}_m^n$ is the error due to roundoff, or a computational mistake. Using the fact that $\bar{u}_m^n$ must also satisfy (7.23), we find that $\bar{\epsilon}_m^n$ satisfies the equation

$$\bar{\epsilon}_m^{n+1} = (1 - 2\lambda)\bar{\epsilon}_m^n + \lambda(\bar{\epsilon}_{m-1}^n + \bar{\epsilon}_{m+1}^n) \tag{7.58}$$

Thus, we conclude that the difference equation itself governs the propagation of errors. We assume that a group of errors is introduced at the initial level $t = 0$ which propagates forward in accordance with (7.58). For difference equations with constant coefficients, the error may be expanded in a finite Fourier series. The error can be written as

$$\bar{\epsilon}_m^0 = \sum_j A_{j0} \exp(i\beta_j mh) \tag{7.59}$$

where $\beta_j = j\pi/(b - a)$ and the number of terms is equal to the number of nodal points on the line $t = 0$. We now seek a solution of (7.58) such that it reduces to (7.59) at time $t = 0$. We assume

$$\bar{\epsilon}_m^n = \sum_j A_j \xi^n(j) \exp(i\beta_j mh) \tag{7.60}$$

where ξ is an arbitrary real or complex number. Since, for linear equations, the sum of independent solutions is also a solution, it is enough if we consider a single term

$$\bar{\epsilon}_m^n = A\xi^n e^{i\beta mh} \tag{7.61}$$

where β is a real number and A is an arbitrary constant. In order that the original error $\bar{\epsilon}_m^0$ shall not grow as n increases, it is necessary and sufficient that

$$|\xi| \leqslant 1 \tag{7.62}$$

If this condition is satisfied then the difference scheme is said to be **stable**. This method of stability analysis is known as the **Fourier series method** or the **Von Neumann method**. The number ξ is called the **amplification factor** of the scheme. Substituting (7.61) into (7.58) and simplifying we get

$$\xi = 1 + \lambda(e^{i\beta h} - 2 + e^{-i\beta h})$$
$$= 1 + 2\lambda(\cos \beta h - 1) = 1 - 4\lambda \sin^2\left(\frac{\beta h}{2}\right)$$

The condition (7.62) yields

$$-1 \leqslant 1 - 4\lambda \sin^2\left(\frac{\beta h}{2}\right) \leqslant 1$$

The right inequality is satisfied trivially and the left inequality will be satisfied for

$$\lambda \leqslant \frac{1}{2 \sin^2\left(\frac{\beta h}{2}\right)} \tag{7.63}$$

Since $0 \leqslant \sin^2 \frac{\beta h}{2} \leqslant 1$, we have the stability condition $0 < \lambda \leqslant \frac{1}{2}$. Thus $\lambda = \frac{1}{2}$ separates the region of stability, where errors decay, from the region of instability where some errors grow. Using the same analysis we find the amplification factor of the scheme (7.24) to be

$$\xi = \frac{1}{1 + 4\lambda \sin^2\left(\frac{\beta h}{2}\right)} \tag{7.64}$$

Since $\lambda > 0$, we find that $|\xi| \leqslant 1$ always. Therefore, the Laasonen scheme (7.24) is stable for all values of the mesh ratio parameter λ. Such a scheme is called an **unconditionally stable scheme**. It is easily shown that the Crank-Nicolson scheme is also an unconditionally stable scheme.

The amplification factor of the Richardson scheme is given by

$$\xi = \xi^{-1} - 8\lambda \sin^2\left(\frac{\beta h}{2}\right) \tag{7.65}$$

or

$$\xi^2 + 8\lambda \sin^2\left(\frac{\beta h}{2}\right)\xi - 1 = 0 \tag{7.66}$$

The roots of this equation are

$$\xi_{1,2} = -4\lambda \sin^2\left(\frac{\beta h}{2}\right) \pm \sqrt{\left(1 + 16\lambda^2 \sin^4\left(\frac{\beta h}{2}\right)\right)} \tag{7.67}$$

We find that the root $|\xi_2| > 1$ for all λ. Hence, the Richardson method is always **unstable**. Finally, we discuss the stability of the DuFort-Frankel method (7.43). The amplification factor ξ is governed by the equation

$$(1 + 2\lambda)\xi^2 - 4\lambda \cos(\beta h)\xi - (1 - 2\lambda) = 0 \tag{7.68}$$

The roots of this equation are

$$\xi_{1,2} = [2\lambda \cos(\beta h) \pm \sqrt{(1 - 4\lambda^2 + 4\lambda^2 \cos^2 \beta h)}]/(1 + 2\lambda)$$
$$= [2\lambda \cos(\beta h) \pm \sqrt{(1 - 4\lambda^2 \sin^2(\beta h))}]/(1 + 2\lambda) \tag{7.69}$$

The values $|\xi_{1,2}|$ for $2\lambda < 1$ and $2\lambda > 1$ are given in Figs. 7.7. We find

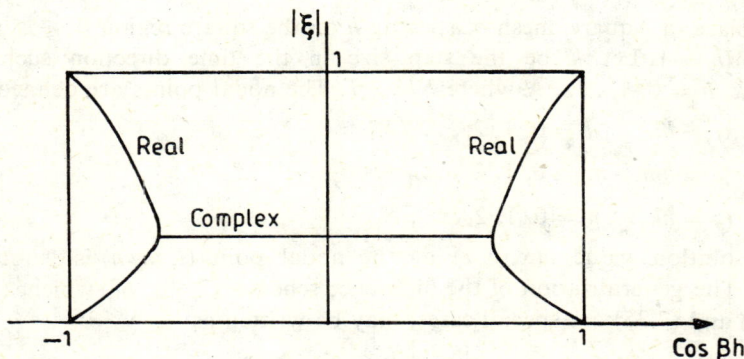

Fig. 7.7(a) Representation of $|\xi_{1,2}|$ for $2\lambda < 1$

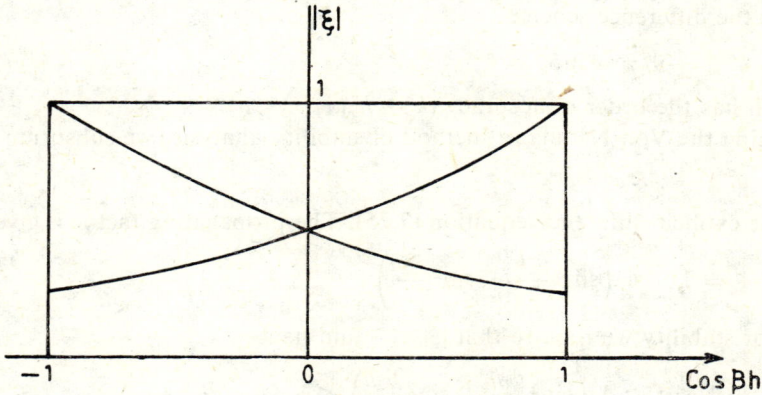

Fig. 7.7(b) Representation of $|\xi_{1,2}|$ for $2\lambda > 1$

that $|\xi_1| \leqslant 1$ and $|\xi_2| < 1$. Hence, the DuFort-Frankel scheme is unconditionally stable.

7.3.3 Two Space Dimensions

We can readily extend the difference schemes in Sec. 7.3.1 to higher space dimensions especially when the region is rectangular. The two dimensional heat flow equation in the rectangular region bounded by

$$R = [0 \leqslant x, y \leqslant 1] \times [0, T]$$

is

$$\frac{\partial u}{\partial t} = \frac{\partial^2 u}{\partial x^2} + \frac{\partial^2 u}{\partial y^2} \quad (7.70)$$

subject to the initial condition

$$u(x, y, 0) = f(x, y) \quad (7.71)$$

and the boundary conditions

$$u(0, y, t) = g_1(y, t), \quad u(1, y, t) = g_2(y, t)$$
$$u(x, 0, t) = h_1(x, t), \quad u(x, 1, t) = h_2(x, t) \quad (7.72)$$

We place a square mesh of spacing h on the square region $0 \leqslant x, y \leqslant 1$ with $Mh = 1$. Let k be the step size in the time direction such that $t = nk$, $n = 0, 1, \ldots, N$ where $Nk = T$. The nodal points are defined by

$$x_l = lh, \quad l = 0, 1, 2, \ldots, M$$
$$y_m = mh, \quad m = 0, 1, 2, \ldots, M$$
$$t_n = nk, \quad n = 0, 1, 2, \ldots, N \quad (7.73)$$

The solution value $u(x, y, t)$ at the nodal point (l, m, n) is denoted by $U_{l,m}^n$. The generalization of the difference scheme (7.40) of which (7.23), (7.29) and (7.35) are special cases, may be written as

$$u_{l,m}^{n+1} = u_{l,m}^n + \lambda(\delta_x^2 + \delta_y^2)[\theta u_{l,m}^{n+1} + (1-\theta)u_{l,m}^n] \quad (7.74)$$

where $u_{l,m}^n$ is an approximate value of $U_{l,m}^n$. For example, the value $\theta = 0$ gives the difference scheme

$$u_{l,m}^{n+1} = u_{l,m}^n + \lambda(\delta_x^2 + \delta_y^2)u_{l,m}^n \quad (7.75)$$

which has the order of accuracy $(k + h^2)$.

Using the Von-Neumann method of stability analysis, we substitute

$$u_{l,m}^n = A\xi^n e^{i\theta_1 lh} e^{i\theta_2 mh} \quad (7.76)$$

in the explicit difference equation (7.75). The propagating factor is given by

$$\xi = 1 - 4\lambda\left(\sin^2\frac{\theta_1 h}{2} + \sin^2\frac{\theta_2 h}{2}\right) \quad (7.77)$$

For stability we require that $|\xi| \leqslant 1$ and hence

$$-1 \leqslant 1 - 4\lambda\left(\sin^2\frac{\theta_1 h}{2} + \sin^2\frac{\theta_2 h}{2}\right) \leqslant 1$$

Since $0 \leqslant \sin^2 \frac{\theta_1 h}{2}, \sin^2 \frac{\theta_2 h}{2} \leqslant 1$, the stability condition is obtained as $0 < \lambda \leqslant \frac{1}{4}$.

Again, for $\theta = \frac{1}{2}$, we may write the difference scheme (7.74) as

$$u_{l,m}^{n+1} = u_{l,m}^n + \frac{\lambda}{2}(\delta_x^2 + \delta_y^2)(u_{l,m}^{n+1} + u_{l,m}^n)$$

or

$$\left[1 - \frac{\lambda}{2}(\delta_x^2 + \delta_y^2)\right]u_{l,m}^{n+1} = \left[1 + \frac{\lambda}{2}(\delta_x^2 + \delta_y^2)\right]u_{l,m}^n \qquad (7.78)$$

which is of order $k^2 + h^2$.

Using the Von-Neumann method, we find

$$\xi = \frac{1 - 2\lambda\left(\sin^2 \frac{\theta_1 h}{2} + \sin^2 \frac{\theta_2 h}{2}\right)}{1 + 2\lambda\left(\sin^2 \frac{\theta_1 h}{2} + \sin^2 \frac{\theta_2 h}{2}\right)}$$

Since, $0 \leqslant \sin^2 \frac{\theta_1 h}{2}, \sin^2 \frac{\theta_2 h}{2} \leqslant 1$ and $\lambda > 0$, the condition $|\xi| \leqslant 1$ is always satisfied. Hence, the method (7.78) is unconditionally stable.

On each time level, a system of linear algebraic equations is to be solved. The coefficient matrix of this system is a band matrix, whose total band width is $2M - 1$ as shown in Fig. 7.8. We number the unknowns in the

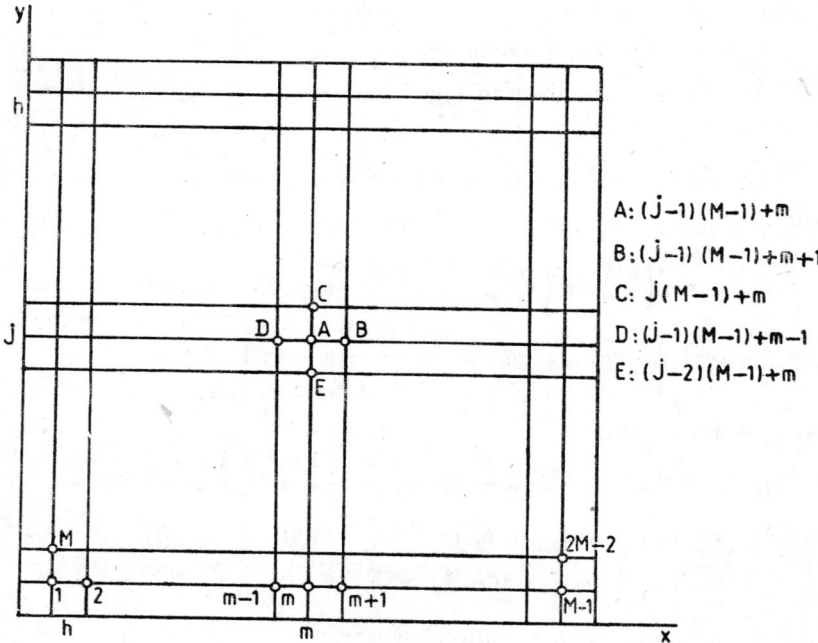

Fig. 7.8 Application of the implicit method on the level $n + 1$

interior starting from left to right in the x-direction and from bottom to top in the y-direction. When we apply (7.78) at A, the five points A, B, C, D, E (and hence the unknowns at these nodes) enter the scheme. The points E and C are $M - 1$ units away from A while B and D are one unit away from A. Hence, the total band width of this system of equations is $2M - 1$. If h is small, then the band width is very large and the solution of this system of equations takes a lot of computational time. To avoid this difficulty we use the **Alternating Direction Implicit methods**.

Example 7.5 Find the solution of the two dimensional heat conduction equation

$$\frac{\partial u}{\partial t} = \frac{\partial^2 u}{\partial x^2} + \frac{\partial^2 u}{\partial y^2}$$

subject to the initial condition

$$u(x, y, 0) = \sin \pi x \sin \pi y, \qquad 0 \leqslant x, y \leqslant 1$$

and the boundary conditions

$$u = 0, \text{ on the boundaries, } t \geqslant 0$$

using the explicit method

$$u_{l,m}^{n+1} = (1 - 4\lambda) u_{l,m}^n + \lambda(u_{l-1,m}^n + u_{l+1,m}^n + u_{l,m-1}^n + u_{l,m+1}^n)$$

with $h = \frac{1}{3}$ and $\lambda = \frac{1}{8}$. Integrate upto two time levels.
Compare the results with the exact solution

$$u(x, y, t) = e^{-\pi^2 t} \sin \pi x \sin \pi y$$

The nodal points are given in Fig. 7.9.
For $\lambda = \frac{1}{8}$, we have

$$u_{l,m}^{n+1} = \tfrac{1}{2} u_{l,m}^n + \tfrac{1}{8}(u_{l-1,m}^n + u_{l+1,m}^n + u_{l,m-1}^n + u_{l,m+1}^n)$$

The initial and boundary conditions become

$$u_{l,m}^0 = \sin\left(\frac{\pi l}{3}\right) \sin\left(\frac{\pi m}{3}\right), \qquad l, m = 0, 1, 2, 3$$

$$u_{l,0}^n = u_{0,m}^n = u_{3,m}^n = u_{l,3}^n = 0, \qquad n = 0, 1$$
$$l, m = 0, 1, 2, 3$$

We get, for $n = 0$,

$$u_{l,m}^1 = \tfrac{1}{2} u_{l,m}^0 + \tfrac{1}{8}(u_{l-1,m}^0 + u_{l+1,m}^0 + u_{l,m-1}^0 + u_{l,m+1}^0)$$

$$u_{1,1}^1 = \tfrac{1}{2} u_{1,1}^0 + \tfrac{1}{8}(u_{0,1}^0 + u_{2,1}^0 + u_{1,0}^0 + u_{1,2}^0)$$

$$= \tfrac{3}{8} + \tfrac{1}{8}(0 + \tfrac{3}{4} + 0 + \tfrac{3}{4}) = \tfrac{9}{16}$$

$$u_{2,1}^1 = \tfrac{1}{2} u_{2,1}^0 + \tfrac{1}{8}(u_{1,1}^0 + u_{3,1}^0 + u_{2,0}^0 + u_{2,2}^0)$$

$$= \tfrac{3}{8} + \tfrac{1}{8}(\tfrac{3}{4} + 0 + 0 + \tfrac{3}{4}) = \tfrac{9}{16}$$

Partial Differential Equations 337

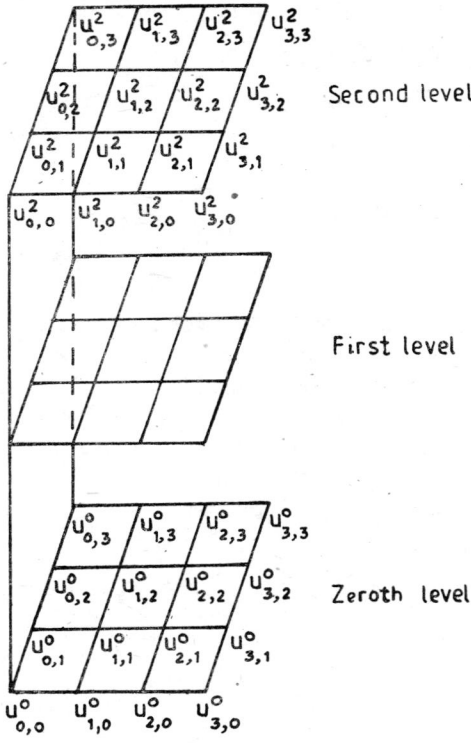

Fig. 7.9 Representation of nodal points

$$u^1_{1,2} = \tfrac{1}{2}u^0_{1,2} + \tfrac{1}{8}(u^0_{0,2} + u^0_{2,2} + u^0_{1,1} + u^0_{1,3})$$
$$= \tfrac{3}{8} + \tfrac{1}{8}(0 + \tfrac{3}{4} + \tfrac{3}{4} + 0) = \tfrac{9}{16}$$
$$u^1_{2,2} = \tfrac{1}{2}u^0_{2,2} + \tfrac{1}{8}(u^0_{1,2} + u^0_{3,2} + u^0_{2,1} + u^0_{2,3})$$
$$= \tfrac{3}{8} + \tfrac{1}{8}(\tfrac{3}{4} + 0 + \tfrac{3}{4} + 0) = \tfrac{9}{16}$$

The exact solution is

$$u(\tfrac{1}{3}, \tfrac{1}{3}, \tfrac{1}{72}) = u(\tfrac{2}{3}, \tfrac{1}{3}, \tfrac{1}{72}) = u(\tfrac{1}{3}, \tfrac{2}{3}, \tfrac{1}{72})$$
$$= u(\tfrac{2}{3}, \tfrac{2}{3}, \tfrac{1}{72}) = 0.6539$$

For $n = 1$

$$u^2_{l,m} = \tfrac{1}{2}u^1_{l,m} + \tfrac{1}{8}(u^1_{l-1,m} + u^1_{l+1,m} + u^1_{l,m-1} + u^1_{l,m+1})$$
$$u^2_{1,1} = \tfrac{1}{2}u^1_{1,1} + \tfrac{1}{8}(u^1_{0,1} + u^1_{2,1} + u^1_{1,0} + u^1_{1,2})$$
$$= \tfrac{9}{32} + \tfrac{1}{8}(0 + \tfrac{9}{16} + 0 + \tfrac{9}{16}) = \tfrac{27}{64}$$
$$u^2_{2,1} = \tfrac{1}{2}u^1_{2,1} + \tfrac{1}{8}(u^1_{1,1} + u^1_{3,1} + u^1_{2,0} + u^1_{2,2})$$
$$= \tfrac{9}{32} + \tfrac{1}{8}(\tfrac{9}{16} + 0 + 0 + \tfrac{9}{16}) = \tfrac{27}{64}$$

$$u_{1,2}^2 = \tfrac{1}{2}u_{1,2}^1 + \tfrac{1}{8}(u_{0,2}^1 + u_{2,2}^1 + u_{1,1}^1 + u_{1,3}^1)$$
$$= \tfrac{9}{32} + \tfrac{1}{8}(0 + \tfrac{9}{16} + \tfrac{9}{16} + 0) = \tfrac{27}{64}$$
$$u_{2,2}^2 = \tfrac{1}{2}u_{2,2}^1 + \tfrac{1}{8}(u_{1,2}^1 + u_{3,2}^1 + u_{2,1}^1 + u_{2,3}^1)$$
$$= \tfrac{9}{32} + \tfrac{1}{8}(\tfrac{9}{16} + 0 + \tfrac{9}{16} + 0) = \tfrac{27}{64}$$

The exact solution is

$$u(\tfrac{1}{3}, \tfrac{1}{3}, \tfrac{1}{36}) = u(\tfrac{2}{3}, \tfrac{1}{3}, \tfrac{1}{36}) = u(\tfrac{1}{3}, \tfrac{2}{3}, \tfrac{1}{36}) = u(\tfrac{2}{3}, \tfrac{2}{3}, \tfrac{1}{36}) = 0.5702$$

Alternating Direction Implicit (ADI) Methods

These methods are two step methods involving the solution of tridiagonal systems of equations along lines parallel to the x and y axis at the first and the second step respectively.

In the **Peaceman-Rachford** ADI method, in the first step we advance from t_n to $t_{n+\frac{1}{2}}$ and use implicit differences for $\partial^2 u/\partial x^2$ and explicit differences for $\partial^2 u/\partial y^2$. In the second step we advance from $t_{n+\frac{1}{2}}$ to t_{n+1} and use explicit differences for $\partial^2 u/\partial x^2$ and implicit differences for $\partial^2 u/\partial y^2$. The ADI method may be written as

$$\frac{u_{l,m}^{n+\frac{1}{2}} - u_{l,m}^n}{k/2} = \frac{1}{h^2}\delta_x^2 u_{l,m}^{n+\frac{1}{2}} + \frac{1}{h^2}\delta_y^2 u_{l,m}^n \tag{7.79}$$

and

$$\frac{u_{l,m}^{n+1} - u_{l,m}^{n+\frac{1}{2}}}{k/2} = \frac{1}{h^2}\delta_x^2 u_{l,m}^{n+\frac{1}{2}} + \frac{1}{h^2}\delta_y^2 u_{l,m}^{n+1} \tag{7.80}$$

These equations may also be written as

$$\left(1 - \frac{\lambda}{2}\delta_x^2\right)u_{l,m}^{n+\frac{1}{2}} = \left(1 + \frac{\lambda}{2}\delta_y^2\right)u_{l,m}^n \tag{7.81}$$

$$\left(1 - \frac{\lambda}{2}\delta_y^2\right)u_{l,m}^{n+1} = \left(1 + \frac{\lambda}{2}\delta_x^2\right)u_{l,m}^{n+\frac{1}{2}} \tag{7.82}$$

The intermediate values $u_{l,m}^{n+\frac{1}{2}}$ are obtained by solving the system of equations produced by (7.81). Adding (7.81) and (7.82) we get

$$u_{l,m}^{n+\frac{1}{2}} = \frac{1}{2}\left(1 + \frac{\lambda}{2}\delta_y^2\right)u_{l,m}^n + \frac{1}{2}\left(1 - \frac{\lambda}{2}\delta_y^2\right)u_{l,m}^{n+1} \tag{7.83}$$

The boundary conditions for obtaining the solution of (7.81) are obtained as

$$u_{l,m}^{n+\frac{1}{2}} = \frac{1}{2}\left(1 + \frac{\lambda}{2}\delta_y^2\right)u_{l,m}^n + \frac{1}{2}\left(1 - \frac{\lambda}{2}\delta_y^2\right)u_{l,m}^{n+1}, \quad l = 0, M \tag{7.84}$$

The solutions on the level $n+1$ are now obtained by solving the system of equations produced by (7.82). The boundary conditions for the solution of this system are given by the prescribed boundary conditions. Eliminating $u_{l,m}^{n+\frac{1}{2}}$ from (7.81) and (7.82), we get

$$\left(1 - \frac{\lambda}{2}\delta_x^2\right)\left(1 - \frac{\lambda}{2}\delta_y^2\right)u_{l,m}^{n+1} = \left(1 + \frac{\lambda}{2}\delta_x^2\right)\left(1 + \frac{\lambda}{2}\delta_y^2\right)u_{l,m}^n \tag{7.85}$$

which is an extension of the Crank-Nicolson scheme to two dimensions. Using, the Von Neumann method to (7.85) we find

$$\xi = \frac{[1 - 2\lambda \sin^2(\theta_1 h/2)][1 - 2\lambda \sin^2(\theta_2 h/2)]}{[1 + 2\lambda \sin^2(\theta_1 h/2)][1 + 2\lambda \sin^2(\theta_2 h/2)]} \tag{7.86}$$

Since $0 \leq \sin^2(\theta_1 h/2), \sin^2(\theta_2 h/2) \leq 1$ and $\lambda > 0$, the condition $|\xi| \leq 1$ is always satisfied. Hence the scheme (7.85) or the ADI method (7.81), (7.82) is unconditionally stable.

Alternatively, (7.85) may be written in the **D'yakonov split form** as

$$\left(1 - \frac{\lambda}{2}\delta_x^2\right) u_{l,m}^{*n+1} = \left(1 + \frac{\lambda}{2}\delta_x^2\right)\left(1 + \frac{\lambda}{2}\delta_y^2\right) u_{l,m}^{n} \tag{7.87}$$

$$\left(1 - \frac{\lambda}{2}\delta_y^2\right) u_{l,m}^{n+1} = u_{l,m}^{*n+1} \tag{7.88}$$

The intermediate boundary conditions in this case are obtained from (7.88)

$$u_{0,m}^{*n+1} = \left(1 - \frac{\lambda}{2}\delta_y^2\right)(g_1)_m^{n+1} \tag{7.89}$$

and

$$u_{M,m}^{*n+1} = \left(1 - \frac{\lambda}{2}\delta_y^2\right)(g_2)_m^{n+1} \tag{7.90}$$

Example 7.6 Find the solution of the two dimensional heat conduction equation

$$\frac{\partial u}{\partial t} = \frac{\partial^2 u}{\partial x^2} + \frac{\partial^2 u}{\partial y^2}$$

subject to the initial condition

$$u(x, y, 0) = \sin \pi x \sin \pi y, \quad 0 \leq x, y \leq 1$$

and the boundary conditions

$$u = 0, \text{ on the boundary for } t \geq 0$$

using the Peaceman-Rachford ADI method. Assume $h = \frac{1}{4}, \lambda = \frac{1}{8}$ and integrate for one time step.

The nodal points are given by

$$x_l = \frac{l}{4}, \quad 0 \leq l \leq 4$$

$$y_m = \frac{m}{4}, \quad 0 \leq m \leq 4$$

The initial and boundary conditions become

$$u_{l,m}^0 = \sin\left(\frac{\pi l}{4}\right) \sin\left(\frac{\pi m}{4}\right)$$

$$u_{0,m}^{n+\frac{1}{2}} = u_{4,m}^{n+\frac{1}{2}} = 0, \quad 0 \leq m \leq 4$$

$$u_{l,0}^{n+1} = 0 = u_{l,4}^{n+1}, \quad 0 \leq l \leq 4$$

The Peaceman-Rachford ADI method is given by

(i) $(1 - \frac{1}{16}\delta_x^2)u_{l,m}^{n+\frac{1}{2}} = (1 + \frac{1}{16}\delta_y^2)u_{l,m}^n$

(ii) $(1 - \frac{1}{16}\delta_y^2)u_{l,m}^{n+1} = (1 + \frac{1}{16}\delta_x^2)u_{l,m}^{n+\frac{1}{2}}$

For $n = 0$, we have for the solution of $u_{l,m}^{1/2}$

$$-\tfrac{1}{16}u_{l-1,m}^{1/2} + \tfrac{9}{8}u_{l,m}^{1/2} - \tfrac{1}{16}u_{l+1,m}^{1/2} = \tfrac{1}{16}u_{l-1,m}^0 + \tfrac{7}{8}u_{l,m}^0 + \tfrac{1}{16}u_{l+1,m}^0$$

On the first mesh line $m = 1$, we get for $l = 1, 2, 3$

$$-\tfrac{1}{16}u_{0,1}^{1/2} + \tfrac{9}{8}u_{1,1}^{1/2} - \tfrac{1}{16}u_{2,1}^{1/2} = \tfrac{1}{16}u_{0,1}^0 + \tfrac{7}{8}u_{1,1}^0 + \tfrac{1}{16}u_{2,1}^0$$

$$-\tfrac{1}{16}u_{1,1}^{1/2} + \tfrac{9}{8}u_{2,1}^{1/2} - \tfrac{1}{16}u_{3,1}^{1/2} = \tfrac{1}{16}u_{1,1}^0 + \tfrac{7}{8}u_{2,1}^0 + \tfrac{1}{16}u_{3,1}^0$$

$$-\tfrac{1}{16}u_{2,1}^{1/2} + \tfrac{9}{8}u_{3,1}^{1/2} - \tfrac{1}{16}u_{4,1}^{1/2} = \tfrac{1}{16}u_{2,1}^0 + \tfrac{7}{8}u_{3,1}^0 + \tfrac{1}{16}u_{4,1}^0$$

Using the boundary conditions, we get

$$\begin{bmatrix} \tfrac{9}{8} & -\tfrac{1}{16} & 0 \\ -\tfrac{1}{16} & \tfrac{9}{8} & -\tfrac{1}{16} \\ 0 & -\tfrac{1}{16} & \tfrac{9}{8} \end{bmatrix} \begin{bmatrix} u_{1,1}^{1/2} \\ u_{2,1}^{1/2} \\ u_{3,1}^{1/2} \end{bmatrix} = \begin{bmatrix} (7\sqrt{2}+1)/(16\sqrt{2}) \\ (14+\sqrt{2})/(16\sqrt{2}) \\ (7\sqrt{2}+1)/(16\sqrt{2}) \end{bmatrix}$$

The solution of this system is

$$u_{1,1}^{1/2} = u_{3,1}^{1/2} = \frac{127 + 16\sqrt{2}}{322} = 0.46468$$

$$u_{2,1}^{1/2} = 0.65716$$

On the second mesh line $m = 2$, we get for $l = 1, 2, 3$

$$\begin{bmatrix} \tfrac{9}{8} & -\tfrac{1}{16} & 0 \\ -\tfrac{1}{16} & \tfrac{9}{8} & -\tfrac{1}{16} \\ 0 & -\tfrac{1}{16} & \tfrac{9}{8} \end{bmatrix} \begin{bmatrix} u_{1,2}^{1/2} \\ u_{2,2}^{1/2} \\ u_{3,2}^{1/2} \end{bmatrix} = \begin{bmatrix} (7\sqrt{2}+1)/16 \\ (14+\sqrt{2})/16 \\ (7\sqrt{2}+1)/16 \end{bmatrix}$$

The solution of the system is

$$u_{1,2}^{1/2} = u_{3,2}^{1/2} = \frac{127\sqrt{2} + 32}{322} = 0.65716$$

$$u_{2,2}^{1/2} = 0.92936$$

On the third mesh line $m = 3$, we get for $l = 1, 2, 3$

$$\begin{bmatrix} \tfrac{9}{8} & -\tfrac{1}{16} & 0 \\ -\tfrac{1}{16} & \tfrac{9}{8} & -\tfrac{1}{16} \\ 0 & -\tfrac{1}{16} & \tfrac{9}{8} \end{bmatrix} \begin{bmatrix} u_{1,3}^{1/2} \\ u_{2,3}^{1/2} \\ u_{3,3}^{1/2} \end{bmatrix} = \begin{bmatrix} (14+\sqrt{2})/32 \\ (7\sqrt{2}+1)/16 \\ (14+\sqrt{2})/32 \end{bmatrix}$$

which gives

$$u_{1,3}^{1/2} = u_{3,3}^{1/2} = \frac{127 + 16\sqrt{2}}{322} = 0.46468$$

$$u_{2,3}^{1/2} = 0.65716$$

We have for the solution of $u^1_{l,m}$

$$-\tfrac{1}{16}u^1_{l,m-1} + \tfrac{9}{8}u^1_{l,m} - \tfrac{1}{16}u^1_{l,m+1} = \tfrac{1}{16}u^{1/2}_{l,m-1} + \tfrac{7}{8}u^{1/2}_{l,m} + \tfrac{1}{16}u^{1/2}_{l,m+1}$$

For $l = 1$ and $m = 1, 2, 3$ we get

$$\begin{bmatrix} \tfrac{9}{8} & -\tfrac{1}{16} & 0 \\ -\tfrac{1}{16} & \tfrac{9}{8} & -\tfrac{1}{16} \\ 0 & -\tfrac{1}{16} & \tfrac{9}{8} \end{bmatrix} \begin{bmatrix} u^1_{1,1} \\ u^1_{1,2} \\ u^1_{1,3} \end{bmatrix} = \begin{bmatrix} 0.44767 \\ 0.63310 \\ 0.44767 \end{bmatrix}$$

which gives

$$u^1_{1,1} = u^1_{1,3} = 0.43186, \quad u^1_{1,2} = 0.61074$$

The coefficient matrix for integration along each mesh line is same as in the previous cases. We find the solutions

$$u^1_{2,1} = u^1_{2,3} = 0.61074, \quad u^1_{2,2} = 0.86371$$

and

$$u^1_{3,1} = u^1_{3,3} = 0.43186, \quad u^1_{3,2} = 0.61074$$

7.4 HYPERBOLIC EQUATIONS

7.4.1 One Space Dimension

Consider the wave equation

$$\frac{\partial^2 u}{\partial t^2} = \frac{\partial^2 u}{\partial x^2} \qquad (7.91)$$

subject to the initial conditions

$$u(x, 0) = f(x), \quad \frac{\partial u}{\partial t}(x, 0) = g(x), \quad -\infty < x < \infty$$

The exact solution of this problem, called the **D'Alembert solution**, is

$$u(x, t) = \tfrac{1}{2}[f(x - t) + f(x + t)] + \frac{1}{2}\int_{x-t}^{x+t} g(\tau)\, d\tau \qquad (7.92)$$

The characteristics are $\xi = x + t =$ constant and $\eta = x - t =$ constant. The transformation $\xi = x + t$ represents a translation of the coordinate system to the left by an amount t. Since this translation is proportional to time t, a point $\xi =$ constant moves to the left with speed unity. That is, a solution of the form $u(x, t) = f(x + t)$, represents a wave with velocity -1 without changing its shape. Similarly, $u(x, t) = f(x - t)$ represents a wave with velocity $+1$. Hence, the solution consists of two waves, one travelling to the left and another travelling to the right with the same speed. At any nodal point (x_m, t_n), we have from (7.92)

$$u(x_m, t_n) = \tfrac{1}{2}[f(x_m - t_n) + f(x_m + t_n)] + \frac{1}{2}\int_{x_m - t_n}^{x_m + t_n} g(\tau)\, d\tau \qquad (7.93)$$

This means that the solution at (x_m, t_n) is determined by the prescribed values of $f(x)$ at the end points of the interval $(x_m - t_n, x_m + t_n)$ and by the prescribed values of $g(x)$ over this interval. The characteristics $\xi = x + t = x_m + t_n$ and $\eta = x - t = x_m - t_n$ intersect the initial line $t = 0$ at $x = x_m + t_n$ and $x = x_m - t_n$. The characteristics are plotted in Fig. 7.10. The region inside the triangle PRS is called the **domain of dependence** of the analytic solution at the nodal point (x_m, t_n). The initial data outside the interval $(x_m - t_n, x_m + t_n)$ does not influence the solution at the nodal point (x_m, t_n). While constructing difference schemes for the numerical solution of the hyperbolic equations this particular aspect of the solutions should be kept in mind.

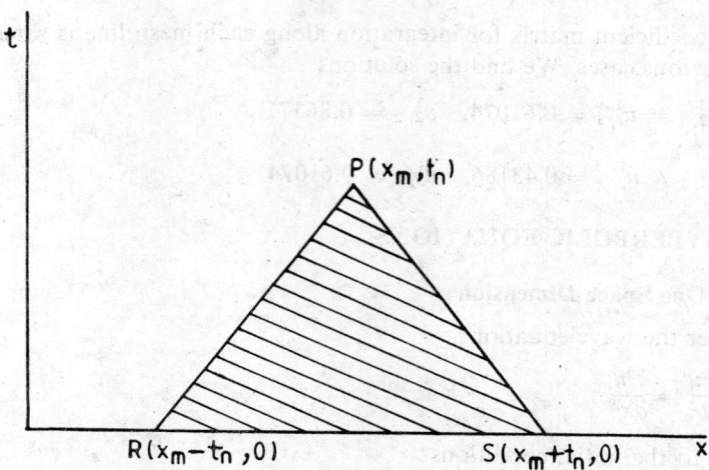

Fig. 7.10 Characteristics of the wave equation

Consider the difference approximations

$$\frac{\partial^2 u}{\partial t^2} = \frac{1}{k^2} \delta_t^2 u + O(k^2) \tag{7.94}$$

and

$$\frac{\partial^2 u}{\partial x^2} = \frac{1}{h^2} \delta_x^2 u + O(h^2) \tag{7.95}$$

Applying these approximations in (7.91), at the nodal point (x_m, t_n), we get

$$\frac{1}{k^2} \delta_t^2 u_m^n = \frac{1}{h^2} \delta_x^2 u_m^n \tag{7.96}$$

or

$$u_m^{n+1} = 2(1 - r^2) u_m^n + r^2 (u_{m+1}^n + u_{m-1}^n) - u_m^{n-1} \tag{7.97}$$

where $r = (k/h)$. This is an explicit three level formula. Note that the minimum number of time levels that can be used is three. In order to start computations, we need the data on two lines $t = 0$ and $t = k$. The information required on the line $t = k$ is obtained by using a suitable approximation to the initial condition $\partial u(x, 0)/\partial t = g(x)$. If we use the approximation

$$\frac{\partial u}{\partial t} = \frac{1}{k} \mu_t \delta_t u + O(k^2) \tag{7.98}$$

we get

$$u_m^{-1} = u_m^1 - 2kg_m \tag{7.99}$$

The extraneous points u_m^{-1} introduced in using the difference scheme (7.97) for $n = 0$ are eliminated by using (7.99). By using Taylor Series expansions, we can easily show that (7.97) is of order $(k^2 + h^2)$. We can study the stability of (7.97) by using the Von-Neumann method. Substituting

$$u_m^n = A\xi^n e^{i\beta mh} \tag{7.100}$$

in (7.97) and simplifying, we get

$$\xi^2 - \left(2 - 4r^2 \sin^2 \frac{\beta h}{2}\right) \xi + 1 = 0 \tag{7.101}$$

which gives

$$\xi_{1,2} = \left(1 - 2r^2 \sin^2 \frac{\beta h}{2}\right) \pm \sqrt{\left(1 - 2r^2 \sin^2 \frac{\beta h}{2}\right)^2 - 1}$$

If $\left|1 - 2r^2 \sin^2 \frac{\beta h}{2}\right| > 1$ then $|\xi_1| > 1$. If $-1 \leqslant \left(1 - 2r^2 \sin^2 \frac{\beta h}{2}\right) \leqslant 1$, then the stability condition $|\xi| \leqslant 1$ is satisfied. Hence, the scheme (7.97) is stable, when

$$-1 \leqslant 1 - 2r^2 \sin^2 \frac{\beta h}{2} \leqslant 1$$

or

$$r \leqslant 1 \tag{7.102}$$

The domain of dependence of the finite difference solutions can also be studied. For the scheme (7.97), lines through the nodal point (x_m, t_n) with slopes $\pm r$ play the role of characteristics. The characteristics, in this case are

$$t - rx = t_n - rx_m$$

and $\tag{7.103}$

$$t + rx = t_n + rx_m$$

They meet the initial line $t = 0$ at $(x_m - t_n/r, 0)$ and $(x_m + t_n/r, 0)$ whose distance is $2t_n/r$. This distance depends on the mesh lengths k and h. In the case of the analytic solution, this distance is $2t_n$. If $r > 1$, that is when the stability condition is violated, then the domain of dependence of the solution at (x_m, t_n) lies inside the domain of dependence of the analytic solution. The data between these two regions is not being utilized in finding the difference solution. In such cases, the convergence as $h \to 0$, $k \to 0$ of the difference solution to the exact solution is not possible. If $r \leqslant 1$, then the domain of dependence of the difference solution is either larger or equal to the domain of dependence of the analytic solution. The difference solutions are obtained using data more or equal to the initial data as used in the case of the analytic solution. The contribution due to the extra initial data tends to zero as $h \to 0$, $k \to 0$, r remaining constant. This stability condition $r \leq 1$, for all explicit formulas, is known as **Courant, Friedrichs and Lewy (CFL) condition**.

If we replace u_m^n by a weighted sum

$$\theta u_m^{n+1} + (1 - 2\theta) u_m^n + \theta u_m^{n-1}$$

on the right hand side of (7.96), then we get

$$\delta_t^2 u_m^n = r^2 \delta_x^2 [\theta u_m^{n+1} + (1 - 2\theta) u_m^n + \theta u_m^{n-1}] \tag{7.104}$$

where $0 \leq \theta \leq 1$. The difference scheme is of order $(k^2 + h^2)$.

For studying the stability, we substitute

$$u_m^n = A \xi^n e^{i\beta m h} \tag{7.105}$$

in (7.104). Simplifying, we get

$$\xi^2 - 2B\xi + 1 = 0 \tag{7.106}$$

where

$$B = 1 - \frac{2r^2 \sin^2 \frac{\beta h}{2}}{1 + 4\theta r^2 \sin^2 \frac{\beta h}{2}}$$

The stability condition $|\xi| \leq 1$ requires that $|B| \leq 1$. We find

(i) $\theta \geq \frac{1}{4}$, unconditional stability

(ii) $0 < \theta < \frac{1}{4}$, $0 < r^2 \leq \frac{1}{1 - 4\theta}$, conditional stability

The value $\theta = \frac{1}{4}$ gives the **Von-Neumann method**

$$(1 - \tfrac{1}{4} r^2 \delta_x^2) \delta_t^2 u_m^n = r^2 \delta_x^2 u_m^n \tag{7.107}$$

Example 7.7 Find the solution of the initial boundary value problem

$$\frac{\partial^2 u}{\partial t^2} = \frac{\partial^2 u}{\partial x^2}, \quad 0 \leq x \leq 1$$

subject to the initial conditions

$$u(x, 0) = \sin \pi x, \quad 0 \leq x \leq 1$$

$$\frac{\partial u}{\partial t}(x, 0) = 0, \quad 0 \leq x \leq 1$$

and the boundary conditions

$$u(0, t) = 0, \quad u(1, t) = 0, \quad t > 0$$

by using

   (i) the explicit scheme (7.97)

   (ii) the implicit scheme (7.104), $\theta = \frac{1}{2}$

Assume $h = \frac{1}{4}$, $r = \frac{3}{4}$ and integrate for 5 time steps. Compare with the exact solution

$$u(x, t) = \sin \pi x \cos \pi t$$

For $h = \frac{1}{4}$, $r = \frac{3}{4}$, we have $k = rh = \frac{3}{16}$. The nodal points are given by

$$x_m = mh, \quad m = 0, 1, 2, 3, 4$$
$$t_n = nk, \quad n = 0, 1, 2, \ldots$$

The initial conditions give

and
$$u_m^0 = \sin\left(\frac{\pi m}{4}\right)$$
$$u_m^{-1} = u_m^1$$
, $m = 0, 1, 2, 3, 4$

The boundary conditions give

$$u_m^n = 0, \quad m = 0, 4$$

(i) The explicit method (7.97) gives

$$u_m^{n+1} = \tfrac{7}{8} u_m^n + \tfrac{9}{16}(u_{m-1}^n + u_{m+1}^n) - u_m^{n-1}, \quad m = 1, 2, 3$$

For $n = 0$, we have

$$u_m^1 = \tfrac{7}{8} u_m^0 + \tfrac{9}{16}(u_{m-1}^0 + u_{m+1}^0) - u_m^{-1}$$

Hence,

$$u_m^1 = \tfrac{7}{16} u_m^0 + \tfrac{9}{32}(u_{m-1}^0 + u_{m+1}^0)$$

$m = 1$, $u_1^1 = \tfrac{7}{16} u_1^0 + \tfrac{9}{32}(u_0^0 + u_2^0) = 0.59061$

$m = 2$, $u_2^1 = \tfrac{7}{16} u_2^0 + \tfrac{9}{32}(u_1^0 + u_3^0) = 0.83525$

$m = 3$, $u_3^1 = \tfrac{7}{16} u_3^0 + \tfrac{9}{32}(u_2^0 + u_4^0) = 0.59061$

For $n \geqslant 1$, we have

$$u_m^{n+1} = \tfrac{7}{8} u_m^n + \tfrac{9}{16}(u_{m-1}^n + u_{m+1}^n) - u_m^{n-1}$$

The solutions are obtained as

$u_1^2 = u_3^2 = 0.27951, \quad u_2^2 = 0.39528$

$u_1^3 = u_3^3 = -0.12369, \quad u_2^3 = -0.17493$

$u_1^4 = u_3^4 = -0.48614, \quad u_2^4 = -0.68750$

$u_1^5 = u_3^5 = -0.68840, \quad u_2^5 = -0.97354$

The exact solution is

$$u(\tfrac{1}{4}, \tfrac{15}{16}) = u(\tfrac{3}{4}, \tfrac{15}{16}) = -0.69352, \quad u(\tfrac{1}{2}, \tfrac{15}{16}) = -0.98079$$

(ii) The implicit scheme (7.104) for $\theta = \frac{1}{2}$ gives

$$-\tfrac{9}{32}u_{m-1}^{n+1} + \tfrac{25}{16}u_m^{n+1} - \tfrac{9}{32}u_{m+1}^{n+1} = 2u_m^n + \tfrac{9}{32}u_{m-1}^{n-1} - \tfrac{25}{16}u_m^{n-1} + \tfrac{9}{32}u_{m+1}^{n-1}$$

$$m = 1, 2, 3$$

For $n = 0$, we get

$$-\tfrac{9}{16}u_{m-1}^1 + \tfrac{25}{8}u_m^1 - \tfrac{9}{16}u_{m+1}^1 = 2u_m^0$$

For $m = 1, 2, 3$, we have the system of equations

$$\begin{bmatrix} \tfrac{25}{8} & -\tfrac{9}{16} & 0 \\ -\tfrac{9}{16} & \tfrac{25}{8} & -\tfrac{9}{16} \\ 0 & -\tfrac{9}{16} & \tfrac{25}{8} \end{bmatrix} \begin{bmatrix} u_1^1 \\ u_2^1 \\ u_3^1 \end{bmatrix} = \begin{bmatrix} 2u_1^0 \\ 2u_2^0 \\ 2u_3^0 \end{bmatrix}$$

whose solution is

$$u_1^1 = u_3^1 = 0.60709, \quad u_2^1 = 0.85855$$

The solutions at the other time levels are

$$u_1^2 = u_3^2 = 0.33533, \quad u_2^2 = 0.47422$$
$$u_1^3 = u_3^3 = -0.03130, \quad u_2^3 = -0.04426$$
$$u_1^4 = u_3^4 = -0.38907, \quad u_2^4 = -0.55022$$
$$u_1^5 = u_3^5 = -0.63677, \quad u_2^5 = -0.90053$$

7.4.2 Two Space Dimensions

Consider the two dimensional wave equation

$$\frac{\partial^2 u}{\partial t^2} = \frac{\partial^2 u}{\partial x^2} + \frac{\partial^2 u}{\partial y^2} \tag{7.108}$$

subject to the initial conditions

$$u(x, y, 0) = f(x, y), \quad 0 \leqslant x, y \leqslant 1$$

$$\frac{\partial u}{\partial t}(x, y, 0) = g(x, y), \quad 0 \leqslant x, y \leqslant 1$$

and the boundary conditions prescribed on the boundary of the unit square.

An explicit difference scheme can be written as

$$\delta_t^2 u_{l,m}^n = r^2(\delta_x^2 + \delta_y^2) u_{l,m}^n$$

or

$$u_{l,m}^{n+1} = 2(1 - 2r^2)u_{l,m}^n + r^2(u_{l-1,m}^n + u_{l,m-1}^n + u_{l,m+1}^n + u_{l+1,m}^n) - u_{l,m}^{n-1} \tag{7.109}$$

It is easily verified that the method is of order $(k^2 + h^2)$.
Substituting

$$u_{l,m}^n = A\xi^n e^{i\beta_1 l h} e^{i\beta_2 m h} \tag{7.110}$$

in (7.109) and simplifying, we get

$$\xi^2 - 2[1 - 2r^2(\sin^2\theta_1 + \sin^2\theta_2)]\xi + 1 = 0 \tag{7.111}$$

where $\theta_1 = \beta_1 h/2$ and $\theta_2 = \beta_2 h/2$.
Solving (7.111), we have

$$\xi_{1,2} = B \pm \sqrt{B^2 - 1} \tag{7.112}$$

where $B = 1 - 2r^2(\sin^2\theta_1 + \sin^2\theta_2)$.
If $|B| > 1$, then $|\xi_1| > 1$ and the method is unstable. If $|B| < 1$, then $\xi_{1,2}$ are a complex pair and $|\xi_{1,2}| = 1$. The stability condition becomes

$$-1 \leqslant 1 - 2r^2(\sin^2\theta_1 + \sin^2\theta_2) \leqslant 1$$

or

$$0 < r \leqslant \frac{1}{\sqrt{2}} \tag{7.113}$$

The implicit difference schemes for solving (7.108) can be obtained by generalizing the implicit scheme (7.104) in one dimension. We have

$$\delta_t^2 u_{l,m}^n = r^2[\theta(\delta_x^2 + \delta_y^2)u_{l,m}^{n+1} + (1 - 2\theta)(\delta_x^2 + \delta_y^2)u_{l,m}^n + \theta(\delta_x^2 + \delta_y^2)u_{l,m}^{n-1}]$$

or

$$[1 - \theta r^2(\delta_x^2 + \delta_y^2)]\delta_t^2 u_{l,m}^n = r^2(\delta_x^2 + \delta_y^2)u_{l,m}^n \tag{7.114}$$

It can be verified that the order of method (7.114) is $(k^2 + h^2)$.
Substituting (7.110) in (7.114) and simplifying, we get

$$\xi^2 - 2\xi\left[1 - \frac{2r^2(\sin^2\theta_1 + \sin^2\theta_2)}{1 + 4\theta r^2(\sin^2\theta_1 + \sin^2\theta_2)}\right] + 1 = 0$$

The stability condition $|\xi| \leqslant 1$ is satisfied, if

$$-1 \leqslant 1 - \frac{4r^2}{1 + 8\theta r^2} \leqslant 1 \tag{7.115}$$

This gives

(i) $\theta \geqslant \frac{1}{4}$, unconditional stability

(ii) $0 < \theta < \frac{1}{4}$, $r^2 \leqslant \frac{1}{2(1 - 4\theta)}$, conditional stability
$$\tag{7.116}$$

Alternating Direction Implicit (ADI) Methods

We now construct ADI methods using the implicit methods (7.114). We write (7.114) as

$$[1 - \theta r^2 \delta_x^2][1 - \theta r^2 \delta_y^2]\delta_t^2 u_{l,m}^n = r^2(\delta_x^2 + \delta_y^2)u_{l,m}^n \tag{7.117}$$

The addition of the term $\theta^2 r^4 \delta_x^2 \delta_y^2 \delta_t^2 u_{l,m}^n$ to the left hand side of (7.114) does not affect the order of the method as it is of higher order than that of the difference scheme. The order of the method (7.117) is same as that of (7.114). It can be verified that the stability condition is satisfied if

348 *Numerical Methods*

$$-1 \leqslant 1 - \frac{4r^2}{(1+4\theta r^2)^2} \leqslant 1 \qquad (7.118)$$

The right hand inequality is trivially satisfied.
The left hand inequality is satisfied if

$$1 + 2r^2(4\theta - 1) + 16\theta^2 r^4 \geqslant 0$$

or

$$[1 + r^2(4\theta - 1)]^2 + r^4(8\theta - 1) \geqslant 0$$

Hence, we have

(i) $\theta > \frac{1}{8}$, unconditional stability

(ii) $0 < \theta \leqslant \frac{1}{8}$, $r^2 \leqslant \frac{1}{(1 - 4\theta) + \sqrt{(1 - 8\theta)}}$, conditional stability
$\qquad (7.119)$

We may write the scheme (7.117) as a two step method in the following forms:

(i) **D'yakonov Split form**

$$(1 - \theta r^2 \delta_x^2) u_{l,m}^{*n+1} = r^2(\delta_x^2 + \delta_y^2) u_{l,m}^n + (1 - \theta r^2 \delta_x^2)(1 - \theta r^2 \delta_y^2)$$
$$\times (2u_{l,m}^n - u_{l,m}^{n-1}) \qquad (7.120)$$

$$(1 - \theta r^2 \delta_y^2) u_{l,m}^{n+1} = u_{l,m}^{*n+1} \qquad (7.121)$$

(ii) **First Lees ADI method**

$$(1 - \theta r^2 \delta_x^2) u_{l,m}^{*n+1} = r^2(\delta_x^2 + \delta_y^2) u_{l,m}^n + (1 - \theta r^2 \delta_x^2)(2u_{l,m}^n - u_{l,m}^{n-1}) \quad (7.122)$$

$$(1 - \theta r^2 \delta_y^2) u_{l,m}^{n+1} = u_{l,m}^{*n+1} - 2\theta r^2 \delta_y^2 u_{l,m}^n + \theta r^2 \delta_y^2 u_{l,m}^{n-1} \qquad (7.123)$$

(iii) **Second Lees ADI method**

$$(1 - \theta r^2 \delta_x^2) u_{l,m}^{*n+1} = r^2(\delta_x^2 + \delta_y^2) u_{l,m}^n \qquad (7.124)$$

$$(1 - \theta r^2 \delta_y^2) u_{l,m}^{n+1} = u_{l,m}^{*n+1} + (1 - \theta r^2 \delta_y^2)(2u_{l,m}^n - u_{l,m}^{n-1}) \qquad (7.125)$$

where $u_{l,m}^{*n+1}$ are called the intermediate values.

For the solution of the system of equations given by (7.120) or (7.122) or (7.124), we require boundary conditions for $u_{l,m}^{*n+1}$. These conditions are obtained from (7.121) or (7.123) or (7.125) respectively. The boundary conditions for the solution of the equations obtained from (7.121) or (7.123) or (7.125) are given by the prescribed boundary conditions. For the implementation of the ADI methods we require the solution on the zeroth and first time levels. The solutions on the zeroth level are obtained from the prescribed initial condition, $u(x, y, 0) = f(x, y)$. The solutions on the first time level may be obtained as follows:

$$u_{l,m}^1 = u_{l,m}^0 + k\left(\frac{\partial u}{\partial t}\right)_{l,m}^0 + \frac{k^2}{2}\left(\frac{\partial^2 u}{\partial t^2}\right)_{l,m}^0 + O(k^3)$$

$$= u_{l,m}^0 + k g_{l,m} + \frac{k^2}{2}\left(\frac{\partial^2 u}{\partial x^2} + \frac{\partial^2 u}{\partial y^2}\right)_{l,m}^0 + O(k^3).$$

A difference approximation of order $(k^2 + h^2)$ is given by

$$u^1_{l,m} = u^0_{l,m} + kg_{l,m} + \frac{r^2}{2}(\delta_x^2 + \delta_y^2)u^0_{l,m}$$

Example 7.8 Solve the initial boundary value problem

$$\frac{\partial^2 u}{\partial t^2} = \frac{\partial^2 u}{\partial x^2} + \frac{\partial^2 u}{\partial y^2}$$

$$u(x, y, 0) = \sin \pi x \sin \pi y$$

$$\frac{\partial u}{\partial t}(x, y, 0) = 0, \quad 0 \leqslant x, y \leqslant 1$$

$$u(x, y, t) = 0, \quad \text{on the boundary, } t \geqslant 0$$

using the D'yakanov split form

$$\left(1 - \frac{r^2}{2}\delta_x^2\right)u^{*n+1}_{l,m} = r^2(\delta_x^2 + \delta_y^2)u^n_{l,m} + \left(1 - \frac{r^2}{2}\delta_x^2\right)\left(1 - \frac{r^2}{2}\delta_y^2\right)$$

$$\times (2u^n_{l,m} - u^{n-1}_{l,m})$$

$$\left(1 - \frac{r^2}{2}\delta_y^2\right)u^{n+1}_{l,m} = u^{*n+1}_{l,m}$$

with $h = \frac{1}{3}$ and $r = \frac{1}{3}$. Perform the integration for one time step.

The nodal points are

$$x_l = lh, \quad y_m = mh, \quad t_n = nk, \quad 0 \leqslant l, m \leqslant 3, \quad n = 0, 1$$

The initial and boundary conditions become

$$u^0_{l,m} = \sin \pi lh \sin \pi mh = \sin \frac{\pi l}{3} \sin \frac{\pi m}{3}$$

$$u^n_{0,m} = u^n_{l,0} = 0, \quad 0 \leqslant l, m \leqslant 3$$

We also have

$$u^1_{l,m} = u^0_{l,m} + k \times 0 + \frac{r^2}{2}(\delta_x^2 + \delta_y^2)u^0_{l,m}$$

$$u^1_{l,m} = u^0_{l,m} + \tfrac{1}{18}(u^0_{l-1,m} + u^0_{l+1,m} + u^0_{l,m-1} + u^0_{l,m+1} - 4u^0_{l,m})$$

which gives

$$u^1_{1,1} = u^0_{1,1} + \tfrac{1}{18}(u^0_{0,1} + u^0_{2,1} + u^0_{1,0} + u^0_{1,2} - 4u^0_{1,1})$$

$$= \tfrac{3}{4} + \tfrac{1}{18}(0 + \tfrac{3}{4} + 0 + \tfrac{3}{4} - 4 \times \tfrac{3}{4}) = \tfrac{2}{3}$$

$$u^1_{1,2} = u^1_{2,1} = u^1_{2,2} = \tfrac{2}{3}$$

For $n = 1$, we get

$$(1 - \tfrac{1}{18}\delta_x^2)u^{*2}_{l,m} = \tfrac{1}{9}(\delta_x^2 + \delta_y^2)u^1_{l,m} + (1 - \tfrac{1}{18}\delta_x^2)(1 - \tfrac{1}{18}\delta_y^2)$$

$$\times (2u^1_{l,m} - u^0_{l,m})$$

$$(1 - \tfrac{1}{18}\delta_y^2)u^2_{l,m} = u^{*2}_{l,m}$$

The intermediate boundary conditions are obtained as

$$u_{l,m}^{*2} = (1 - \tfrac{1}{18}\delta_y^2)0 = 0, \quad l = 0, 3$$

We have

$$-\frac{1}{18}u_{l-1,m}^{*2} + \frac{10}{9}u_{l,m}^{*2} - \frac{1}{18}u_{l+1,m}^{*2} = \frac{1}{162}\Big[328 u_{l,m}^1 + u_{l-1,m-1}^1$$
$$+ u_{l+1,m-1}^1 + u_{l-1,m+1}^1 + u_{l+1,m+1}^1 - 2(u_{l,m-1}^1 + u_{l-1,m}^1 + u_{l+1,m}^1$$
$$+ u_{l,m+1}^1)\Big] - \frac{1}{324}[400 u_{l,m}^0 - 20(u_{l-1,m}^0 + u_{l+1,m}^0 + u_{l,m-1}^0 + u_{l,m+1}^0)$$
$$+ (u_{l-1,m-1}^0 + u_{l+1,m-1}^0 + u_{l-1,m+1}^0 + u_{l+1,m+1}^0)]$$

$$l = 1, m = 1; \quad \frac{10}{9}u_{1,1}^{*2} - \frac{1}{18}u_{2,1}^{*2} = \frac{1951}{3888}$$

$$l = 2, m = 1; \quad -\frac{1}{18}u_{1,1}^{*2} + \frac{10}{9}u_{2,1}^{*2} = \frac{1951}{3888}$$

This gives

$$u_{1,1}^{*2} = u_{2,1}^{*2} = \frac{1951}{4104}$$

We also find

$$u_{1,2}^{*2} = u_{2,2}^{*2} = \frac{1951}{4104}$$

The second set of equations become

$$-\frac{1}{18}u_{l,m-1}^2 + \frac{10}{9}u_{l,m}^2 - \frac{1}{18}u_{l,m+1}^2 = u_{l,m}^{*2}$$

we obtain

$$l = 1, m = 1; \quad \frac{10}{9}u_{1,1}^2 - \frac{1}{18}u_{1,2}^2 = \frac{1951}{4104}$$

$$l = 1, m = 2; \quad -\frac{1}{18}u_{1,1}^2 + \frac{10}{9}u_{1,2}^2 = \frac{1951}{4104}$$

We get

$$u_{1,1}^2 = u_{1,2}^2 = \frac{1951}{4332}$$

Similarly, we find

$$u_{2,1}^2 = u_{2,2}^2 = \frac{1951}{4332}$$

7.4.3 First Order Equation

Consider the first order hyperbolic equation

$$\frac{\partial u}{\partial t} + \frac{\partial u}{\partial x} = 0 \tag{7.126}$$

subject to the initial condition $u(0, x) = f(x)$.

The characteristic of this hyperbolic equation is
$$\xi = x - t = \text{constant} \tag{7.127}$$
which is a straight line inclined to the x axis at an angle $\pi/4$.

The analytical solution is
$$u(x, t) = f(x - t) \tag{7.128}$$

Superimpose on the given region a rectangular grid with mesh spacing h and k in the x and t directions respectively. The nodal points are given by
$$x_m = mh, \quad m = 0, 1, 2, \ldots$$
$$t_n = nk, \quad n = 0, 1, 2, \ldots$$

We now consider the approximations

$$\frac{\partial u}{\partial x} = \begin{vmatrix} \frac{1}{h} \Delta_x u + O(h) & (7.129) \\ \frac{1}{h} \nabla_x u + O(h) & (7.130) \\ \frac{1}{h} \mu_x \delta_x u + O(h^2) & (7.131) \end{vmatrix}$$

Explicit difference formulas

Using the forward difference approximation for the derivative $\partial u/\partial t$ and (7.129) at (x_m, t_n) we get
$$\Delta_t u_m^n = -r \Delta_x u_m^n$$
or
$$u_m^{n+1} = (1 + r)u_m^n - r u_{m+1}^n \tag{7.132}$$

where $r = k/h$. Expanding (7.132) in Taylor series we find that the order of the formula is $(k + h)$. Substituting
$$u_m^n = A\xi^n e^{i\beta mh} \tag{7.133}$$

in (7.132) and simplifying, we obtain
$$\xi = (1 + r) - re^{i\beta h} = (1 + r - r\cos\beta h) - ir \sin \beta h$$
$$|\xi|^2 = (1 + r - r\cos\beta h)^2 + r^2 \sin^2 \beta h$$
$$= 1 + 4r(1 + r) \sin^2 \frac{\beta h}{2}$$

Hence $|\xi| > 1$ always and the scheme (7.132) is unstable. This instability can be explained by the Courant-Friedrichs-Levy (CFL) condition as discussed in the previous section. The CFL condition requires that the characteristic through P, which is making an angle $\pi/4$ with the x-axis, must intersect the previous mesh line (level n) within the range of points considered by the formula at the previous time level. In the present case, the information between the points A and B on the level n is not being used and hence convergence cannot take place as $h \to 0$, r being fixed.

Using the approximation (7.130) at (x_m, t_n) we get

$$\Delta_t u_m^n = -r \nabla_x u_m^n \tag{7.134}$$

or

$$u_m^{n+1} = (1 - r)u_m^n + r u_{m-1}^n \tag{7.135}$$

The difference scheme (7.135) is of order $(k + h)$. Substituting

$$u_m^n = A\xi^n e^{i\beta mh}$$

in (7.135) and simplifying, we get

$$\xi = (1 - r) + r e^{-i\beta h}$$

$$|\xi|^2 = (1 - r + r \cos \beta h)^2 + r^2 \sin^2 \beta h$$

$$= 1 - 4r(1 - r) \sin^2 \frac{\beta h}{2} \tag{7.136}$$

Maximizing $\sin^2 \frac{\beta h}{2}$, we get $|\xi|^2 = (1 - 2r)^2$. Hence $|\xi| \leqslant 1$ for $0 < r \leqslant 1$. The scheme (7.135) is stable for $0 < r \leqslant 1$. The characteristic AP at P falls inside the nodes on the nth time level. We find

$$\tan \frac{\pi}{4} = \frac{k}{AB}$$

or

$$k = AB \tag{7.137}$$

The CFL condition requires $0 \leqslant AB \leqslant h$ or $0 \leqslant k \leqslant h$ or $0 \leqslant r \leqslant 1$. This is the maximum possible stability region for a two level explicit formula according to the CFL condition.

Consider the Taylor expansion

$$u_m^{n+1} = u_m^n + k \left.\frac{\partial u}{\partial t}\right|_m^n + \frac{k^2}{2} \left.\frac{\partial^2 u}{\partial t^2}\right|_m^n + \cdots$$

Neglecting the higher order terms, we may write

$$u_m^{n+1} = u_m^n - k \left.\frac{\partial u}{\partial x}\right|_m^n + \frac{k^2}{2} \left.\frac{\partial^2 u}{\partial x^2}\right|_m^n \tag{7.138}$$

Using the approximations (7.131) and (7.138), we get

$$u_m^{n+1} = u_m^n - r \mu_x \delta_x u_m^n + \frac{r^2}{2} \delta_x^2 u_m^n$$

or

$$u_m^{n+1} = u_m^n - \frac{r}{2}(u_{m+1}^n - u_{m-1}^n) + \frac{r^2}{2}(u_{m+1}^n - 2u_m^n + u_{m-1}^n) \tag{7.139}$$

The schematic representation is given in Fig. 7.11.

The equation (7.139) is called the **Lax-Wendroff** formula. The order of the formula (7.139) is $(k^2 + h^2)$. It is the most commonly used formula

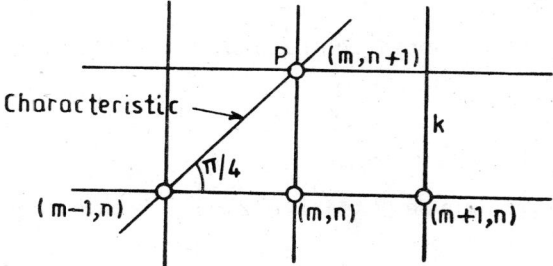

Fig. 7.11 Schematic representation of (7.139)

for the solution of first order hyperbolic equations. Substituting

$$u_m^n = A\xi^n e^{i\beta mh}$$

in (7.139) and simplifying, we get

$$\xi = (1 - r^2) - ir \sin \beta h + r^2 \cos \beta h \tag{7.140}$$

Hence,

$$|\xi|^2 = \left(1 - 2r^2 \sin^2 \frac{\beta h}{2}\right)^2 + r^2 \sin^2 \beta h$$

$$= 1 - 4r^2(1 - r^2) \sin^4 \frac{\beta h}{2}$$

Maximizing $\sin \frac{\beta h}{2}$, we get $|\xi|^2 = (1 - 2r^2)^2$. Hence, $|\xi| \leqslant 1$ for $0 < r \leqslant 1$ and the scheme (7.139) is stable for $0 < r \leqslant 1$. Note that the characteristic at P intersects the previous mesh line within the nodes being used on that line and the CFL condition is satisfied.

If we approximate

$$\frac{\partial u}{\partial t} = \frac{1}{k} \mu_t \delta_t u + 0(k^2)$$

and substitute (7.131) in (7.126) at the nodal point (x_m, t_n), we get

$$\mu_t \delta_t u_m^n = -r \mu_x \delta_x u_m^n$$

or

$$u_m^{n+1} = u_m^{n-1} - r(u_{m+1}^n - u_{m-1}^n) \tag{7.141}$$

This is a three level explicit formula of order $(k^2 + h^2)$. The scheme (7.141) is called the **leap frog scheme**. The scheme is stable for $0 < r \leqslant 1$ and coincides with the CFL condition.

Implicit schemes

Implicit difference schemes can only be applied to initial boundary value problems. Consider the solution of (7.126) in the quarter plane $0 \leqslant x < \infty$, $t \geqslant 0$ subject to the boundary condition

$$u(0, t) = g(t) \tag{7.142}$$

Using the approximations (7.20ii) and (7.130) at (x_m, t_{n+1}) we get

$$u_m^{n+1} = u_m^n - r(u_m^{n+1} - u_{m-1}^{n+1})$$

or

$$(1 + r)u_m^{n+1} - ru_{m-1}^{n+1} = u_m^n \tag{7.143}$$

Even though the formula (7.143) is implicit, it is effectively explicit, since (7.143) can be written as

$$u_m^{n+1} = \frac{1}{1+r}(u_m^n + ru_{m-1}^{n+1}) \tag{7.144}$$

Since the boundary condition on $x = 0$ is prescribed, the solution u_{m-1}^{n+1} is known from the boundary condition when the solution u_m^{n+1} is being determined. Proceeding to the right, all the solutions on this mesh line are determined explicitly. The amplification factor of (7.143) is

$$\xi = \frac{1}{(1+r) - re^{-i\beta h}} = \frac{1}{\left(1 + 2r\sin^2\frac{\beta h}{2}\right) + ir\sin\beta h}$$

$$|\xi|^2 = \frac{1}{\left(1 + 2r\sin^2\frac{\beta h}{2}\right)^2 + r^2\sin^2\beta h} \leq 1$$

Hence, the scheme (7.143) is unconditionally stable. The Crank-Nicolson type formula can be used if boundary conditions are given at $x = 0$ as well as at another boundary, say $x = 1$. However, this problem is over determined, because theoretically the solutions on $x = 1$ can be obtained from the initial condition and the boundary condition at $x = 0$. Such a formula should be used with great care.

Example 7.9 Find the solution of

$$\frac{\partial u}{\partial t} + \frac{\partial u}{\partial x} = 0$$

subject to the initial condition

$$u(x, 0) = 0, \quad x < 0$$
$$= x, \quad 0 \leq x \leq 1$$
$$= 2 - x, \quad 1 \leq x \leq 2$$
$$= 0, \quad x > 2$$

using the Lax-Wendroff formula. Assume $h = \frac{1}{2}$ and $r = \frac{1}{2}$ and compute upto two time steps.

The Lax-Wendroff formula becomes

$$u_m^{n+1} = \tfrac{3}{4}u_m^n - \tfrac{1}{8}u_{m+1}^n + \tfrac{3}{8}u_{m-1}^n$$

Since $u(0, x) = 0$ for $x < 0$, it is sufficient to start using the formula with $n = 0$ and $x = 0$.

For $n = 0$, we have

$$u_m^1 = \tfrac{3}{4}u_m^0 - \tfrac{1}{8}u_{m+1}^0 + \tfrac{3}{8}u_{m-1}^0$$

For $m = 1, 2, \ldots, 6$, we obtain

x_m	0	$\tfrac{1}{2}$	1	$\tfrac{3}{2}$	2	$\tfrac{5}{2}$
u_m^1	$-\tfrac{1}{16}$	$\tfrac{1}{4}$	$\tfrac{7}{8}$	$\tfrac{3}{4}$	$\tfrac{3}{16}$	0

On the second time level, that is $n = 1$, we have

$$u_m^2 = \tfrac{3}{4}u_m^1 - \tfrac{1}{8}u_{m+1}^1 + \tfrac{3}{8}u_{m-1}^1$$

We find the solutions as

x_m	$-\tfrac{1}{2}$	0	$\tfrac{1}{2}$	1	$\tfrac{3}{2}$	2	$\tfrac{5}{2}$	3
u_m^2	$\tfrac{1}{128}$	$-\tfrac{5}{64}$	$\tfrac{7}{128}$	$\tfrac{21}{32}$	$\tfrac{111}{128}$	$\tfrac{27}{64}$	$-\tfrac{9}{128}$	0

7.4.4 System of Equations

Consider the system of equations

$$\frac{\partial \mathbf{u}}{\partial t} + \mathbf{A}\frac{\partial \mathbf{u}}{\partial x} = 0 \tag{7.145}$$

where $\mathbf{u} = [u_1 \; u_2 \ldots u_N]^T$ is an N-component vector and $\mathbf{A}$ a constant $N \times N$ matrix with real and distinct eigenvalues and N linearly independent eigenvectors. The explicit schemes (7.132), (7.135), (7.139) and (7.141), and the implicit scheme (7.143) may easily be written for (7.145). For example, the Lax-Wendroff difference scheme (7.139) becomes

$$\mathbf{u}_m^{n+1} = \mathbf{u}_m^n - \tfrac{1}{2}r\mathbf{A}(\mathbf{u}_{m+1}^n - 2\mathbf{u}_{m-1}^n) + \tfrac{1}{2}r^2\mathbf{A}^2(\mathbf{u}_{m+1}^n - 2\mathbf{u}_m^n + \mathbf{u}_{m-1}^n) \tag{7.146}$$

Further the Lax-Wendroff scheme may also be written as a two-step method

(i) $\quad \mathbf{u}_m^{n+1} = \tfrac{1}{2}(\mathbf{u}_{m+1}^n + \mathbf{u}_{m-1}^n) - \tfrac{1}{2}r\mathbf{A}(\mathbf{u}_{m+1}^n - \mathbf{u}_{m-1}^n)$ \hfill (7.147)

(ii) $\quad \mathbf{u}_m^{n+2} = \mathbf{u}_m^n - r\mathbf{A}(\mathbf{u}_{m+1}^{n+1} - \mathbf{u}_{m-1}^{n+1})$ \hfill (7.148)

Substituting (7.147) in (7.148) we get the composite scheme

$$\mathbf{u}_m^{n+2} = \mathbf{u}_m^n - \tfrac{1}{2}r\mathbf{A}(\mathbf{u}_{m+2}^n - \mathbf{u}_{m-2}^n) + \tfrac{1}{2}r^2\mathbf{A}^2(\mathbf{u}_{m+2}^n - 2\mathbf{u}_m^n + \mathbf{u}_{m-2}^n)$$

which is exactly the Lax-Wendroff scheme (7.146) for a network with spacing $2h$ and $2k$. The procedure (7.147) and (7.148) is called the **Richtmyer two-step method**. The Lax-Wendroff scheme (7.146) is of order $(k^2 + h^2)$ and is stable if $r\lambda \leqslant 1$, where $\lambda = \rho(\mathbf{A})$.

The hyperbolic equation (7.91) can also be written as a system of two first order equations and then difference schemes to these first order equations can be written. We may write (7.91) as

$$\frac{\partial v}{\partial t} = \frac{\partial w}{\partial x} \tag{7.149}$$

and

$$\frac{\partial w}{\partial t} = \frac{\partial v}{\partial x} \tag{7.150}$$

where $w = \partial u/\partial x$ and $v = \partial u/\partial t$. The initial conditions for w and v become

$$w(x, 0) = f'(x) \quad \text{and} \quad v(x, 0) = g(x)$$

Let a difference scheme be written as

$$\Delta_t v_m^n = r\delta_x w_m^n$$
$$\nabla_t w_{m-\frac{1}{2}}^{n+1} = r\delta_x v_{m-\frac{1}{2}}^{n+1}$$

or

$$v_m^{n+1} = v_m^n + r(w_{m+\frac{1}{2}}^n - w_{m-\frac{1}{2}}^n) \tag{7.151}$$

$$w_{m-\frac{1}{2}}^{n+1} = w_{m-\frac{1}{2}}^n + r(v_m^{n+1} - v_{m-1}^{n+1}) \tag{7.152}$$

This is a set of explicit schemes of first order. When v_m^{n+1} are determined from (7.151), these values are used on the right hand side of (7.152). Substituting

$$v_m^n = v_0^n e^{i\beta mh} \tag{7.153}$$

and

$$w_m^n = w_0^n e^{i\beta mh} \tag{7.154}$$

in (7.151) and (7.152) and simplifying, we get

$$v_0^{n+1} = v_0^n + ipw_0^n$$
$$w_0^{n+1} = w_0^n + ipv_0^{n+1} = w_0^n + ip(v_0^n + ipw_0^n)$$
$$= ipv_0^n + (1 - p^2)w_0^n$$

or

$$\begin{bmatrix} v_0^{n+1} \\ w_0^{n+1} \end{bmatrix} = \begin{bmatrix} 1 & ip \\ ip & 1 - p^2 \end{bmatrix} \begin{bmatrix} v_0^n \\ w_0^n \end{bmatrix} \tag{7.155}$$

where $p = 2r \sin \beta h/2$.

The amplification matrix is

$$G(k, \beta) = \begin{bmatrix} 1 & ip \\ ip & 1 - p^2 \end{bmatrix} \tag{7.156}$$

The necessary condition for stability is that the eigenvalues of the amplification matrix **G** be less than or equal to one in magnitude. This condition is called the **Von-Neumann necessary condition** for stability.

The eigenvalues of $G(k, \beta)$ are given by

$$\begin{vmatrix} 1 - \mu & ip \\ ip & 1 - p^2 - \mu \end{vmatrix} = 0$$

or
$$\mu^2 - (2-p^2)\mu + 1 = 0 \tag{7.157}$$

We obtain from (7.157).
$$\mu_{1,2} = \tfrac{1}{2}[(2-p^2) \pm p\sqrt{(p^2-4)}] \tag{7.158}$$

If $p^2 > 4$, $|\mu_1| > 1$ and the difference scheme is unstable. If $p^2 = 4$, then $\mu_{1,2} = -1$. In this case, the scheme may or may not be stable. If $p^2 < 4$, then $\mu_{1,2}$ are the complex pair and $|\mu_{1,2}| = 1$. This gives the stability condition

$$4r^2 \sin^2 \frac{\beta h}{2} < 4$$

or
$$r < 1 \tag{7.159}$$

Conservation form

Here we assume that the matrix $\mathbf{A}$ in (7.145) is the Jacobian matrix of the derivatives of the vector function $\mathbf{f}$ with respect to the components of the unknown vector $\mathbf{u}$. The equation (7.145) becomes

$$\frac{\partial \mathbf{u}}{\partial t} + \mathbf{A}(\mathbf{u}) \frac{\partial \mathbf{u}}{\partial x} = 0 \tag{7.160}$$

which may be written in the form

$$\frac{\partial \mathbf{u}}{\partial t} + \frac{\partial \mathbf{f}(\mathbf{u})}{\partial x} = 0 \tag{7.161}$$

where
$$\mathbf{A}(\mathbf{u}) = \frac{\partial \mathbf{f}}{\partial \mathbf{u}}$$

The equation (7.161) is said to be in conservation form. The Lax-Wendroff method for (7.161) may be derived as follows:

$$\frac{\partial \mathbf{u}}{\partial t} = -\frac{\partial \mathbf{f}(\mathbf{u})}{\partial x}$$

$$\frac{\partial^2 \mathbf{u}}{\partial t^2} = \frac{\partial}{\partial t}\left(-\frac{\partial \mathbf{f}}{\partial x}\right) = -\frac{\partial}{\partial x}\left(\frac{\partial \mathbf{f}}{\partial t}\right)$$

$$= -\frac{\partial}{\partial x}\left(\frac{\partial \mathbf{f}}{\partial \mathbf{u}} \frac{\partial \mathbf{u}}{\partial t}\right)$$

$$= \frac{\partial}{\partial x}\left(\mathbf{A}(\mathbf{u}) \frac{\partial \mathbf{f}}{\partial \mathbf{u}} \frac{\partial \mathbf{u}}{\partial x}\right) = \frac{\partial}{\partial x}\left(\mathbf{A}(\mathbf{u}) \frac{\partial \mathbf{f}}{\partial x}\right) \tag{7.162}$$

We also have

$$\mathbf{u}(x_m, t_{n+1}) = \mathbf{u}(x_m, t_n) + k\left(\frac{\partial \mathbf{u}}{\partial t}\right)_m^n + \tfrac{1}{2}k^2\left(\frac{\partial^2 \mathbf{u}}{\partial t^2}\right)_m^n + \cdots \tag{7.163}$$

Substituting (7.162) into (7.163) we get

$$\mathbf{u}(x_m, t_{n+1}) = \mathbf{u}(x_m, t_n) - k\left(\frac{\partial \mathbf{f}}{\partial x}\right)^n_m + \tfrac{1}{2}k^2\left(\frac{\partial}{\partial x}\left(\mathbf{A}(\mathbf{u})\frac{\partial \mathbf{f}}{\partial x}\right)\right)^n_m + O(k^3)$$

Replacing the derivatives by the difference expressions, the Lax-Wendroff method is given by

$$\mathbf{u}_m^{n+1} = \mathbf{u}_m^n - \tfrac{1}{2}r(\mathbf{f}_{m+1}^n - \mathbf{f}_{m-1}^n) + \tfrac{1}{2}r^2[\mathbf{A}_{m+\frac{1}{2}}^n(\mathbf{f}_{m+1}^n - \mathbf{f}_m^n)$$
$$- \mathbf{A}_{m-\frac{1}{2}}^n(\mathbf{f}_m^n - \mathbf{f}_{m-1}^n)] \tag{7.164}$$

where

$$\mathbf{A}_{m+\frac{1}{2}}^n = \mathbf{A}(\mathbf{u}_{m+\frac{1}{2}}^n) \quad \text{and} \quad \mathbf{f}_m^n = \mathbf{f}(\mathbf{u}_m^n)$$

We further approximate

$$\mathbf{A}_{m+\frac{1}{2}}^n = \tfrac{1}{2}(\mathbf{A}_{m+1}^n + \mathbf{A}_m^n), \quad \mathbf{A}_{m-\frac{1}{2}}^n = \tfrac{1}{2}(\mathbf{A}_m^n + \mathbf{A}_{m-1}^n)$$

When we solve (7.160) as an initial boundary value problem, we require an additional numerical boundary condition at the right boundary. This boundary condition at $m = M$ may be written as

$$\mathbf{u}_M^{n+1} = \mathbf{u}_M^n - \frac{r}{2}(3\mathbf{f}_M^n - 4\mathbf{f}_{M-1}^n + \mathbf{f}_{M-2}^n)$$
$$+ \frac{r^2}{2}[\mathbf{A}_m^n(\mathbf{f}_m^n - \mathbf{f}_{m-1}^n) - \mathbf{A}_{m-1}^n(\mathbf{f}_{m-1}^n - \mathbf{f}_{m-2}^n)] \tag{7.165}$$

Example 7.10 Find the solution to the differential equation

$$\frac{\partial u}{\partial t} + \frac{\partial}{\partial x}(\tfrac{1}{2}u^2) = 0$$

subject to the conditions

$$u(0, t) = 0, \quad u(x, 0) = x, \quad 0 \leqslant x \leqslant 1$$

using the Lax-Wendroff formula. Assume $h = 0.2$, $r = 0.5$ and integrate for one time step.

We have $f(u) = u^2/2$ and $A = u$. The Lax-Wendroff scheme (7.164) with $r = \tfrac{1}{2}$ becomes

$$u_m^{n+1} = u_m^n - \tfrac{1}{8}[(u_{m+1}^n)^2 - (u_{m-1}^n)^2] + \tfrac{1}{32}[(u_{m+1}^n + u_m^n)\{(u_{m+1}^n)^2 - (u_m^n)^2\}$$
$$- (u_m^n + u_{m-1}^n)\{(u_m^n)^2 - (u_{m-1}^n)^2\}]$$

On the right boundary, we get from (7.165)

$$u_M^{n+1} = u_M^n - \tfrac{1}{8}[3(u_M^n)^2 - 4(u_{M-1}^n)^2 + (u_{M-2}^n)^2]$$
$$+ \tfrac{1}{16}[u_M^n\{(u_M^n)^2 - (u_{M-1}^n)^2\} - u_{M-1}^n\{(u_{M-1}^n)^2 - (u_{M-2}^n)^2\}]$$

For $n = 0$, we have

$$u_m^1 = u_m^0 - \tfrac{1}{8}[(u_{m+1}^0)^2 - (u_{m-1}^0)^2] + \tfrac{1}{32}[(u_{m+1}^0 + u_m^0)\{(u_{m+1}^0)^2 - (u_m^0)^2\}$$
$$- (u_m^0 + u_{m-1}^0)\{(u_m^0)^2 - (u_{m-1}^0)^2\}]; \quad m = 1, 2, 3, 4$$

$$u_5^1 = u_5^0 - \tfrac{1}{8}[3(u_5^0)^2 - 4(u_4^0)^2 + (u_3^0)^2] + \tfrac{1}{16}[u_5^0\{(u_5^0)^2 - (u_4^0)^2\}$$
$$- u_4^0\{(u_4^0)^2 - (u_3^0)^2\}]$$

We have

$$u_0^0 = 0, \quad u_1^0 = 0.2, \quad u_2^0 = 0.4, \quad u_3^0 = 0.6, \quad u_4^0 = 0.8 \quad \text{and} \quad u_5^0 = 1$$

Substituting and simplifying, we get

$$u_1^1 = 0.182, \quad u_2^1 = 0.364, \quad u_3^1 = 0.546, \quad u_4^1 = 0.728 \quad \text{and} \quad u_5^1 = 0.9085$$

7.5 ELLIPTIC EQUATIONS

Consider the solution of the linear partial differential equation

$$A\frac{\partial^2 u}{\partial x^2} + C\frac{\partial^2 u}{\partial y^2} + D\frac{\partial u}{\partial x} + E\frac{\partial u}{\partial y} + Fu = F^* \tag{7.166}$$

in the region **R** with boundary $\partial \mathbf{R}$, where A, C, D, E, F and F^* are continuous functions of x and y in $\mathbf{R} + \partial \mathbf{R}$. Assume that $A > 0, C > 0$ and $F \leq 0$ so that the ellipticity condition and the weak max-min principle are satisfied. Elliptic partial differential equations always occur as **boundary value problems**. The boundary conditions can be of the following three types:

(i) The Dirichlet or First Boundary Condition

We have

$$u(x, y) = g(x, y), \quad (x, y) \in \partial \mathbf{R} \tag{7.167}$$

where $g(x, y)$ is a prescribed function which is defined and continuous on $\partial \mathbf{R}$. The condition (7.167) is called the **Dirichlet condition**. If $g(x, y) = 0$ on $\partial \mathbf{R}$ then the boundary condition is called the homogeneous Dirichlet condition.

(ii) The Neumann or Second Boundary Condition

We have

$$\frac{\partial u}{\partial \mathbf{n}} = g(x, y), \quad (x, y) \in \partial \mathbf{R} \tag{7.168}$$

where $g(x, y)$ is a prescribed function defined and continuous on $\partial \mathbf{R}$ and $\mathbf{n}$ is the outwardly directed normal. The condition (7.168) is called the **Neumann condition**. We may also have homogeneous or inhomogeneous Neumann boundary conditions.

(iii) The Third or Mixed Boundary Condition

We may have

$$\frac{\partial u}{\partial \mathbf{n}} + \alpha(x, y)u = g(x, y), \quad (x, y) \in \partial \mathbf{R} \tag{7.169}$$

where $\alpha(x, y) > 0$ and $g(x, y)$ are defined and continuous on $\partial \mathbf{R}$. Equation (7.169) is called the **mixed boundary condition**.

In order to find the numerical solution of (7.166) with appropriate boundary conditions, we superimpose on **R**, a rectangular network with mesh lengths h and k in the x and y directions respectively. The nodal points are given by

$$x_l = x_0 + lh, \quad l = 0, \pm 1, \pm 2, \ldots$$
$$y_m = y_0 + mh, \quad m = 0, \pm 1, \pm 2, \ldots \quad (7.170)$$

Denote the nodal point (x_l, y_m) by (l, m) and the numerical solution at this nodal point by $u_{l,m}$. If all the four neighbouring nodes of a nodal point are inside **R**, then this node is called an **internal node** or **pivot**. Otherwise, it is called a **boundary node** or pivot (see Fig. 7.12).

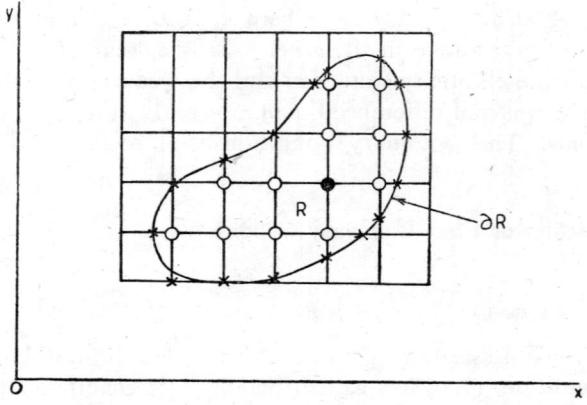

Fig. 7.12 Internal and boundary nodes

If (l, m) is an internal pivot, then we may use the approximations

$$\frac{\partial u}{\partial x} = \frac{1}{h} \mu_x \delta_x u + O(h^2)$$

$$\frac{\partial u}{\partial y} = \frac{1}{k} \mu_y \delta_y u + O(k^2)$$

$$\frac{\partial^2 u}{\partial x^2} = \frac{1}{h^2} \delta_x^2 u + O(h^2)$$

$$\frac{\partial^2 u}{\partial y^2} = \frac{1}{k^2} \delta_y^2 u + O(k^2) \quad (7.171)$$

Substituting (7.171) in (7.166), we get the difference scheme as

$$\frac{1}{h^2}\left[A_{l,m}\delta_x^2 + \frac{h^2}{k^2} C_{l,m}\delta_y^2 + hD_{l,m}\mu_x\delta_x + \frac{h^2}{k} E_{l,m}\mu_y\delta_y \right.$$
$$\left. + h^2 F_{l,m} \right] u_{l,m} = F^*_{l,m}$$

or

$$\frac{1}{h^2}[a_{l,m}u_{l+1,m} + b_{l,m}u_{l-1,m} + c_{l,m}u_{l,m+1} + d_{l,m}u_{l,m-1} - e_{l,m}u_{l,m}] = F^*_{l,m} \quad (7.172)$$

where

$$a_{l,m} = A_{l,m} + \frac{h}{2}D_{l,m}; \quad b_{l,m} = A_{l,m} - \frac{h}{2}D_{l,m}$$

$$c_{l,m} = p^2 C_{l,m} + \frac{ph}{2}E_{l,m}; \quad d_{l,m} = p^2 C_{l,m} - \frac{ph}{2}E_{l,m}$$

$$e_{l,m} = 2A_{l,m} + 2p^2 C_{l,m} - h^2 F_{l,m}; \quad p = \frac{h}{k}$$

The order of the difference scheme (7.172) is $(k^2 + h^2)$. While writing the difference scheme at any internal pivot, if any of the neighbouring points lie on the boundary ∂R, then its value is substituted from the boundary conditions. If the Neumann type or a mixed boundary condition is prescribed at this neighbouring point, then the unknown at this neighbouring point enters the difference scheme.

If the points of intersection of the mesh lines and the boundary of the given region R are not nodal points, then the difference scheme (7.172) needs to be modified. Consider the general case when the mesh lengths on all four sides of the nodal point are different as given in Fig. 7.13. Let the

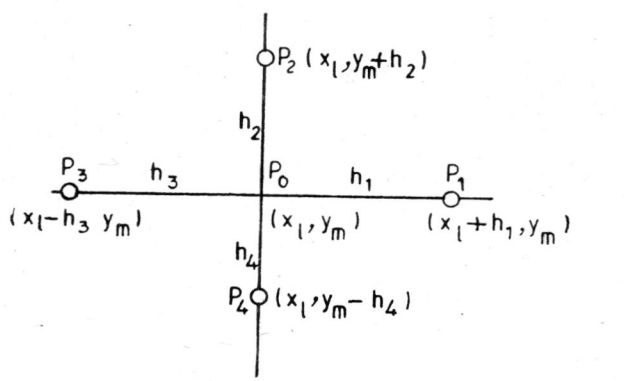

Fig. 7.13 Nodal points with unequal mesh lengths

four neighbouring points be $P_0(x_l, y_m)$, $P_1(x_l + h_1, y_m)$, $P_2(x_l, y_m + h_2)$, $P_3(x_l - h_3, y_m)$ and $P_4(x_l, y_m - h_4)$ where $0 < h_1, h_2, h_3, h_4 \leq h$. We write the difference approximation as

$$\left(A\frac{\partial^2 u}{\partial x^2} + C\frac{\partial^2 u}{\partial y^2} + D\frac{\partial u}{\partial x} + E\frac{\partial u}{\partial y} + Fu\right)_{l,m} = \alpha_0 u(x_l, y_m)$$
$$+ \alpha_1 u(x_l + h_1, y_m) + \alpha_2 u(x_l, y_m + h_2) + \alpha_3 u(x_l - h_3, y_m)$$
$$+ \alpha_4 u(x_l, y_m - h_4) \quad (7.173)$$

Expanding in Taylor series and comparing the various order derivative terms, we get

$$\alpha_0 + \alpha_1 + \alpha_2 + \alpha_3 + \alpha_4 = F_{l,m}$$

$$h_1\alpha_1 - h_3\alpha_3 = D_{l,m}$$

$$h_2\alpha_2 - h_4\alpha_4 = E_{l,m}$$

$$h_1^2\alpha_1 + h_3^2\alpha_3 = 2A_{l,m}$$

$$h_2^2\alpha_2 + h_4^2\alpha_4 = 2C_{l,m}$$

The solution of this system is

$$\alpha_0 = F_{l,m} - \left[\frac{1}{h_1h_3}\{2A_{l,m} + (h_3 - h_1)D_{l,m}\}\right.$$
$$\left. + \frac{1}{h_2h_4}\{2C_{l,m} + (h_4 - h_2)E_{l,m}\}\right]$$

$$\alpha_1 = \frac{2A_{l,m} + h_3 D_{l,m}}{h_1(h_1 + h_3)}; \quad \alpha_2 = \frac{2C_{l,m} + h_4 E_{l,m}}{h_2(h_2 + h_4)}$$

$$\alpha_3 = \frac{2A_{l,m} - h_1 D_{l,m}}{h_3(h_1 + h_3)}; \quad \alpha_4 = \frac{2C_{l,m} - h_2 E_{l,m}}{h_4(h_2 + h_4)}$$

The difference scheme becomes

$$\alpha_0 u(x_l, y_m) + \alpha_1 u(x_l + h_1, y_m) + \alpha_2 u(x_l, y_m + h_2) + \alpha_3 u(x_l - h_3, y_m)$$
$$+ \alpha_4 u(x_l, y_m - h_4) = F^*_{l,m} \tag{7.174}$$

It can be verified that the difference scheme (7.174) is of order (h_i). If $h_1 = h_3 = h$ and $h_2 = h_4 = k$, then (7.174) simplifies to (7.172) whose order of approximation is $(k^2 + h^2)$. When the difference scheme (7.172) or (7.174) is used at all the pivots, we get a system of linear algebraic equations with the same number of equations as the number of unknowns. This system can be written as

$$\mathbf{Au} = \mathbf{g} \tag{7.175}$$

which can be solved by using any iterative method. We find that $\alpha_1, \alpha_2, \alpha_3$ and α_4 are positive if

$$2A_{l,m} > \max\,[|D_{l,m}|h_3,\,|D_{l,m}|h_1]$$
and
$$2C_{l,m} > \max\,[|E_{l,m}|h_2,\,|E_{l,m}|h_4] \tag{7.176}$$

It may be noted that if (7.176) is satisfied, then $\alpha_0 < 0$. Thus, the off-diagonal elements of **A** are positive, the diagonal elements are negative and **A** has diagonal dominance. The matrix **A** is also irreducible. This guarantees the convergence of the solutions of the system of equations (7.175).

The nodal points in the interior of the region **R** and on the boundary $\partial\mathbf{R}$, should be numbered from left to right in the positive x direction and from bottom to top in the positive y direction. Any other systematic numbering of the nodal points may be followed. However, if the nodal points are numbered arbitrarily, the diagonal dominance of **A** in (7.175) may be destroyed and convergence cannot be guaranteed.

7.5.1 Dirichlet Problem

Putting $A = C = 1$, $D = E = F = F^* = 0$ in (7.166) we get the Laplace equation

$$\nabla^2 u = \frac{\partial^2 u}{\partial x^2} + \frac{\partial^2 u}{\partial y^2} = 0 \tag{7.177}$$

Equation (7.177) along with the Dirichlet boundary conditions is called the **Dirichlet problem**.

The difference equation (7.172) for $h = k$ (square network) becomes

$$u_{l-1,m} + u_{l+1,m} - 4u_{l,m} + u_{l,m-1} + u_{l,m+1} = 0 \tag{7.178}$$

which in schematic form is shown in Fig. 7.14. Equation (7.178) is referred to as the five point formula.

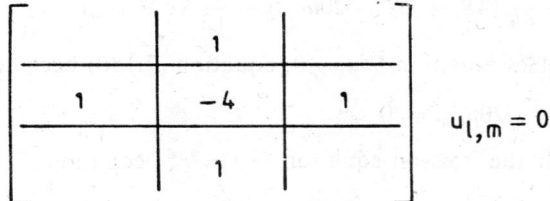

Fig. 7.14 Five point difference scheme

If $A = C = 1$, $D = E = F = 0$ and $F^* \neq 0$ in (7.156), then we get the **Poisson equation**

$$\frac{\partial^2 u}{\partial x^2} + \frac{\partial^2 u}{\partial y^2} = F^* \tag{7.179}$$

The corresponding difference scheme for $h = k$ becomes

$$u_{l-1,m} + u_{l+1,m} - 4u_{l,m} + u_{l,m-1} + u_{l,m+1} = h^2 F^*_{l,m} \tag{7.180}$$

Incorporating the boundary condition (7.167) in (7.178) or (7.180) we get a system of equations of the form (7.175). The approximations (7.178) and (7.180) are of order h^2. Higher order difference schemes on the square network for the Laplace and Poisson equations can be obtained as follows. We know that

$$u(x_l \pm h, y_m) = E_x^{\pm 1} u_{l,m} = e^{\pm h D_x} u_{l,m}$$
$$u(x_l, y_m \pm h) = E_y^{\pm 1} u_{l,y} = e^{\pm h D_y} u_{l,m} \tag{7.181}$$

where $D_x = \partial/\partial x$ and $D_y = \partial/\partial y$.

We find
$$S_1 = u_{l+1,m} + u_{l-1,m} + u_{l,m-1} + u_{l,m+1}$$
$$= 4u_{l,m} + h^2 \nabla^2 u_{l,m} + \frac{h^4}{12}(D_x^4 + D_y^4)u_{l,m} + O(h^6) \qquad (7.182)$$

$$S_2 = u_{l+1,m+1} + u_{l+1,m-1} + u_{l-1,m-1} + u_{l-1,m+1}$$
$$= 4u_{l,m} + 2h^2 \nabla^2 u_{l,m} + \frac{h^4}{6}(D_x^4 + 6D_x^2 D_y^2 + D_y^4)u_{l,m} + O(h^6) \qquad (7.183)$$

$$4S_1 + S_2 = 20u_{l,m} + 6h^2 \nabla^2 u_{l,m} + \frac{h^4}{2} \nabla^4 u_{l,m} + O(h^6) \qquad (7.184)$$

where ∇^4 is the biharmonic operator $\nabla^4 = \partial^4/\partial x^4 + 2\partial^4/\partial x^2 \partial y^2 + \partial^4/\partial y^4$.

Hence
$$\nabla^2 u_{l,m} = \frac{1}{6h^2}(4S_1 + S_2 - 20u_{l,m}) - \frac{h^2}{12} \nabla^4 u_{l,m} + O(h^4) \qquad (7.185)$$

Neglecting the truncation error, we get the nine point scheme
$$\nabla^2 u_{l,m} = \frac{1}{6h^2}(4S_1 + S_2 - 20u_{l,m}) - \frac{h^2}{12} \nabla^2(\nabla^2 u_{l,m}) \qquad (7.186)$$

For the Laplace equation $\nabla^2 u = 0$, equation (7.186) becomes
$$4S_1 + S_2 - 20u_{l,m} = 0$$

Similarly, for the Poisson equation $\nabla^2 u = F^*$, equation (7.186) becomes
$$4S_1 + S_2 - 20u_{l,m} = 6h^2 \left(F_{l,m}^* + \frac{h^2}{12} \nabla^2 F_{l,m}^* \right) \qquad (7.187)$$

which is of order h^4.

Example 7.11 Find the solution of $\nabla^2 u = 0$ in **R** subject to the Dirichlet condition $u(x, y) = x - y$ on $\partial \mathbf{R}$, where **R** is the region inside the triangle with vertices $(0, 0)$, $(7, 0)$, $(0, 7)$ and $\partial \mathbf{R}$ is its boundary. Assume the step length $h = 2.0$.

There are three nodes in **R** as shown in Fig. 7.15. At the node numbered 1, equation (7.178) can be used and at nodes 2 and 3 the modified formula (7.174) can be used. The numbering of the nodes on the boundary is immaterial. Applying the difference formulae and using the known boundary values, we obtain the system of equations

$$\begin{bmatrix} -1 & \frac{1}{4} & \frac{1}{4} \\ \frac{1}{3} & -2 & 0 \\ \frac{1}{3} & 0 & -2 \end{bmatrix} \begin{bmatrix} u_1 \\ u_2 \\ u_3 \end{bmatrix} = \begin{bmatrix} 0 \\ -4 \\ 4 \end{bmatrix}$$

The solution of this system is
$$u_1 = 0, \quad u_2 = 2, \quad u_3 = -2$$

These are the exact values, since the exact solution is $u(x, y) = x - y$.

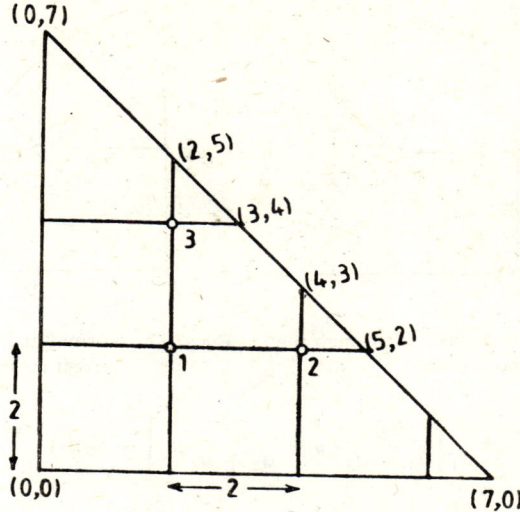

Fig. 7.15 Nodal points in example 7.11

7.5.2 Neumann Problem

If

$$\int_{\partial R} g\, dS = 0 \tag{7.188}$$

then the Neumann problem has at least one solution. If $u(x, y)$ is a solution, then every solution of the form $u(x, y) + c$ where c is an arbitrary constant is also a solution. Thus, the Neumann problem has a one parameter family of solutions. If some other conditions are provided, then the solution can be obtained uniquely. But, then the problem can be considered as a mixed boundary value problem.

7.5.3 Mixed Problem

In the interior of **R**, the difference approximations derived earlier can be used. We need, now, approximations for the derivative condition on the boundary ∂R so that they can be used in the mixed boundary conditions, at the nodes on the boundary. Two cases may arise. The normal at the node P on the boundary may pass through the nearest boundary point Q as in Fig. 7.16 or it may intersect the nearest mesh line as in Fig. 7.17. In the first case, we have

$$\left(\frac{\partial u}{\partial \mathbf{n}}\right)_P = \frac{u_P - u_Q}{d_1} + 0(d_1) \tag{7.189}$$

In the second case, we obtain

$$\left(\frac{du}{\partial \mathbf{n}}\right)_P = \frac{u_P - u_S}{d_1} + 0(d_1)$$

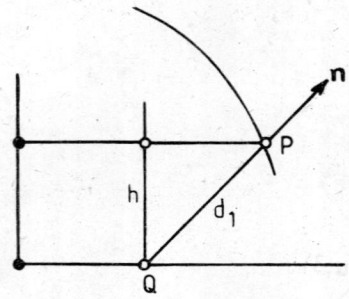

Fig. 7.16 Normal at P passing through the nearest mesh point

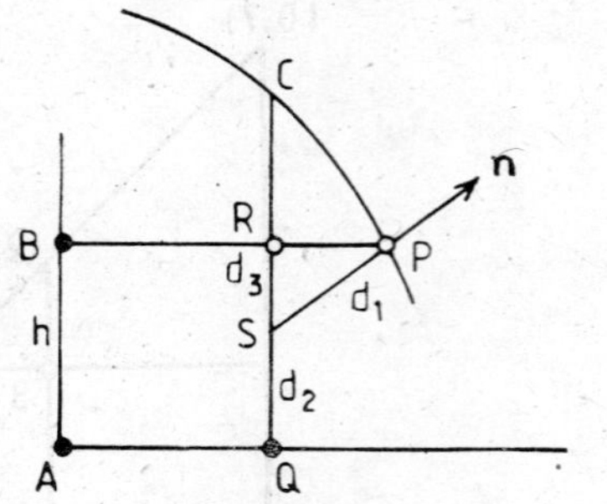

Fig. 7.17 Normal at P intersecting the nearest mesh

$$= \frac{u_P}{d_1} - \frac{1}{d_1}\left[\frac{d_2}{d_2+d_3}u_R + \frac{d_3}{d_2+d_3}u_Q\right] + 0(d_1)$$

$$= \frac{u_P}{d_1} - \frac{1}{d_1(d_2+d_3)}(d_2 u_R + d_3 u_Q) + 0(d_1) \qquad (7.190)$$

Both the equations (7.189) and (7.190) are first order approximations. Higher order approximations involving the nodes at A, B, C, P, Q and R can be derived. However, these approximations often destroy the diagonal dominance of the system of the equations.

Example 7.12 Solve the mixed boundary value problem

$$\nabla^2 u = 0, \quad 0 \leqslant x, y \leqslant 1$$

$$\left.\begin{array}{l} u = 2x, \quad y = 0 \\ u = 2x - 1, \quad y = 1 \end{array}\right\} \quad 0 \leqslant x \leqslant 1$$

$$\left.\begin{array}{l} u_x + u = 2 - y, \quad x = 0 \\ u = 2 - y, \quad x = 1 \end{array}\right\} \quad 0 \leqslant y \leqslant 1$$

The analytical solution is $u(x, y) = 2x - y$. Use the five point formula with $h = k = \frac{1}{3}$.

The nodal points are shown in Fig. 7.18.

At the node numbered 1, we approximate

$$\frac{\partial u}{\partial x} = \frac{u_2 - u_1}{h} + 0(h)$$

and hence at 1, the difference approximation is

$$-2u_1 + 3u_2 = \frac{5}{3}$$

Similarly, at 4 we have

$$-2u_4 + 3u_5 = \frac{4}{3}$$

Using the five point formula at the nodes 2, 3, 5 and 6 we get the system of equations

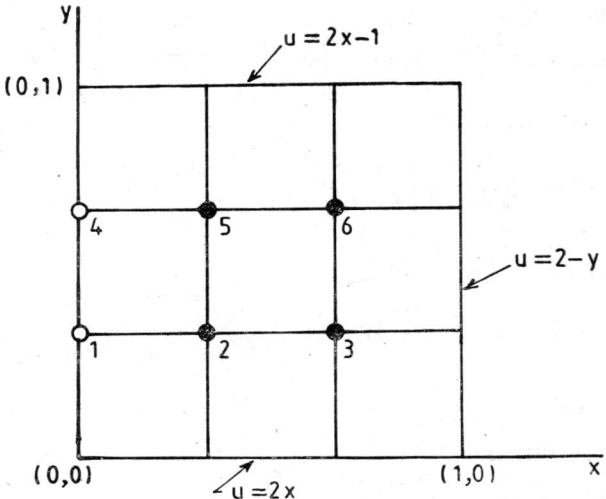

Fig. 7.18 Nodal points in the example 7.12

$$\begin{bmatrix} -2 & 3 & 0 & 0 & 0 & 0 \\ 1 & -4 & 1 & 0 & 1 & 0 \\ 0 & 1 & -4 & 0 & 0 & 1 \\ 0 & 0 & 0 & -2 & 3 & 0 \\ 0 & 1 & 0 & 1 & -4 & 1 \\ 0 & 0 & 1 & 0 & 1 & -4 \end{bmatrix} \begin{bmatrix} u_1 \\ u_2 \\ u_3 \\ u_4 \\ u_5 \\ u_6 \end{bmatrix} = \begin{bmatrix} \frac{5}{3} \\ -\frac{2}{3} \\ -3 \\ \frac{4}{3} \\ \frac{1}{3} \\ -\frac{5}{3} \end{bmatrix}$$

The solution of the system of equations is

$u_1 = -\frac{1}{3}, \quad u_2 = \frac{1}{3}, \quad u_3 = 1, \quad u_4 = -\frac{2}{3}, \quad u_5 = 0, \quad u_6 = \frac{2}{3}$

which is same as the exact solution.

Example 7.13 Solve the mixed boundary value problem

$\nabla^2 u = 0, \quad 0 \leq x^2 + y^2 \leq 1; \quad x \geq 0, y \geq 0$

$u = 0 \quad \text{on } x = 0 \text{ and } y = 0$

$\dfrac{\partial u}{\partial \mathbf{n}} = x - y \quad \text{on } x^2 + y^2 = 1$

Use the five point formula with $h = \frac{1}{2}$.
The nodal points are shown in Fig. 7.19.
The coordinates of the nodes are $1(\frac{1}{2}, \frac{1}{2})$, $2(\sqrt{3}/2, \frac{1}{2})$, $3(\frac{1}{2}, \sqrt{3}/2)$.
The coordinates of P and Q are $(\frac{1}{2}, 1/2\sqrt{3})$ and $Q(1/2\sqrt{3}, \frac{1}{2})$.
Hence $PM = 1/2\sqrt{3}, PS = (\sqrt{3}-1)/2\sqrt{3}, NQ = 1/2\sqrt{3}, QS = (\sqrt{3}-1)/2\sqrt{3}$,
$SD = (\sqrt{3}-1)/2 = SC, PD = QC = (\sqrt{3}-1)/\sqrt{3}$.
At 1, we use the formula (7.174). We find

$-2\sqrt{3}u_1 + u_2 + u_3 = 0$

At the nodes 2 and 3 we use the formula (7.190).

We obtain the equations

$$\frac{\sqrt{3}}{\sqrt{3}-1}u_2 - \frac{2\sqrt{3}}{\sqrt{3}-1}\left[\frac{1}{2\sqrt{3}}u_1 + \frac{\sqrt{3}-1}{2\sqrt{3}}\cdot 0\right] = \frac{\sqrt{3}-1}{2}$$

or

$-u_1 + \sqrt{3}u_2 = \frac{1}{2}(2-\sqrt{3})$

and

$-u_1 + \sqrt{3}u_3 = -\frac{1}{2}(2-\sqrt{3})$

The solution of this system is

$$u_1 = 0; \quad u_2 = -u_3 = \frac{\sqrt{3}(2-\sqrt{3})}{3}$$

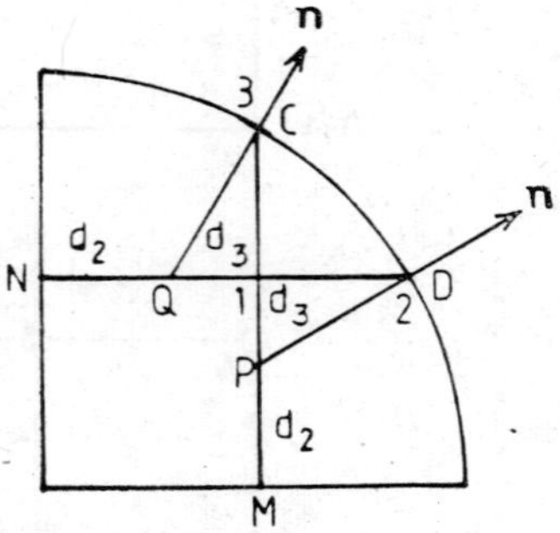

Fig. 7.19 Nodal points in example 7.13

PROBLEMS

1. (a) Calculate the characteristics of
   $u_{xx} - t^2 u_{tt} = 0$
   (b) For which values of x at the level $t = 1$ is the solution of
   $u_{xx} - t^2 u_{tt} = 0$
   $u(x, 0.1) = f(x)$
   $u_t(x, 0.1) = g(x)$
   influenced by a perturbation of $f(x)$ at $x = 1$?
   (Uppsala Univ., Sweden, BIT 13(1973), 493)

2. In which parts of the (x, y)-plane is the following equation elliptic?
   $u_{xx} + 4u_{xy} + (x^2 + 4y^2)u_{yy} = \sin(xy)$
   (Uppsala Univ., Sweden, BIT 14(1974), 366)

3. The heat conduction equation $u_t = u_{xx}$ is approximated by
   $(V_{i, j+1} - V_{i, j-1})/2k = (V_{i-1, j} - 2V_{i, j} + V_{i+1, j})/h^2$
   (a) Determine the truncation error.
   (b) Investigate the stability using Von-Neumann's method.
   (Uppsala Univ., Sweden, BIT 17(1977), 294)

4. Consider the heat equation
   $u_t = u_{xx}, \quad 0 \leqslant x \leqslant 1, \quad 0 \leqslant t \leqslant T$
   $u(0, t) = u(1, t) = 0$
   $u(x, 0) = f(x)$

Investigate the stability conditions on k and h for $0 \leqslant \alpha \leqslant 1$ when the equation is approximated by

$$\frac{v(x, t+k)-v(x, t)}{k} = \alpha \frac{v(x+h, t+k)-2v(x, t+k)+v(x-h, t+k)}{h^2}$$

$$+(1-\alpha)\frac{v(x+h, t)-2v(x, t)+v(x-h, t)}{h^2}$$

(Gothenburg Univ., Sweden, BIT 14(1974), 254)

5. The heat conduction equation $\frac{\partial u}{\partial t} = \frac{\partial^2 u}{\partial x^2}$ is approximated with the difference equation

$$\frac{u_{j,n+1}-u_{j,n}}{k} = \frac{1}{2h^2}[u_{j+1,n}-2u_{j,n}+u_{j-1,n}+u_{j+1,n+1}-2u_{j,n+1}$$

$$+ u_{j-1,n+1}]$$

where $u_{j,n} = u(x+jh, t+nk)$. The solution $u(x, t)$ is supposed to possess continuous partial derivatives of sufficiently high order. Show that the truncation error is $O(h^2) + O(k^2)$.

(Lund Univ., Sweden, BIT 7(1967), 170)

6. One wants to use the difference approximation

$$\frac{u_{i,j+1}-(1-\theta)u_{i,j}-\theta u_{i,j-1}}{(1+\theta)k} = \frac{u_{i-1,j}-2u_{i,j}+u_{i+1,j}}{h^2}$$

for approximating a solution of the partial differential equation $\frac{\partial u}{\partial t} = \frac{\partial^2 u}{\partial x^2}$ in the domain $0 \leqslant x \leqslant 1$, $t \geqslant 0$. Investigate the stability properties for different values of $\theta (0 \leqslant \theta \leqslant 1)$ and $r = \frac{k}{h^2}$.

(Uppsala Univ., Sweden, BIT 7(1967), 247)

7. Discuss the stability of a method to integrate the heat equation

$$\frac{\partial u}{\partial t} = \frac{\partial^2 u}{\partial x^2}, \quad t \geqslant 0, \quad 0 \leqslant x \leqslant 1$$

$$u(0, t) = u(1, t), \quad t \geqslant 0$$

$u(x, 0)$ given, $0 \leqslant x \leqslant 1$

based on a central difference approximation of $\partial^2 u/\partial x^2$ and the trapezoidal rule for the time integration.

(Stockholm Univ., Sweden, BIT 5(1965), 68)

8. Use at least two suitable difference schemes to determine $u(0.4, 0.2)$ where $u(x, t)$ denotes the solution of

$u_t = u_{xx}$

$$u(x, 0) = \begin{cases} 2x, & x \in [0, \tfrac{1}{2}] \\ 2(1-x), & x \in [\tfrac{1}{2}, 1] \end{cases}$$

$$u(x+1, t) = u(x, t)$$

Use the mesh size $h = 0.2$ and an appropriate time step k. Discuss the results.

Hint. Use the symmetry properties of the problem and the solution.

(Gothenburg Univ., Sweden, BIT 13(1973), 375)

9. A difference formula for the partial differential equation
$$\frac{\partial u}{\partial t} = \frac{\partial^2 u}{\partial x^2}, \quad t > 0$$
is
$$[1 - \tfrac{1}{2}(r - \tfrac{1}{6})\delta_x^2] u_j^{n+1} = [1 + \tfrac{1}{2}(r + \tfrac{1}{6})\delta_x^2] u_j^n$$
where $r = k/h^2$. Determine the truncation error and examine the stability of the formula.

10. A three level difference formula for the heat conduction equation
$$\frac{\partial u}{\partial t} = \sigma \frac{\partial^2 u}{\partial x^2}, \quad \sigma > 0, \; 0 \leqslant x \leqslant 1, \; t \geqslant 0$$
can be written as
$$(1 + \theta)\frac{u_m^{n+1} - u_m^n}{k} - \theta \frac{u_m^n - u_m^{n-1}}{k} = \sigma \frac{\delta_x^2 u_m^{n+1}}{h^2}$$
where $\theta = \text{constant} \geqslant 0$ and $(\sigma k / h^2) = \text{constant}$. Examine the stability of this scheme.

11. A difference method for the numerical solution of
$$\frac{\partial u}{\partial t} = \frac{\partial^2 u}{\partial x^2}, \quad t > 0$$
under suitable initial and boundary conditions, may be written as
$$u_j^{n+1} = \tfrac{4}{3} u_j^n - \tfrac{1}{3} u_j^{n-1} + \tfrac{2}{3} r \delta_x^2 u_j^{n+1}$$
where $r = (k/h^2)$. Find the truncation error and examine the stability of the formula. If the initial condition is
$$u(x, 0) = x, \qquad 0 \leqslant x \leqslant \tfrac{1}{2}$$
$$= 1 - x, \quad \tfrac{1}{2} \leqslant x \leqslant 1$$
and the boundary conditions are
$$u(0, t) = 0 = u(1, t)$$
find the finite difference solutions when $h = \tfrac{1}{4}$ and $r = \tfrac{1}{2}$. Integrate upto two time steps.

12. The parabolic partial differential equation
$$\frac{\partial u}{\partial t} = \frac{\partial^2 u}{\partial x^2}$$
with values of u specified for $0 \leqslant x \leqslant 1$ when $t = 0$ and for $x = 0$

and $x = 1$ when $t > 0$ is to be solved by using the following difference schemes:

(a) $u_m^n - u_m^{n-1} = p(u_{m+1}^n - 2u_m^n + u_{m-1}^n)$

(b) $u_m^{n+1} - u_m^n = p(u_{m+1}^n - 2u_m^n + u_{m-1}^n)$

where $u_m^n = u(mh, nk)$ and $p = \dfrac{k}{h^2}$.

Find the truncation error of the difference schemes.
Derive also the stability conditions of these difference schemes.

13. Obtain a difference approximation to the partial differential equation

$$\frac{\partial u}{\partial t} = \frac{\partial}{\partial x}\left(u \frac{\partial u}{\partial x}\right)$$

with initial and boundary conditions

$u(x, 0) = 300 + 3(x - 4)^2$, $0 \leq x \leq 4$
$u(0, t) = 348$; $u(4, t) = 300$, $t \geq 0$

Use the formula valid for $\rho(x) \in C^2$, $u(x) \in C^4$

$$\frac{\partial}{\partial x}\left(\rho(x)\frac{\partial u}{\partial x}\right)_{x=x_\mu} = \frac{1}{2h^2}[(\rho_{\mu+1} + \rho_\mu)(u_{\mu+1} - u_\mu)$$
$$- (\rho_\mu + \rho_{\mu-1})(u_\mu - u_{\mu-1})] + O(h^2)$$

where $u(x_0 + \mu h) = u_\mu$ etc.
Choose $x_0 = 0$, $h = \Delta x = 1$, $k = \Delta t = 0.001$ and integrate until $t = 0.003$.

(Inst. Tech., Stockholm, Sweden, BIT 4(1964), 197)

14. The **Douglas-Rachford** scheme for solving the parabolic equation in two dimensions

$$\frac{\partial u}{\partial t} = \frac{\partial^2 u}{\partial x^2} + \frac{\partial^2 u}{\partial y^2}$$

is given by

$$\left(1 - \frac{r}{2}\delta_x^2\right)u_{l,m}^{*n+1} = \left(1 + \frac{r}{2}\delta_x^2\right)u_{l,m}^n + r\delta_y^2 u_{l,m}^n$$

$$\left(1 - \frac{r}{2}\delta_y^2\right)u_{l,m}^{n+1} = u_{l,m}^{*n+1} - \frac{r}{2}\delta_y^2 u_{l,m}^n$$

Find the truncation error of the formula and examine its stability.

15. The function $u(x, y, t)$ satisfies the parabolic equation

$$\frac{\partial u}{\partial t} = \frac{\partial^2 u}{\partial x^2} - \frac{\partial^2 u}{\partial y^2}$$

with the initial and boundary conditions specified as

$u(x, y, 0) = \sin(2\pi x)\sin(2\pi y)$, $0 \leq x, y \leq 1$
$u(x, y, t) = 0$, $t > 0$, on the boundary.

Obtain the numerical solution by Douglas-Rachford ADI method with $h = \tfrac{1}{3}$, $r = 1.0$ and integrate for one time step.

16. For the solution of the differential equation

$$\frac{\partial u}{\partial t} = \frac{\partial^2 u}{\partial x^2} + \frac{\partial^2 u}{\partial y^2} + cu, \quad c \text{ constant}$$

the following difference schemes were written:

(i) $u_{l,m}^{n+1} = [1 - r(\delta_x^2 + \delta_y^2) + crh^2]u_{l,m}^n$

(ii) $[1 - crh^2 - r(\delta_x^2 + \delta_y^2)]u_{l,m}^{n+1} = u_{l,m}^n$

where $r = k/h^2$. Show that (i) is stable for $0 < r \leqslant 1/(4 - ch^2/2)$, $c < 0$ and (ii) is unconditionally stable for $c \leqslant 0$.

17. The heat flow equation in two dimensions is given by

$$\frac{\partial u}{\partial t} = \frac{\partial^2 u}{\partial x^2} + \frac{\partial^2 u}{\partial y^2}$$

valid in $0 \leqslant x, y \leqslant 1$, $t \geqslant 0$. The initial and boundary conditions are given as

$u(x, y, 0) = \sin 2\pi x \sin 2\pi y, \quad 0 \leqslant x, y \leqslant 1$

$u(x, y, t) = 0, \quad t > 0$, on the boundary

Solve the above equation using the Peaceman-Rachford ADI method with $h = \tfrac{1}{3}$, $r = 1$. Integrate for one time step.

18. The differential equation

$$\frac{\partial^4 u}{\partial x^4} + K \frac{\partial^2 u}{\partial t^2} = 0$$

is to be solved for $0 \leqslant x \leqslant a$, $0 \leqslant t$, K is a given constant and the boundary values

$u(x, 0), u_t(x, 0)$

$u(0, t), u_x(0, t)$

$u(a, t), u_x(a, t)$

are also given.

The differential equation is approximated with the following difference equation:

$$\frac{u_{i+2, j} - 4u_{i+1, j} + 6u_{i, j} - 4u_{i-1, j} + u_{i-2, j}}{h^4}$$

$$+ K \frac{u_{i, j+1} - 2u_{i, j} + u_{i, j-1}}{k^2} = 0$$

where h and k are chosen so that $Kh^4 = k^2$.

(a) Determine the truncation error.

(b) Suppose that the values $u_{i,1}$ ($i = 1, 2, \ldots$) are known and that the values $u_{i,2}, u_{i,3}$ etc. are calculated from the difference equation. Show that this method is unstable.

(Inst. Tech., Copenhagen, Denmark, BIT 11(1971), 225)

19. Show that the partial differential equation

$$\frac{\partial^2 u}{\partial t^2} = \frac{\partial^2 u}{\partial x^2}$$

with $u = f(x)$ and $\frac{\partial u}{\partial t} = g(x)$, $t = 0$ may be replaced by the difference scheme

$$u_m^{n+1} = -u_m^{n-1} + 2(1 - p^2)u_m^n + p^2(u_{m+1}^n + u_{m-1}^n)$$

$$u_m^0 = f_m$$

$$u_m^1 = kg_m + (1 - p^2)f_m + \tfrac{1}{2}p^2(f_{m+1} + f_{m-1})$$

where $u_m^n = u(mh, nk)$, $f_m = f(mh)$, $g_m = g(mh)$ and $p = \frac{k}{h}$. Find an expression of the local truncation error of each of the finite difference expressions. Using

$$f(x) = x, \qquad 0 \leq x \leq \tfrac{1}{2}$$
$$= 1 - x, \quad \tfrac{1}{2} \leq x \leq 1$$
$$g(x) = 0, \qquad 0 \leq x \leq 1$$

and $u(0, t) = 0 = u(1, t)$, find the solution upto two time steps with $h = 0.2$, $p = \tfrac{1}{2}$.

20. The following initial boundary value problem is given:

$$\frac{\partial^2 u}{\partial t^2} = \frac{\partial^2 u}{\partial x^2} + \frac{\partial^2 u}{\partial y^2}, \quad 0 < x, y < 1, \quad t > 0$$

$$u(x, y, 0) = xy, \quad 0 \leq x, y \leq \tfrac{1}{2}$$
$$= (1 - x)(1 - y), \quad \tfrac{1}{2} \leq x, y \leq 1$$
$$= (1 - x)y, \quad \tfrac{1}{2} \leq x \leq 1, 0 \leq y \leq \tfrac{1}{2}$$
$$= x(1 - y), \quad 0 \leq x \leq \tfrac{1}{2}, \tfrac{1}{2} \leq y \leq 1$$

$$\frac{\partial u}{\partial t}(x, y, 0) = 0, \quad 0 \leq x, y \leq 1$$

$u(x, y, t) = 0$ on the boundary for $t > 0$

Use an $0(k^2 + h^2)$ ADI method to find solutions with $h = \tfrac{1}{3}$, $r = 1/2$. Integrate upto two time steps.

21. The difference equation

$$\frac{u_{i, j+1} - u_{i, j-1}}{2k} = \frac{u_{i+1, j} - (u_{i, j+1} + u_{i, j-1}) + u_{i-1, j}}{h^2}$$

is a consistent difference approximation to different partial differential equations depending on how h and k approach zero. Investigate what partial differential equations are intended and give the corresponding conditions on h and k.

(Uppsala Univ., Sweden, BIT 8(1968), 59)

22. Find the solution of the initial value problem

$$\frac{\partial u}{\partial t} + \frac{\partial u}{\partial x} = 0$$

together with the initial condition

$$u = \begin{cases} 0 & x < 0 \\ \frac{x}{2}, & 0 \leqslant x \leqslant 2 \\ 2 - \frac{x}{2}, & 2 \leqslant x \leqslant 4 \\ 0, & x > 4 \end{cases}$$

using (i) Lax-Wendroff method, (ii) Leap-frog scheme with $h = 0.5$, $p = 0.5$, where $p = k/h$. Integrate for two time steps.

23. The differential equation

$$\frac{\partial u}{\partial t} + c \frac{\partial u}{\partial x} = 0$$

is approximated by the implicit method

$$\left[(\alpha + \beta) - (\beta - \alpha) \frac{p^2}{4} \delta_x^2 + p\alpha\mu_x\delta_x \right] u_m^{n+1}$$
$$= \left[(\alpha + \beta) + (\beta - \alpha) \frac{p^2}{4} \delta_x^2 - p\beta\mu_x\delta_x \right] u_m^n$$

where $p = \frac{ck}{h}$ and α and β are arbitrary parameters. Determine

(i) the local truncation error for arbitrary α and β,
(ii) the stability conditions.

24. Let

$$v_{p,q+1} = v_{p,q} + \tfrac{1}{2}\lambda(4v_{p+1,q} - v_{p+2,q} - 3v_{p,q})$$
$$+ \tfrac{1}{2}\lambda^2(v_{p+2,q} - 2v_{p+1,q} + v_{p,q})$$

where $\lambda = \frac{k}{h}$, $v_{p,q} = v(ph, qk)$ be a difference approximation of

$$\frac{\partial u}{\partial t}(x, t) = \frac{\partial u}{\partial x}(x, t)$$

(a) What does the computation molecule look like?
(b) How big is the local truncation error?

$$\left(\text{The equation shall be written as } \frac{v_{p,q+1} - v_{p,q}}{k} = \ldots \right)$$

(c) State some value of λ for which the difference approximation is stable, and some for which it is unstable.

(Uppsala Univ., Sweden, BIT 9(1969), 400)

25. Solve the differential equation $\nabla^2 u = 16$ for a square of side 2 with $u = 0$ on the boundary.

(a) Formulate the corresponding difference equation with mesh size h in both directions.

(b) Solve the difference equation exactly for $h = 1$ and $h = \frac{1}{2}$.

(c) Give the formula for a convergent iteration procedure for the solution of the difference equation.

(d) Give the formula for the solution by over-relaxation.

(e) The mesh should be made finer by successive divisions of the mesh size by two in both directions. How would you use the values which were obtained with the coarser grid in order to get starting values for the iteration with the finer grid?

(Norwegian Inst. Tech., Trondheim, Norway, BIT 4(1964), 61)

26. Apply the nine point formula to solve the partial differential equation

$$\frac{\partial^2 u}{\partial x^2} + \frac{\partial^2 u}{\partial y^2} = x^2 + y^2; \quad 0 < x < 3, 0 < y < 3$$

$u(x, y) = 0$, on the boundary.
Assume the step length $h = 1$.

27. Consider the boundary value problem

$$\nabla^2 u = \frac{\partial^2 u}{\partial x^2} + \frac{\partial^2 u}{\partial y^2} = 4, \; |x| \leqslant 1, |y| \leqslant 1$$

$u = 0$ on the boundary
Write a five point and nine point difference formulation with mesh spacing h in both directions. Solve the difference equations exactly for $h = \frac{1}{2}$.

28. Solve the partial differential equation

$$\nabla^2 u = x^2 - 1, \; |x| \leqslant 1, |y| \leqslant 1$$

with $u = 0$ on the boundary of the square.
Formulate the five point and nine point difference schemes with mesh size h in both directions. Solve the difference equations for $h = \frac{1}{2}$.

29. Let **R** be the set of points for which $x > 0, y > 0, x^2 + y^2 < 9$ and $\partial \mathbf{R}$ be its boundary. In **R** solve the Dirichlet problem

$$\nabla^2 u = 4 \text{ in } \mathbf{R}$$

$u(x, y) = x^2 + y^2$ on $\partial \mathbf{R}$

using a five point difference formula, assuming the origin at $(0, 0)$ and $h = 1$.

30. Find the solution of the Dirichlet problem
$$\nabla^2 u = x^2 - y^2; \quad y > 0, \; x^2 + y^2 < 4$$
$$u(x, y) = \tfrac{1}{12}(x^4 - y^4) \text{ on the boundary}$$
using a five point formula with $h = 1.0$.

31. Find the solution of the mixed boundary value problem
$$\nabla^2 u = 0, \quad x > 0, \; y > 0, \quad x^2 + y^2 < 1$$
$$u(x, y) = x^2 - y^2; \quad \text{on } x = 0 \quad \text{or} \quad y = 0$$
$$\frac{\partial u}{\partial n}(x, y) = 2(x^2 - y^2), \quad \text{on } x^2 + y^2 = 1$$
with step length $h = \tfrac{1}{2}$ and a five point difference scheme.

32. Find a five point difference scheme of the form (7.174) for the elliptic partial differential equation
$$4u_{xx} + 6u_{yy} + 7u_x - 5u_y - 9u = e^{-x}$$
Determine an upper bound for the step length h such that
$$\alpha_1, \alpha_2, \alpha_3, \alpha_4 > 0 \quad \text{and} \quad \alpha_0 < 0.$$

33. An elastic membrane satisfies the Laplace equation
$$\frac{\partial^2 u}{\partial x^2} + \frac{\partial^2 u}{\partial y^2} = 0$$
The boundary conditions are $u = 0$ on the inner boundary and $u = 1$ on the outer boundary (see Fig. 7.20).

(a) Give the linear equation system for the u-values at the nodes when the five-point ∇_5^2-operator is used to approximate the Laplace operator. Use symmetry to get as few unknowns as possible.

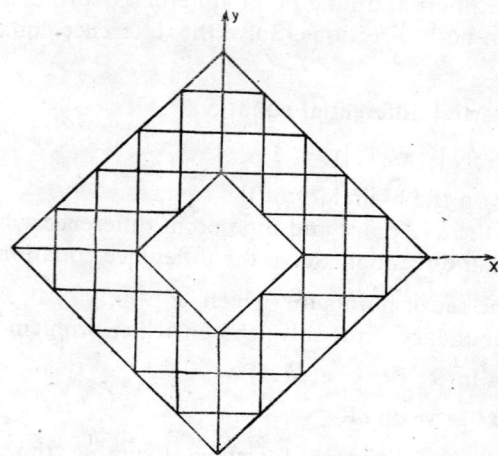

Fig. 7.20 Nodal points

(b) Solve the system to 3 correct decimal places.

(Royal Inst. Tech., Stockholm, Sweden, BIT 18(1978), 366)

34. The function $u(x, y)$ satisfies the differential equation

$$\frac{\partial^2 u}{\partial x^2} + \frac{\partial^2 u}{\partial y^2} + \frac{1 + e^u}{2} = 0$$

for $|x| \leq 1$, $|y| \leq 1$, and the boundary condition $u(x, y) = 0$ for $|x| = 1$ and for $|y| = 1$. Determine an approximate value of $u(0, 0)$ using a difference approximation. Use $h = 1$ and $h = \frac{1}{2}$ and perform Richardson extrapolation.

Observe $u(x, y) = u(-x, y) = u(y, x)$. The problem has two solutions. Only one with the smallest values of u is required here.

(Lund Univ., Sweden, BIT 8(1968), 343)

35. Compute the smallest eigenvalue of the differential equation

$$\nabla^2 u = -\lambda u$$

with $u = 0$ on the boundary of a square of size 3 by 3, by substituting $\nabla^2 u = -\lambda u$ in both sides of the relation given in (7.187). Choose $h = \frac{3}{2}$, $h = 1$ and use Richardson extrapolation under the assumption that the error in λ is $O(h^4)$. Compare the result with the exact value $2\pi^2/9$.

(Inst. Tech., Stockholm, Sweden, BIT 5(1965), 142)

APPENDIX-I

We now give the flowcharts for a few frequently used numerical methods.

Flowchart 1. Newton-Raphson Method

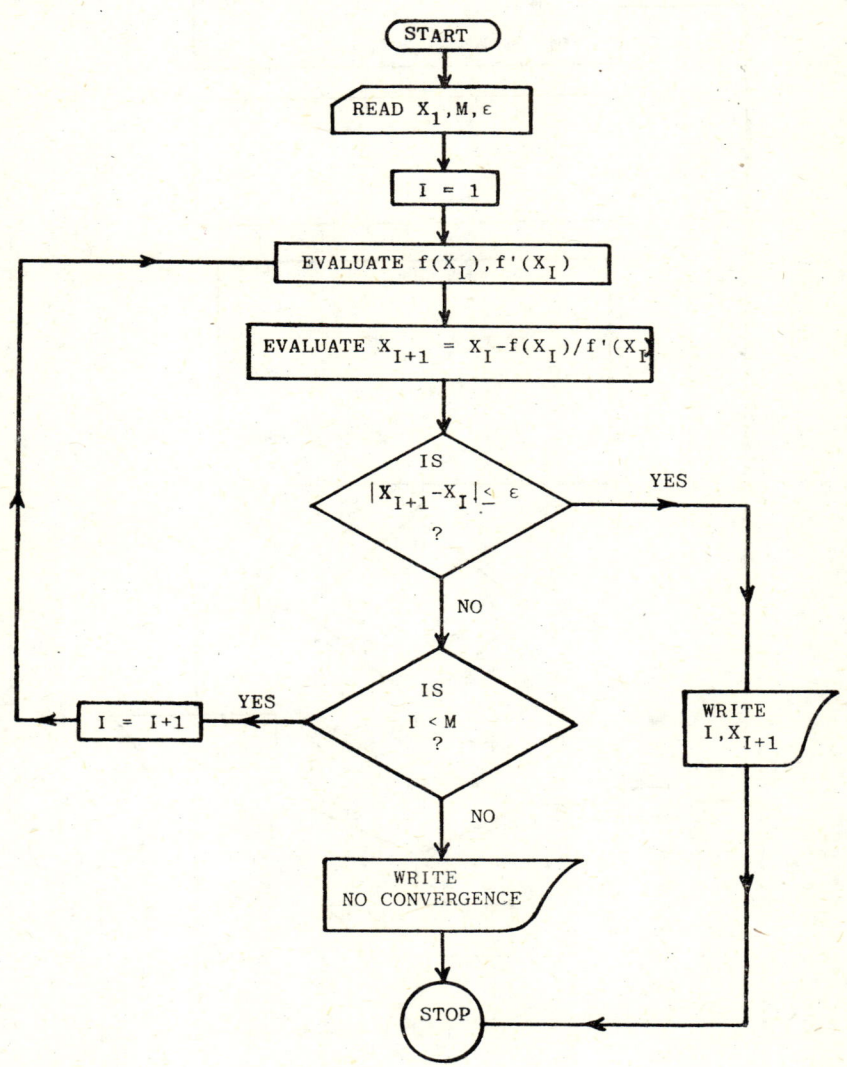

380 Numerical Methods

Flowchart 2. Bairstow's Method

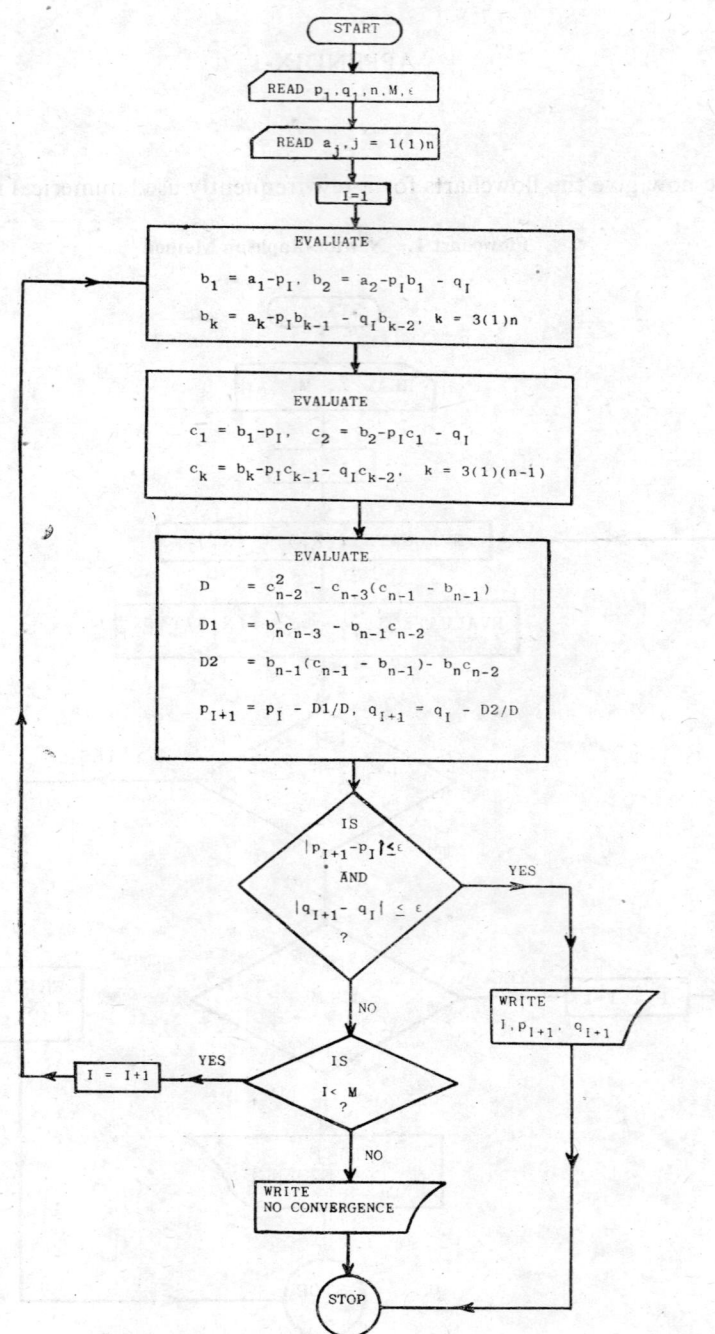

Appendix 381

Flowchart 3. Gauss Elimination Method

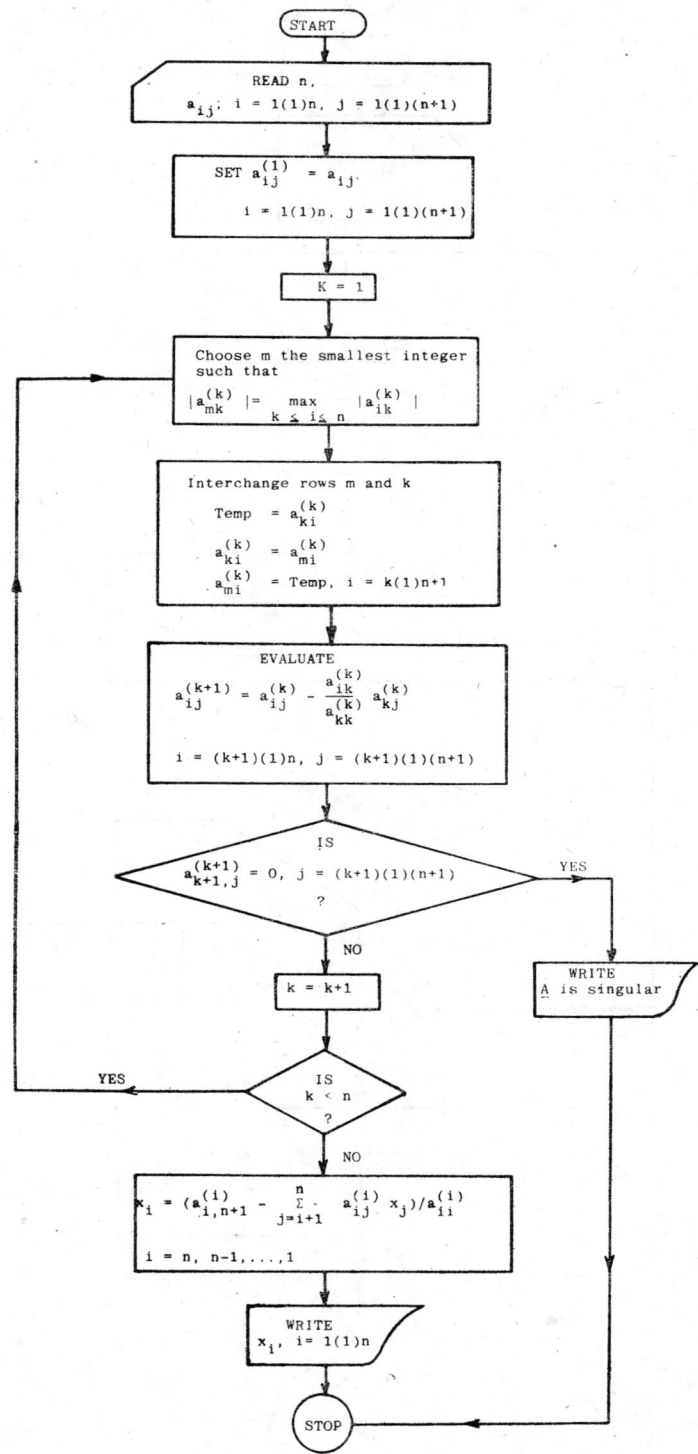

Flowchart 4. Gauss-Seidel Iterative Method

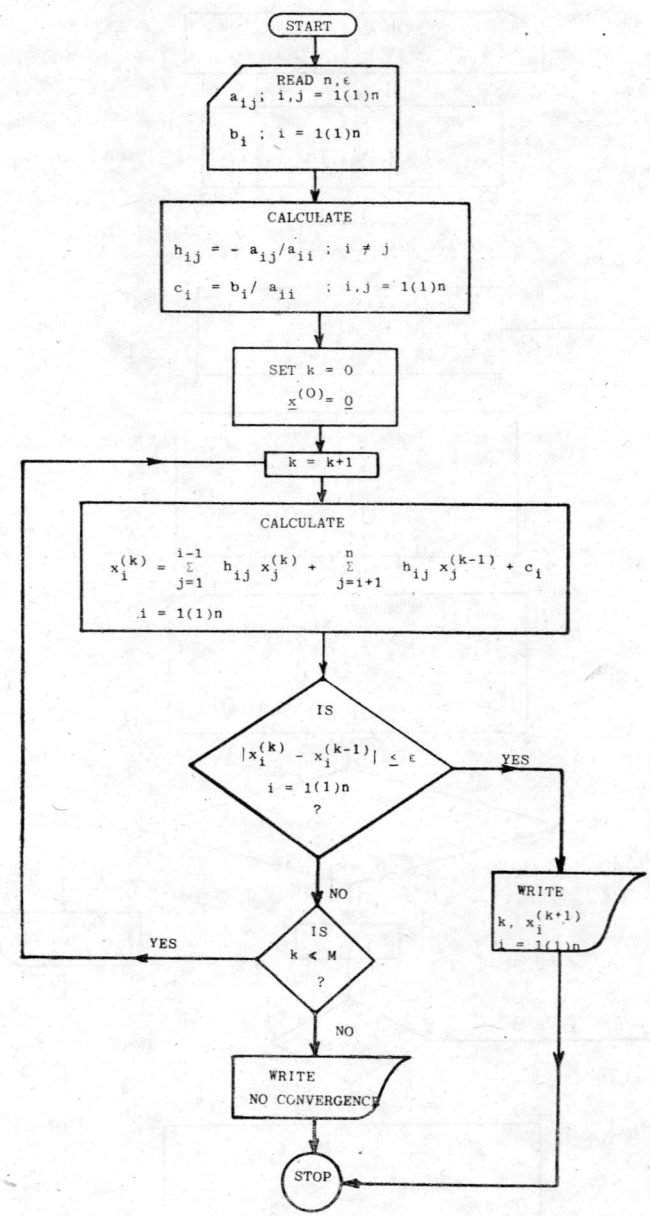

Flowchart 5. Power Method

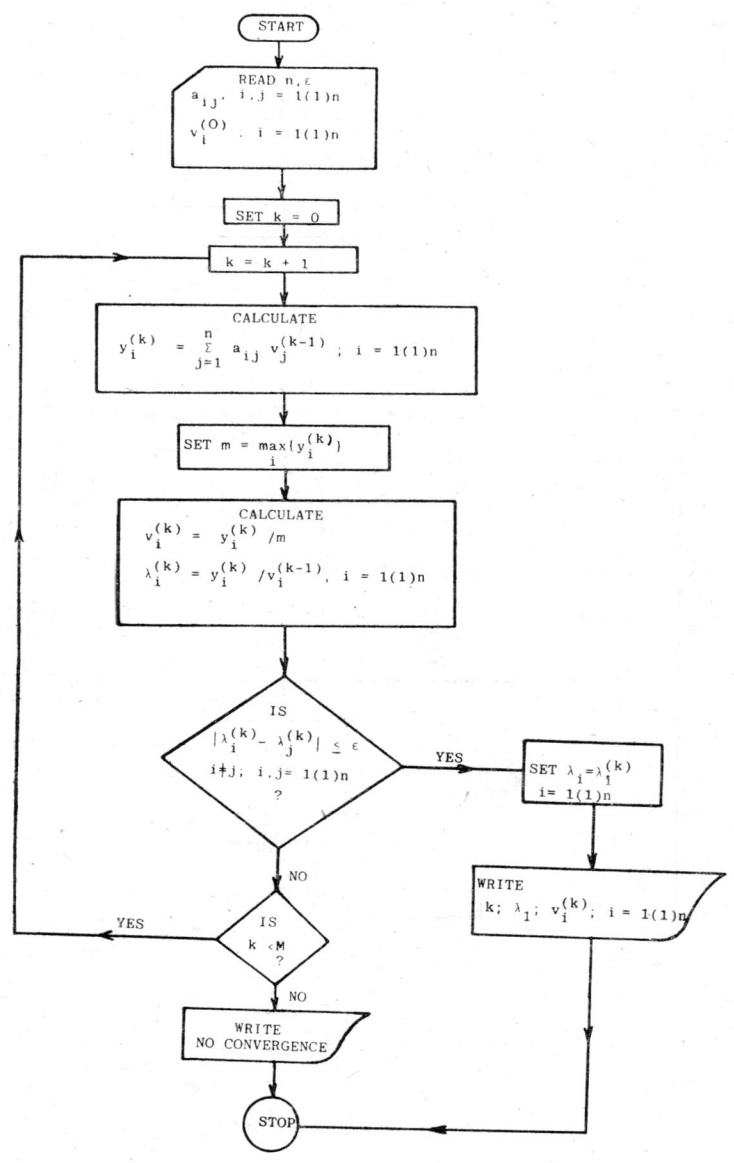

Flowchart 6. Runge-Kutta Fourth Order Method

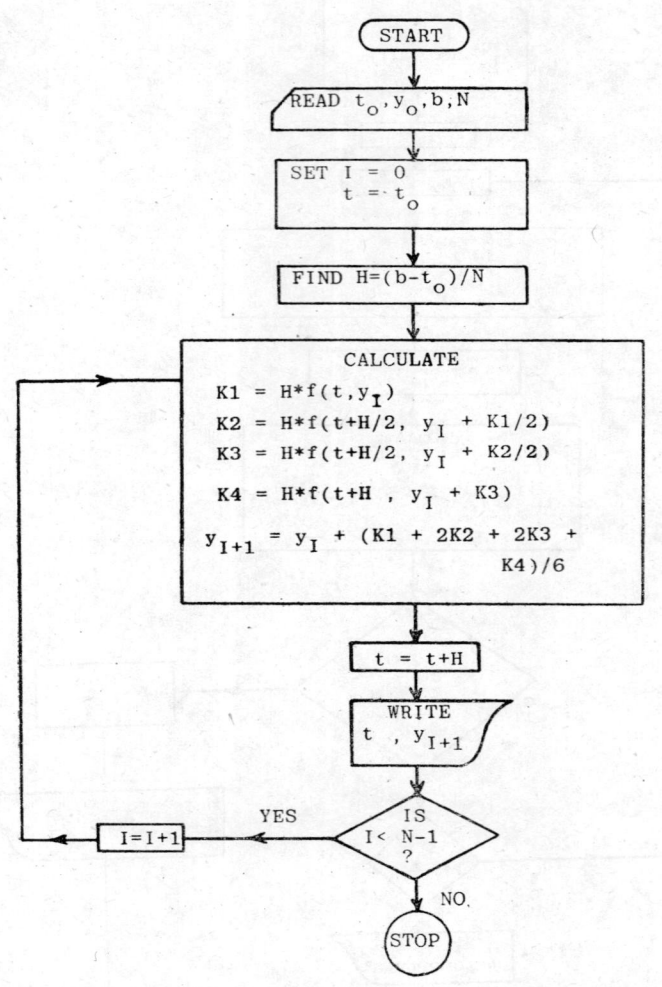

Answers and Hints to the Problems

Chapter 2

1. (a) (1.5, 1.75), 1.57. The root lies in (1.5703, 1.5742).
   (b) (1, 2), 1.80. The root lies in (1.7959, 1.7969).
2. Linear. $\epsilon_{k+1} = (f''(\xi)/2f'(\xi))\epsilon_0\epsilon_k + \cdots$
3. (ii) $x_{n+1} = (x_n^2 + N)/(2x_n)$
   $x_{n+1} = (2x_n^2 + N)/(3x_n^2)$
   $x_{n+1} = x_n(2 - Nx_n)$
   4.24, 0.06, 2.62
4. (i) 1.5, 1.856
   (ii) 0.75, 0.567
6. Order of convergence = 2 (when $f'(a) \neq 1$ or 0). Asymptotic error constant is $f(a)(1 - f'(a))/f'(a)$. In Newton-Raphson's method the error constant is $f''(a)/(2f'(a))$.
7. Three; 1.856
8. $p = \frac{5}{9}$, $q = \frac{5}{9}$, $r = -\frac{1}{9}$; third order
   $\epsilon_{k+1} = 5\epsilon_k^3/(3\xi^2) + \cdots$
9. $x_0 = 2$, $x_1 = 2.4375$, $x_2 = 2.4495$
   $x_0 = 2$, $x_1 = 2.4167$, $x_2 = 2.4495$
11. (i) $x_{k+1} = x_k + \dfrac{(x_k - x_{k-1})(x_k - x_{k-2})f_k(f_{k-1} - f_{k-2})}{(x_k - x_{k-1})(f_{k-2} - f_k)f_{k-1} + (x_k - x_{k-2})(f_k - f_{k-1})f_{k-2}}$
    (ii) $\epsilon_{k+1} = \left(\dfrac{c_2^2}{c_1^2} - \dfrac{c_3}{c_1}\right)\epsilon_k\epsilon_{k-1}\epsilon_{k-2}$; $c_r = \dfrac{f^{(r)}(\xi)}{r!}$; 1.84
    (iii) 1.615, 1.485
12. $x_{k+1} = x_k - \dfrac{(x_k - x_{k-1})f_k[f_{k-1}(f_k - f_{k-1}) - (x_k - x_{k-1})f_k f'_{k-1}]}{2f_k f_{k-1}(f_k - f_{k-1}) - (x_k - x_{k-1})(f_k^2 f'_{k-1} + f_{k-1}^2 f'_k)}$
    rate of convergence = 2.732; 1.856, 1.856
13. Second; 0.797
14. (ii) (a) 0.7676, 0.7929
    (b) 1.9531, 3.1353

16. (i) $x_{n+1} = \epsilon/(x_n^3 + 1)$
    (ii) $22(\epsilon^{10})$

17. $|x_{n+1} - \xi| \leq \dfrac{c}{1-c} |x_{n+1} - x_n|;\ \phi'(\xi) \leq c$

    $|x_2 - \xi| \leq \dfrac{c}{1-c} |x_2 - x_1| \leq 0.03001$

    $|x_{n+2} - \xi| \leq c^n |x_2 - \xi| \leq (0.53)^n (0.03001) < 5 \times 10^{-5}$
    which gives $n = 11$

18. 1.2959

19. 22, three iterations of Δ^2 process

20. Third order: $\gamma_1 = \dfrac{1}{f_k'},\ \gamma_2 = \tfrac{1}{2}\dfrac{f_k''}{(f_k')^3},\ \gamma_3 = 0$

    Fourth order: $\gamma_1 = \dfrac{1}{f_k'},\ \gamma_2 = \tfrac{1}{2}\dfrac{f_k''}{(f_k')^3}$,

    $\gamma_3 = \left[\tfrac{1}{2}\dfrac{(f_k'')^2}{(f_k')^2} - \tfrac{1}{6}\dfrac{f_k'''}{f_k'}\right]\Big/(f_k')^3$

    0.259

21. $k = 12,\ 4.223,\ \pm 0.005$

22. $\alpha = 0.63$

23. (b) $x_0 < \alpha_1$, no convergence
    $x_0 > \alpha_1$, convergence to α_2
    (c) $x_0 < \gamma$, convergence to α_1
    $x_0 > \gamma$, convergence to α_2
    where γ is the zero of $1 - 2xe^{x^2-1}$

24. $n = 4$

25. (i) 0.362 and 1.309
    (ii) 0.8858

28. $\alpha = 1,\ \beta = \tfrac{1}{2}$, third order

29. -1.667

30. (i) $(x, y) = (-0.25, 0.75),\ (0.075, 0.975),\ (-0.0017, 0.9973)$
    (ii) $(x, y) = (-0.5080, -0.3864),\ (-0.50878, -0.38655),$
    $(-0.50883, -0.38666)$

31. (i) all roots
    (ii) 2
    (iii) 0.67

32. $x = 1.09,\ y = 1.94$

33. $x = 2.41225 \pm 10^{-5},\ y = -0.64386 \pm 0.5 \times 10^{-5}$

34. (x, y) or $(y, x) = \pm(1.035, 0.222)$
35. The roots of the quadratic equation
$$\lambda^2 - (f_x + g_y + f_y g_x)\lambda + f_x g_y = 0$$
are less than unity in modulus.
If $f_x = g_y = 0$, then the roots of the first quadratic are $|\lambda| = \sqrt{|f_y g_x|}$ and the roots of the second quadratic are $|\lambda| = 0, |f_y g_x|$.
36. $a = -g_y/D$, $b = f_y/D$, $c = g_x/D$, $d = -f_x/D$,
$D = f_x g_y - g_x f_y$; $x_0 = 0$, $y_0 = \sqrt{2}$
37. (a) From the starting point $(\frac{1}{10}, \frac{1}{4}, \frac{1}{12})$ one Newton-Raphson step yields
$x = 0.068960$, $y = 0.246443$, $z = 0.076929$
   (b) Iteration formulas
$x_{k+1} = (1 - \sin(x_k + y_k))/10$
$y_{k+1} = (1 + \cos^2(z_k - y_k))/8$
$z_{k+1} = (1 - \sin z_k)/12$
Final solution: $x = 0.068978$, $y = 0.246442$, $z = 0.076929$
Five iterations are required.
38. $x = 2.999820$, $y = 2.962681$, $z = 2.030252$
39. Three; 0, 2, 2
40. Two; $P'(2) = 53$, $P''(2) = 110$
41. (i) 0.931
   (ii) 1.404
   (iii) -1.167
42. Chebyshev method: 1.2222, 1.2599
   Method (2.29): 1.2449, 1.2597
   Method (2.31): 1.2099, 1.2597
43. (a) ± 0.12450, ± 0.90114, ± 1.99310
   (b) $1.3h$
44. (i) $p = 1.9413$, $q = 1.9538$
   (ii) $p = 1.1446$, $q = 1.4219$
   (iii) $p = 2.3923$, $q = 3.13$
45. (i) $1 \pm i$ (4 squarings)
   (ii) $\alpha_1 = \alpha_2 = 2.000002$, $\alpha_3 = 0.999998$ (4 squarings)
   (iii) 1.0912, -0.9172, $0.413 \pm 1.956i$ (3 squarings)
46. 0.3347 (Birge-Vieta method, two iterations, $x_0 = 0.0$)
   1.2146 (Birge-Vieta method, two iterations, $x_0 = 1.2$)
   -1.3888 (Birge-Vieta method, two iterations, $x_0 = -1.4$)
   $-0.0802 \pm 1.3283i$ (deflated quadratic polynomial)

47. $-0.689752, 0.770091$

48. $2.363376 \pm 0.5 \times 10^{-6}$

49. $x_{k+1} = x_k - \dfrac{f_k}{f_k'} - \dfrac{1}{2}\dfrac{f_k'' f_k^2}{(f_k')^3}$ (third order)

    $x_{k+1} = x_k - \dfrac{f_k}{f_k'} - \dfrac{1}{2}\dfrac{f_k'' f_k^2}{(f_k')^3} - \left[\dfrac{1}{2}\left(\dfrac{f_k''}{f_k'}\right)^2 - \dfrac{1}{6}\dfrac{f_k'''}{f_k'}\right]\dfrac{f_k^3}{(f_k')^3}$ (fourth order)

50. Second order:
    $x_{k+1} = x_k + A_1(x_k, y_k); \ y_{k+1} = y_k + B_1(x_k, y_k)$
    $A_1(x_k, y_k) = -(fg_y - gf_y)_k/D; \ B_1(x_k, y_k) = -(gf_x - fg_x)_k/D$
    where $D = (f_x g_y - f_y g_x)_k$

    Third order:
    $x_{k+1} = x_k + A_1(x_k, y_k) + A_2(x_k, y_k)$
    $y_{k+1} = y_k + B_1(x_k, y_k) + B_2(x_k, y_k)$
    $A_2 = -(f_2 g_y - g_2 f_y)_k/D, \ B_2 = -(g_2 f_x - f_2 g_x)_k/D$
    $f_2 = \tfrac{1}{2}(A_1^2 f_{xx} + 2A_1 B_1 f_{xy} + B_1^2 f_{yy})_k$
    $g_2 = \tfrac{1}{2}(A_1^2 g_{xx} + 2A_1 B_1 g_{xy} + B_1^2 g_{yy})_k$

Chapter 3

3. Use a diagonal matrix to transform A into symmetric form

4. $1\Big/\left(2 \cos \dfrac{3\pi}{7}\right) \approx 2.247$

5. $f(\mathbf{A}) = \mathbf{A} \sinh(2\sqrt{13})/\sqrt{13}$

6. (a) and (c)

7. (a) $x_1 = 1, \ x_2 = \tfrac{1}{2}, \ x_3 = -\tfrac{1}{2}$
   (b) $x_1 = 1, \ x_2 = 0, \ x_3 = -1$

8. $-\dfrac{1}{16}\begin{bmatrix} 8 & -7 & -5 \\ -8 & 1 & 3 \\ -8 & 5 & -1 \end{bmatrix}$

9. $\mathbf{A}^{-1} = (a_{ij})$, where
    $a_{ij} = \begin{vmatrix} 1 & \text{if } i = j \\ -x & \text{if } i = j+1 \\ 0 & \text{otherwise} \end{vmatrix}$

10. $\mathbf{L}^{-1} = \begin{bmatrix} 1 & & & & & \\ \tfrac{1}{2} & 1 & & & \bigcirc & \\ \tfrac{1}{3} & \tfrac{2}{3} & 1 & & & \\ \vdots & \vdots & \vdots & & & \\ \tfrac{1}{n} & \tfrac{2}{n} & \cdots & \tfrac{(n-1)}{n} & 1 \end{bmatrix}$

11. $\mathbf{LL}^T$, where
$$L = \begin{bmatrix} 1 & 0 & 0 & 0 & 0 \\ -1 & 1 & 0 & 0 & 0 \\ 0 & -1 & 1 & 0 & 0 \\ 0 & 0 & -1 & 1 & 0 \\ 0 & 0 & 0 & -1 & 1 \end{bmatrix}$$

12. $-\frac{1}{14} \begin{bmatrix} -5 & 1 & 7 \\ 1 & -3 & -7 \\ 7 & -7 & -7 \end{bmatrix}$

13. $\frac{1}{5} \left[\begin{array}{cc|cc} 4 & -3 & 2 & -1 \\ -3 & 6 & -4 & 2 \\ \hline 2 & -4 & 6 & -3 \\ -1 & 2 & -3 & 4 \end{array} \right]$

14. (a) Jacobi method
15. (a) $|k| < \sqrt{2}/2$
    (b) $\omega \approx 1.03$
16. (a) $\rho(\mathbf{J}) = \sqrt{0.41}$, $\rho(\mathbf{G}) = 0.41$
    (b) $w_{opt} \approx 1.13$
17. (a) Jacobi: $[1 \quad 7/5 \quad 1]^T$, $[3/20 \quad 3/5 \quad 1/5]^T$, $[3/4 \quad 127/100 \quad 3/4]^T$
    Gauss-Seidel: $[1 \quad 4/5 \quad 2/5]^T$, $[3/5 \quad 24/25 \quad 12/25]^T$,
    $[13/25 \quad 124/125 \quad 62/125]^T$
    Exact solution: $[1/2 \quad 1 \quad 1/2]^T$
    (b) $\rho(\mathbf{J}) = 0.7359$, $\nu_J = 0.3067$; $\rho(\mathbf{G}) = 0.4$, $\nu_G = 0.9163$
18. (b) $\rho(\mathbf{J}) = \sqrt{11/18}$, $w_{opt} = 1.23183$, $\rho(\mathrm{SOR}) = 0.23183$, $\nu = 1.4618$
    (c) $[0.51652 \quad 1.49008 \quad 0.49642]^T$
    Exact: $[0.5 \quad 1.5 \quad 0.5]^T$
19. $\alpha = -0.4$
20. Let $\mathbf{M} = \mathbf{D}^{-1}(\mathbf{A} - \mathbf{D})$, where $\mathbf{D}$ is the diagonal part of $\mathbf{A}$.
    Then $w_{opt} = 2/(1 + \sqrt{1 - \rho(\mathbf{M})})$; $w_{opt} = 2/(1 + \sin \pi/6) = \frac{4}{3}$.
21. (a) $\rho(\mathbf{B}^{-1}\mathbf{C}) < 1$
    (b) The eigenvector is $[1 \quad 1 \quad \ldots \quad 1]^T$.
23. 2^{n-1}
24. The limit is the null matrix.
25. (a) $\|\mathbf{A}\| = \rho(\mathbf{A}) = 5$
    (b) $\mathbf{x} = [\frac{1}{2} \quad -\frac{1}{2} \quad -\frac{1}{2} \quad \frac{1}{2}]^T$

26. (a) $|\Delta x_1| \leq 492\epsilon$, $|\Delta x_2| \leq 1860\epsilon$, $|\Delta x_3| \leq 1500\epsilon$
    (b) $|\Delta y| \leq 132\epsilon$

27. The condition number is minimized for $\alpha = 12.5$. cond $(A(12.5)) = 11$

28. $\|\delta x\| \leq \sqrt{0.4} \times 10^{-4}$

29. (c) $\text{cond}_\infty (A) \geq \frac{3}{2}\epsilon^{-1}$ [True value: $\text{cond}_\infty (A) = 6(1 + \epsilon^{-1})$]

30. The eigenvalues of A are 1, -1 and 0. If ϵ is small, we have the corresponding estimates 4ϵ, 6ϵ and 12ϵ.

31. $\lambda_1 = 1$, $\lambda_2 = -1$, $\lambda_3 = 2$
    $x_1 = [1 \quad 3 \quad 2]^T$, $x_2 = [0 \quad 3 \quad 2]^T$, $x_3 = [1 \quad 4 \quad 3]^T$

32. (a) There is no limit.
    (b) $|1 + \beta_1 + \beta_2| < 1$

34. $-\frac{1}{2} < \alpha < 0$

35. (a) Make a similarity transformation with a suitable diagonal matrix.
    (b) (i) $|\Delta\lambda_1/\lambda_1| < \delta + 0(\epsilon^{1/2})$
        (ii) $|\Delta\lambda_1/\lambda_1| < \delta\epsilon^{-2} + 0(\epsilon^{-3/2})$

37. $\lambda_{1,2} = \frac{1}{2}(a_1 + a) \pm \sqrt{\frac{1}{4}(a_1 - a)^2 + a_2^2 + \ldots + a_n^2}$
    $\lambda_3 = \lambda_4 = \ldots = \lambda_n = a$

38. (a) $\begin{bmatrix} 1 & 0 \\ 1 & 1 \end{bmatrix} \begin{bmatrix} 0 & 10^{-2} \\ 0 & 0 \end{bmatrix} \neq \begin{bmatrix} 0 & 10^{-2} \\ 0 & 0 \end{bmatrix} \begin{bmatrix} 1 & 0 \\ 1 & 1 \end{bmatrix}$
    (b) $\begin{bmatrix} (a+b)/2 & (a-b)/2 \\ (a-b)/2 & (a+b)/2 \end{bmatrix}$, where $a = (1.1)^{10}$, $b = (0.9)^{10}$

39. $A^{10} = \frac{1}{9} \begin{bmatrix} -1 & 8 & -4 \\ 8 & -1 & -4 \\ -4 & -4 & -7 \end{bmatrix}$

40. $(-3.66, 3.66)$, $\|A\| \leq 3.66(\sqrt{5} + \sqrt{2})$

41. Transform A to diagonal form
    $\lambda_1 = \frac{1}{2}$, $|\lambda_1 - \bar{\lambda}_1| \leq 5 \times 10^{-2}$; $\lambda_2 = \frac{5}{2}$, $|\lambda_2 - \bar{\lambda}_2| = 0$;
    $\lambda_3 = -1$, $|\lambda_3 - \bar{\lambda}_3| \leq 5 \times 10^{-2}$

42. Put $a - \lambda = 2 \cos \phi$.
    $|\lambda| \leq |a| + 2$, $|\lambda - a| \leq 2$, $\lambda_k = a - 2\cos k\phi$, $x_{ik} = \sin(ik\phi)$
    where $\phi = \pi/(n+1)$ and $i, k = 1, 2, \ldots, n$

43. $\lambda_1 = 0.38563$, $\lambda_2 = -1.31256$, $\lambda_3 = 5.92693$
    $x_1 = \begin{pmatrix} 0.56540 \\ -0.29488 \\ -0.82429 \end{pmatrix}$, $x_2 = \begin{pmatrix} 0.54566 \\ -0.73605 \\ 0.40061 \end{pmatrix}$, $x_3 = \begin{pmatrix} -0.61853 \\ -0.67629 \\ -0.40007 \end{pmatrix}$

Answers and Hints to the Problems 391

44. $\mathbf{A}' = \begin{bmatrix} 1 & 2\sqrt{2} & 0 \\ 2\sqrt{2} & 3 & 0 \\ 0 & 0 & -1 \end{bmatrix}; \lambda = 5, \mathbf{x} = \begin{bmatrix} 1 \\ 1 \\ 1 \end{bmatrix}$

45. $\mathbf{A}' = \begin{bmatrix} 1 & 2\sqrt{2} & 0 & 0 \\ 2\sqrt{2} & 0 & \sqrt{2} & 0 \\ 0 & \sqrt{2} & -1/2 & 5/2 \\ 0 & 0 & 5/2 & -5/2 \end{bmatrix}$

46. -2.37228; 2; 3.37228; Exact: 2, $(1 \pm \sqrt{33})/2$
47. (i) $-20.00606, -10.11194, 5.11800$ (6 iterations)
    Exact: $-20, -10, 5$
    (ii) 3.4138, 0.5862 (5 iterations); Exact: $2 \pm \sqrt{2}$
48. $\lambda = 3.561553$
49. $\lambda = 20.124$, $\mathbf{x} = [0.062 \quad 1.000 \quad 0.062]^T$
50. $\lambda = 2.4812$

Chapter 4

1. $1 + \frac{1}{2}x - \frac{1}{8}x^2$, 1.0246875, 0.0625
2. $\log 2 + \frac{1}{2}(x-1) - \frac{1}{8}(x-1)^2 + \frac{1}{24}(x-1)^3 - \frac{1}{64}(x-1)^4$, 0.000002
3. 12
4. 12, 7
5. 0.33961, 0.0003125
6. 0.00091
7. 0.03836, 0.142
9. $(-1)^n x_0 x_1 \ldots x_n$
10. $8x^2 - 19x + 12$, 1.5
11. 9
13. $(-1)^n / (x_0 x_1 x_2 \ldots x_n)$
16. $p(1.0) = -0.445$, $p(2.0) = -4.840$
17. 10
18. (a) 3.9 ± 0.06
    (b) $3.9585 \pm 2 \times 10^{-4}$
19. $a = 1 - s$, $b = s$, $c = -s(s-1)(s-2)/6$, $d = s(s^2-1)/6$
20. $a = (s-1)(s^2+s-2)/4$, $b = (s+1)(2+s-s^2)/4$
    $c = (s-1)^2(s+1)/4$, $d = (s-1)(s+1)^2/4$
    error term $= \dfrac{(s^4 - 2s^2 + 1)}{4!} y^4(\xi)$, $x_0 - h < \xi < x_0 + h$

21. $a_0 = \dfrac{2}{(b-a)^3}[f(a) - f(b)] + \dfrac{1}{(b-a)^2}[f'(a) + f'(b)]$

$a_1 = \dfrac{3}{(b-a)^2}[f(b) - f(a)] - \dfrac{1}{(b-a)}[2f'(a) + f'(b)]$

$a_2 = f'(a), \ a_3 = f(a)$

22. $f(x) = a_0 + a_1(x - x_0) + a_2(x - x_0)^2 + a_3(x - x_0)^3 + a_4(x - x_0)^4 + a_5(x - x_0)^5$

$a_0 = 1, \ a_1 = 2, \ a_2 = \tfrac{1}{2}, \ a_3 = \dfrac{1}{2h^3}(40 - 24h - 5h^2),$

$a_4 = \dfrac{1}{2h^4}(7h^2 + 32h - 60), \ a_5 = \dfrac{3}{2h^5}(8 - 4h - h^2)$

$P\left(\dfrac{x_0 + x_1}{2}\right) = (128 + 20h - h^2)/64$ where $h = x_1 - x_0$

23. $y(x) = \begin{cases} (17x^3 - 51x^2 + 94x - 15)/15, & 1 < x \leqslant 2 \\ (-55x^3 + 381x^2 - 770x + 531)/15, & 2 \leqslant x \leqslant 3 \\ (38x^3 - 456x^2 + 1651x - 1380)/45, & 3 \leqslant x < 4 \end{cases}$

$y(1.5) = 103/40, \ y'(2) = 94/15$

24. $P_3(x) = \begin{cases} -24x^3 + 68x^2 - 66x + 23, & \tfrac{1}{2} \leqslant x \leqslant 1 \\ \tfrac{1}{27}(-40x^3 + 188x^2 - 310x + 189), & 1 \leqslant x \leqslant \tfrac{3}{2} \\ \tfrac{1}{108}(-28x^3 + 184x^2 - 427x + 369), & \tfrac{3}{2} \leqslant x \leqslant 2 \end{cases}$

25. $P_2(x) = \begin{cases} \dfrac{1}{\sqrt{10}}[(2\sqrt{5}-8\sqrt{2}+2\sqrt{10})x^2+(\sqrt{5}-8\sqrt{2}+3\sqrt{10})x+\sqrt{10}], \\ \hspace{6cm} -1\leqslant x\leqslant 0 \\ \dfrac{1}{\sqrt{10}}[(2\sqrt{5}-8\sqrt{2}+2\sqrt{10})x^2-(\sqrt{5}-8\sqrt{2}+3\sqrt{10})x+\sqrt{10}], \\ \hspace{6cm} 0\leqslant x\leqslant 1 \end{cases}$

$|f - P_2(x)| \leqslant 0.0155$

$P_4(x) = \dfrac{1}{6\sqrt{5}}[2(2\sqrt{10}+12\sqrt{5}-32)x^4 - (\sqrt{10}+30\sqrt{5}-64)x^2 + 6\sqrt{5}],$

$\hspace{8cm} -1 \leqslant x \leqslant 1$

$|f - P_4(x)| \leqslant 0.00505.$

28. 0.019

29. $N_0 = (x^2 - h^2)(y^2 - k^2)/d, \quad N_1 = x(x - h)y(y + k)/4d,$
$N_2 = -(x^2 - h^2)y(y + k)/2d, \quad N_3 = x(x + h)y(y + k)/4d,$
$N_4 = -x(x - h)(y^2 - k^2)/2d, \quad N_5 = -x(x + h)(y^2 - k^2)/2d,$
$N_6 = x(x - h)y(y - k)/4d, \quad N_7 = -(x^2 - h^2)y(y - k)/2d$
$N_8 = x(x + h)y(y - k)/4d, \quad d = h^2k^2$

30. $\tfrac{1}{2}(x - 1)(x - 2)(-2y^2 + 7y + 1) - x(x - 2)(y^2 + 2y + 3) + \tfrac{1}{2}x(x - 1)(y^2 + 3y + 7)$

32. $a = 13.77, \ b = 4.06, \ c = -2.53$

33. $\frac{3}{4}(2\pi - 5) + \frac{15}{4}(3 - \pi)x^2$
34. $1.143x^2 - 0.971x + 1.286$
35. $a = 0.6853, b = 0.3059$
37. $a = 1.31281, b = 0.752575$
38. (i) $a = 0.784976, c = -0.733298$
    (ii) $b = 0.9995$.
39. $F'(0) = 3 \sum_{i=-n}^{n} iY_i/[n(n+1)(n+2)]$

40. $\frac{1}{60}\begin{bmatrix} 2 & 1 & 0 & 0 & . & . & 0 \\ 1 & 4 & 1 & 0 & . & . & 0 \\ 0 & 1 & 4 & 1 & . & . & 0 \\ . & . & . & . & . & . & . \\ 0 & 0 & . & . & 1 & 4 & 1 \\ 0 & 0 & . & . & 0 & 1 & 2 \end{bmatrix}$

41. $p_{k+1}(x) = (x - b_k)p_k(x) - c_k p_{k-1}(x)$
    $b_k = \sum_{j=0}^{n} x_j[p_k(x_j)]^2 \Big/ \sum_{j=0}^{n} [p_k(x_j)]^2$
    $c_k = \sum_{j=0}^{n} [p_k(x_j)]^2 \Big/ \sum_{j=0}^{n} [p_{k-1}(x_j)]^2$
    $p_{-1}(x) = 0, p_0(x) = 1, p_1(x) = x, p_2(x) = x^2 - \frac{1}{2}$
    Write the given equation as
    $\sum_{j=0}^{4} [(1 + x_j^2)^{-1} - \{d_0 p_0(x_j) + d_1 p_1(\ ) + d_2 p_2(x_j)\}]$
    $a_0 = 166/175, a_1 = 0, a_2 = -16/35$

42. $a_0 = \frac{1}{\pi}\int_0^1 \frac{T_0^*(x)\,dx}{(1+x)\sqrt{x(1-x)}}, a_r = \frac{2}{\pi}\int_0^1 \frac{T_r^*(x)\,dx}{(1+x)\sqrt{x(1-x)}}, r = 1, 2, 3, 4$
    $a_0 = 1/\sqrt{2}, a_1 = 4 - 3\sqrt{2}, a_2 = 17\sqrt{2} - 24, a_3 = 140 - 99\sqrt{2},$
    $a_4 = 577\sqrt{2} - 816$
    maximum error: $3363\sqrt{2} - 4756$

43. $a = -0.75, b = 1.66, |\text{max. error}| = 0.1$
45. $-\frac{9}{8}x^2 + 2x + \frac{1}{16}$
47. $a \approx 0.54$
48. $-3x^2 - 29x + 7, \delta = 2$ for $x = 0, 1, 2, 3, 4$
49. $a = c = 0, b = 1, d = \frac{1}{8}$
50. $0.75x^4 + 0.97x^2 + 0.96$
51. $(1 + e)/2$ (in both cases)

394 Numerical Methods

52. $a = 0.862, b = 0.995, c = 0.305$
54. $(160x^2 - 168x + 131)/128$
55. $\frac{1}{4}$
56. $\frac{4799}{4800} + \frac{3455}{13824} x + \frac{103}{1800} x^2 + \frac{37}{3456} x^3$
57. $0.9967 - 0.1389x^2$
58. $0.999729x - 0.164451x^3 + 0.020382x^5$
59. (a) $g(x) = x$, not unique
    (b) $g(x) = -\frac{10}{3}x^2 + 3x$, unique
60. $197x^3 - 1200x^2 + 1801x - 480$

Chapter 5

1. $\alpha_0 = -3/2h, \alpha_1 = 2/h, \alpha_2 = -1/2h$
   Error term: $-(h^2/3)f'''(\xi), x_0 < \xi < x_2$
2. $f'(6.0) = -3.748, f''(6.3) = 0.25$
4. (a) $c_1 = y'''(x)/3$
   (b) $1.080 \pm 3 \times 10^{-3}$
5. $\alpha = -1/h, \beta = 1/h, \gamma = h/24, \delta = -h/24, h = b - a$
   The truncation error tends to $-\{(b - a)^4/2880\}y^{(5)}((a + b)/2)$
6. $a_{2n+1} = (-1)^n(n!)^2/(2n + 1)!, a_{2n} = 0$
7. (a) $k_i = 2i, i = 1, 2, \ldots$
   (b) $h = 0.4 \quad 1.289393$
   $\qquad\qquad\qquad\qquad 1.257469$
   $h = 0.2 \quad 1.265450 \qquad\qquad 1.256595$
   $\qquad\qquad\qquad\qquad 1.256650$
   $h = 0.1 \quad 1.259600$
8. (a) $f''_{i,h}(x) = f''_{i-1,h}(x) + \dfrac{f''_{i-1,h}(x) - f''_{i-1,2h}(x)}{4^i - 1}$
   where i denotes the ith iterate
   (b) $h = 0.2, f''(0.3) \approx -0.74875$
   $h = 0.1, f''(0.3) \approx -0.75000$
   Extrapolated $f''(0.3) \approx -0.75008$
   Rounding error $\leqslant 0.0025$, Truncation error $\leqslant 0.00008$
9. $f'(1) = 0.54030$
10. $p = 0$, **Trapezoidal rule:**
    $\qquad 0.09710999$
    $\qquad\qquad\qquad 0.09763533$
    $\qquad 0.09750400 \qquad\qquad 0.09763357$
    $\qquad\qquad\qquad 0.09763368$
    $\qquad 0.09760126$

Simpson's rule:
0.09766180
               0.09763357
0.09763533   0.09763357
               0.09763357
0.09763368

$p = 1$, **Trapezoidal rule:**
0.04741863
               0.04811455
0.04794057   0.04811657
               0.04811645
0.04807248

Simpson's rule:
0.04807333
               0.04811730
0.04811455   0.04811656
               0.04811658
0.04811645

11. $A_0 = 12/15$, $A_1 = 16/15$, $A_2 = 2/15$, $R = (8\sqrt{2}/315)h^{7/2}f'''$
12. $a = 5/12$, $b = 2/3$, $c = -1/12$, $k = h^4/24$, $n = 3$
13. $a = 0$, $b = 3/4$, $c = 1/4$, Truncation error: $O(h^4)$
14. $p = 1/12$; $R = (h^5/720)f^{(4)}(\xi)$, $x_0 < \xi < x_1$

$$\int_a^b f(x)\,dx = \frac{h}{2}[(f_0+f_n) + 2(f_1+f_2+ \cdots +f_{n-1})] + \frac{h^2}{12}[f_0'-f_n'] + O(h^4)$$

15. $a = 7/5$, $b = 16/5$, $p = 1/15$, $R = (h^7/4725)f^{(6)}(\xi)$, $x_0 < \xi < x_2$

$$\int_a^b f(x)\,dx = \frac{h}{15}[7(f_0+f_{2n}) + 16(f_1+f_3+f_5+ \cdots +f_{2n-1})$$
$$+ 14(f_2+f_4+ \cdots +f_{2n-2})] + \frac{h^2}{15}[f_0' - f_{2n}'] + O(h^6)$$

16. $\alpha = 1/2$, $\beta = 1/6$, $p = 3$
17. (a) 0.002365
    (b) Step length $= 0.005$, 5 decimals
18. 0.33036
19. $\alpha_1 = \pi/4$, $\alpha_2 = \pi/2$, $\alpha_3 = \pi/4$, 2.6233, Exact value: 2.62205755
20. -0.916
21. 1.57210
22. -0.16 ± 0.23
23. $0.3517 \pm 2 \times 10^{-5}$

24. $A_1 = A_4 = 1/6$, $A_2 = A_3 = 5/6$, $x_2 = -1/\sqrt{5}$, $x_3 = 1/\sqrt{5}$
Error term of sixth order

25. 0.203508, 0.202714

26. $A_{-1} = \frac{5}{9}(a + \sqrt{\frac{3}{5}})$, $A_0 = \frac{8}{9}$, $A_1 = \frac{5}{9}(a - \sqrt{\frac{3}{5}})$, $x_1 = \sqrt{\frac{3}{5}}$
The order is 5 and is independent of a.

27. $A_1 = A_3 = 1.107$, $A_2 = 1.119$, $x_1 = -x_3 = -0.8138$, $x_2 = 0$, $\sigma = 5$

28. 1.18266

29. 0.25435

30. (i) 0.57143, 0.58824
(ii) 0.43246, 0.49603
(iii) 1.181636, 1.417963
(iv) 1.772454, 1.772454

31. (a) 1.39113, 1.38684
(b) 1.38642, 1.38243

32. $x_1 = 1/3$, $H_1 = 1.5$, $B_1 = 0.5$, $R = \frac{2}{27}f'''(\xi)$, $-1 < \xi < 1$

33. $x_1 = -x_2 = 1/\sqrt{5}$, $H_1 = H_2 = 5/6$, $B_1 = B_2 = 1/6$,
$R = -(2/23625)f^{(6)}(\xi)$, $-1 < \xi < 1$

34. (a) 1.32471, 1.31135
(b) 1.30795, 1.31261
(c) 1.46584, 1.29661

35. (i) $n = 3$, $I = 0.1663$, $n = 4$, $I = 0.1832$; $n = 5$, $I = 0.1918$.
(ii) $n = 2$, $I = 0.2605$; $n = 3$, $I = 0.2595$; $n = 4$, $I = 0.2592$

36. $A_{1,2} = (3 \mp \sqrt{6})/6$, $x_{1,2} = (3 \pm \sqrt{6})/2$, $c = 53/128$, $n = 4$.

37. 6.494, Exact value $= \pi^4/15$

38. $\alpha = 1/\sqrt{3}$

39. (i) 0.31233, Exact: 0.31756
(ii) 0.31772

40. $n = 2$ 4.134
    0.952
$n = 4$ 3.997

Chapter 6

1. $y_n = (1/160)[63(-3)^n + 32(2)^n - 40n^2 - 60n - 95]$

2. $y_n = \frac{1}{3}\left[1 + \frac{(-1)^{n+1}}{2^n}\right]$
Instability (due to a term proportional to n in the general solution).

4. $u_n = \frac{1}{2}(i)^{n+1}[(-1)^n e^{inx} - e^{-inx}]$
   or $u_n = -\sin(nx - s\pi/2)$, where $n = 4k + s$ and $s = 0, 1, 2, 3$
6. $y(0.1) = 1.355$, $y(0.2) = 1.85548$, $y(0.3) = 2.55161$, $y(0.4) = 3.51092$
   (a) $t < 0.0437$
   (b) $n = 10$ (using the exact solution)
7. $y(0.6) = 0.61035$, $y(0.8) = 0.84899$
8. (i) $y_{n+1} = y_n + (1 - e^{-h})y'_n$
   (ii) $y_{n+1} = y_n + \left(\dfrac{1 - \cos(h)}{\sin(h)}\right)(y'_n + y'_{n+1})$
11. $h \leq a/b$
12. $A = \begin{bmatrix} 1 - \dfrac{\alpha^2}{2} + \dfrac{\alpha^4}{24} & -\alpha + \dfrac{\alpha^3}{6} \\ \alpha - \dfrac{\alpha^3}{6} & 1 - \dfrac{\alpha^2}{2} + \dfrac{\alpha^4}{24} \end{bmatrix}$
    $\alpha = kh$, $\alpha < 2\sqrt{2}$
13. $y(0.1) = 1.07137$, $y(0.2) = 1.09109$, $u(0.1) = 0.44113$,
    $u(0.2) = -0.03237$
14. $\lambda = 0.84673$
15. (b) $\epsilon < 10^{-3}/6$
17. Stability if $s > 0$ or $s < -\frac{1}{2}$
18. (b) It is a first order method, stable and A-stable
    (c) $h = 0.2$, $y(0.2) = [0.75 \quad 0.018]^T$
        $h = 0.1$, $y(0.2) = [0.72 \quad 0.026]^T$
        Extrapolated value: $y(0.2) = [0.69 \quad 0.034]^T$
19. $a = 1/3$, $w_1 = 1/4$, $w_2 = 3/4$, Third order; $-6 \leq \lambda h < 0$
20. Third order; $-6 \leq \lambda h < 0$
21. $y(0.1) \approx 0.9145$, $y(0.2) \approx 0.85405$, $y(0.3) \approx 0.81146$
22. $a = 1$, $b = d = 1/3$, $c = 4/3$. Weakly stable, unstable for $\lambda < 0$
23. $a = 1$, $b = 21/8$, $c = -9/8$, $d = 15/8$, $e = -3/8$
    Truncation error: $(81/240)h^5 y^{(5)}(\xi)$, $x_{n-3} < \xi < x_{n+1}$
24. Weak stability. The difference equation has the approximate roots
    $(1 + h/8)(-1/2 \pm i\sqrt{(3/2)}$ and $1 - h$.
25. $\alpha = 28$, $\beta = 12$, $\gamma = 48$, unstable
26. $a < 1$
27. (a) Weakly stable
28. $\lim\limits_{n \to \infty} |y_n| = \infty$, $\lim\limits_{n \to \infty} |y(x_n)| = 0$

29. (a) inherent instability
    (b) stability when $ah < 2.785$
    (c) strong instability

30. (a) $-(2/19)h^5 y^{(5)}(\xi_n)$, $x_n - 3h < \xi_n < x_n + h$,
    (b) Yes

31. (a) $y_n = A + B\left[\dfrac{1 - Vh/2p}{1 + Vh/2p}\right]^n$
    (b) $y_n = A + \dfrac{B}{(1 + Vh/p)^n}$
    (c) 1

32. $\xi^2 - (1 + \mu + 0.75\mu^2)\xi + 0.25\mu^2 = 0$
    $\xi^2 - (1 + \mu + 0.5\mu^2 + 0.375\mu^3)\xi + 0.125\mu^3 = 0$, $\mu = \lambda h$

33. $T_1 = (251/720)h^5 y^{(5)}(\xi)$, $x_{n-3} < \xi < x_{n+1}$
    $T_2 = -(19/720)h^5 y^{(5)}(\xi_1)$, $x_{n-2} < \xi_1 < x_{n+1}$

34. (a) $C = -(1/24)(\alpha - 9)$, $D = (1/12)(7\alpha + 9)$, $E = (1/24)(17\alpha - 9)$.
    (b) Stable for $-0.6 < \alpha < 1$

35. Applicable: (b) and (c)
    Stable: (b) always
                (c) if $k < \sqrt{2}/h$

36. (a) $\mathbf{B}(h) = \mathbf{I} + h\mathbf{A} + \tfrac{1}{2}h^2\mathbf{A}^2$
    (b) $-2/a$
    (c) $p = 2$

37. $x_n = -(n^3 - 10^6 n)/6000$, $n = 1(1)999$

38. (c) With central difference approximations of the derivatives we arrive at a linear 3-diagonal system for $y_0, y_1, \ldots, y_{N-1}$ approximating $y(0), y(h) = y(-h), \ldots, y(2 - h) = y(-2 + h)$ respectively. With $h = 1$, there are two unknowns $y(0) \approx y_0 = 1.1684$, $y(1) = y(-1) \approx y_1 = 0.8487$

39. $y(0) = -19.8523$, $y(0.25) = -14.6080$, $y(0.5) = -9.5294$, $y(0.75) = -4.6861$

40. (a) Second order method and extrapolation:

	$y(0)$		$y(0.5)$
$h = 0.5$	0.967213		0.721311
		0.932953		0.691803
$h = 0.25$	0.941518		0.699180

Fourth order method and extrapolation:

	$y(0)$		$y(0.5)$
$h = 0.5$	0.927585		0.688044
		0.932067		0.691235
$h = 0.25$	0.931787		0.691035

(b) **Extrapolated values: Second order:** $y(0) = 0.850433$
$y(0.5) = 0.623968$, **Fourth order:** $y(0) = 0.850815$
$y(0.5) = 0.624243$

41. (a) $a = 1/20$, $b = 2/15$
    (b) $y_{n+2} = -16y_{n+1} + 34y_n - 16y_{n-1} - y_{n-2} + \frac{4}{3}h^2(2f_{n+1} + 11f_n - 2f_{n-1})$

42. (a) $n = 5$
    (b) $f(12) = 2.3756$

43. $y(x) \approx 1 + \alpha x + 6(1 + \alpha)x^5$ where $\alpha = 9/55$

44. $y(0) \approx -0.308$

45. (a) $a = 1/6$, $b = -1/720$, $a_1 = -1/6$, $a_2 = 1/12$, $a_3 = 1/30$, $a_4 = 1/72$
    (b) For the $N+3$ unknowns $y(kh) = y_k$, $k = -2, -1, 1, \ldots, N-1$, $N+1$, $N+2$, we get the $N+3$ equations
$$h^{-4}\delta^4 y_k = (1 + a\delta^2 + b\delta^4)(p(x_k)y_k + q(x_k)),\ k = 0(1)N$$
$$hy'(kh) = \mu\delta y_k + a_1\Delta^3 E^{-1}y_k + h^4(a_2 + a_3\mu\delta + a_4\delta^2)$$
$$\times (p(x_k)y_k + q(x_k)),\ k = 0, N$$

46. $a = -\sinh(1)/24\cosh^2(1) = -0.020565$
    $b = 12a = -0.24678$

47. $h = \frac{1}{4}$: $y(\frac{1}{4}) = y(\frac{3}{4}) = 2.4$, $y(\frac{1}{2}) = 3.2$, $y'(0) = 5.6$
    $h = \frac{1}{6}$: $y(\frac{1}{6}) = y(\frac{5}{6}) = 1.88$, $y(\frac{1}{3}) = y(\frac{2}{3}) = 2.65$, $y(\frac{1}{2}) = 2.98$,
        $y'(0) = 5.26$
    Extrapolated value: $y'(0) = 5.00$

48. $u_1' = u_2$, $u_1(0) = 0$
    $u_2' = u_1 + x$, $u_2(0) = \alpha$
    Use $\alpha = \frac{1}{2}$ and $-\frac{1}{2}$. We find $c_1 = 0.357645$, $c_2 = 0.642355$.
    $u(0.2) \approx -0.028471$, $u(0.4) \approx -0.050081$, $u(0.6) \approx -0.057763$,
    $u(0.8) \approx -0.043895$, $u(1) \approx 4 \times 10^{-7}$

49. With $u'(2) = 0.5$ and -0.5, we find $c_1 = 0.71081$, $c_2 = 0.289189$,
    $u(2.25) = 0.037887$, $u(2.5) = 0.048746$, $u(2.75) = 0.035475$,
    $u(3.0) = -1.4 \times 10^{-7}$

50. Starting with $u'(0) = \alpha_0 = -1.8$, $u'(0) = \alpha_1 = -1.9$, we get
    $\alpha_2 = -2.0035$. We find
    $u(0.1) = 0.82614$, $u(0.2) = 0.69405$, $u(0.3) = 0.59132$, $u(0.4) = 0.50985$,
    $u(0.5) = 0.44414$

Chapter 7

1. (a) $t = c_1 e^x$ and $t = c_2 e^{-x}$
   (b) $1 - \log 10 \leqslant x \leqslant 1 + \log 10$
2. The equation is elliptic outside the ellipse $(x/2)^2 + y^2 = 1$
3. (a) $0(k^2 + h^2)$
   (b) Always unstable
4. $0 \leqslant \alpha < 1/2$, stable if $k/h^2 \leqslant 1/(2(1-2\alpha))$
   $1/2 \leqslant \alpha < 1$, always stable
6. Stable if $2r \leqslant (1-\theta)/(1+\theta)$
7. Crank-Nicolson scheme: unconditionally stable
8. Crank-Nicolson scheme: $k = 0.2$ gives $u(0.4, 0.2) = 0.3$
   Laasonen scheme: $k = 0.2$ gives $u(0.4, 0.2) = 0.52$
9. $0(k^2 + h^4)$, unconditionally stable
10. Unconditionally stable for $\theta \geqslant 0$
11. $0(k^2 + h^2)$, unconditionally stable
    $u_1^1 = u_3^1 = 15/68$, $u_2^1 = 11/34$ (Crank-Nicolson)
    $u_1^2 = u_3^2 = 269/1564$, $u_2^2 = 89/391$
12. (a) $0(k + h^2)$, unconditionally stable
    (b) $0(k + h^2)$, $0 < p \leqslant 1/2$
13. $u(0, 0.003) = 348$, $u(1, 0.003) = 332$, $u(2, 0.003) = 318$
    $u(3, 0.003) = 307$, $u(4, 0.003) = 300$
14. $0(k^2 + h^2)$, unconditionally stable
15. $u_{1,1}^1 = u_{2,2}^1 = 0.03$, $u_{1,2}^1 = u_{2,1}^1 = -0.03$
17. $u_{1,1}^1 = u_{2,2}^1 = 0.03$, $u_{1,2}^1 = u_{2,1}^1 = -0.03$
18. (a) $0(h^2)$
19. $0(k^2 + h^2)$, $0(k^2 + h^2)$, stable for $0 < p \leqslant 1$
    $u_1^1 = u_4^1 = 0.2$, $u_2^1 = u_3^1 = 0.375$,
    $u_1^2 = u_4^2 = 0.19375$, $u_2^2 = u_3^2 = 0.30625$
20. $u_{1,1}^1 = u_{2,1}^1 = u_{1,2}^1 = u_{2,2}^1 = 1/12$
    $u_{1,1}^2 = u_{2,1}^2 = u_{1,2}^2 = u_{2,2}^2 = 11/486$
21. $\dfrac{k}{h^2} = r : \dfrac{\partial u}{\partial y} = \dfrac{\partial^2 u}{\partial x^2}$
    $\dfrac{k}{h} = r : \dfrac{\partial u}{\partial y} = \dfrac{\partial^2 u}{\partial x^2} - r^2 \dfrac{\partial^2 u}{\partial y^2}$

22. (i)

x_m	$-\frac{1}{2}$	0	$\frac{1}{2}$	1	$\frac{3}{2}$	2	$\frac{5}{2}$	3	$\frac{7}{2}$	4	$\frac{9}{2}$	5
u_m^1	0	$-\frac{1}{32}$	$\frac{1}{8}$	$\frac{3}{8}$	$\frac{5}{8}$	$\frac{15}{16}$	$\frac{7}{8}$	$\frac{5}{8}$	$\frac{3}{8}$	$\frac{3}{32}$	0	0
u_m^2	$\frac{1}{256}$	$-\frac{5}{128}$	$\frac{9}{256}$	$\frac{1}{4}$	$\frac{63}{128}$	$\frac{53}{64}$	$\frac{119}{128}$	$\frac{3}{4}$	$\frac{129}{256}$	$\frac{27}{128}$	$\frac{9}{256}$	0

(ii) Evaluate the solution values at the first step by Lax-Wendroff scheme

x_m	0	$\frac{1}{2}$	1	$\frac{3}{2}$	2	$\frac{5}{2}$	3	$\frac{7}{2}$	4	$\frac{9}{2}$
u_m^2	$-\frac{1}{16}$	$\frac{3}{64}$	$\frac{1}{4}$	$\frac{15}{32}$	$\frac{7}{8}$	$\frac{29}{32}$	$\frac{3}{4}$	$\frac{33}{64}$	$\frac{3}{16}$	$\frac{3}{64}$

23. (i) $0(k^2 + h^2)$

(ii) unconditionally stable for $\beta^2 \geq \alpha^2$

24. (b) $0(k^2)$

(c) stable for $0 < \lambda \leq 2$, unstable for $\lambda > 2$

25. (b) Using five point formula, we get

$h = 1.0; u(0, 0) = -4,$

$h = 0.5, u(0, 0) = -4.5, u(\frac{1}{2}, 0) = 3.5, u(\frac{1}{2}, \frac{1}{2}) = -2.75$

26. $u(1, 1) = -158/77, u(2, 1) = -224/77$

$u(2, 2) = -290/77, u(1, 2) = -224/77$

27. Five point formula:

$u(0, 0) = -9/8, u(\frac{1}{2}, 0) = -7/8, u(\frac{1}{2}, \frac{1}{2}) = -11/16$

Nine point formula:

$u(0, 0) = -636/539, u(\frac{1}{2}, 0) = -495/539, u(\frac{1}{2}, \frac{1}{2}) = -783/1078$

28. Five point formula:

$u(0, 0) = \frac{1}{4}, u(\frac{1}{2}, 0) = 23/128, u(\frac{1}{2}, \frac{1}{2}) = 9/64, u(0, \frac{1}{2}) = 25/128$

Nine point formula:

$u(0, 0) = 0.25, u(\frac{1}{2}, 0) = 0.178977, u(\frac{1}{2}, \frac{1}{2}) = 0.140625,$

$u(0, \frac{1}{2}) = 0.196023$

29. $u(1, 1) = 1.6277, u(2, 1) = u(1, 2) = 4.2553, u(2, 2) = 8.0269$

30. $u(0, 1) = -1/12, u(1, 1) = u(-1, 1) = 0$

31. $u(\frac{1}{2}, \frac{1}{2}) = 0, u\left(\frac{\sqrt{3}}{2}, \frac{1}{2}\right) = -u\left(\frac{1}{2}, \frac{\sqrt{3}}{2}\right) = \frac{5(\sqrt{3} - 1)}{4\sqrt{3}}$

32. $h < 8/7$

33. (a) $\begin{bmatrix} -4 & 1 & & & 0 \\ 1 & -4 & 2 & & \\ & 1 & -4 & -1 & \\ & & 1 & -4 & 1 \\ 0 & & & 2 & 4 \end{bmatrix}$, $\mathbf{u} = \begin{bmatrix} 0 \\ -1 \\ 0 \\ -2 \\ 0 \end{bmatrix}$

when the nodes are appropriately ordered.

(b) $\mathbf{u}^T = [0.102 \quad 0.407 \quad 0.263 \quad 0.647 \quad 0.323]$

34. $u(0, 0) = 0.334$
35. $\lambda = 2.1917$, exact: 2.1932

Index

abscissas, 213
absolute error, 7
absolute norm, 89
absolutely stable, 281, 286
A-stable methods, 286
Adams-Bashforth method, 249, 273
Adams-Moulton method, 253, 273
adjoint matrix, 71
Aitken's Δ^2-method, 37
Aitken's interpolation, 139, 142
algorithm, 13
Alternating direction implicit method, 338
amplification factor, 3 32
arguments, 135
asymptotic error constant, 32
augmented matrix, 71

back substitution method, 74
backward Euler method, 253
Bairstow method, 47
base, 1
  ernstein polynomials, 175
Bessel interpolation, 150
best approximation, 165
binary, 1
Birge-Vieta method, 46
bisection method, 22
bits, 2
boundary node, 360
boundary value problem, 289, 359

canonical representation, 243
Cauchy-Riemann equations, 42
Cayley-Hamilton theorem, 103
characteristics, 314
characteristic equation, 72, 255
characteristic value problem, 71, 290
Chebyshev equioscillation theorem, 177
Chebyshev method, 30
Chebyshev polynomial, 173
Cholesky method, 83
chord method, 24
closed type method, 218
composite trapezoidal rule, 228

composite Simpson's rule, 229
condition number, 90
conditionally stable, 285
convergence, 32
convergent method, 248, 329
coordinate functions, 165
corrector method, 271
Cotes numbers, 215
Courant-Friedrichs-Levy condition, 344
Cramer's rule, 74
Crandall method, 324
Crank-Nicolson method, 323

D'Alembert solution, 341
decomposition method, 81
difference equations, 247, 248
Dirichlet condition, 359
Dirichlet problem, 363
domain of dependence, 342
Douglas-Rachford scheme, 371
DuFort and Frankel method, 325
D'Yakonov split form, 339, 348

eigenfunction, 71
eigenvalues, 71
eigenvalue problem, 71, 290
elliptic partial differential equation, 315
equivalence theorem, 319
error constant, 215
error equation, 33
error of approximation, 197, 329
error tolerance, 19
euclidean norm, 166
Euler method, 249
Euler-Cauchy method, 262
explicit method, 249, 258, 321
explicit Runge-Kutta method, 262
extraneous roots, 285
extrapolation problem, 183

Faddeev-Leverrier method, 102
finite elements, 155
first divided difference, 139
fixed point, 5

Index

floating point, 5
flowchart, 14
forward substitution method, 73
Fourier series method, 332
Fox's example, 121
Frobenius norm, 89

Gauss-Chebyshev integration methods, 224
Gauss-elimination method, 77
Gauss-Hermite integration methods, 226
Gauss-Jordan method, 80
Gauss-Laguerre method, 225
Gauss-Legendre integration method, 220
Gauss-Seidel iteration method, 93
Gaussian integration methods, 219
Gerschgorin bounds, 105
Gerschgorin circles, 105
Givens method, 111
Graeffe's root squaring method, 50
Gregory-Newton backward difference interpolation, 150
Gregory-Newton forward difference interpolation, 148
grid points, 247
growth parameter, 285

Hermite interpolating polynomial, 137, 154
Hermite polynomial, 226
Heun method, 265, 309
hexadecimal, 1
Hilbert norm, 89
homogeneous equation, 314
Householder method, 113
Householder transformations, 114
hyperbolic partial differential equation, 315

illconditioned, 91
implicit method, 253, 258, 322
increment function, 258
inherent error, 7
inhomogeneous equation, 314
initial conditions, 244
initial boundary value problem, 319
initial value problem, 244
instructions, 13
integral surface, 314
integration rule, 213
intermediate value theorem, 21
interpolation problem, 135
internal node, 360
inverse power method, 120
iterated interpolation, 139
iteration function, 34

Jacobi iteration method, 92

Jacobi method, 109
Jacobian, 42

knots, 155
Kutta method, 265

L^p-norm, 166
Laasonen method, 322
Lagrange bivariate interpolation, 163
Lagrange fundamental polynomial, 139
Lagrange interpolating polynomial, 137
Lagrange interpolation, 138
Laguerre polynomial, 226
Lanczos economization, 179
Lax-Wendroff method, 352, 358
layer, 320
leap frog scheme, 353
least square approximation, 166
Lees ADI method, 348
Legendre polynomial, 172, 220
level, 320
linear differential equation, 243
Lipschitz constant, 244
Lobatto integration method, 221
local discretization error, 249
local truncation error, 249, 321

machine epsilon, 7, 8
Maclaurin series expansion, 181
mantissa, 6
matrix norm, 89
maximum norm, 89
mesh points, 247
mesh ratio parameter, 321
mesh spacing, 247
midpoint method, 254
Milne-Simpson methods, 274
minimax property, 174
mixed boundary conditions, 359
modified predictor-corrector method, 277
Muller method, 29
Multiple root, 18
multipoint iteration method, 31
multistep method, 248, 283

nearly optimal, 265
natural spline, 161
Neumann condition, 359
Newton bivariate interpolation, 164
Newton-cotes integration methods, 215
Newton interpolation with divided differences, 140, 143
Newton-Raphson method, 25
nodes, 135, 213
nonlinear differential equation, 243

non-periodic spline, 161
normal equations, 166
normal form, 6
numerical error, 280
numerical solution, 247
numerically unstable, 10
Numerov method, 292
Nyström method, 254, 273

octal, 1
open-type methods, 218
optimal formula, 264
optimal value, 206
order, 32, 243, 258, 321
ordinary differential equation, 243
orthogonal functions, 170
overshoot, 299

Padé-approximation, 181
parabolic partial differential equation, 315
partition method, 86
Peaceman-Rachford ADI method, 338
permanence property, 184
periodic spline, 161
periodic stability, 281
piecewise cubic Hermite interpolation, 157
piecewise cubic interpolation, 157
piecewise interpolation, 155
piecewise linear interpolation, 155
pivot, 360
pivot elements, 77
pivoting, 77
Poisson equation, 363
power method, 117
predictor method, 271
predictor-corrector method, 277
principal root, 285
property A, 71
pure initial value problem, 319

quadrature formula, 213
quasilinear, 314

Radau integration method, 222
radix, 1
rate of convergence, 32
rational approximation, 181
Rayleigh quotient, 104
reduced characteristic equation, 285
regula falsi method, 24
relative error, 7
relatively stable, 281
relaxation parameter, 94
remainder, 136
residual vector, 91

Richardson's extrapolation, 208
Richardson method, 325
Richtmyer two step method, 355
Romberg integration, 231
root, 18
root condition, 272
roundoff error, 7
Routh-Hurwitz criterion, 286
Runge's example, 185
Runge-Kutta method, 262
Rutishauser method, 116

Schmidt method, 321
secant method, 24
shape function, 157
shooting method, 298
significant digits, 9
similar matrices, 104
similarity matrix, 104
Similarity transformation, 104
simple root, 18
Simpson's rule, 217
Simpson's $\frac{3}{8}$th rule, 217
singlestep methods, 248
SOR method, 94
spectral norm, 89
spectral radius, 72
spiral solution, 35
spline in compression, 189
spline in tension, 189
spline interpolation, 159
square norm, 166
square root method, 83
stable, 248, 280, 285, 332
staircase solution, 35
Steffenson's method, 61
step length, 247
Stirling interpolation, 150
Sturm function, 44
Sturm sequence, 44, 111
Sturm's theorem, 44

tabular points, 135
Taylor series method, 259
three level formula, 325
trapezoidal rule, 215
triangularization method, 81
truncation error, 8, 136, 140
two level formula, 321
two step method, 248

unconditionally stable scheme, 332
under relaxation, 94
undershoot, 299
uniform approximation, 174

uniform norm, 166
uniform (minimax) polynomial approximation, 175
unstable, 280, 285

vector norm, 89
Von Neumann condition for stability, 356
Von Neumann method, 332, 344

Weierstrass approximation theorem, 165
weight function, 213
weights, 213
well posed, 318
Wilkinson's example, 10
words, 5

zero, 18